I0036272

English Farming
Past and Present

English Farming
Past and Present

Lord Ernle

New (Sixth) Edition

With Introductions By
G. E. Fussell
and
O. R . McGregor

Routledge
Taylor & Francis Group

First published in 1912 by Longmans, Green & Co. Ltd
New (Sixth) Edition first published by Heinemann Educational
Books Ltd and Frank Cass and Company Ltd, 1961

This edition first published in 2018 by Routledge
2 Park Square, Milton Park, Abingdon, Oxon, OX14 4RN
and by Routledge
52 Vanderbilt Avenue, New York, NY 10017, USA

Routledge is an imprint of the Taylor & Francis Group, an informa business

© 1961 by G. E. Fussell and O. R. McGregor

Publisher's Note
The publisher has gone to great lengths to ensure the quality of this reprint but
points out that some imperfections in the original copies may be apparent.

Disclaimer
The publisher has made every effort to trace copyright holders and welcomes
correspondence from those they have been unable to contact.
A Library of Congress record exists under ISBN:

ISBN 13: 978-1-138-39208-3 (hbk)
ISBN 13: 978-1-138-39212-0 (pbk)
ISBN 13: 978-0-429-42235-5 (ebk)

ENGLISH FARMING
PAST AND PRESENT

LORD ERNLE

English Farming

Past and Present

NEW (SIXTH) EDITION

WITH INTRODUCTIONS BY

G. E. FUSSELL

AND

O. R. McGREGOR

HEINEMANN

LONDON MELBOURNE TORONTO

FRANK CASS & CO. LTD

KINGSWOOD BOOKS ON SOCIAL HISTORY

General Editors: H. L. Beales and O. R. McGregor

First published by Longmans, Green & Co. Ltd, 1912

First edition 1912
Second edition 1917
Third edition 1922
Fourth edition 1927
Fifth edition 1936
(edited by Sir A. D. Hall)

New (Sixth) Edition first published by Heinemann Educational
Books Ltd and Frank Cass and Company Ltd, 1961

Introductions copyright © 1961
G. E. Fussell and O. R. McGregor

Published by Heinemann Educational Books Ltd
15–16 Queen Street, Mayfair, London, W.1
and
Frank Cass and Company Ltd
91 Southampton Row, London, W.C.1

*This book has been printed in Great Britain by
litho-offset at Taylor Garnett Evans & Co. Ltd.,
Watford, Herts, and bound by them*

PREFACE TO SIXTH EDITION

THE edition of *English Farming Past and Present* here reprinted is the fifth, published in 1936, edited and revised by Sir Daniel Hall, and out of print for many years. As far as Chapter XVI, the text stands as written by Lord Ernle; the extent of Sir Daniel Hall's additions and alterations is explained in his preface.

We have not attempted to bring this classic up-to-date by textual exegesis or by amending the appendices. Our aim has been to supply critical and bibliographical introductions in the form of commentaries both upon the sources used by Lord Ernle and upon subsequent writing. Thus the essays are not devotional exercises. But we hope that the sacrifice of piety will provide the present-day reader with the twin advantages of Lord Ernle's and Sir Daniel Hall's text and a guide to the findings and direction of recent scholarship.

Each introduction stands on its own. Neither of us is responsible for what the other has written.

G. E. FUSSELL
O. R. McGREGOR

January 1961.

PREFACE TO FIFTH EDITION

WHEN Lord Ernle and Messrs. Longmans did me the honour of asking me to prepare a new edition of *English Farming, Past and Present* I agreed with some trepidation. I knew that I had neither the desire nor the knowledge to alter what Lord Ernle had written of " the past." But what in 1912 he wrote of " the present " has since taken on a different colour, and the post-war period has witnessed revolutionary changes in the practices of agriculture and in the attitude of the State, of which the student of agriculture might well desire some summary account.

Except for an occasional footnote I have, therefore, left untouched Lord Ernle's text as far as Chapter XVI, but to that chapter I have added a section bringing the history of tithe down to the present time. I have left the earlier part of Chapter XVIII as Lord Ernle wrote it, but have substituted for the latter part a new section dealing with the practice of agriculture in the period from the beginning of the century to the outbreak of the war in 1914. Lord Ernle had touched upon education and research, but their growth has been so rapid that I have treated both the pre-war and the later period in one new chapter. Lord Ernle's account of the " War and State Control " remains ; it is his own record of actions for which he bore so large a responsibility. From that point I have taken up the tale with two chapters, one dealing with the more important items of legislation affecting agriculture, and a second discussing the changes in the methods of farming that have seemed to me to be significant in this latter period. As in these chapters I have perforce had to deal with matters of opinion, I can only express beforehand my regrets for such errors of omission or commission as I must have made. I have many people to thank for their assistance in providing information, notably Mr. C. S. Orwin of Oxford and Dr. V. E. Wilkins, and several of my old colleagues in the Ministry.

Sequels are notoriously colourless and unsatisfactory and I cannot hope to have provided an exception. But it is necessary that Lord Ernle's classic should be furnished with the data concerning the more recent events, and I take pride in being associated with a book from which I and my contemporaries have learned to look upon English agriculture as something continuously growing, whose present status and future development cannot be understood without an appreciation of its roots in the past.

A. D. HALL

February 1936

AUTHOR'S PREFACE

English Farming Past and Present is based on an article which appeared in the *Quarterly Review* for 1885. The article was subsequently expanded into a book, published in 1888 by Messrs. Longmans under the title of *The Pioneers and Progress of English Farming.*

The book has been out of print for twenty years. Written with the confidence of comparative youth and inexperience, it expressed as certainties many opinions which might now be modified, if not withdrawn. But its motives were two convictions, which time has rather strengthened than weakened. One was, that the small number of persons who owned agricultural land might some day make England the forcing-bed of schemes for land-nationalisation, which countries, where the ownership of the soil rested on a more democratic basis, repudiated as destructive of all forms of private property. The other was, that a considerable increase in the number of peasant ownerships, in suitable hands, on suitable land, and in suitable localities, was socially, economically, and agriculturally advantageous.

Since 1888, the whole field of economic history has been so carefully and skilfully cultivated, that another work on a branch of the subject might appear superfluous. But there still seemed to be room for a consecutive history of English agriculture, written from a practical point of view, and tracing the influence of the progress of the industry on the social conditions of those engaged in its pursuit. Great economic changes have resulted from small alterations in the details of manufacturing processes. Similar changes may often be explained by some little-noticed alterations in farming practice. The introduction of the field-cultivation of turnips, for example, was as truly the parent of a social revolution as the introduction of textile machinery. The main object of *The Pioneers and Progress of English Farming,* and, in greater detail, of *English Farming Past and Present,* is to suggest that advances in agricultural skill, the adoption of new methods, the application of new

A*—E F P P

resources, the invention of new implements, have been, under the pressure of national necessities, powerful instruments in breaking up older forms of rural society, and in moulding them into their present shape.

Students of economic and social questions—and at the present day most people are interested in these subjects—will decide whether the influence of these simple and natural causes has been greater or less than is suggested. Even those who consider that their importance is exaggerated, may find in the record of their progress a useful commentary on the political explanations which they themselves prefer to adopt. The book may still serve another purpose. It touches rural life at many different points and at many different stages. Dwellers in the country are surrounded by traces of older conditions of society. They may perhaps find, through *English Farming Past and Present*, a new interest in piecing together the fragments of an agricultural past, and in reconstructing, as in one of the fashionable occupations of the day, a picture of the Middle Ages or of the eighteenth century in the midst of their own familiar surroundings.

Now that the book is in print and on the eve of publication, I feel more acutely than ever the disadvantages under which it has been prepared. *English Farming Past and Present* is the by-product of a life occupied in other pursuits than those of literature. It has been impossible to work upon it for any continuous period of time. Written in odd half-hours, it has been often laid aside for weeks and even months. My thanks are therefore due, in more abundant measure, to Professor Ashley, Sir Ernest Clarke, and Mr. H. Trustram Eve, who have kindly read the proof-sheets and helped me with corrections, and above all to Mr. G. H. Holden, who has also verified the references and prepared the Index.

ROWLAND E. PROTHERO

September 6, 1912

CONTENTS

CHAPTER VII

JETHRO TULL AND LORD TOWNSHEND : 1700-1760

CHAPTER VIII

THE STOCK-BREEDER'S ART AND ROBERT BAKEWELL : 1725-1795

CHAPTER IX

ARTHUR YOUNG AND THE DIFFUSION OF KNOWLEDGE : 1760-1800

CHAPTER XIV

THE RURAL POPULATION: 1780-1813

CHAPTER XV

AGRICULTURAL DEPRESSION AND THE POOR LAW: 1813-1837

CHAPTER XVI

TITHES

APPENDIX III

THE CORN LAWS

APPENDIX IV

APPENDIX V

APPENDIX VI

APPENDIX VII

APPENDIX VIII

APPENDIX IX

APPENDIX X

INTRODUCTION

PART ONE: BEFORE 1815

by

G. E. FUSSELL

I

ERNLE'S SOURCES AND OTHERS

S I N C E Lord Ernle, then Mr. R. E. Prothero, first published his *English Farming, Past and Present* in 1912, a great many students have written innumerable essays and books on the varied aspects of English agrarian and agricultural history. Before then and between the time when the first version of his book, *The Pioneers and Progress of English Farming*, was published in 1888 many works on the economic history of farming, the history of systems of land tenure, the social relations of noble, franklin and serf, appeared, but in these works there is little about the story of farming practice and its development. A notable exception is Russell M. Garnier's *History of the Landed Interest* (2 vols., 1892–3), a work that should certainly be read in parallel with Ernle by all students, although much of its interest is placed in Scotland.

A few attempts have been made to produce textbooks that might supersede Ernle's work, but so far none of these has done so. It has its shortcomings, discernible in the light of modern research, but it remains and will, I believe, remain the classic handbook of the subject. For one thing the style of writing is highly picturesque, at times vying with Macaulay or with Trevelyan. Many of the thumb-nail sketches, like that of the settlement of Glendale, Northumberland, in the late eighteenth century, are quite unforgettable, and remain firmly fixed in the reader's mind. A rather pedestrian style of writing often favoured by the more meticulous mind is much less of a mnemonic, though it may possibly present the facts more accurately, but a discussion of this abstruse question would lead into a very tangled forest, and need not be pursued here. I doubt that it would be adequately overtaken.

Nearly three-quarters of a century has passed since Ernle wrote the little book which is the foundation of his larger work, and it would be astonishing if the intensive research into the subject which has been carried out by numerous scholars should not have turned up new classes of evidence, both of direct and indirect value. Some of these I have used myself, and other writers have developed their use, and indicated other founts of information. For himself Ernle relies in the main upon the printed farming textbooks from Fitzherbert onwards, and for what is really the beginning of his story upon Walter of Henley and other treatises written at about the same time, which were printed by Cunningham and Lamond as *Walter of Henley's Husbandry*, etc., in 1890. These works he studied with the greatest thoroughness, and he expanded his reading by perusing some of the controversial pamphlets and other literature of Tudor and later times. Other material that was available he entirely neglected. Though he had read Maitland he does not seem to have been acquainted with the other great medievalists, Vinogradoff, Round and the rest. He used the farming textbooks, but not the historians who were his contemporaries : nor was he over-critical of the sources he did use.

The early farming textbooks, on which I, too, have mainly relied, contain many anachronisms, and are a strange mixture of practical wisdom with ancient superstition derived from classical and other sources. Many repeat almost verbally what their predecessors said, and often they proposed the most remarkable proceedings. They must however have been used as handbooks by the few people who read them or could read them, and to this degree must have seemed reasonable to the contemporary public. So far as their theories are concerned they are often quite absurd to a present-day reader, but this is because their wild speculations were the beginnings of the enquiries that have developed into modern scientific farming. When they state that a method was that of farmers working in a particular area, or something that an individual farmer was doing, there is no reason to suppose that they were not speaking the truth; at the most they are only suspect when they are plagiarists of earlier writers, or the context is generally unreliable. They are, in fact, the prime source of information about farming practice, and certainly for agricultural theory. To a degree therefore Ernle was justified in his choice of sources.

But the textbooks need to be supplemented, and some of the material which Ernle could have used in this way was available when he was preparing his book : the balance has since been discovered. The first and most obvious is the immense number of apparently trivial, but often surprisingly useful, local histories of parishes, hundreds and counties. No one could hope to read them all. Legions have been printed since the seventeenth century, and many of them are now of little interest. Some are just the opposite, and, amongst a medley of amateur antiquarianism, documents of the greatest value are sometimes included in whole or in part. Probably Ernle had to neglect this source on the score of time. He was not only a historian, but a practising land agent and politician, and a writer on other subjects. Maybe the possibilities of these works did not occur to him. Details of these productions filed in the British Museum before 1881 can be found in John P. Anderson's *The Book of British Topography*, 1881. Such things in countless numbers and of the most variable quality have been issued since then, and will doubtless continue to appear.

Again the great topographers from Leland through Camden and their imitators are informative, and the works of the foreign and other tourists who travelled in England in search of information about political and social organisation, or merely on pleasure bent, seeking the picturesque, contain incidental remarks that are highly illuminating —see my *The Exploration of England. A select bibliography of travel and topography, 1570–1815* (1935). The greatest exemplar of these tourists was Arthur Young, and he had many apostles, but this kind of writing only became popular towards the end of the eighteenth century. Earlier " tours and topography " do however contain both economic and technical notes about farming, and a great deal about the social conditions their authors encountered.

Most of the descriptions of journeys are naturally in the form of diaries, day-to-day records of events. Keeping diaries of the uneventful round of rustic daily life is an occupation that can almost be said to be a human failing. Very ordinary country squires and some farmers kept them, fortunately, and, like those of better known and more famous personalities in a wider field, supply what must be accepted as definite and accurate evidence of the way of life of the writers and their locality. Ernle made substantial use of only one of these, that of Henry Best, which was printed by the Surtees Society just over a hundred years ago under the title *Rural Economy in Yorkshire in 1641* (vol. XXXIII, 1857). Similarly he only used one or two of the county descriptions, Robert Reyce's *Breviary of Suffolk*, 1618, and the two books on Oxford and Stafford by Robert Plot.

Best's record was an account book as well as a diary, and some other account books have since been printed and widely used. Among them are Eleanor C. Lodge's " Account Book of a Kentish Estate, 1616–1704 ", *Proceedings of the British Academy, IV*, 1927, and my " Robert Loder's Farm Accounts, 1610–1620 ", *Camden Society 3rd Series*, vol. LIII, 1936. Only a year before Best's work was published, the Chetham Society produced *The house and farm accounts of the Shuttleworths of Gawthorpe Hall . . . Lancashire* (1856). There must be many more collected in the repositories of county archives that have been established in the past few decades. Indeed extracts from such documents have been printed in the proceedings and transactions of similar societies, and county and more local archaeological and antiquarian societies, some of which have acquired outstanding reputations, while others have been content with rather undiscriminating trivialities. All can be useful, as I tried to show when I extracted sample material from the inventories of domestic goods and live and dead stock made by the executors of wills, and for other purposes, as " Farmer's Goods and Chattels, 1500–1800 " in *History N.S.*, XX, Dec. 1935, a source that has been exploited in greater detail by Dr. W. G. Hoskins, Francis W. Steer, F. G. Emmison, G. H. Kenyon and others.

One other valuable contribution several of these societies have made is to print exegeses upon and details of the 1801 and other crop censuses made when invasion was threatened during the French Wars. These documents were, I believe, discovered first by Dr. Hoskins, and have since formed the subject of study by several scholars, their results having been published by local societies, and by the Royal Geographical Society in its Journal.

The essays in the different Victoria County Histories are of a more comprehensive character, and deal with a longer period of time than these documents and cannot be neglected : but valuable as most are, they are not all of an equally high standard. In this connection I may perhaps be allowed to mention the dozen or so county societies which have been good enough to publish my local studies under the title of *Four centuries of farming systems in . . .* individual counties. These I fear have many defects, but were an attempt to gather together information about each county discussed. They can, of course, be supplemented by the large number of contributions to the learned journals, both national and foreign, that are concerned with the subject.

So far I have only discussed printed sources, because it was upon printed sources that Ernle chiefly relied. My remarks should be amplified by reference to Dr. Joan Thirsk's " Content and sources of English agrarian history after 1500 ", *Agricultural History Review*, III, 1955.

The basic sources for any history earlier than the era of the printed book is the written document, and I am certain that chance has played a large part in the survival of particular papers. Necessarily, too, so far as rural life is concerned, the legal aspect of land holding was of the highest importance to great landowners, the church and the king. Inevitably such things as royal laws, charters and leases of land must have some relation to the contemporary processes of farming, but they are rarely definitive enough to give precise information. Custumals only begin to take the solid form of the written word at a comparatively late period, and accounts of royal and ecclesiastical estates only become fairly numerous in the thirteenth century and later. Land surveys and terriers, too, of medieval date are late rather than early.

Many scholars have worked on these sources in the past century. Little of their work beyond that of Seebohm's *English Village Community*, 1883, was known to Ernle. Much of it was done after his book was written. A recent summary of both documents and modern works has been provided by R. H. Hilton, " The Content and Sources of English Agrarian History before 1500 ", an essay that also appeared in the *Agricultural History Review*, vol. III, 1955. I shall refer to them further in the attempt that follows to expand the first few pages of Ernle's book in the light of this research, and shall also set out something of the results obtained by modern archaeological investigation, which has made a material contribution to our knowledge of prehistoric and pre-Norman Conquest agriculture in this country, although many questions relating to this time remain as yet unanswered, or at the best are the subject of speculation.

II

PREHISTORY AND ROMAN OCCUPATION

ARABLE farming was practised on small plots in very ancient times in England : in other countries it may have been more ancient, and developed on a larger scale combined closely with livestock husbandry. A great deal of work has been done by the archaeologists on prehistoric village sites, and on Roman and Anglo-Saxon remains. Of this I cannot pretend to make a critical assessment ; it is far outside the realm of my special knowledge, which is indeed largely confined to the modern period. The most I can do is to indicate the nature of the work that has been done, and the publications in which it can be studied. If I offer any speculation on the technical questions involved to which firm answers founded on firm evidence have yet to be discovered, it must be understood that they are no more than speculations.

The settlements of early man in England were on the hills, Salisbury Plain, Dartmoor, the South Downs, where traces of their activities remain, or have been clearly defined by the new process of air photography. Two modern writers have summarised what is known about the technique of farming in the days before the coming of the Romans. They are M. E. Seebohm (now Mrs. Christie) in *The Evolution of the English Farm*, 2nd edition, 1952, and R. Trow-Smith in *English Husbandry from the earliest times to the present day*, 1951. To these may 'be added E. Cecil Curwen, *Plough and Pasture*, 1946: and, of course, the authorities on which these writers rely.

The farmers of the Neolithic, Bronze and Early Iron Ages all cultivated wheat for bread corn, and kept livestock, cattle, sheep and pigs. Cultivation is believed to have been by hoe in the very earliest times, and by a light two-ox plough without a coulter or a mouldboard, known to the Romans as the aratrum, and in Scandinavian countries as the ard. This was an ancient implement used by the Sumerians some 3,500 B.C., but is said to have been brought to this country about 1,000 B.C. by immigrants of Celtic stock. It is the implement which ploughed the so-called Celtic rectangular fields that have been studied so carefully from the air of late. Light ploughs of this pattern have survived in the Mediterranean region until the present day, and are effective enough in the kind of dry land farming usual in warm dry climates, where the soil must be brought to a fine tilth, and kept in that state to retain as much soil moisture as possible and prevent undue evaporation. Its use in Northern Europe in the Bronze Age is further evidence that the climate was then warm and dry (see F. G. Payne, " The British Plough : some stages in its development ", *Agricultural History Review*, V, 1957, and the authorities there cited). In the eighteenth century several eminent agriculturists made, or had made, models of the ard and other ploughs then in use, and these are now housed in the Royal Swedish Academy of Agriculture Museum, Stockholm. Others are housed in the Swedish Institute of Agricultural Engineering, Ultuna, Uppsala. Both collections have been described by Ragnar Jirlow, the first in the Academy's Report No. 6, 1951, *Äldre Plogar och Årder i Kungl. Land-bruksacademiens Museum*, and the second in the Institute Report No. 213, *Plogmodeller fran 1700 Talet vid Landbrukshögskolan*, 1948. The actual models doubtless give a better idea of what the ard was, but the pictures in these reports are illuminating. There are slight modifications between local types, but in general there is great similarity to other ancient implements. Others in Gothenburg Historical Museum are the same. I understand, too, that the Museum of English Rural Life have made an ard, and done some ploughing with it.

Air photography and even ground examination can indicate little of how the ploughing of the Celtic fields was done, but since the plough used was very like, if not exactly the same as the aratrum of the classical authors, the preparation of the seedbed may have been along the classical lines, and the cropping may have been on the two-course of crop and fallow rotation of the Roman writers. For the crop the ground was prepared by two ploughings, occasionally three, in the Greece of Homer, Hesiod and Xenophon: Theophrastus recommended four. When three ploughings were given they were carried out in April, July and September. Virgil advised his Roman readers to plough and sow before the cold of winter, and to plough four times, twice in summer and twice in winter, but Varro, Columella and Palladius adopted the wise mean of three, which was usual in the open fields, and even in enclosures in England until the eighteenth century. Since the aratrum was the plough used in the Mediterranean countries as it was in north-western Europe during the Bronze and Early Iron Age, it is not unreasonable to suppose that the recorded procedure had been that of the peoples who left no written records—but this is, of course, pure speculation.

The Roman fields, like those of the Celtic peoples in this country, were rectangular, and it is now generally believed that the shape of these fields was determined by the tools used, and the soil and climatic conditions. In order to obtain a good tilth with a not too efficient implement that must be pressed to one side by the driver in order to invert the furrow slice, if such it could be called, the second ploughing must be done crossways to the first at right angles in order to break up the surface as completely as possible. It would be inconvenient to do this if the plot were long and narrow ; nearly square is the obvious shape so that each run of the plough before turning might always be much the same. This is, in fact, roughly the shape of the Celtic fields, and the two factors, plough and field design, combine to make this belief reasonable.

On a seedbed of this kind wheat, barley, oats and beans were cultivated. The types of wheat were Emmer (*Triticum diococcum*) ; Spelt (*Triticum spelta*) and a primitive form of *Triticum vulgare* called by Professor Perceval, Breadwheat. The barley was Bere (*Hordeum vulgare*), a type that has been grown ever since, especially in what are called backward parts of the country, hilly and poor soils. Oats were of two types, *Avena strigosa*, and *A. brevis*. The bean was *Vicia Fabia*, small, about the size of the pea, and commonly cultivated by Bronze Age and Neolithic people in Central Europe. Already by 300 B.C. wheat was " abundant in the south-east, and thrashing was done under cover on barn floors ". (John Perceval, *Wheat in England*, 2nd edition, 1948. J. R. B. Arthur, " Prehistoric wheats in Sussex ", *Sussex Archaeological Collection*, XCII.) All the domestic animals were kept by Iron Age men, cattle, sheep, goats, pigs and horses, the latter being a necessity for warriors who fought in chariots. R. Trow-Smith in his *History of British Livestock Husbandry to 1700 A.D.*, 1957, has discussed the types. Since all we know of these animals is learned from their skeletal remains from natural death, slaughter or the cooking fire, nothing can be gleaned about the methods of breeding, feeding and herding of the different classes. They must have lived as well, or as ill, as they could on natural vegetation, and very much in a state of nature except that some efforts were made to protect them from the onslaughts of wild carnivores and night raiders.

The social organisation of these peoples, our so remote ancestors, will probably never be definitely known. All that the ages have preserved besides the outline of their fields is some indication of their methods of dealing with their physical environment, their housing and household goods, but these are subjects which are adequately dealt with in the several works on prehistory, and need not engage us here, especially as all I could do would be to reproduce their results.   Woollen fabrics and perhaps linen were woven, pottery was made, so that men wore clothing and their women were able to cook in other ways than by grilling or roasting against the fire.   Grain, after being threshed in one or other primitive fashion, was parched or partly dried by heat and stored in pit silos to preserve it for future use.   It was ground in a hand quern.   (E. C. Curwen, " Prehistoric agriculture in Britain," *Antiquity*, I, p. 261.   Gordon Childe, *Prehistoric Communities in the British Isles*, 1947.)

At some time towards the end of the Early Iron Age the climate of north-west Europe including, of course, our island, began to deteriorate. The mild dry Mediterranean type of weather changed to the modern sort, wet and cold, an atmosphere carrying a heavy charge of moisture through which the sun penetrated with difficulty and only in especially good conditions.   The soil, too, especially the lowland with its extensive forest cover, was sodden, the rivers inclined to spread out into riparian marshes.   These new conditions were not created in any cataclysmic way.   The men of the time must often have said to one another, as we have been doing for the past few years, that they could not remember such a long series of bad winters and poor summers.   The ploughmen found that working with the ard in these changing conditions was growing more and more difficult.   Mr. F. G. Payne (op. cit.) believes they added a coulter to split the earth before the share, and a mouldboard to invert the furrow slice behind it.   Both these " improvements " added to the tractive effort required to drive the plough through the soil.   To ease this increased draught the fore-end of the ploughbeam may have been supported on a framework consisting of an axle carrying a pair of wheels, and so became what has been called the Belgic plough. The period when this plough was first designed is not known.   It has been placed in the fourth century B.C. by Curwen, who is convinced that it was an unmistakable heavy wheeled plough, though wheels were not essential, and over a large part of England swing ploughs, or ploughs without wheels, were in use until the end of animal traction. They are, if I am not mistaken, still used in the archaic open fields at Laxton (or Lexington), Nottinghamshire.

Clapham (*Concise Economic History of Britain*, 1949) attributes this agricultural revolution, caused by changes in climate and an onslaught on heavier wet soils, to the Belgae: Childe (*What Happened in History*, Penguin, 1957), to the Germans, who having learned the secrets of iron working from the Celts, apparently invented " a system of tillage appropriate to the heavy clay lands of the Northern European forests— deep cultivation with a heavy plough drawn by eight oxen, and equipped with a mouldboard and coulter to turn over the sod instead of just scratching the soil as the Mediterranean and Celtic ploughs did ". He suggests that these people and their plough reached south-eastern England about 75 B.C., and that they were such excellent farmers that they were exporting grain from the cultivation of their newly won lands twenty-five years later. If the long, heavy plough, with or without wheels, and drawn by more than two, but not necessarily eight oxen, was introduced then, it was the agent of a great change in the layout of arable land. No longer was it best to have the fields nearly square so that they could equally well be ploughed parallel to any side, but the length of the plough and its team as well as its capacity for work led the farmers to form their " lands " in long narrow strips, a procedure that finally created the well-known open-field strip system of Midland England in Medieval times, and retained it there until the enclosures of the eighteenth and nineteenth centuries. Various technical reasons are adduced in explanation of this change : but, if the plough was first introduced by Belgic immigrants, its use was extended after the Roman occupation with the arrival of the Anglo-Saxon invaders, and it may be discussed more fully in that connection.

Meanwhile the invention of the scythe shows that Early Iron Age men were trying to improve the conditions under which their cattle lived. The scythe enabled them to cut grass from the wild pastures or the marshy riparian flood meadows to make hay for feeding their animals in the winter, but to what extent they did this cannot now be known. Here again it is only when the classical writers begin to describe the farming of their times that reasonable suppositions (for they are no more) can be formulated. The Romans came to Britain at the opening of the Christian era, and remained for some four centuries. They must have brought with them both practical experience of farming, and of the theoretical views held by their sages. Both are discussed by Raymond Billiard in *L'agriculture dans l'antiquité d'après les Georgiques de Virgile*, Paris, 1928. It is, of course, unlikely that the Roman methods would be adopted *in toto*. Factors of climate, soil, etc., would make this practically impossible. Practical farming necessarily varied through the widespread Roman Empire, and it might be that a fruitful line of research would be to identify the legions stationed at different places in England. Their origin might suggest the adoption of their national farming technique in the stations, for example, the legion that was settled as *coloni* in the Lincolnshire Fens, and whose farming may have supplied the garrisons at Lincoln and York by water transport along the canals cut by the Romans through the Fens. This is however nothing more than a passing thought, and must be left to someone younger and better equipped than myself, possibly to Mr. Simon Applebaum, whose " Agriculture in Roman Britain ", *Agricultural History Review*, VI, 1958, is the most detailed study of this era that has so far been made.

There were changes in the approximately four centuries of the Roman occupation, but one thing is very important—the land was nearly empty. The total population of the " province " was probably never more than half a million, and the greater part of England remained in the state to which nature had then brought it. The patches of settlement were small, and the new towns laid out by the Romans were more frequent in the south and south-west than elsewhere. Great areas were given up to forest ; fens covered much of eastern England and of Somerset ; wild moorland, almost if not wholly uninhabited, stretched across England in the north and even the peopled parts of chalk downs were not worried with any traffic problems. The pack trains carrying goods along the ancient ways and new roads, and the small, infrequent parties of travellers only had to face the natural hazards of the journey and the possibility of attack by spoilers.

There was a great extension of corn growing in Roman times in western Europe from the Atlantic to the Rhine, and this applied in England as well. The Belgic settlers had begun to attack the heavier soils, the medium loams, before the Romans, and consequently there was some improvement in crops and livestock. The Roman conquest did not wipe out the previous occupiers, who continued to work their square fields, and to live in pit dwellings or circular houses of wattle and thatch. Progress in the heavier and valley soils was made by Belgic elements, and by the occupants of the new Roman villas, which were isolated country houses standing in the midst of their estates. There is some controversy, which will probably never be finally resolved, about the farming methods and tools used on these estates.

The layout of the Roman villa seems to have made provision for housing livestock, and this must have enabled cattle keepers to select their breeding animals. The use of the scythe for cutting natural hay must have simplified, to a minor degree if no more, the problem of winter keep. The ox and the horse of Roman times was larger than its ancestors. But the sites indicate that they were chosen for their arable possibilities. Applebaum has endeavoured to show that a preference for winter-sown cereals over spring-sown began to appear, " partly under the impact of climatic deterioration " which was cumulative in its effects.

Whether this procedure was also affected by the " new " plough, or whether the " new " plough was the prime cause will never be certainly decided. There are still many questions to be answered about the primitive plough for which the few scattered relics of coulters and shares are inadequate. The ard or araire continued to be used in the Celtic fields. The Romans had applied the coulter to the aratrum. Ears, either on one or both sides of the share, had added to the efficiency of this plough. The famous Pliny passage refers to wheeled ploughs in Rhaetia and in Cis-Alpine Gaul, of which Virgil was a native, but though he was writing in the first century A.D. the text containing the passage is of later date, and is corrupt according to the textual critics. Charles Parain is of the opinion that the mouldboard certainly came much later than wheels. " Perhaps," he writes in " The Evolution of Agricultural Technique " (contributed to vol. I of the *Cambridge Economic History*, 1941), " it was invented in Flanders, a flat wooden plank serviceable only in strong stoneless land such as the Flemish clay."

Applebaum is of opinion that some villas were erected on the sites of earlier improved farms, and were sometimes worked in conjunction with Celtic villages, but Collingwood does not agree. For the latter, *Roman Britain*, 1937, the village was a group of huts of one room, round in shape, perhaps pit dwellings or dry stone hutments built above ground. The old villages and the new villas were in separate places. Villas are found south of a line between Severn and Trent, though a few sites are outside this area. The villages remained on Salisbury Plain, Cranborne Chase and the other hill sites. Taxation or tribute (rent) pressed heavily on the cultivators of the soil. Landownership was profitable, but the cultivator himself was left with only the barest of subsistence (see Charles Stevens, " Agriculture and Rural Life in the later Roman Empire ", *Cambridge Economic History*, I). Nevertheless the inhabitants of the Thames valley villages " in their damp and dismal pits " were not undernourished as their immunity from rickets demonstrates. They had strong bones though they were a trifle shorter than their descendants. Of their corn they had little left when the landlord and the tax collector had taken their "share". All of it was eaten. No rats flourished in their granaries. They ate good meals of pork and mutton, and it may have been the high proportion of flesh in their diet that kept up their health and strength.

The villa system, adopted by the landowners of Romano-British origin was, Collingwood thinks, a Romanised version of a system already existing before the Conquest. The labour was probably slaves housed with the animals, or at least in the same part of the house, if they lived in ; otherwise they were probably *Servi casati*, hutted slaves, each "given a piece of land and told to keep himself ". Such men probably practised a manual hoe culture. Most of the hand tools used today were used in Roman Britain, iron or iron-tipped spades and forks, rakes, mattocks, bill-hooks, and so on.

All the cereals were grown, wheat, barley, rye and oats (J. R. B. Arthur, " British grain in Roman times ", *Agriculture*, 64, 1957). Wheat of several types was predominant ; barley ran it a close second ; and oats, the last being grown to a greater extent in the north, perhaps as a fodder crop for horses. Besides the cereals some vegetables were cultivated, and in the orchards there were several varieties of fruit. The normal Roman villa was a profit-making farm. " Comfortable on the average, luxurious at best, and squalid at worst, the Roman British villas ranged in size from cottage to mansion, and the associated acreage varied accordingly " (I. A. Richmond, *Roman Britain*, Penguin, 1958).

The simplest type of house was a barn-like building half of which housed the family, and the other half the livestock. These included cattle, sheep, pigs and horses, and some poultry, fowls, and the goose, which was a Roman introduction. Later and more developed styles were as elaborate as that at Chedworth, Gloucestershire, where there was some working up of agricultural raw material. The arrangement of the expanding villa, or the (if the term may be used) Romanisation of farms, as on Cranborne Chase, allowed for the better maintenance of livestock over winter, and the consequent accumulation of manure could be used to increase the grain yield. Possibly this may have permitted the introduction of a three-course rotation, suggested by the more advanced classical writers, in place of the normal crop and fallow. Marling was a process well known in classical countries, and the material may have been dug and used in Roman Britain (see my " Marl, an ancient manure," *Nature*, 24 Jan., 1959). The compost heap too must have been made of all that variety of waste and other products the classics had recommended.

There were regional divergencies in the systems of farming. In Southern Roman Britain some evidence has survived of the increase of pastoral or livestock husbandry during the Roman occupation at the expense of arable, but in the north the process was probably reversed (Richmond). There is, however, still so much to be learned about farming and the life of the natives and of the immigrants in the country-side that it is here impossible to do more than indicate the modern sources mentioned above, and give the barest outline of what they relate.

III

THE ANGLO-SAXONS

ARCHAEOLOGY has provided much the greater part of the evidence about the "life and times ", as the phrase used to run, of the Roman occupation of Britain, and of the numerous peoples who preceded them. The arrival of the Teutonic invaders from the valley lands of the Elbe and the Rhine, from Friesia, Scandinavia and Denmark, does not precisely mark the beginning of written history ; it remained unwritten for some centuries, but a few documents have come down to us to supplement such things as Bede's *Ecclesiastical History*, and the *Anglo-Saxon Chronicle*. It is from legal enactments, charters making grants of land, and other documents of this kind that most information can be gained about the rural life and farming of the Anglo-Saxons, as they are called.

This conquest was not cataclysmic. It was not completed in a brief space of time, and there had been Teutonic immigrants before the last legion left. It had indeed been necessary to appoint a Count of the Saxon Shore to defend that coastline against these barbarians. It was bad policy, perhaps, but at the last some were hired to protect the land from their kin. These were paid with land and became settlers. They did not succeed in doing the work they were hired to do, and possibly did not even try very hard : but the whole of England, as it then became, was not occupied until two hundred years had passed.

Since the general occupation proceeded by degrees there was no wholesale destruction or flight of the provincial population followed by the formation of Teutonic communities on a clean slate such as J. R. Green described in his *Short History*. The sporadic warfare of these centuries naturally interrupted the continuity of provincial institutions, and isolated villas may have been burned out, while towns were also affected. (Vinogradoff, *The Growth of the Manor*, 1905, pp. 117–18.)

The Roman towns had been decaying for a long time before the Anglo-Saxons came. The eruption of the barbarians across Europe had broken the economic nexus with Rome, and as their overseas trade dwindled they shrank. The new invaders did not destroy such towns as were still occupied. The townsmen could supply many things they needed. It has been pointed out that there is a difference between sacking a town and destroying it : or again that the Saxons feared the stone-built towns as the work of giants. They lived and were buried outside the towns. Their cemeteries are found outside York, Caistor, Cambridge and Lincoln. As they were farmers their livelihood was in the country, not in the towns, and it was a racial characteristic that they preferred to live each family in splendid isolation. (D. B. Harden, ed. *Dark Age Britain : Studies presented to E. T. Leeds*, 1956, p. 116 ff.)

These new people brought with them the plough over which there has been so much controversy. This was the implement developed in the heavy forest soils of southern Germany, and afterwards brought to the settlements on the coastline from Denmark to northern France. It is the so-called Belgic plough, and had been brought to this country before the Roman Conquest. Probably some of the Romano-British villa owners used it, but its real introduction was by the Anglo-Saxons, who followed the Romans and were the first Englishmen.

This plough, as described by Pliny, was fitted with a share, a coulter and a mouldboard. It may or may not have had wheels : but it was an altogether larger object than the Celtic plough or that used by the Romans on their light Mediterranean soil. If the Piercebridge model is any guide to the size of the plough-oxen then the heavier plough might certainly require eight or even more draught cattle. In the model the animals, supposing their proportion to the driver to be correct, could not have been larger than a small modern Kerry cow. They were nothing approaching the size of the beasts now kept as a curiosity plough team at Cirencester Park. Whether the animals were large or small, the generally used traction unit has been accepted as eight animals. This is based on the Domesday use of this unit, and it is now quite certain that in practice, as opposed to the legal use of it for surveying purposes, the eight-ox team was by no means usual after the conquest. Teams of varying numbers of animals were yoked to the plough, from the two oxen which hauled the aratrum with or without the new attachments, coulter and mouldboard, to more than eight, which hauled the heavier plough through the denser soils of the valleys preferred by the new settlers.

The technicalities involved in clearing and breaking up new land with this plough have been discussed at length by Dr. and Mrs. Orwin in *The Open Fields*, 1st edition, 1938, 2nd edition, 1954: and indeed the ideas of their predecessors carefully considered. It is, of course, impossible to deny that it would be more convenient to use this heavy plough to make a longer furrow than that common in the small square Celtic fields or the similar jugera of the Roman coloni. This factor has been generally accepted as the main reason for the change in the shape from the small square field to the arrangement of long narrow lands or strips, which was the common layout of the open arable fields from some unfixable date before the Norman Conquest until enclosure and amalgamation changed them once more into fields defined by ditch and hedge or other physical boundary.

The origin of the mixed strip system of arable landholding is still unexplained, widespread as it was. Shifting cultivation in the forest lands has been suggested as a reason for experience in co-operation in working the land, but clearly when the Teutonic peoples settled in this country they brought with them a fully developed system. I imagine that they did not invariably attack virgin soil, cleared of brushwood, forest growth or tall weeds, aquatic or otherwise. Some of them must have taken over land previously cultivated. Even if they destroyed the buildings of a Roman villa, they would have been foolish to neglect the opportunity offered to them by the efforts of the Roman cultivator. In this connection Dr. H. P. R. Finberg's *Roman and Saxon Withington*, 1955, is of interest. (See also W. G. Hoskins, " The English Landscape " in *Medieval England*, ed. A. L. Poole, 1958, vol. I.)

In places where they did break new ground either in forest clearings, comparatively open spaces with less tree cover, such as the less steeply inclined parts of downland, or marshy land adjoining rivers, it seems to me very doubtful whether any very large area was laid out at any one time in one place. It is much more likely that a family group, possibly of three generations, with followers, servants and slaves, would settle on a likely spot, clear as much of the most suitable land as possible, and plough it in the long furrows and narrow lands their plough and the need for surface drainage demanded. Such activities would never reach the sum total of 2,000 or 3,000 acres that the Orwins say it would puzzle a modern surveyor to arrange. The quantity, supposing that there were as many as ten heads of families, and allowing a conventional virgate of thirty acres to each, would be only 300 acres in each of three fields, or only one field might be ploughed out in the first year of occupation, the second being broken and fallowed in the following spring and summer and so on. However it was done, an area of 300 acres would present no insuperable obstacle to providing intermixed strips because all of it could be seen. Whatever method of distribution was used, that of one " land " to the owner of the plough, the next to the near ox and so on, on Seebohm's principle, or on some idea of sharing out the good land equally, which presupposes a clear idea of soil values, is immaterial to the actual process. As the cleared and cultivated area was expanded new " shots " or " furlongs " could readily be added. This is, of course, guesswork, but no more so than other suggestions. If the settlement was made by a body of people, owning more than one plough and plough team, the allocation of " lands " would be more complicated unless each plough group opened up a separate though adjacent site, possibly with a piece of unploughed land between the two, that would later be brought into the tilled acreage by some member of the slowly growing population. Based as they are on the technical necessities implicit in the type of plough used, the Orwins' theories are the most likely to be correct, but, in fine, they and all the others remain largely conjectural, and most likely will always remain so.

The cultivated area remained relatively small, as how should it do otherwise, with a population that increased by slow degrees to some two million at the time of Domesday. Miss Dorothy Whitelock (*The Beginnings of English Society*, Pelican *History of England*, vol. 2) points out that there were still great stretches of woodland all over the country, and large areas of undrained marsh. The Anglo-Saxon settlements were relatively small, and surrounded by wood and waste. The extant documents are mainly about the open arable fields, but there were also other methods of working the land. Owners of some small estates held their land in blocks, as they did around the isolated farms in the forest clearings that were steadily being won from the waste.

In spite of their possession of the long plough with mouldboard and coulter, and a fair range of hand tools, scythes, sickles, bill-hooks, hammers, axes, iron-tipped spades, our Anglo-Saxon ancestors' arable technique was very much less productive than animal-powered farming, with no more equipment except harrow and roller, was to become as the centuries passed. This is natural enough, but it had heavy consequences. Famine, as Ernle himself said of Tudor living, trod upon the heels of feasting. A full year's subsistence could only be won by steady work all the year round, and a bad season meant that there was a serious food shortage in the months immediately before the next harvest just as there is in primitive Africa today. Wheat, barley and rye were the cereals, and beans and peas were the main legumes grown, and these crops combined with a full year's fallow produced the three-course so-called wheat and bean rotation, that lasted for more than a millennium, especially on the heavy clays. A few vegetables and herbs might be grown in a garden.

The shortage of winter feed—there was little other than wild grass hay—made it necessary to kill off and salt down many of the livestock in the autumn, though, of course, the breeding animals had to be maintained as best they could. The scanty supplies of hay were supplemented with cereal straw, bean haulm, browse, i.e., branches of trees with withered leaves, anything that the animals would eat. A result of major importance was that the number of livestock was kept low, and such manure as they produced—it was rarely anything like enough—was of little fertilising value.

The open-field system which the Anglo-Saxons brought with them from north-west Germany did not spread over the whole of England. Sir Frank Stenton, our greatest authority on Anglo-Saxon England, whose book of that title was published in 1943, deduced from Ine's law that an open-field system existed in seventh-century Wessex, and also probably in the Midlands, Lindsey and Deira. In east Yorkshire, Lincoln and the east Midlands it seems earlier than the Danish invasions of the ninth century : but a large part of the country was not farmed in this way. It was not practised in the north and north-west, along the Welsh border nor in Devon, though the last has been questioned by Dr. Finberg. It was only fully developed in the country which had been settled before the sixth century, and wide areas in that region escaped it. "In Kent a separate race followed a separate agrarian tradition, and in Essex infertile soil and ancient woodland made open-field cultivation unprofitable. Nothing is known of the rural economy of East Anglia before the end of the eleventh century . . .", but the open-field system of the Midland plan was never used there. Probably the first settlers held compact blocks, which disintegrated by reason of division amongst co-heirs. The interest of the open-field system "has led to some exaggeration of the extent to which it formed the framework of old English agrarian life."

Less attention has been paid to the use of the waste though some scholars have emphasised its importance : Sir Frank Stenton in *Anglo-Saxon England* and Maitland in *Domesday Book and Beyond*. The arable could not, in fact, have existed without the waste, and when by assarting and extension of the arable the waste was reduced in area difficulties often arose, as, for example, at Laxton, where most of the waste had been absorbed by the seventeenth century.

The arable open fields, or enclosures, were the nexus, because cereals were the larger part of diet, but the plough beasts had to eat if they were to keep up their strength and continue to do their work satisfactorily. They found their nutriment on the waste at the borders of the arable, on the green tracks that served for village communications, on the hades and slades, the sikes and unploughed headlands. It must have been a fairly poor diet, tough grass much mixed with weeds, and probably overwhelmingly fibrous. Swine, under the care of a herd, roamed the surrounding woods, and comparatively large flocks of sheep were kept on the marshes. For example, there was pasture for some three hundred sheep in Romney Marsh, part of a royal gift, in A.D. 697. Tenterden, Kent, was originally a swine pasture for the men of Thanet, but blocks of pasture had been assigned to individual manors before the Norman Conquest. Not only was there an enormous area of waste in the weald which stretched from Hampshire to Kent, and was 120 miles long by 30 broad, but it was extensive elsewhere. In the parish of Whalley, Lancashire, which Vinogradoff (*Growth of the Manor*, 1905) estimates as comprising 161 square miles, at least 70 formed the forest of Blackburnshire, and were claimed by no township or manor. Some 33,000 acres here were used as pasture and woodland by the various townships, only 3,500 acres being cultivated as arable. There was similar waste marsh in the Fens, and the east border of Essex, " a sort of fringe often covered by water with no delimitation of ownership and used by all the adjoining villages for sheep." In the south-west and north-west a sparse population followed pastoral trades making indiscriminate use of the wide areas of heath, moor and hill.

The size of the kernel of arable land held by different classes of farmers and their relation to each other, to the king, or other intermediary landlord, does not seem very easy to determine. A variety of terms is used in the scanty documents to describe land area. In Kent the sulung was a unit of cultivation that could be ploughed by a single team of eight oxen. It was estimated to be equal to two hides, and, though this is vague enough, because the area of the hide was not constant, it is a sign of the wealth and importance of the Kentish ceorl. Since the hide was elsewhere considered the area that was necessary to support a family, it is evident that in normal seasons the Kentish ceorl would have produce for sale. At Domesday the hide in the east

Midlands, Cambridge, Essex and the Isle of Ely, was reckoned at 120 acres; in Wessex it deteriorated to only 40 acres, which may have been because of the different soil texture.

The ceorl was a freeman and could be the owner of land that had come to him from his fathers, and would go to his sons. Other rights and duties define his status. His holding, too, might be more than one hide, and, if so, he was on the way to becoming a lord. But as time passed some of these men, from whatever cause, lost their original proud status, and only occupied their land and homestead as tenants of some lord at a rent and services. Gradually for some as yet unexplained reason the social standing of the ceorl degenerated. " The general drift of peasant life," wrote Stenton, " in these centuries before the Norman Conquest was from freedom to servitude." At the beginning of the eleventh century thousands of ceorls were bound by a strict routine of weekly labour to a private lord.

In East Anglia, Lincoln, Nottingham and Derby, where the Danes had penetrated and conquered towards the end of the ninth century, the distribution of the land was different. Vinogradoff puts the holding of the free peasant here at as little as twelve acres. Here, too, land was reckoned in carucates of eight bovates, or oxgangs. Trow-Smith, citing the Rev. O. J. Reichel, gives the land measure in Devon as " furlongs ", four of which comprised a ploughland, about sixty-four acres.

The Celtic pastoralists of Wales lived under a system that may have applied to all the Celtic settlements from an earlier time, but there is, of course, no written evidence to support this supposition. The unit was at one time a " tyddyn ", a patriarchial household that might be occupied by three generations and the domestic animals in a building all under one roof similar to that still found in some of the remoter parts of Holland or Southern Germany and Austria. The family lived in one end, and the animals in the other with no partition to delimit the respective areas. A development may have been a group of several " tyddyns ", occupied by different members of the kindred, each having his cow, yoke of oxen, and share in the common fields. Such a group, often or originally nine in number, formed a " trev " or hamlet. This arrangement obviously lent itself to expansion. Vinogradoff likens it to an amoeba which breaks up after three or four generations: the offshoots formed new " trevs ". Each " tyddyn " was enclosed by a wattle or thorn fence, a necessary precaution to prevent the cattle feeding on the thatch, and the whole " trev " was enclosed by a stockade even as G. L. Gomme found was the practice in some remote parts of the world in the late nineteenth century (*Village Community*, 1890). The Welsh occupiers were assessed by groups, and paid their chief a food rent and a small sum of money.

The " tyddyns " near to the arable in the valleys were of sturdy construction, but the practice of transhumance made it necessary to have a summer dwelling in the mountains. This is said to have been a slighter structure, and it was here that the people lived during the warmer months, milking their ewes and cows and making cheese and butter. Both human beings and their livestock returned to the valleys in the autumn to gather their harvest and to enjoy the comparative shelter of the lower elevation.

These Welsh (or Celtic) farmers seem to have enjoyed the status of freemen, a condition that was fast vanishing from Saxon England before the Norman Conquest. The *Rectitudines Singularum Personarum* and the *Gerefa*, the latter being perhaps an addendum to the former, is the only authority of the early eleventh century. It has been closely analysed by Sir Frank Stenton, and by Dorothy Whitelock (*The Beginnings of English Society*, Pelican *History of England*), and by Peter Hunter Blair (*An Introduction to Anglo-Saxon England*, 1956).

Rural society was then composed of three or four classes in addition to the local magnate. The most substantial of these was the geneat, a peasant who had some of the duties of a mounted retainer. His obligations were to mow and reap for a definite period during the lord's harvest ; to guard the lord's person, to carry messages, and do other more or less personal duties. Sir Frank thinks it probable that the holding of the eleventh-century geneat originated in a gift made by an early lord to one of his servants. The gebur occupied a smaller tenement, and his obligations were of a more servile nature. His holding was a yardland—one-quarter hide, or some thirty acres—varying of course with the district—which became the standard but not necessarily the normal holding of a medieval villein. At least one-sixth of the man's time was spent working on the lord's holding, two days all the year round and three in harvest, as well as extra duties at the spring ploughing and sowing. In addition to all this, the lord could call upon him for boon works whenever he chanced to require the gebur's services. The man also paid rent in kind, and a few pence in money. All this seems very exigent, and it is difficult to see how the gebur produced his own living, or what small proportion of it remained to him when he had done so. But the lord should have provided the initial outlay by supplying two oxen, one cow, six sheep and seven acres already sown on his yardland.

Still humbler was the kotsetla. He was a trifle better off than the later cottager for he should have been occupier of at least five acres. He paid no rent but his service obligations were exacting, though they varied in different places. The fortunate might do week work on Monday only ; others as much as three days a week, in harvest possibly more. A day's work at harvest was estimated at one acre oats, one-half acre of other corn. This man was probably either the younger son of a ceorl, or a descendant of such a younger son.

These three classes contained the majority of peasants, but some estate servants may have been even humbler. Whatever their status, life must have been a continual round of work. The actual procession of the seasons with their necessary routine of ploughing, sowing, weeding and harvesting on the arable, and the life processes in the animals, were, of course, the same as they are today : but the implements were crude, and the traction animals weak and weedy so there was a constant struggle against the ever-encroaching wilderness of unwanted vegetation, against which fallow ploughing, and grazing of the stubble by cattle, sheep and horses was the only remedy.

The year's work was, if that is possible, even more endless than it is on a modern farm. As soon as harvest was over, the fallow had to be ploughed for the final bout, such manure as there was having first been carted and spread, before the winter corn was sowed. Immediately that was finished the fallow ploughings began, and these went on at intervals of some months if the village had two fields. Where there were three the autumn work was doubled ; ploughing of the fallow for the spring corn was followed by ploughing in the barley stubble when the animals were moved away from it. There were all sorts of jobs in the winter, killing animals and curing the meat, making tools and kitchen utensils of wood or horn, preparing skins of animals for use, cutting timber, doing house repairs, threshing grain as needed, tending stock in the house and much more. With Plough Monday work in the fields began again. The spring corn land had to be ploughed again, and so did the barley stubble that was to lie fallow all the year ; the strips from which the wheat had been reaped, cleared of grazing cattle, were broken up. In March or April the spring grain was sowed, and possibly harrowed in. Cows and sheep with their young calves and lambs could begin to go out to graze on the new grass of the meadows. Dairy work became important. In May the hay meadows were closed

to the livestock and put up for hay, the animals going out to the wild
pasture of the waste under the care of cowherd, shepherd and swineherd, the goose girl of legend and folklore watching the poultry on the
common. The summer months were an orgy of work with a multi-
plicity of tasks. Hay must be made, the corn crops reaped, the sheep
shorn, the beans and peas harvested, and in the intervals the busy
husbandman may have done some summer ploughing : the preparation
for each crop was traditionally three ploughings at least, and continued
to be so for hundreds of years. And no sooner had one year's annual
round been completed, than preparations for the next must be commenced. Indeed in a season of late harvest the two must often have
overlapped. It was a dour, hard life by modern standards, by which it
ought not to be judged. The peasant countries of today offer the husbandman no more, but he is often a surprisingly cheerful person, and
our Anglo-Saxon ancestors were, no doubt, equally so, though that is
not to say that some of them were not discontented with their lot, which
there was little opportunity of changing. There was some relief in the
seasonal festivals and the numerous church holy days.

IV

MEDIEVAL ENGLAND

T H E broad sweep of generalisation made in Ernle's first few pages is a surprisingly accurate synopsis of the conclusions reached at the time he was writing, and is confirmed by later research. It is much the same as that stated by Gordon East in his *Historical Geography of Europe*, 1930, p. 57. The general structure of agricultural society and land holding was already stratified, as indeed it seems to have been in what we should call the country of origin of the Anglo-Saxon invaders.

Equality in size of holding was never a reality. There were always different classes of men whose position was defined by the size of their holding, and in the approximately six hundred years that had passed since their first landing changes had come about that had pressed heavily on the subordinate classes. All through this period the rich and powerful had been trying to consolidate and improve their position. Misfortune, which could descend like a thunderbolt upon the farmer, whose toil was wringing a bare subsistence from his few acres, made the unlucky one more dependent than ever upon his lord. A bad season, an outbreak of disease amongst the livestock, an onslaught by outlaws, less frequent perhaps than natural disasters, all could force a freeman to place his hands between those of a noble and become his man. For these and other reasons the germ of the manorial system, if no more, was already present in early eleventh-century society.

The Norman Conquest created conditions in which it was necessary to make records, the greatest being Domesday Book, which reduced everything and everybody to order as the precise Norman mind desired. Indeed it was a necessary preliminary to government and taxation, and if not quite so orderly and uniform as the king might have desired, that must be attributed to the varying conditions the surveyors met with rather than any lack of zeal on their part. A useful and informative list of this and other surveys is provided in " A Classified List of the Agrarian Surveys in the Public Record Office," compiled by one of Hubert Hall's seminars and printed in *Economica* No. 4.

Ernle himself has pointed to the danger of broad and general state-ments, a danger that was made so clear by Maitland : but he does not fail to provide a lengthy and detailed description of life on a thirteenth-century manor. By implication, if not direct statement, this description was intended by Ernle to apply to the social and agrarian organisation of the manor, and consequently the village, all over the country in the age of Walter of Henley. It is not so applicable, as both Vinogradoff and Maitland are careful to say, and it would not have been nationally applicable in the period between the departure of the legions and the Norman Conquest ; only to the Midland belt where the open-field system was usual. Some modifications in detail are necessary in those areas where there were only two fields, the northern slope of the Sussex downs, the northern slope of the Berkshire downs, and the slope from the Lincolnshire highlands towards the North Sea. The area in which the open-field system of farming was practised has been closely defined by Dr. and Mrs. Orwin in *The Open Fields*, 1954. In addition to the great block of open-field farming in the midland and eastern counties, which persisted until the eighteenth century, there were at different times, mainly early times, outliers of open field in the south-west, the north-west and north-east, and Kent and Essex may have had some, though, generally speaking, these counties seem to have been enclosed directly from the waste.

It may be true as Ernle says (p. 5) that all over this area there were nucleated villages, and no isolated farmhouses, but the country was not yet wholly occupied in the eleventh century, and both Maitland and Vinogradoff emphasise that the pattern of settlement was not every-where the same, a conclusion that has been confirmed by recent work done by Dr. Joan Thirsk, but not yet published—in especial relation to East Anglia.

Domesday Book is the document that describes England as it was towards the end of the eleventh century. It is unique, and has been studied by generations of scholars. Maitland's book has become a classic. Parts of Domesday have been translated and printed in the massive collection of *English Historical Documents* edited by D. C. Douglas and printed by Eyre and Spottiswoode during the past few years. H. C. Darby's volumes on *Domesday Geography* are a most valuable analysis. Nothing is more clearly disclosed than the wide variety of conditions. The village was the most common form of settlement over most of the country. In Lincolnshire and Yorkshire there were scattered hamlets and homesteads, but most of Yorkshire had been devastated and the outlines of the arable fields lost. When new settlers came they would begin anew by ploughing small areas, enough for maintenance, and would possibly find an opportunity of making three fields instead of two (Maitland, p. 264). There was one house at Eardisley, Herefordshire, with one plough team, two serfs and a Welshman who paid 3s. rent, but then few ploughlands were recorded in Herefordshire, only five to a square mile in the east of the county, falling to one in the west, and less than one-half in the north-west, suggesting that there was still a good deal of land available for future cultivation. Similarly there was only one plough team to the square mile in the Forest of Dean, but over the Cotswolds there were from three to five, though the density in the Vale of Gloucester was lower. Shropshire was not densely populated with 1,809 teams, and in this county, too, plentiful land was not yet cultivated. Whether the same could be said for Leicestershire, where there were 1,858 teams, is an open question. (Darby and Terrett, *Domesday Geography of Midland England*, 1954.) Some of the great ecclesiastical estates owned a large number of ploughs. Tavistock Abbey, for instance, had twenty-two plough teams at work on their Devonshire lands. (H. P. R. Finberg, *Tavistock Abbey, a study in the social and economic history of Devon*, 1951, p. 86.) This abbey, too, like others, had large herds and flocks, among which were no less than 167 goats.

When the density of plough teams varied so much the numbers of their owners varied equally. Few villages were so large as Elmston, Derby, where there were thirty-six households, presumed by Vinogradoff to comprise 180 persons with some bordarii and cottars in addition, but in that county the number of isolated households was also small. There were thirty-three hamlets of two to five households, fifty-eight small villages of six to eleven, forty-three larger villages of twelve households. The prevailing settlement was therefore, he argued, the village of moderate size. Essex on the contrary was a county of larger settlements. Here they were thirty-six of 381 settlers, 115 hamlets, ninety-seven small villages and 133 large. Examples could be multiplied, but sufficient has been said to indicate the wide diversity between the conditions in which rural people (there were few others) lived in the eleventh century.

Before the Conquest reclamations of forest and waste had been going on sporadically and slowly as human requirements and ambitions demanded. Where every village and indeed " every hide " almost had to be self-sufficient, a slight increase in population enforced an extension of the arable so that the extra mouths could be filled. Place names show that this was being done " before the Danish invasions ". Some names embody the name of the man who was responsible, " one of the driving peasants who throve so that he had full five hides of his own land, church and kitchen, bell house and burhgate ". There was a great deal of land suitable for reclamation—in the sense of being cultivated— at the Norman Conquest, and the new invaders had every reason to encourage agricultural development. The ploughing of new land and the allocation of new rights in the pastures and woodlands meant increased rents. The consequence was that men were expanding the acreage under the plough in the old-established villages and the hamlets that had grown up on their borders. As the village people slowly grew more numerous they built a few more houses and ploughed a few more acres. Those whose needs caused them to break new land on the outskirts built on the edge of the woods, or in a natural or man-made clearing, and so the number of scattered homesteads increased, or small groups of two or three settlers who gathered together for company or protection in tiny hamlets. The process continued for a couple of centuries, and reached its maximum in the thirteenth century. (*Cambridge Economic History*, I, pp. 50, 75; Clapham, *Concise Economic History*, p. 79; Hoskins, " English Landscape " in A. L. Poole, *Medieval England*, 1958, vol. I.)

These new lands need not necessarily have been laid out on the strip system. The long furrow which was a consequence of using the swing plough or wheel plough with coulter and mouldboard did not forbid enclosed fields as eighteenth-century farming clearly demonstrates. Vinogradoff believed, with some show of reason, that the only influence on the distribution of fields was the lie of the country, the quality and quantity of soil at the disposal of the tillers, and the system by which they worked it. Even if compact enclosures were divided through gavelkind, or for some other reason, they could retain their shape on the ground, though Mrs. Thirsk thinks that division in this way in East Anglia may have created a strip system of ownership. The ploughlands could be " rounded off units on the soil ", and occasionally even in village settlements the inhabitants received separate allotments, not intermixed, but by the side of each other, perhaps particularly in Danish counties.

In the eleventh century the meadows were scanty, rarely sufficient to supply the demesne livestock, and the tenants' horses and oxen. The possibility of selling hay off the manor was rare indeed. Wild pasture was still the main resource, as it continued to be for centuries. Pannage for swine, the most common flesh food of the Middle Ages, was an important source of revenue from the woodlands; it competed with the use of timber for building and heating, and tended to open up the forest and expand the natural clearings. Moreover the grazing animal consumed the young shoots and so helped to prevent natural regeneration.

Farming technique changed little or not at all during this time though it is possible that by rearrangement, or extension of the area, or merely by actual use, the three-field layout replaced the two-field in some places. There is no possible measure of this change, if indeed it took place. Two-field villages were still to be found in the seventeenth and eighteenth centuries. Much has been written about the fossilisation of practice that was inherent in the open-field system of land-holding : it was one of the most loudly-voiced arguments of the eighteenth-century enclosers. But half a century ago Maitland pointed out that there was change although the arrangement seemed designed to resist innovation, and to force men who were in favour of newfangled notions to abandon them under the pressure of tradition and their fellows' mental inertia. Nevertheless there is little trace of new processes, new implements or new crops. Maitland added that in villages (or manors) where the arable was laid out in three fields the farmers did not necessarily follow the three-course rotation of two crops and a fallow. Examples where two fields of the three were left idle were to be found in fourteenth-century Yorkshire and as late as the eighteenth century in one Suffolk village.

The standard practice of the sixteenth century and later, according to the textbooks from Fitzherbert onwards, was to plough three times to prepare a seedbed. Maitland doubted whether this was ever general in the times of early open field farming, and after the Norman Conquest. Ernle, who apparently relied upon sixteenth-century textbooks, states the dates when the three ploughings were carried out (p. 9), the third taking place between Easter and Whitsuntide. Walter of Henley advised three, the first in winter, second in Lent, and the third in " lyke seede tyme in wyntur ". Maitland is also sceptical about the universality of fallowing, and, as we know from the Broadbalk field experiments at Rothamsted, wheat can be grown on the same land year after year for a century without the yield falling below twelfth-century standards, so this contention is tenable. " The villein was ", Maitland said, " required to plough between Michaelmas and Christmas, and between Christmas and Ladyday, but nothing is said about ploughing in summer. " Though he was writing half a century ago I can heartily echo his words, " We are only beginning to learn a little about medieval agriculture " (p. 399).

The " unfree " ploughman of Aelfric's *Colloquies* complains of his lot because he was under compulsion to plough an acre a day. He may have been able to do this in some conditions, and upon some soils, but many centuries later Arthur Young thought that an acre a day was only possible to be done by the skilled East Anglian ploughman at the second ploughing. He, like Maitland, was doubtful whether an acre in the " forenoon " was possible when breaking up stubble, a doubt which I share having regard to Young's description of late eighteenth-century work in Gloucester, Monmouth and Glamorgan on " light and middling turnip land ".

Ernle's sketch of the farming of an open-field village between the Conquest and the Tudor age is exciting, full of gusto, picturesque and detailed. The organisation of a manor and its inhabitants and their way of living has all the same qualities. Having read it an unforget-table impression of life on a medieval manor remains. It is, however, a generalised outline, and, as I have said, to some extent relies upon printed books of the sixteenth century, though Ernle mentions F. W. Page's *End of Villeinage in England*, 1900, and makes use of John Smyth's *Lives of the Berkeleys*, 1883, as well as one or two modern studies in local history.

Very simplified as Ernle's sketch is, it may, I think, be accepted as a working hypothesis. It can and will be subjected to elaboration. The manor, in its heyday, took many different shapes. Frequently it did not coincide with the village : and the manor of the Danelaw, of East Anglia, Cheshire, Kent, Northumberland and elsewhere did not, to put it at a low level, always conform to the standard pattern. There are many works that have described its variations, but these cannot be discussed at length here. A few of the relevant treatises are : Sir Frank Stenton, *Types of Manorial Structure in the Northern Danelaw*, 1910, and his *Free Peasantry of the Northern Danelaw*, 1926 ; D. C. Douglas, *Social Structure of medieval East Anglia*, 1927; H. J. Hewitt, *Medieval Cheshire*, 1929 ; the voluminous works of G. G. Coulton are almost too well known to need mention ; and the delightful studies of H. S. Bennett, *Life on the English Manor, 1150–1400*, 1948, and *The Pastons and Their England*, 1932. These all suggest modifications of Ernle in their own particular way. The work of E. A. Kosminsky, though written with a political bias, and professedly does not deal with agricultural technique, attempts " to make a statistical study of variations in the structure of the manor ". Its title is *Studies in the agrarian history of England in the thirteenth century*, 1956. Besides these an infinity of local studies has appeared in the pages of the learned journals both historical and geographical.

The emphasis in Ernle's sketch is upon the arable farming of the manor. The diet of the great majority of the population was mainly cereal, bread and pottage. Salted fish was consumed in some quantity, the consequent thirst assuaged with copious draughts of beer, another derivative of the cereal crop. The dairy contributed in greater or lesser degree to the general table, if table there was. But " fresh butcher's meat was rarely eaten and, if it was, was almost universally grass fed ". Indeed little attention had been paid to the early development of animal farming until R. Trow-Smith worked on the subject and produced his *British Livestock Husbandry to 1700*, 1957. So far as Ernle discussed this subject it was in terms of grazing on the meadows at times when not put up for hay, or on the stubbles of the arable at prescribed times, and at other periods of the year on the waste of the manor. All classes of stock lived sparely and generally in a poor state, breeding by the haphazard union of nobody's son with everybody's daughter. Not only were the animals small and debilitated in health; disease was rife. Sheep suffered from the scab and the rot, cattle from " murrain ", possibly a collective name that covered a variety of diseases.

Here once again it is necessary to emphasise the wide extent of the wastes, moor and forest that existed even in the midst of open-field England. (H. C. Darby, ed., *An historical geography of England before A.D. 1800*, 1936, Chap. V.) The royal forest was an important part of the total of uncultivated land, and has been estimated at as high a figure as one-third of the total area of the country. Windsor Forest covered a large extent, " all Berkshire, parts of Hampshire and extended into Surrey as far as Guildford", wrote Miss Neilson in the *Cambridge Economic History*, I. " Not merely woods such as Sherwood, Selwood, Dean, Andred, Windsor, Arden, and such hill districts as the Chilterns, the Peak, Exmoor, Dartmoor, and the Yorkshire Wolds were subject to forest law, but whole counties—Devon, Cornwall, Essex, Rutland, Northampton, Leicestershire and Lancashire," but some onslaughts on these had already begun in the thirteenth century. (Darby, p. 176 ; Clapham, *Concise Economic History*, p. 86 ff.)

The royal forests were the scene of large-scale cattle and sheep ranching, and similar enterprises were undertaken at the great Cistercian and other granges. Herds of 400 to 900 cows were kept in the royal vaccaries at Eversley and Bagshot, and there were others in Savernake and Blackburnshire. Long-woolled sheep flourished in the Cotswolds and Lincoln, short wool came from the flocks of Hereford and Salop, a coarse grade from the chalk downs. The flocks of the Midlands produced a medium grade. Vast flocks were owned by the great ecclesiastical and lay landlords. Ely Abbey possessed estates carrying no less than 13,400 sheep. Peasants, too, owned numerous sheep : one Wiltshire group of 198 having 3,760, which, if evenly divided, gave them a sizeable flock apiece. Miss Neilson concludes that the " old picture of the static self-contained medieval village must certainly be modified, and place must be made for a more active agrarian life by no means confined within its own narrow limits ".

The effects of the grazing animal on the woodlands must have been marked. Goats, for example, are voracious and omnivorous feeders. They rapidly destroy scrub, and, if they continue to feed over an area, it can soon be reduced to a condition in which tillage is possible, even if not very easy. Swine can root amongst the young growth. The larger animals destroy young shoots, and so prevent natural regeneration in, shall we say, damp oak forest. The result of feeding animals in the beech woods of the chalk hills, the damp oak woods of the Weald and similar country, and amongst the furze and brambles of other types of waste most certainly developed patches of more or less open grassland here and there. These gaps in the wild cover would appeal to the intending coloniser, and very soon after the Conquest some fairly extensive attempts at colonisation were encouraged by Normans both lay and ecclesiastical. (Darby, op. cit., p. 179 ff.) The human use of the woodland, too, must have had an appreciable effect. The forest provided material for building, for furniture and tools and tool handles, and, not least, firewood. Damage to standing timber must have been extensive when a villager collected his firewood by hook or crook. Furze was gathered by all who did their own baking. For all these reasons forest clearing must have been continuous though perhaps on a relatively small scale. The effect was to put some area of land into a state that was preliminary to reclamation for cultivation. Technical progress in the vast Fen of the Eastern Counties is a special study, and has been exhaustively treated by H. C. Darby in *The Medieval Fenland*, 1940.

A good deal of large-scale enterprise between the Conquest and the end of the thirteenth century was carried out by the great ecclesiastics. The Cistercians founded Rievaulx and Fountains in the fourth decade of the twelfth century, and became famous horse and cattle breeders, as well as wool producers, here and elsewhere. The work of Abbot Samson at Bury St. Edmunds has been recognised since the days of Thomas Carlyle in repairing buildings and restoring and expanding arable farms. A. T. M. Bishop has described the grants of waste in devastated Yorkshire, which resulted in the rise of a class of free peasants similar to those of the Central Danelaw. Far away in Devonshire the abbots of Tavistock Abbey were reclaiming land by the age-old process of burnbaiting (Denshiring), manuring with sea-sand and keeping large flocks and herds.

It was not this rather extensive work on what might almost be called the outskirts of civilisation that Ernle meant when he described the reclamation work called assarting ; it was the small scattered extensions of the arable made here and there on the waste of the open-field villages. This was done all over the open-field area as population slowly increased. Improving landlords, like the Berkeleys, encouraged it, and also did something towards consolidating the scattered strips of land belonging to a farm into more compact holdings. This work was important in two ways. The newly tilled lands were outside the ordinary customary control of the open-field farmers, and were subject to no customary manorial services. The growth of a money economy perhaps encouraged manorial lords to allow such encroachments on the waste that would supply them with a money rent. This tended to assist in changing the traditional manorial economy, and helped towards its disintegration. Obviously, too, it reduced the area of the waste, cleared forest and scrub, and offered a means of making a living to those of the population who were not entitled to anything in the open fields, or whose share, by parcelisation or otherwise, had become too small to bear a sufficient food crop for their maintenance. It was a possible recourse for a landless man who had a family to support, and had neither land nor employment to provide for them. In the more densely populated parts of the country this nibbling away at the waste, coupled perhaps with the lord's activity in the same direction, could lead to a shortage of grazing land for the whole village community. H. S. Bennett in *Life on the English Manor*, 1937, p. 51 ff., expands this subject as does H. C. Darby. This second disadvantage was felt earlier and later in different communities. For example at Wigston Magna in Leicestershire it was felt in the Middle Ages (W. G. Hoskins, *Midland Peasant*, 1957), while at Laxton, Nottinghamshire, it only presented a problem in the early seventeenth century (Orwin, *Open Fields*). The enactment of the Statutes of Merton, 1235, and Westminster, 1285, only serve to emphasise the point, though they did not, I think, provide an effective remedy, which was in point of fact impossible. Enough has perhaps been said to indicate the limitations of Ernle's general outline of the working of the manorial organisation of English agriculture in its heyday, the extent to which details can be filled in as a result of modern research, and how much still remains to be learned about it. His discussion of the social pattern in which the people of the time were arranged can be accepted with the same limitations, and can be expanded by reference to the authorities already mentioned. In addition there is Professor Postan's careful study of the " Famulus, the estate labourer in the XIIth and XIIIth centuries ", *Economic History Review*, Supplement No. 2, a class that is only mentioned *en passant* by Ernle, and the use of the parish priest, often a local farmer, as agent by absentee landlords (Bennett), but it would be a work of supererogation to elaborate these questions, and impossible in the small space available. Lady Doris Stenton's *English Society in the early Middle Ages* (Pelican *History of England*) is a useful summary.

Many influences were at work that made for change. A monetary economy was slowly taking the place of what has been called natural economy ; the woollen trade was expanding, which gave an added impetus to this modification ; men were demanding, and slowly obtaining by various devices, improved social conditions ; the area of useful land was being extended with the consequent rise of a class who were outside the range of the traditional and practical restraints of open-field farming ; slaves were known no more, and servile burdens were growing lighter though not as speedily as those who suffered from them wished. All these factors played their part in the gradual disintegration of the manor for, like all others, this social scheme contained within itself the seeds of its own dissolution. It is however remarkable that the same effects were felt all over western Europe in the late thirteenth and fourteenth centuries. Wars and plague and famine accelerated the process.

It is not necessary to add to what Ernle (and others) have said about the effects of the famine years of the fourteenth century, the Black Death, the revolt of 1381 ; the results of these events coupled with those of the Hundred Years War are known to every schoolboy. Perhaps the most important of them was the sudden and disastrous decimation of the population by the plague, which has been estimated at the loss of from one-third to one-half of all the people then living. No doubt this accounts for some of the *Lost Villages of England* described by M. W. Beresford in his book of that title, 1954, but it did not have much, if any, effect on the traditional methods of farming practised by the survivors: see also Hoskins, " English Landscape ", in Poole, *Medieval England*, 1955, vol. I.

As Ernle himself says little or no improvement in farming technique was possible in the fourteenth and fifteenth centuries. Other historians have come to the same conclusion, but the most modern research has revealed, or perhaps I should say confirmed, some ideas that were current, but have not been discussed by Ernle. I have already mentioned Maitland's pronouncement on the improbability of a single pattern of farming over the whole country, and something of what demonstrates the correctness of this mental attitude. Dr. Hoskins has shown that the open arable system was more elastic than was formerly believed. He found that the farmers of Wigston Magna were in the habit of leaving such strips of their arable as they chose to lie fallow, and producing a miscellaneous fodder crop of weeds and grass in the midst of a field under corn (*The Midland Peasant*, 1957), and Dr. Finberg has described the convertible husbandry in Devonshire (Hoskins and Finberg, " Open field in Devon ", *Devonshire Studies*, 1952).

Co-operation in ploughing, i.e. several peasants combining their oxen to make up a sufficiently powerful team for a plough provided by one or more of them, has for long been an accepted doctrine. While few villein occupiers of the standard thirty-acre farm would have been well enough off to make up a plough team of eight oxen there is nothing against their owning a pair of oxen that might have been sufficient to haul a light type of plough in not too heavy soil—though conclusive evidence might be hard to find. Some peasants certainly owned much larger holdings, and may very well be supposed, like Latimer's father, to have owned their own plough and enough oxen to haul it. Rising families like the Pastons certainly did, and work with the lord's plough and team had long been one of the services due from tenants (Bennett, pp. 44 ff.). These remarks suggest that there must have been different designs of plough in use, hauled by teams made up of different numbers of animals. The varying size of teams is confirmed by H. G. Richardson's Historical Revision No. c, "The Medieval Plough Team", *History*, March 1942.

Intrinsically all ploughs are the same in basic pattern. There must be a beam on which the coulter is fixed, and to which the stilts or handles are attached. The share beam is carried below it, and attached to it. The mouldboard, usually on the right-hand side, is fixed behind the share, and carries the furrow slice as it turns over. Within the limits of this pattern there is great scope for differences, differences which Fitzherbert remarked as being regional in the early sixteenth century, and which must therefore have been developed through long years of trial and error. (See D. Chadwick, *Social Life in the days of Piers Plowman*, 1922.)

Recovery after the disasters of the fourteenth century was more rapid than might have been expected. True, the fifteenth century saw the Wars of the Roses, which only ended in 1485 ; but in fact these wars may have helped forward the changes in social organisation that would otherwise have happened more slowly. Ernle did not think ordinary rural life was much affected by these dynastic struggles of the great nobles, an opinion that was shared by Maitland. G. G. Coulton in his *Medieval Village*, 1925, agrees. Bennett, in *The Pastons and their England*, paints an altogether different picture—of terror by night and day—but this is a question I must leave open. I have no grounds for any opinion.

Another point that needs remark is that Ernle appears to have thought that the stock and land lease was a product of the fourteenth century, but it was common form for the gebur of the *Rectitudines* to be provided, not only with sown land and seed and some animals, but even with household goods. What may have been new was the practice of some great landowners, e.g. Merton College, to give up the personal cultivation and management of distant estates, and to let the land to farmers at a money rent on lease, sometimes a lease as long as thirty years, at others for terms as brief as five years. (Thorold Rogers, *Six Centuries of Work and Wages*, 1906, pp. 277–8.) This is a phenomenon that Ernle dealt with (p. 49).

The first two chapters of this great work are, as I see it, largely concerned with the story of English farming from the time of Walter of Henley to the accession of Henry VII. This attempt to expand what Ernle had to say about this period is necessarily brief and discursive. It tends to show that in the main what Ernle wrote was a good generalisation, something that he himself had warned readers that it must necessarily be. Several writers treating of the whole time between the Norman Conquest and the founding of the Tudor dynasty have expanded his brief pages into book length studies—some indeed treat only of a point in that period—and it is these works, many of which I have named, that must guide the reader who wishes to make a more detailed study, which will reveal the charm and variety of medieval English rural life in all its divergent aspects.

V

TUDOR AND EARLY STUART PERIOD

I F it is correct to say that the Wars of the Roses did not particularly affect the majority of rural people, it must be equally true that no cataclysmic change in farming came about as a result of the foundation of the new Tudor dynasty. Sudden change in the agricultural world is not usual, but is sometimes caused by those disasters that can be grouped under the rather inappropriate title " Acts of God ". Nothing of this kind happened in the sixteenth century, but for economic and other reasons the developments that had already been plainly visible for the last several centuries were accelerated, and their results most volubly discussed.

Ernle has summarised the pamphlets and other contemporary literature which complained of or supported enclosure. There was a great outcry against the conversion of arable to pasture for sheep walks, something that had been happening for a good many years in greater or lesser degree. Ernle, after a lengthy disquisition upon the protestations of the pamphleteers, including Sir Thomas More, comes down fairly heavily on the side of the progressive farmers who wished to work their land in severalty. This they could do without restriction, and the change was ultimately inevitable, although, as everyone knows, it was not completed until the mid-nineteenth century. He does not fail to recognise that some enclosures were made in Tudor times, like some of earlier date, for other reasons than increasing the number of sheep carried (p. 64), but most history books accept the complainants' plea that the most damaging enclosures were made in order to replace arable farming by sheep grazing. No doubt this is true. If, as Ernle said, some 35,000 people were ejected from their holdings in the place where they were born, a more important factor in Tudor emotion than it is today, a measurable amount of suffering was caused, and there was some justice in the outcry and the disturbances that took place. Few of these people would have taken any comfort from the fact that they were a small percentage of the total population, but it was a tough age, and the landowners and large farmers seem to have been a pitiless crew, despite the legislation and commissions of enquiry that attempted, Canute like, to stem this tide for a century and a half. Towards the end of the sixteenth and in the early seventeenth century enclosure was made in order to improve and extend arable farming. In the north, e.g. Lancashire, numbers of small intakes were made from the waste in small plots of tillage (see G. H. Tupling, *The Economic History of Rossendale*, 1927). Other similar local histories indicate that the same process was going on in some other places. It was, of course, merely a continuation of the assarting that had been characteristic of an earlier time. The agrarian demands of the rebels of 1536 and 1549 are now readily accessible in *English Economic History ; Select Documents*, ed. by Bland, Brown and Tawney, 1921.

Some twenty-five years ago I expressed doubts about the accepted views of this movement in my essay " Farming methods in the early Stuart period ", *Journal of Modern History* (Chicago, U.S.A.), VII, March, 1935. This and the authorities therein cited has now been confirmed by the researches of Dr. Joan Thirsk (*Tudor Enclosures*, Historical Association pamphlet, General Series No. 41, 1959). If an area of between one-half and three-quarters of a million acres was affected by enclosure during the century and a half before A.D. 1650 it was a small percentage of the total area of the country, but a much larger percentage of the cultivated area which was still encompassed by vast areas of unused, or land unusable in its contemporary condition. Perhaps I should say again that this comparison, had they been able to make it, would have provided no more comfort for the outcasts than the previous one.

The disturbance caused by the enclosing activity was spread over a lengthy period : that caused by the Dissolution of the Monasteries, and the distribution of their estates, mentioned only *en passant* by Ernle, possibly for politic reasons, was much more cataclysmic, and concerned a much greater area, sometimes estimated as highly as one-fifth of the whole of England. Few villages or towns could have been free from the impact of this large-scale change in landownership, and by the subsequently frequent changes in ownership from one hand to another. It brought a new set of men into control, and gave opportunities to the thrifty and farseeing tenant to become the owner of his holding, to add to it, and to rise in the social scale—as well as the chance of his being ousted in favour of someone who would pay more highly for the privilege of working the land.

These internal processes of change were rendered easier by the influx of Spanish gold from the newly discovered New World. This and other precious metals at once increased the amount of currency in circulation, and reduced the value of money. It made financial transactions simpler, facilitated borrowing, and the issue of credit, and, as Ernle puts it, when the moral scruples about usury had relaxed in approval of legitimate interest, the fabric of modern financial economy had begun to be woven. All this combined to form the " Agrarian Problem in the 16th century " of which no more exhaustive study has yet been made than the work with this title published in 1912 by Professor R. H. Tawney.

The rise of substantial farmers in one county has been described by W. G. Hoskins in " The Leicestershire Farmer in the 16th century ", first published in the *Transactions of the Leicestershire Archaeological Society*, 1941–2, and expanded in his *Essays in Leicestershire History*, 1950, while other aspects in that county are discussed in *Studies in Leicestershire Agrarian History* published by the Society in 1949. It is expanded to national dimensions by Mildred Campbell in *The English Yeoman under Elizabeth and the early Stuarts*, Yale University Press, 1942. The basis of Dr. Hoskins's study of the Leicestershire farmer was his use of inventories made by executors of their wills. Another series of *Elizabethan Inventories*, a kind of evidence the value of which had not been appreciated when Ernle was writing, has been edited by C. E. Freeman, and printed in the *Publications of the Bedfordshire Historical Record Society*, vol. XXXII. Slight sidelights on the condition of some rural people are given in a short paper of my own, " The social and agrarian background of the Pilgrim Fathers ", *Agricultural History* (U.S.A.), VII, October, 1933.

The classes above and below them were in different case, as Ernle pointed out. The smaller landowners may have found themselves with a practically fixed income in a world of rising prices, and in difficulties because of their increased conspicuous expenditure. The old families may have wasted or lost their inheritance, and been substituted by new men, but this class steadily grew more powerful and prosperous. " The Rise of the Gentry " has been the subject of some controversy, the arguments being advanced by Professor Tawney in the *Economic History Review*, XI, 1941, and by H. R. Trevor-Roper in *Economic History Review*, Supplement No. 1, " The Gentry 1540–1640 ", to which a postscript was added by Professor Tawney in *Economic History Review*, 2nd series, VII, August, 1954. Clearly they did come to play a larger part in the affairs of the nation than they had in the past. The dispossessed, Ernle's vagrom men, whose condition he makes clear enough, have been the subject of a later and special study by A. V. Judges with the title, *The Elizabethan Underworld,* 1930 ; the physical position of the day labourer in my *English Rural Labourer, his home, furniture, clothing and food from Tudor to Victorian days,* 1949.

When he turned to the actual practice of farming in Tudor times Ernle said that the new printed textbooks, which had been appearing since the beginning of the sixteenth century, did not indicate any distinct advance on thirteenth-century methods : indeed there were signs of a retrograde movement : but I doubt whether any human activity could have remained static for 300 years. Fundamentally, of course, there was no great difference in what the farmer did to produce the same crops and the same livestock as his ancestors. There was still the problem of winter keep for the livestock, and of shortage of manure for the corn crops, but the theorists were beginning to envisage new methods. Ernle has discussed these men and their writings at some length, and his very real knowledge of modern practical farming was the effective yardstick by which he measured them.

If there had been little change in the implements used, or the ordinary ways of doing the jobs that make up the farming year, the specialist— in a very ill-defined way as yet—was beginning to emerge. Ernle knew very well that there were regional types (as we have come to know them) of farming in his own day, and recognised that they already existed in Elizabethan England if not before then. The sources from which the food supply of London was drawn emphasise this development. A few examples will suffice. Cereal produce was drawn from the Home Counties by road and river, and from the ports of the south-east coast. Butter and cheese came from Suffolk and the more distant counties of Cheshire and thereabouts. Livestock could walk to market from the grazing counties of the north and north-west. F. J. Fisher's essay, " The Development of the London Food Market, 1540–1640 ", *Economic History Review,* V, 1935, clarifies the point, and possibly mine, " Traffic in farm produce in 17th century England ", *Agricultural History,* XX, April, 1946, may expand it a trifle. If the material provided by R. Trow-Smith in *History of British Livestock Husbandry to 1700,* 1957, is added, little more need be said. While these writings may support and expand Ernle's thesis, he had in fact said much the same thing in less detail, or perhaps I should say, in a more general way.

Some regional methods were of the greatest importance. They were the initial stages of activities that became widespread. Fitzherbert mentioned the smoke of the fires seen in Devon when the farmers were paring and burning the land, a practice, as Dr. Finberg said, of unknown antiquity. The use of sea-sand and sometimes seaweed as manure in some of the coastal counties, notably Sussex and Cornwall, could not be adopted by inland farmers, but they were advised to use pond and river mud by the self-appointed experts of the day. In France, Bernard Palissy had begun to examine different soils mainly in order to obtain proper material for his pottery, but incidentally to note their different qualities and various requirements in manure. I gather Ernle did not approve of green manuring. Growing buckwheat to plough in was, he wrote, miserable practice. Marling was revived according to him, but I do not think it had ever been completely interrupted. Leland found it general on the sandy soils of Salop, Cheshire and Lancashire, and it had been known to Walter of Henley. The special crops of certain districts, and the degree of attention given to particular classes of livestock in others, was remarked upon by Leland, and later by Camden and the topographers who followed in his wake and copied him. (See E. G. R. Taylor's " Leland's England " and " Camden's England " both in Darby's *Historical Geography*.) In this connection A. L. Rowse's *The England of Elizabeth*, 1950, is instructive.

Ernle's only comment upon the farm implements of this century and a half is that different ploughs had been developed to meet the necessities of varying soils, but he does not give any details. He added that more iron was used in plough and waggon building, that waggon wheels were now shod with iron, but this does not take us very far. I have attempted to expand this subject in *The Farmer's Tools A.D. 1500–1900 : a history of British farm implements and machines*, 1952, but there is, in fact, very little to be said of this period. A light plough was certainly used in Lincolnshire, and perhaps in East Anglia, which was the precursor of the two-horse plough so much praised by Arthur Young some three centuries later. For the rest the plough was the cumbersome affair of traditional lore, though its parts were modified in size and shape at the discretion of local wisdom. The harrow was used. It might be a stout one with ash teeth, or it might be a thorn bush weighted with a log. Carts, waggons and hand tools completed the list of the farmer's equipment, and, though the contemporary assessment of needs appears lengthy and comprehensive, it amounts only to the barest equipment for manual work, sowing and harvesting crops, etc. Ernle remarked Barnaby Googe's description of the so-called Gaulish reaper of classical times, an imaginary reconstruction of which is shown in an illustration in Raymond Billiard's book. Why this apparatus vanished in the fifteen hundred years between classical times and the nineteenth century cannot be explained, but disappear it apparently did.

Another point in which one might join issue with Ernle is his statement that the art of gardening had declined after the fifteenth century, an opinion that was certainly not shared by the Hon. Alicia Amherst, whose *History of Gardening in England* appeared in 1895, and was therefore to hand when our author was writing. Fitzherbert made the care of a garden a part of a wife's duties, and Tusser paid the subject a good deal of attention. New fruits were imported and grown on ; herbs and roots for the London food market were produced in market gardens on the verge of the city. Orchards were developed. As people grew more wealthy in the reign of Elizabeth they spent more money on their gardens, and Bacon's princely garden would have covered thirty acres. Around Norwich refugees from the Low Countries are said to have brought with them their intensive cultivation of vegetables, including turnips and possibly clover for improved grassland, although clover is an indigenous plant in many of our natural pastures.

Many industrious authors in the early seventeenth century added to the few textbooks published in the sixteenth. Ernle provided a slight bibliography of these books, and analysed their theories, many of which sound surprisingly modern. Rather more comprehensive details of these works are supplied in my *Old English Farming Books : Fitzherbert to Tull, 1523–1730*, 1947 : earlier bibliographies are mentioned in it.

Ernle has written no truer sentence in the whole of his book than the words which open Chapter V. Prospects were bright for farmers and rural society, as they were in so many other respects, but also there were bad seasons, outbreaks of plague, and finally political disputes that led to the Civil War. The movements of the armies discomposed local residents and their demands for food and equipment were often rather more than a nuisance because animals wanted for breeding and the dairy might be taken, and supplies of grain essential to local households carried away, but on the whole the rural population was not driven away from its homes, nor were many villages and towns destroyed. Except for the furore caused by the passage of the contending armies through a neighbourhood, ordinary life in the countryside went on much as usual. The composition of the armies, and especially the officers, was more important. On both sides were great landowners, and greater and lesser gentry, the people who were most literate, and whose ambition, it may be conceived, was to increase the value and extent of their estates. To this end they were inclined, as their descendants were, to try out new ideas, to encourage the cultivation of new crops, to select animals for breeding, to experiment with unusual manures and so on. When they deserted their homes to join the armies these things went by the board, and the energetic priests of the new agricultural doctrines addressed themselves to a vacant auditorium. The tension did not discourage these writers, and they continued production throughout the Commonwealth even more intensively.

The value of turnips and clover had already been recognised in theory by the end of the sixteenth century. Some writers suggest that these crops were actually grown in the field then ; e.g. Russell M. Garnier, in " The Introduction of Forage Crops into Great Britain ', *J.R.A.S.E.*, 1896, pp. 77–97 : but this needs further evidence before it can be accepted. The early seventeenth century writers did not neglect these plants, but the real foundation of the alternate husbandry was laid by Sir Richard Weston of Sutton Place, Surrey, not only in his *Discours of Husbandrie used in Brabant and Flanders*, 1645 (?), but in practical cultivation on his estates at Worplesdon, Surrey. This is most handsomely recognised by Ernle, but has been somewhat obscured by the better-known attribution of turnip culture to the eminent " Turnip " Townshend. These writers, too, from Sir Hugh Plat onwards, all advised other things besides the customary animal and human waste products used for manure, not neglecting lime, chalk, marl, and other " earths ", as well as industrial wastes, though the latter were not experimented with until the end of the seventeenth century. Gadgets to speed up sowing the seed were recommended. Dibbling boards were actually made, and Gabriel Plattes set out elaborate instructions for making a seed drill : but these instructions are almost, if not quite, impossible for a modern mind to understand, and the production of this, the first actual agricultural machine, had to wait till after the Restoration.

Another thing that exercised these writers as much, or possibly more than the farmers, was the control of soil moisture. Efficient drainage is one of the most important elements in successful crop growing. The natural flooding of rivers made the riparian meadows lush and nutritious, and led to the systems of controlled irrigation of grassland that came to be known as water meadows. But arterial drainage was even more important in the Eastern Counties, in Romney Marsh, in Somerset and elsewhere. In the Caroline and Commonwealth period the vast undertaking of draining was begun once again. Of the initial series of operations Ernle provides an enthusiastic account, as he does of the results. It need not be expanded here because H. C. Darby in *The Drainage of the Fens*, 1940, has done that in a manner not likely to be superseded, and his work may be supplemented by L. C. Harris, *Vermuijden and the Fens*, 1953, and A. K. Astbury, *The Black Fens*, 1958. The story is one that finds a place in simple outline in most history textbooks, but deserves close attention because the peculiar circumstances and conditions led to what was and is one of the most striking examples of regional farming in this country. The farming of the Lincolnshire Fens takes on a very different appearance in the recent work by Joan Thirsk, *English Peasant Farming*, 1957, from what it has in Ernle's account of the area, and this is one of the most striking examples of the study of a region so far made. It is not possible to make a close comparison between its findings and those of Ernle in this introduction ; but it is clear that there was much more actual farming in the Fens, and indeed denser population than Ernle supposed. Surplus dairy and livestock products were exported and cereals and other crops were cultivated.

The examination and collation of inventories has steadily become a more popular form of research in the last twenty-five years. One of these projects, that of F. G. Emmison in " Jacobean Household Inventories ", *Publications of the Bedfordshire Historical Record Society*, XX, contains a large number of these documents relating to deaths at the end of the second decade of the seventeenth century when plague was raging. Unfortunately, Mr. Emmison's careful analysis does not include any attempt to deduce the system of farming followed by these unfortunates. Analysis of this kind might be fruitful. It was made by Francis W. Steer in his *Farm and Cottage Inventories of mid-Essex, 1635–1749*, 1950. Here during the period covered the farming was the old wheat and bean course with a bare fallow, turnips being " scarcely more than a culinary crop by the time this series of inventories comes to an end ". The usual livestock were kept, and the farming was of the traditional " mixed " type. At Kirdford, Sussex, the traditional " mixed farming " was usual. G. H. Kenyon has made this clear in his " Kirdford Inventories, 1611 to 1776 with particular reference to Weald Clay Farming ", *Sussex Archaeological Collection*, XCIII, and the other papers cited. Here there was but little change excepting the introduction of some ley farming at the end of the seventeenth century.

It was as true at A.D. 1650 as it was at the Norman Conquest and before that the arable farmer of the open field could not carry on his business without livestock. He must have oxen, or rarely horses, to haul his plough and his wain ; he must have cows to provide replacements and to supply some liquid milk and dairy products ; he must have sheep to provide some addition to the milk supply, to provide wool and some meat if he was sufficiently well-to-do to eat some mutton and beef in place of the universal pig meat. All these animals were also producers and often transporters of manure. The arable farmer was completely dependent on his livestock, and Ernle's contention (p. 121) that " Tudor farmers had treated arable and pasture farming as two distinct branches, which could not be combined " is not strictly accurate, nor is it rendered so by his following sentence. The open-field arable farmers did not, it is true, pay any great attention to their livestock, which can almost be said to have picked up their living by the way, but they had to have them nevertheless. Outside the great central block of open-field villages there were areas where pasture farming was predominant, and where possibly only subsistence cereals were cultivated, but this again is rather proof of the growing regionalism of English farming, which was being steadily more closely adapted to its environment, than anything else.

VI

FROM THE RESTORATION TO WATERLOO

FARMING under the later Stuarts was for Ernle a period of quiescence, when little progress was made, when there was no leadership from the great landowners, when people were satisfied to accept the limitations of their environment, and to continue in the traditional ways of their forefathers. This is really rather less than just to the farmers of this time, some of whom were willing to adopt the new crops. He does give a word of appreciation to the few who began then to work out the rotation, which can be called the alternate husbandry, on the basis of what was being learned from the advanced farming of the Low Countries.

One thing he did not know. Reginald Lennard had not yet discovered amongst the papers of the Royal Society the few reports that had been made in reply to the questionnaire sent out by the Society's Georgical Committee. These reports were analysed in Mr. Lennard's essay, " English Agriculture under Charles II ", *Economic History Review*, IV, 1932, and recently a brief new appraisal of the " First Agricultural Survey " has been made by Nigel Harvey in *Agriculture*, April, 1959.

The new rotation, when it became more or less consolidated, came to be known as the Norfolk four-course, but this is quite inexplicable unless it was correlated to Townshend becoming the eponymous hero of the turnip. The inclusion of turnips and clover, or the cultivation of these crops, not necessarily in that particular succession, was not initiated in Norfolk only. Farmers in other counties were growing both quite as early as Norfolk men, but, of course, the proportion who did so was very small. As Marshall declared 150 years ago, clover was grown in Kent in the middle of the seventeenth century, a remark that is confirmed by Miss E. C. Lodge's *Account Book of a Kentish Estate, 1616–1704*, 1927. Clover seed was on sale in the shops of London seedsmen in the 1650's. Turnips were grown as a fodder crop at the same date in a few places in Suffolk as Mr. Kerridge has proved.

By the end of the seventeenth century both crops were cultivated, not only in Norfolk, but in Suffolk, Essex, Northampton, Kent, Surrey,

Hampshire, the Isle of Wight, Berkshire, Wiltshire and Devon ; in addition clover had been adopted in Cambridge, Oxford, Stafford, Worcester, Hereford, Dorset and Sussex, and in some of these counties rye or rye grass was mixed with clover seed, or grown as a crop by itself : see my " Adventures with Clover ", *Agriculture*, October, 1955. When writing of " Turnip " Townshend's agricultural activities Ernle said that, after his retirement from political life, he devoted himself " to improvements in the rotation of crops, and to the field cultivation of turnips and clover, *which in the preceding half century* [my italics] had been successfully introduced into the country ". G. C. Broderick had earlier made the plain statement (*English Land and English Landlords*, 1881) that the four-course rotation was worked out in the seventeenth century, and its first introduction followed the close of the Civil War. He does not support this with any documentation, but undoubtedly it is correct.

Broderick's conclusion has been confirmed by later research, both my own and other people's. The point has been carried even further, and it has been suggested that the field cultivation of clover and turnips was learned by unidentified farmers near Norwich from Flemish refugees in Elizabeth I's reign, and that these crops became common from early in the seventeenth century, presumably before the Civil War, but this needs further confirmation. The writers who make this assertion are J. F. Benese, *Anglo-Dutch relations from the earliest times to the death of William III*, 1925 : J. Arnold Fleming, *Flemish Influence in Britain*, 1930, vol. I : see also J. Thorold Rogers, *Economic Interpretation of History*, 1888, but it must be accepted with reserve.

The first specific reference to turnips in Norfolk known to Miss Naomi Riches (*The Agricultural Revolution in Norfolk*, University of North Carolina Press, Chapel Hill, U.S.A., 1937) was at Shropham in 1681. This village was on the Townshend estates.   Colonel Walpole at Houghton was even earlier in the van of agricultural improvement.   He grew large quantities of turnips as early as 1673, and clover was a crop that was part of his regular routine, apparently being sown in April, a very proper time.   He had adopted the correct technique of cultivating root crops, a process that was quite unusual then.   He weeded turnips regularly, and double hoed, which must, I think, mean that he first bunched and then singled the plant, exactly the modern procedure. Others in the neighbourhood who were growing clover at the same date were Walpole's cousin, Ruding, a yeoman farmer at Rougham ; Mrs. Armiger, a small landowner at Burnham Thorpe ; Allen of Ingoldisthorpe, one of the lesser gentry ; and Kent, steward to Lord Townshend. It seems unlikely that they were the only ones. (J. H. Plumb, ed., *Studies in Social History : the Walpoles, father and son*, 1955.) Clearly there was a measurable development in the use of the new crops by the end of the seventeenth century, though we cannot measure it, and to these things must be added the potatoes that were grown in Lancashire, which were remarked upon by Ernle, but have not been much regarded by other writers.

The ancient methods of adding to the still inadequate supplies of animal manure were continued in the areas where they had been practised for so long ; burn beating was practised in the south-west where seasand, seaweed and shells were used ; paring and burning in Cumberland and Westmorland, in Kent and on the moss land of Hereford. Marling was traditional in Lancashire, Cheshire and Stafford, and liming and chalking in various places. Woollen rags were bought in London and tried by some farmers in Berkshire. (See my " Pioneer farming in the late Stuart age ", *J.R.A.S.E.*, vol. 100, iii, 1940.)

These efforts were all directed towards the improvement of arable farming. The new crops, of course, were forage crops in the main, such potatoes as were grown in the south being recommended as pig rather than human food, a prejudice Cobbett shared 150 years later. But despite extra home-grown feed available on the small number of outstanding farms there seems to have been little improvement in breeding. Lord Scudamore is said to have imported Dutch cattle as a useful cross on the Hereford, but precise evidence of this is lacking. The legislation against Irish imports was designed to help the English breeders ; this prohibition may have had good economic effects, but there is little to show any of a technical kind.

Equally with the introduction of the new crops the interest in mechanical aids was towards facilitating the work on the arable land. Worlidge's design for a drill was produced almost immediately after the Restoration. Ernle accepts Bradley's statement that he tried to make this machine but failed. A modern farmer, Mr. Averill, of Shrawley, Worcestershire, on the other hand, found it possible to follow Worlidge's instructions, and built a model. The same thing has been done in Chicago : but Worlidge's ingenuity has been obscured in the greater fame of Jethro Tull's apparatus. Neither of these was the first seed drill ever made, but these were the earliest in this country (see my *Farmer's Tools*, Chap. III).

For long enough it has been realised that this type of progress was most applicable to the lighter soils, ranging from the sandy soils of the eastern counties to some loams, and the chalk and limestone downlands. It was on these soils that so-called improving farmers worked, and Ernle does not fail to emphasise the greater specialisation, the basis of modern regional types of farming, that continued to develop during this time, and thenceforward, for this and other reasons. Moreover, he points out that until the sweeping enclosures of the late eighteenth century the area of uncultivated, almost unused, land in the country was still very large (p. 153). It is, however, unnecessary in this introduction to discuss the enclosure movement, or the arguments for and against open fields ; it has already been done in many easily accessible works.

Since the work of " Turnip " Townshend and Coke of Norfolk has played so large a part in history textbooks, not to mention Ernle's work, it is necessary to say something more about what happened in Norfolk.

Coke of Norfolk was not the first member of his family to undertake land reclamation on the Holkham estate. In 1660 John Coke, son of the Lord Chief Justice, had reclaimed 360 acres from the sea ; in 1722 Thomas Coke, afterwards Lord Leicester, another 400 acres. He struggled to improve this barren sweep of country. This was a substantial translation of land that had been coastal marshes, or worse, into land fit for farming. (C. W. James, *Chief Justice Coke, his family and descendants at Holkham*, 1929 ; A. M. W. Stirling, *Coke of Norfolk and his friends*, 1912.) At about the same date as the second piece of land reclamation, seventeen acres of turnips were being hoed on the Holkham estate on 23 September, 1723.

Only seven years later Charles, Viscount Townshend, retired and consoled himself for his loss of political prestige by personal attention to his hereditary estates at Raynham. He, according to Ernle, devoted himself to experimenting in farming practices he had observed abroad, but these were already known in Norfolk, having been conducted by Colonel Walpole and others at Houghton some sixty years before, and possibly also on Townshend's own estate. This Ernle did not, and possibly could not, have known, and he described the Raynham estate when Townshend returned to it as mainly rush-strewn marshes or sandy wastes. He gives Townshend the credit of being the initiator of the Norfolk four-course rotation, a statement that is palpably inexact. Townshend, as Ernle's story shows, took a firm grasp of improvements already spreading through the county, probably extended the area under the system on his own estates so far as he farmed it, and his great prestige made him a remarkably fine publicist for it : but I must add that Defoe believed the new crops to have spread from Norfolk all over the south of England by 1724. James Beeverell, a Frenchman, had been impressed by the Norfolk arable sheep economy in 1707 (*Les delices de la Grande Bretagne*, 1707, vol. I).

The further development of Norfolk farming in the second half of the eighteenth century has come to be inextricably bound up with Coke of Norfolk's work : but he did not enter upon his estate until 1776. Later writers have said that the place was a desert when he came there, Ernle that not an acre of wheat was grown between Holkham and Lynn, but this was not so. Young had made his three most famous farming tours some years before this date, and since he saw the district, it is conceivable that what he reported is correct. He was rather a tyro then, and must have learned a great deal from his " Tours " : he was no longer a tyro thirty-five years later when he wrote his *General View* of the county in 1804. Then he wrote " For thirty years from 1730 to 1760 the great improvement in the northwestern part of the county took place, and . . . rendered the county in general famous. For the next thirty years to about 1790, I think they nearly stood still ; they *reposed on their laurels* . . ." Then drilling was introduced and caused a second revolution. Coke began his sheep shearings.

Between 1730 and 1760 the improvement in north-west Norfolk was made by tenant farmers and small landowners. Their names and their farms are catalogued by Young. They may have been encouraged by what Townshend was doing, but they had had earlier examples as I have said. Young attributed the revival of marling to his wife's great-great-grandfather, Mr. Allen of Lyng House, and to Townshend, but supplies no reason why the process was begun by these landowners, if they did, in fact, re-establish it. They may have been inspired by classical reading, or by reading the textbook encomiums of this method of supplying fertility to light soil.

The sheep shearings at Holkham begun by Coke, like those of the Duke of Bedford at Woburn, speedily became very famous, and were important factors in the spread of new knowledge : but it would appear that Ernle relied too much on Rigby for his account of Coke. Dr. R. A. C. Parker (" Coke of Norfolk and the Agrarian Revolution ", *Economic History Review*, 2nd series, VIII, December, 1955) has recently cast some doubt upon the orotund rise in rent from £2,000 to £20,000 that legend has attributed to his efforts, and, as I hope I have shown, his estate and the area in general was by no means a desert when he inherited it.

Coke's achievement may have been exaggerated, but the force of his example, and that of the other great landowners all over the country whose names adorn Young's pages, must have been substantial. They used and displayed the new machines, the seed drill, the Rotherham plough, the winnowing box and threshing machine, and all the rest : they grew experimental crops, and tried out new manures : and what they did must not be underestimated, nor should it be exaggerated. Many farmers kept them company along the road towards increased production, and some may even have preceded them. Ernle has made an analysis of the *General View* of each county prepared under the auspices of the Board of Agriculture at the end of the century, but these are of variable value and require criticism based upon the qualifications of the writers, some of whom obtained their jobs by patronage in the usual manner of that day for reasons that were not entirely founded upon agricultural experience. The great personal survey of the nation's farming made by William Marshall is not open to this criticism : Marshall had his faults, but ignorance of farming was not one of them. With the help of Mrs. Constance Goodman I tried to expand this and extend it in time in an essay on " Crop Husbandry in 18th century England " that appeared in *Agricultural History* (U.S.A.), October, 1941, and January, 1942. A similar analysis of the county reports was made by J. H. Clapham in his *Economic History of Modern Britain*, vol. I, 1926.

The main purpose of growing clover and turnips was to supply the livestock with more adequate winter feed, the problem of all the previous centuries. When they had it the farmers could maintain their flocks and herds in good condition over the winter, and with good fortune keep them in flourishing condition for the pail or the butcher. The enclosures that made the cultivation of these crops possible made the breeder's task easier and caused him to demand still more fenced fields, in which he could keep selected livestock. Among the progressive breeders Ernle gives the palm, not unjustly, to Robert Bakewell, whose work he has fully described as well as that of his predecessors and successors : but is a trifle carried away by his enthusiasm for the great improvers. He says Bakewell's success raised up a host of imitators, which was no doubt true, but rather gives the impression that the majority of breeders adopted the improved methods, which is very doubtful because the enclosures of this century were at first certainly made to provide opportunity for improving arable land.

Better graziers and flockmasters there were, and they succeeded in developing carcass weight of both sheep and cattle, milk yield and so on, but there is little evidence to support any exact statistical measure of these achievements. Ernle accepted as a criterion the average weights of carcasses sold at Smithfield in 1710 and 1795 as supplied by Sir John Sinclair, with the very proper caveat that Sir John is not always a reliable witness. These figures I questioned in two essays : " The size of English cattle in the 18th century ", *Agricultural History*, III, October, 1929, and in collaboration with Constance Goodman, " Eighteenth century estimates of English sheep and wool production ", ibid., IV, October, 1930. Even if Sinclair's figures truly represent the average carcass weights at Smithfield Market, they need not reflect a general development, partly because the best meat (the " improved ") animals were sent to London, and partly because the large dairy Shorthorns of the suburbs would unduly increase the average. Moreover the cattle and sheep sold in Smithfield Market formed a very small proportion of the total livestock population of the whole country.

R. Trow-Smith's *British Livestock Husbandry 1700–1900* (1959) is a specialist work, a continuation of his earlier volume, and is the only study of its kind. This book supports Ernle in his placing of Bakewell as the mainspring of animal breeding in the eighteenth century, but I do not think it adds anything to our knowledge of Bakewell's methods, nor indeed does H. C. Pawson's work, *Robert Bakewell, Pioneer Livestock Breeder*, 1957, which contains an interesting selection of Bakewell's correspondence discovered by the author. Mr. Trow-Smith deplores the lack of material for the earlier part of the eighteenth century, and has limited himself to a few outstanding contemporary works, except for Scotland which is outside my scope. In the result the book becomes almost entirely an analysis of the *General Views* and Marshall's *Rural Economy* series in dealing with the eighteenth century. It is nevertheless a valuable commentary upon and supplement to Ernle. There is indeed little appropriate contemporary material for a work on this subject to be found relating to the early eighteenth century, but what there is of it I examined some years ago in an essay " Animal Husbandry in 18th century England ", *Agricultural History*, XI, April and July, 1937. There were serious outbreaks of animal disease in the second and fifth decades of the century, which may have delayed progress. All the work that had been done to extend the arable area, to produce fodder crops including such things as cabbages and rape as well as the better known clover and turnips, and so to make it possible to keep more and better livestock, the invention of new machines, seed drills and threshers, chaff and root cutters, and better designs of ploughs and cultivators, had made enterprising farmers and enthusiastic landlords richer men by the end of the eighteenth century. The establishment of the Board of Agriculture in 1793 was intended to synthesise all this, and by the publication of a national survey to inform everyone of the progress and practice of everyone else. Its constitution and the work it did have been reconsidered by Rosalind Mitchison in " The old Board of Agriculture, 1793–1822 ", *English Historical Review*, January, 1959. Legislation carried out by the landowning representatives of the people and of themselves, e.g., the Corn Laws, had been carefully designed to support all these efforts. The progress of enclosure had helped, but had not been an unmixed blessing, as Ernle is careful to point out. It may have made for the engrossing of farms in some parts of the country, but not everywhere, and recent research by Dr. G. E. Mingay, not published at the time of writing, shows that engrossing was not, in fact, so widespread as some contemporary writers and some modern historians would have us believe. The best known of later studies are A. H. Johnson's *The disappearance of the small landowner*, 1909, and Hermann Levy's *Large and small holdings*, 1911. It is one of the many facets of agrarian history in which there is no clear definition, and Ernle has very correctly not committed himself, while admitting that there was some engrossing, the limits of which he had no means of computing. In this connection it may perhaps be pointed out that England is still mainly a country of medium and small farms. Ernle discusses the relation between progress in farming methods and enclosing and engrossing in some detail, and it would be rash to say that he does not treat the subject justly. Naturally he is strongly in favour of enclosure for all the practical reasons that support it, and of the large-scale enterprise supported by adequate capital. Reasonable students of modern conditions cannot fail to have the same bias, but there is always room for the holding of small areas on which intensive production of a highly capitalised and specialised character is conducted, and which may be regarded as a large-scale undertaking.

The small farms, frequently open-field farms, that were operated at the end of the eighteenth century were not of this kind. They were very often the homes of poor farmers and of backward farming. The outbreak of the Revolutionary and Napoleonic wars made it necessary, as war always does, to increase the national output of food if the country was to remain within reasonable touch of being self-supporting. Prices rose, enclosure proceeded at a more rapid rate, the enthusiasm for improvement was intensified, and a great deal of marginal land was brought under cultivation that would in other circumstances have remained waste. Ernle discusses all this in a thorough and almost unprejudiced manner. He does not forget the famine years that came with bad harvests, nor the necessity for corn imports, about which W. Freeman Galpin (*The grain supply of England during the Napoleonic period*, 1925, Macmillan and Co., New York), is fully informative. This is another subject which is too large for discussion within the limits prescribed for me here.

It is common knowledge that the farm worker was growing more and more depressed as the farmer and landowner became more well-to-do. Despite a rise in the nominal rate of wages during the war the labourer was brought, if he had ever lived otherwise, to a bare subsistence level of living, and his wages were almost everywhere supplemented by a dole paid out of the Poor Rate. Those of the tenant farmers and small owner-occupiers who had been reduced in the social scale by the turn of events (some would say the rapacity of other men) were forced into the ranks of the wage-earners. Some became successful in the new towns ; others starved there as successfully as they would have done if they had stayed in the country : but J. L. and Barbara Hammond (*The Village Labourer*, 1911) and the authorities therein cited have demonstrated that there was a great deal of suffering amongst the poorest of rural society both before, during and after the wars. Ernle does not dispute this palpable truism, but a corrective for a too extreme view is an essay by J. H. Clapham : " The growth of an agricultural proletariat " (*Cambridge Historical Journal*, January, 1923). Some discussion of the subject will be found in my *English Rural Labourer*, and something about the farmer and his wife at this and other times in the two books with which my wife helped me, *The English Countryman, his life and work, 1500–1900 A.D.*, 1955, and *The English Countrywoman, a farmhouse social history, 1500–1900 A.D.*, 1953. All alike felt the cold blasts of adversity when the wars were over, but that is a subject which I must leave to Mr. McGregor, who will deal with the nineteenth century.

Before I cease there is one other point. Ernle does not discuss in any detail the efforts that were made to arrive at an understanding of the scientific processes involved in plant nutrition, or what contemporaries regarded as agricultural chemistry. Some progress had been made before the end of the eighteenth century, but the nascence of " science with practice " was for Ernle, unless I do him an injustice, the work of Humphry Davy. He does mention Francis Home, who wrote in 1757 ; but the attempts to apply science to agrciulture were much older than that as I have shown in my contribution to the fourth volume of the *History of Technology*. The early writers on this subject concerned themselves mostly with crop nutrition, but there was speculation and empiric experiment on animal nutrition. (See Cyril Tyler, " The development of feeding standards for livestock ", *Agricultural History Review*, IV, 1956.) It is doubtful whether very much of the scientific writing on these subjects in the early days of their pursuit ever reached the practising farmer, and they may possibly be considered to belong to the history of science rather than to the history of agriculture, but the intimate nexus between them and farming, the modern farmer's necessary preoccupation with a wide range of sciences and with mechanics, suggest that the history of science applied to agriculture might well form a subject or a series of subjects for future students.

I have no more to say except that the close reading of Ernle's book that I have been forced to make in order to write the foregoing has left me with greater admiration for him than I have had for a good many years. He is open to some criticism, but his descriptive powers are unequalled by any later writer on the subject, and his ability to make broad and accurate generalisations from a mass of evidence is remarkable. It is indeed so remarkable that a great deal of the research done in the past thirty years has largely confirmed him, although it has made some corrections : otherwise it would have been completely worthless, and no one could be foolish enough to make any such suggestion. I must however say again that it will be a very long time before *English Farming Past and Present* is finally superseded, the ultimately inevitable fate of all such treatises.

INTRODUCTION

PART TWO: AFTER 1815

by

O. R. McGREGOR

I

THE POLITICS OF
ENGLISH FARMING PAST AND PRESENT

ROWLAND EDMUND PROTHERO's long life began in the years
of Victorian optimism and ended as the tawdry thirties were crumbling
into Hitler's war. Born in the year of the Great Exhibition, the third
of the four sons of the Reverend George Prothero, he came to maturity
with the Ballot Act and outlived George V by one year. His education
and early career[1] ran in the conventional grooves of the upper middle
class to which his family belonged. He went up from a public school,
Marlborough, to Balliol in the second year of Jowett's mastership. A
second class in classical greats and a first in modern history led, in 1875,
to a fellowship of All Souls which he held until 1891. He ate his
dinners in the Middle Temple and joined the Oxford Circuit in 1878
only to be robbed of the prospect of a legal career when failing eyesight made reading impossible. Relief and restored vision came
from tramping through France in 1881 and 1882. The affectionate
knowledge of the French countryside gained during these and later
travels was distilled into the elegantly nostalgic survey of *The Pleasant
Land of France* first published in 1908.

In 1883, a chance meeting on the Southampton-Havre boat with
Henry Reeve, at that time conducting the *Edinburgh Review,* led to a
request for an article on Alexander Pope and started Prothero on his
fifteen years' career as a literary journalist. His next essay appeared
in the *Quarterly Review* for April, 1885, under the title " The Pioneers
and Prospects of English Agriculture " and was expanded, three years
later, into his first book, *The Pioneers and Progress of English Farming.*
There followed a series of writings characteristic of a conservatively-
minded *littérateur* of the period whose long suit was industry rather
than sensibility. By 1891, he had placed some forty-two articles in
the *Edinburgh* and *Quarterly* and twelve in the *Church Quarterly,* besides

---

[1] The details are available in his autobiography, *Whippingham to Westminster,*
posthumously published in 1938, which regrettably carries the account of his life
only to 1917. Something of his later career may be gleaned from the obituary
by Dr. C. S. Orwin in the *Dictionary of National Biography* (1931–1940).

many occasional pieces in the *Athenaeum, Guardian* and *Blackwood's*. James Knowles made him assistant editor of the *Nineteenth Century* and soon, as he recounts in his autobiography, "I had in my pocket, not only the reversion to Reeve in the editorship of the *Edinburgh Review*, but . . . the reversion to Smith in the editorship of the *Quarterly*.[1]" Prothero early established his reputation for sound and predictable judgment which ensured a generous welcome for his formidable output of biographical compilations from the respectable literary world of the nineties. The commissioned *Life and Correspondence of Arthur Penrhyn Stanley* of 1893 was followed, two years later, by *Letters and Verses of Arthur Penrhyn Stanley*. Immediately afterwards, he brought out two volumes of *The Private Letters of Edward Gibbon* to which he added, between 1898 and 1901, six volumes of *The Letters and Journals of Lord Byron*. The next four years saw only *The Psalms in Human Life* and *The Letters of Richard Ford, 1797–1858*. He records that, when he was appointed to the editorial chair of the *Quarterly* in 1893, a newspaper remarked : "Mr. Prothero is unknown except as a devotee of cricket."[2] There was truth in the gibe. " My own circle of literary friends," he wrote, "was, I suppose, old-fashioned . . . I had never even met any of the men who filled the pages of *The Yellow Book* or *The Savoy*."[3] He was not equipped to restore the waning influence of his periodical. The narrowness of his artistic sympathies and the social exclusiveness which determined his friendships shut him off from (it is his own description) " the large and new public thirsting for information and for the gratification of their un-formed tastes."[4] They shut him off, too, from the creative writers of that period. When he accepted an offer from the eleventh Duke of Bedford in 1898 to become agent-in-chief for the estates, he had carried his literary career as far as Balliol connections could take it, and rather further than his mid-Victorian susceptibilities warranted.

Prothero's new career had its origin in family influence and in his own polemical writing on the politics of land. In 1878, at the request of the Duchess of Bedford to his mother, he took Lord Herbrand Russell, who succeeded to the dukedom in 1893, as a pupil for a term. For this service, the Duke pressed on him a cheque for £300. At the end of his life Prothero gratefully recalled this incident with the

---

[1] *Whippingham to Westminster* (1938), p. 125.
[2] *ibid.*, p. 171.
[3] *ibid.*
[4] *ibid.* p., 170.

comment : " Small boys are said never to forget those of their elders who give them a tip. From my own experience I can only say that the gratitude of a young man, who is similarly tipped, remains even fresher in the memory as long as life lasts."[1] Twenty years later the tipped tutor became his former pupil's estate agent. He was well fitted for the post by knowledge, if not by practical experience, of the land ; and by his position as the leading critic within the Conservative Party of Liberal land policies. He never deviated from the principles outlined in *The Pioneers and Prospects of English Agriculture* and elaborated in *The Pioneers and Progress of English Agriculture*, a present-minded book written " to apply the results of history to the present conditions of English farming,"[2] and to counter " wild talk about State-ownership, ransom, and natural rights, societies to nationalize the land, [and] heroic remedies of illogical half-disciples of Mr. George."[3] In the early eighties, Prothero was the self-appointed champion of agricultural landlords threatened on many fronts. The transcontinental railways in India and North America had begun to feed the steam steel ships which were unloading growing quantities of cheap grain on the British market. The rising rents and comfortable prosperity which landlords and farmers had enjoyed in their golden age after the repeal of the Corn Laws were thus being undermined. The impetus generated by the anti-Corn Law campaign had passed over to radical, urban middle-class Financial and Administrative Reform associations organised up and down the country in order to expose and to eliminate the inefficiencies of a political system dominated by the territorial aristocracy. Criticisms of the English land system accumulated from these and other sources (from, for example, the legislation which Irish misery and intransigence extorted from the imperial parliament) and, by the seventies, they had swelled into variously articulated demands for free trade in land. The enlarged electorate, the long-resisted concession of a secret ballot, the first statutory and compulsory interference with the contractual freedom of landlord and tenant in 1883, and the warnings which Joseph Arch wrote on cottage walls, all pointed to an early and inescapable redefinition of the social functions of land and its owners.

These new determinants were given a cutting edge by the widely disseminated and eagerly discussed doctrines of Henry George, by the revival of socialism and by the enthusiasm of some liberal politicians

---

[1] *Whippingham to Westminster*, p. 192.
[2] Preface, p. v.
[3] *Quarterly Review*, vol. 159, no. 318, p. 359.

too easily persuaded that three acres and a cow would provide their party with a fresh infusion of loyal voters in the counties. Prothero's concern for the prospects of English agriculture derived from his hostility to such forces and from his fears that they would erode the very foundations of social stability. In his interpretation of the politics of the early eighties, the Atlas of land which bore English society on its back was being so debilitated by depression that it could no longer hold against the attacks of menacing agitators. " Mr. Chamberlain's language," he wrote, ". . . is allowed to permeate society to the infinite danger of agricultural interests. But . . . those who pander to the lawless passions of the mob do so at their peril. If labourers are stimulated to assert their so-called right to a part of the land, workmen will not be slow to claim with equal justice the lion's share in the profits of the capitalist."[1] Prothero made no intellectual effort to grasp the implications of doctrines that he instinctively abhorred and he failed to discriminate between the aims of different land reformers. He never understood, for example, the source of Chamberlain's concern for land. In an article of 1883, Chamberlain wrote : " The wide circulation of such books as the *Progress and Poverty* of Mr. Henry George, and the acceptance which his proposals have found among the working classes, are facts full of significance and warning. If something be not done quickly to meet the growing necessities of the case we may live to see theories as wild and unjust as those suggested by the American economist adopted as the creed of no inconsiderable portion of the electorate."[2]

For the young Prothero, the pressing question was : " On which line is future land-legislation to proceed? "[3] First posed in the *Quarterly*

---

[1] *Quarterly Review, op. cit.*, p. 359.

[2] *Fortnightly Review*, vol. 40, p. 762, quoted Elsie E. Gulley, *Joseph Chamberlain and English Social Politics* (1926), p. 210. This book contains an excellent account (ch. VII) of " Chamberlain and Land Reform ". Prothero lacked the political sophistication to distinguish vigour of language from the content of proposals for action. Chamberlain, described in " The Pioneers and Prospects of English Agriculture " as pandering " to the lawless passions of the mob ", thus wrote to Lady Dorothy Nevill in 1883 : " Have you read two books lately published— *Progress and Poverty*, by H. George, and *Land Nationalisation*, by A. Wallace? They come to the same conclusion, ' *l'ennemi c'est le propriétaire* ', and they advocate the same remedy, namely, confiscation of property in land. I am told that these books are being eagerly read by the working classes in London, and that the feeling in favour of drastic measures is growing. In all seriousness, if I were a large landowner I should be uneasy. They are so few, and the landless are so many. There is only one way of giving security to this kind of property, and that is to multiply the owners of it. Peasant proprietorship in some form or other, and on a large scale, is the antidote to the doctrines of confiscation which are now making converts ". Dorothy Nevill, *Under Five Reigns* (1910), pp. 206-7.

[3] *Quarterly Review, op. cit.*

Review article, it was answered at length in his *Pioneers and Progress of English Agriculture*. That book surveyed, first, the history of farming and concluded that improvements had always been effected by private capital and enterprise ; and, second, the range of policies then advocated to meet the circumstances of the eighties. He dismissed two popular nostrums. Although " many farmers slumber in their empty corn bins, dreaming of Protection,"[1] this solution was unrealistic and politically impracticable. On the other hand, peasant proprietorship would be socially and politically advantageous if it resulted from evolution, not from revolution. " A peasant proprietary increases the number of those who have something to lose and nothing to gain by revolution, encourages habits of thrift and industry, gives the owner of land, however small his plot, a stake in the country, and a vested interest which guarantees his discharge of the duties of a citizen. Combined with the *partage forcé*, it checks population, for *la plupart des Normands n'ont pas lu Malthus, mais ils pratiquent instinctivement ses conseils.* . . . It affords a training to the rural population for which we in England have found no substitute. It checks the centralisation of pauperism, the overgrowth of population, and the migration into towns. The element of stability which it contributes to the State is more valuable to the French than ourselves. There the towns are inflammable as touchwood, while the country ignites more slowly. Yet even here it is useful to have a class of slow-thinking men, who will answer political firebrands with *Cela est bien, mais il faut cultiver notre jardin.*"[2] But legislation which attempted to establish such a class would be " opposed to natural laws " and therefore " as effective as the Pope's bull against a comet".[3] Moreover, the economic prospects of a peasantry would be no better than those facing tenant farmers and, in any case, the structure of agriculture could not be re-shaped overnight. An English peasantry offered the best hope of countering the lawless passions of the industrial mob, but its establishment as a stable element in the countryside could result only from judicious encouragement of " natural " forces. " For immediate relief of agricultural distress it is folly to look to peasant proprietorship."[4] Thus the future had to lie with an acceptance of the existing land system and a-willingness to modify it. " The course of agricultural history has on the whole been governed by natural

---

[1] *The Pioneers and Progress of English Agriculture*, p. 239.
[2] *ibid.*, p. 138.   [3] *ibid.*, p. 140.
[4] *ibid.*

economic laws. . . . But the record of the growth of large estates is not so free from the taint of oppression that landlords can appeal with entire confidence to their moral title-deeds. The paucity of their numbers, and the exceptional nature of their position and property, should add no insecurity to their possessions ; yet they warn landlords to think less of their rights than their duties ; they render it essential that no individual should fall below the highest standard of his class."[1]

Prothero suggested two main respects in which insistence upon rights was inhibiting the discharge of duties. First, the "farmers' real grievance against the landlord is the absence of security for unexhausted improvements under a lease."[2] Secondly, the privileged legal position of landowners required equitable reforms in the law of real property : the abolition of primogeniture, entail and settlement ; the assimilation of the laws relating to realty and personalty ; and the cheapening and easing of procedures for the transfer of land. He thought such reform urgent because " at present land lies so far beyond the general public that the masses have little respect for, or desire to guard, the sanctity of real property ".[3] He urged landlords to a cool appraisal of their future. " The crisis is indisputably grave ; revolutionary legislation is powerfully advocated, and the position of landlords completely isolated. Agricultural labourers possess the franchise at a stage in their civilisation which renders them an easy prey to unscrupulous agitators. The fatal Land Act of 1881 has already born fruit in the demands of the Farmers' Alliance, and in the land bills . . . put forward by English, Scotch and Welsh farmers. At the first gleam of agricultural prosperity the cry for tenant right will be renewed. Landlords have now the opportunity of removing legitimate grounds of discontent . . . of striking from the hands of socialistic theorists weapons which are dangerous to the safety of society. The distinction between giving and giving up is vital. But here there is not even a question of giving. Changes such as are suggested entail no surrender of rights, no sacrifice of pecuniary interests. On the contrary, they are dictated to landlords not merely by political prudence, but by commercial self-interest."[4] The lurid language in which the Prothero of *The Pioneers and Progress of English Agriculture* dressed his anxieties serves only to emphasise the triviality of his political analysis and the inconsistencies of his thinking. He began by denying the premises of land reformers through

---

[1] *The Pioneers and Progress of English Agriculture*, pp. 146–7.
[2] *ibid.*, p. 216.   [3] *ibid.*, p. 238.
[4] *ibid.*, pp. 241–2.

an appeal to history in which he discerned the operation of natural laws but ended, nevertheless, by accepting their conclusions concerning practical policies. The application of his principles to the problems of landownership in the eighties yielded nothing more systematic or sophisticated than proposals which united expediency with an appeal to political prudence and financial advantage. It could not be otherwise because he was concerned to sustain the past that he admired as a barrier against a future that he feared. He had no understanding of the dynamic of an industrial society, and no sympathy with the democratic aspirations of its urban citizens. His social ideal was the benevolent paternalism of a territorial aristocracy balanced and buttressed by small cultivators.

When Prothero left the editorship of *The Quarterly* for the agency-in-chief of the Bedford estates, he went to an employer who not only shared his views but also personally embodied his ideal of the beneficently enlightened landowner ruling over a little sovereign commonwealth peopled by dependants. In the introductory to *The Story of a Great Agricultural Estate*, which the eleventh Duke of Bedford published in 1897 to "[prove] that the system of land tenure which allows a great estate to descend unimpaired from one generation to another, secures to those dwelling on the soil material and moral advantages greater than any that are promised under any alternative system ",[1] there are clear traces of Prothero's influence in the form of phrases taken, but not acknowledged, from his writings. Both owner and new agent alike shared a horror of " class legislation which scares capital from the land " and " behind which looms the ominous prospect of confiscation ".[2]

Prothero's new work was thus congenial, and the terms of his employment were generous. Salary and pension apart, they included free occupancy of a country house, Oakley, in Bedfordshire, staffed at the Duke's expense with three indoor and seven outdoor servants and with three gamekeepers. The house, ominously furnished in the French style of the end of the eighteenth century, was heated and lighted at his cost. Game, garden produce, cream, milk and butter as well as whisky and port were free allowances.[3] He also enjoyed as an additional perquisite the facilities of another country house, Thorney in Cambridgeshire, on the same terms. If Prothero was fortunate in taking up his

[1] Duke of Bedford, *op. cit.*, p. 3.
[2] *ibid.*, pp. 10–11.
[3] *Whippingham to Westminster, op cit.*, pp. 191–3.

agency at a time when the worst effects of the agricultural depression were over, his life was darkened by the death of his first wife only six weeks after they had settled at Oakley. Three years later, he married again ; and began to add public to his professional work. As the acreage of the Bedford estate dwindled (sales of land reduced it by half in ten years), so Prothero's political activities widened. He accepted the first chairmanship of the Bedfordshire County Council's Higher Education Committee, was soon elected an alderman and appointed to the membership of several committees. At the Duke's request, he fought two unsuccessful elections as Unionist candidate for North Bedfordshire but finally entered the House of Commons in the early summer of 1914 as one of the members for the University of Oxford after the death of Sir William Anson, Warden of All Souls. His only major publication in this period of his life was *English Farming Past and Present* in 1912.

War brought new responsibilities and wider opportunities for public service. The conventional round of recruiting speeches was followed by work on a number of committees of which the most important was that, with Lord Milner as chairman, appointed in 1915 " to consider and report what steps should be taken, by legislation or otherwise, for the sole purpose of maintaining and, if possible, increasing the present production of food in England and Wales, on the assumption that the war may be prolonged beyond the harvest of 1916 ". Although its influential *Interim Report* was not immediately acted on, the measures to secure increased food production adopted after Lloyd George became Prime Minister at the end of 1916 mainly followed the lines there laid down. The reconstructed coalition government had to tackle a perilous food situation, and the wily poacher who had, in his time, filched more from landowners than their game, offered the Presidency of the Board of Agriculture, with a seat in the Cabinet, to the Duke of Bedford's agent. Prothero discharged his ministerial duties with efficiency and distinction, using his great authority as an agriculturist to persuade landowners and farmers to accept in the national interest a policy that created new and, as some thought, dangerous precedents. " There is no precedent ", wrote Walter Long (at that time a member of the Cabinet as Secretary of State for the Colonies) to Lloyd George in the course of a forceful denunciation of the Corn Production Bill, 1917, " for giving tremendous powers of this kind ... and I fully share the view that a wholly unfair advantage has been taken of the military situation to pass land legislation which would in a quieter time be absolutely impossible. . . . . The fact that land legislation of the future is to be of this very drastic kind is causing a profound amount of irritation among men who have been among the most loyal and devoted supporters of the War from its very commencement."[1]

Prothero ·was always sensitive to the charge of helping to sell the pass that he had defended for a quarter of a century. Accordingly, when the war was over and he had been raised to the peerage as Baron Ernle of Chelsea, and forced into premature retirement by a recurrence of his old eye trouble which led to increasing blindness, he surveyed his war-time record. "The Food Campaign, 1916–18 " was written to demonstrate that the agricultural policies of peace could not be based upon the expedients of war. " The control ", he wrote, " which, in the last stages of the war, the State exercised over industries, has also profoundly influenced the trend of contemporary politics. It has stimulated the growth of Socialism. Of this control of industry by the State agriculture was an example, and it is significant that the Land Policy of the Independent Labour Party closely follows, and, in all its main details, adopts, the organization created in the course of the Food Campaign. For these reasons, it is desirable that the story should be authoritatively told by the Minister . . ."[2]   That story issued, he insisted, in the conclusion that " no sound argument in favour of State control of the agricultural industry can be based upon the Food Production Campaign " because " patriotism was, throughout, the strong incentive to the efforts of farmers ".[3]

The chapters of *English Farming Past and Present* which survey the history of the recent period must be read as a partisan contribution to the late-nineteenth and early-twentieth century debate on the function of land in English society. They are *The Pioneers and Prospects of English Agriculture* come to maturity and express a deepened faith in the beliefs that underlay *The Pioneers and Progress of English Farming* of 1888. In the Preface to the first edition of *English Farming Past and Present*, Prothero clearly stated his " two convictions, which time has strengthened rather than weakened. One was, that the small number of persons who owned agricultural land might some day make England the forcing-bed of schemes for land-nationalisation, which countries, where the ownership of the soil rested on a more democratic basis, repudiated as destructive of all forms of private property. The

---

[1] Quoted in D. Lloyd George, *War Memoirs* (Odhams Press edition, 1938), vol. I, pp. 767–8.
[2] Lord Ernle, *The Land and its People* (n.d.), p. 100.
[3] *ibid.*, p. 168.

other was, that a considerable increase in the number of peasant owner-ships, in suitable hands, on suitable land, and in suitable localities, was socially, economically, and agriculturally advantageous." He was forging weapons manufactured with ore mined from the past for his attack upon the present. Political conviction and present-mindedness gave his book vitality and vigour still unsapped after half a century. When it appeared it was acclaimed as a classic and as the standard history of the subject : it has been so regarded ever since. Its historiographical place is with the two other great contributions to the same debate : with the book of J. L. and Barbara Hammond, *The Village Labourer, 1760–1832*, published in 1911 ; and with that of R. H. Tawney, *The Agrarian Problem in the Sixteenth Century*, published in 1912. The Hammonds were passionate radicals, Tawney was then, as he has since remained, a democratic socialist. Like Prothero, they were immersed in the political controversies of that day, and were writing of the past in their present.

Tawney studied " the appearance, or at any rate the extension, of the tripartite division into landlord, capitalist farmer, and land-less agricultural labourer, the peculiar feature of English rural society which has been given so much eulogy in the eighteenth century and so much criticism in our own ".[1]

The Hammonds made a searing commentary on the antithesis of De Tocqueville who had seen, in England, an aristocracy with power and no privileges and, in France, one with privileges and no power, and had admiringly compared the blending of classes and the calm of liberty and order which, he held, that condition had produced in England, with the impoverishment of wealth, responsibility and initiative in his own country. They examined the social history of England " in the days when the great oaks were in the fulness of their vigour and strength " and traced " what happened to some of the classes that found shelter in their shade ".[2] Their characterisation of the territorial aristocracy did not ignore the record of literature, art and politics, the glittering surface of the age of reason, but revealed also the governing attitude of mind that " left dim and meagre records of the disinherited peasants that are the shadow of its wealth ; of the exiled labourers that are the shadow of its pleasures ; of the villages sinking in poverty and

---

[1] *The Agrarian Problem in the Sixteenth Century*, p. 1.
[2] J. L. and B. Hammond : *The Village Labourer* (2nd edition, 1913), p. 25. It is regrettable that the Hammonds permitted the first chapter of their book, " The Concentration of Power ", which provides the reader with a clear statement of their approach, to be omitted from the 3rd (1919) and subsequent editions.

crime and shame that are the shadow of its power and its pride."[1]
The overmastering importance of the politics of land in the Edwardian
era was such that Tawney and the Hammonds, social critics who found
in the study of history the clue to their understanding of industrial
society, took land in the sixteenth and eighteenth centuries as the
theme of their first major works. But their critique of that society
pointed to the future ; Prothero's was stuck in the past. Nothing
reveals more clearly the social irrelevance and political myopia of the
later chapters of *English Farming Past and Present* than a comment
in the chapter on " Peace-Time Farming, 1919–1927 ", added to the
fourth edition. Lord Ernle there contrasted the aftermath of
the great war that ended in 1815 with that ending in 1918. " Even the
worst feature of the nineteenth century has been reproduced in an
altered form," he wrote. " If for the administration of the old Poor
Law is substituted the unemployment benefit of to-day, a parallel is
established. A similar danger threatens the country. Should a
generation arise which has not known the self-respecting and independent
traditions of the industrious poor, voluntary pauperism would become a
permanent characteristic. Experience of the twenty years that followed
the close of the Napoleonic wars offers a warning of the direction towards
which the nation is heading."[2] Indeed, the later section of the book
that left Lord Ernle's pen can be regarded only as an anti-socialist tract
akin to *The Pioneers and Progress of English Farming* in so far as the
arguments it seeks to rebut are neither stated nor examined.

English Farming Past and Present has retained its reputation as a
classic partly because its author's political convictions did not seriously
distort his vision of the period discussed in Mr. Fussell's Introduction,
and, partly, because in the new edition of 1936, edited by Sir Daniel Hall
and now here reprinted, very little of Lord Ernle's original text relating
to the nineteenth and twentieth centuries remains. Sir Daniel shared
few of Ernle's presuppositions about the land and its ownership. He
was one of the leading exponents of land nationalisation as the essential
basis of rural reconstruction,[3] and the survey of the history of small
holdings which he wrote for his edition of *English Farming Past and
Present* concluded : ". . . it is difficult to suppose that a peasantry can
be recreated in England or that English farming can be revived by a
wholesale division of the large farms. Every year science, machinery,

---

[1] Hammond, *op. cit.*, p. 332.
[2] *English Farming Past and Present* (4th edition, 1927), pp. 417–18.
[3] His mature views can be most easily discovered in his final testament, *Reconstruction and the Land* (1941).

and the art of organisation are advancing and widening the gap in efficiency which separates the large from the small holding. Economic pressure will have its way in this as in preceding centuries, more rapidly, indeed, in the modern world, as science is really beginning to obtain some control over the processes of life which determine production from the land."[1] Ernle's Preface to the first edition still stands in the fifth ; forlorn, neglected and irrelevant. " I knew ", explained Sir Daniel, " that I had neither the desire nor the knowledge to alter what Lord Ernle had written of ' the past '. But what in 1912 he wrote of ' the present ' has since taken on a different colour . . ."[2] Thus, in the present edition, Lord Ernle's historical justification for a territorial aristocracy has disappeared : in its place there is an account of such changes in organisation, techniques and legislation as were relevant to the interests of an agricultural scientist and administrator.

The critical bibliography that follows does not aim to be exhaustive. It has been written in order to suggest lines of approach to recent agricultural history that are neglected or ignored in Sir Daniel Hall's edition, written almost a quarter of a century ago.

[1] p. 430.
[2] Preface to 5th edition, p. v.

II

THE HISTORIOGRAPHY OF ENGLISH FARMING

E R N L E ' S work was strongly influenced by his social and political convictions. Such involvement does not conflict with the purposes of the historian ; indeed, there are those who hold it to be the very breath of living history. But there is a clear distinction between the scholarship of the historian and the propaganda of the politician. This distinction was too often blurred in Ernle's writing. The modern reader is more conscious of the deft pen of a learned journalist than of the careful craftmanship of a scholar. He prided himself on his knowledge of the early writers on agriculture,[1] but even here his reading was narrower than that of earlier writers ; and he resolutely ignored the research of his contemporaries. As Mr. Fussell points out, he had read Maitland but not Vinogradoff or Round ; in none of the editions is there reference to the book of Russell M. Garnier, *History of the English Landed Interest : Its Customs, Laws and Agriculture* (2 vols., 1908), a distinguished study now inexplicably neglected, or to the work of Brodrick, Cunningham, Gonner, or Curtler. The findings of new research were never incorporated in the successive editions which, for example, do not refer to the books of Tawney or the Hammonds and dismiss the book of W. Hasbach, *A History of the English Agricultural Labourer* (1908) in a footnote to a statistical table in an appendix. Indeed, the list of books that Ernle did not consult when preparing his first and later editions includes nearly all the important contributions to his subject.

A narrow bibliography reflected a narrow conception of agricultural history. Ernle followed the tradition, well established by 1912 and still observed today, of writing about England in isolation from the rest of the country. The distortions that result from neglecting developments in Scotland and Ireland are as serious for agricultural as for other branches of economic history. It is not the purpose of this bibliography to attempt a survey of the literature relating to these other areas of the United Kingdom. But the student who is unaware of the circumstances

[1] *Whippingham to Westminster*, p. 251.

which, on the one hand, led to the great famine and made the problems of Irish land a continuous preoccupation of the imperial parliament, and, on the other, produced the brutal oppressions of the highland clearances as well as the magnificent farming of the Lothians, is not likely to make much of his study of the causes and course of rural change in England.

The brilliant and impressive book of an "amateur" historian (it is the author's own, modest description), Redcliffe N. Salaman, *The History and Social Influence of the Potato* (1949) is indispensable. Its main conclusions for Ireland may be quickly obtained from Dr. Salaman's Finlay Memorial Lecture, "The Influence of the Potato on the Course of Irish History" (1943). The essential background is comprehensively surveyed in the penetrating volumes of Gustave de Beaumont, *L'Irlande: Sociale, Politique et Religieuse* (1st edition, 1839, 7th and best revised edition, 1863, Paris)[1] and studied in the suggestive books of K. H. Connell, *The Population of Ireland, 1750–1845* and T. W. Freeman, *Pre-Famine Ireland* (1957). The famine will be the central theme of a forthcoming volume from Mrs. Cecil Woodham-Smith which will supplement and extend the symposium edited by R. D. Edwards and T.D. Williams, *The Great Famine: Studies in Irish History, 1845–52* (1956). Different aspects of Irish experience after the forties may be pursued in the essay of D. V. Glass, "Malthus and the Limitation of Population Growth" in *Introduction to Malthus*, ed. D. V. Glass (1953), in the definitive history of J. L. Hammond, *Gladstone and the Irish Nation* (1938) which relates the later land legislation to political movements and attitudes in England and Ireland, and in the sociological analysis of C. M. Arensberg and S. T. Kimball, *Family and Community in Ireland* (Cambridge, Mass., 1940). The remarkably erudite synthesis of R. D. Collison Black, *Economic Thought and the Irish Question* (1960) is indispensable.

For Scotland, the eighteenth-century developments are summarised in the book of James E. Handley, *Scottish Farming in the Eighteenth Century*, and set in the context of general industrial development by that of Henry Hamilton, *The Industrial Revolution in Scotland* (1932). The story is carried to the middle of the nineteenth century in Part I of the perceptive, penetrating, original (but curiously titled) social history of L. J. Saunders, *Scottish Democracy 1815–1840* (1950). A detailed description of changing farming practice is given in the article of John Dudgeon, "Account of the Improvements which have taken

[1] A translation of the 1st edition, ed. W. Cooke Taylor, was published in 1839.

place in the Agriculture of Scotland since . . 1783 ", *J.R.A.S.E.*, vol. I, 1840. Something may be gleaned of the later period from the short, introductory outline of T. Bedford Franklin, *A History of Scottish Farming* (1952). There is much useful information in the compendium of Alexander Ramsay, *History of the Highland and Agricultural Society of Scotland* (1879), and in the symposium, prepared under the auspices of the Society, *Report on the Present State of the Agriculture of Scotland* (1878). The *Memoir of George Hope of Fenton Barns* (1881), compiled by his daughter, not only describes the career of the greatest of nineteenth-century farmers but is also an essential source for the understanding of the attitudes and opinions of an articulate tenant towards the land system—it may be supplemented by the nostalgic and charmingly written reminiscences of one of Hope's pupils, A. G. Bradley, *When Squires and Farmers Thrived* (1927) ; and the past is brought into relation with the present by the ten-year-old enterprise from which will one day emerge a complete *Third Statistical Account of Scotland*. The pioneer volume of John Strawhorn and William Boyd, *Ayrshire*, was published in 1951. Students of English agriculture may with advantage refer to the already published volume of Catherine P. Snodgrass, *County of East Lothian* (1953). The terrible fate of the highlanders still awaits a historian capable of translating an analysis of human degradation into the poetry of high tragedy. Some of the necessary analysis is contained in the book of Malcolm Gray, *The Highland Economy* (1957) ; other points of view and different emphases may be sought in those of John Stuart Blackie, *The Scottish Highlander and the Land Laws* (1885), Alexander MacKenzie, *The History of the Highland Clearances* (2nd revised edition, 1914) and Thomas Johnston, *The History of the Working Classes in Scotland* (n.d.).

The distinctiveness of Welsh experience is as difficult to assimilate as it is easy to neglect. Perhaps the best approach to the Principality's elusive agrarian history is through the survey of A. W. Ashby and I. L. Evans, *The Agriculture of Wales and Monmouthshire* (1944), and the references there cited. Modern writers have done little to correct the undue concentration upon matters English in Ernle's book. The major exception is the weighty trilogy of Sir John Clapham, *An Economic History of Modern Britain* (3 vols., 1926, 1932, 1938), which remains unchallenged as containing the best, systematic account of the economic history of agriculture in the modern period. More succinct, but nevertheless valuable, accounts are those contained in the text-books of W. H. B. Court, *A Concise Economic History of Britain from 1750 to Recent*

Times (1954) and W. Ashworth, *An Economic History of England* (1960). The use which these authors make of *English Farming Past and Present* is noteworthy. Although Sir John Clapham refers in a footnote to " Lord Ernle's classic ",[1] there is no point at which his narrative relies upon it ; and it is not treated as an authoritative source by Professors Ashworth and Court.

In approaching recent studies devoted exclusively to the history of English agriculture, the student must guard against an insular as well as a regional bias. For this purpose there is the survey of Sir John Russell, *World Population and World Food Supplies* (1954), and the comparative statistical analysis of E. M. Ojala, *Agriculture and Economic Progress* (1951). The pre-1914 European background is surveyed in the book of E. A. Pratt, *The Organization of Agriculture* (1904), and the post-war circumstances are described and explained in those of H. Hessell Tiltman, *Peasant Europe* (1936), Doreen Warriner, *The Economics of Peasant Farming* (1939), P. Lamartine Yates, *Food Production in Western Europe* (1940), and Doreen Warriner and P. Lamartine Yates, *Food and Farming in Post-War Europe* (1943), a short but admirable survey. The publications of the Food and Agriculture Organization of the United Nations provide a running commentary on post-war developments and prospects. The short, comparative survey of *Land Reform : Defects in Agrarian Structure as Obstacles to Economic Development* (United Nations, Department of Economic Affairs, New York, 1951) is very useful. Such reading is necessary if English problems are to be kept within a sensible perspective.

A formidable list of books has been published since 1900 on the general history of English farming either during the last century and a half or some shorter part of it. The book of E. A. Pratt, *The Transition in Agriculture* (1906), describes the agricultural structure that emerged from the years of depression, and that of W. H. R. Curtler, *A Short History of English Agriculture* (1909), followed Prothero's book of 1888 in its attempt to cover the whole period from the Middle Ages to the twentieth century in a continuous narrative.[2] The very large book of J. A. Venn, first published in 1923 with an extended and re-written second edition, *The Foundations of Agricultural Economics Together with An Economic History of British Agriculture During and After the Great War* (1933) is rewarding to the leisured and persistent

---

[1] *An Economic History of Modern Britain* (vol. III, 1938), p. 76, fn. 2.
[2] The earliest such attempt known to the present writer was that of C. Wren Hoskyns, *A Short Inquiry into the History of Agriculture in Ancient, Medieval, and Modern Times* (1849).

reader who can relax after his exertions by turning to the elegant and delightful collection of essays of James A. Scott Watson and May Elliott Hobbes, *Great Farmers* (1st ed. 1937, revised and enlarged, 1951). The results of an enquiry organised by Viscount Astor and B. Seebohm Rowntree were published under the title *British Agriculture : The Principles of Future Policy* (1938). They provide the most elaborate and extensive analysis of the industry available for the 1930's and may usefully be supplemented by the findings of the report of L. Dudley Stamp, *The Land of Britain : Its Use and Misuse* (1948). The essays presented to Sir Daniel Hall in 1939, *Agriculture in the Twentieth Century*, contain important summaries of scientific, technical and administrative changes written by leading authorities. The mechanisms and results of governmental control over agriculture is the main theme of the study of Edith H. Whetham, *British Farming : 1939–1949*. The end of Hitler's war saw the publication of a number of useful books : that of C. S. Orwin, *A History of English Farming* (1949), is a short and elegantly written introduction ; those of Victor Bonham-Carter, *The English Village* (1952), and of Robert Trow-Smith, *English Husbandry* (1951) and *Society and the Land* (1953) are intended for the general reader for whom they provide attractive accounts, first, of the changing practice of husbandry and, second, of the declining social and economic importance of the countryside. The essay in bucolic military history of E. W. Martin, *The Secret People : English Village Life after 1750* (1954), recounts the changes in village life as warfare against urban enemies. Mr. Martin is not alone in feeling that agricultural historians record the lives of balanced men " in a world growing increasingly unbalanced ".[1] That phrase of Dr. Hoskins points the outlook which underlies his extensive contributions to English topography and local history : perhaps the most interesting for the recent period is his pioneering exploration of *The Making of the English Landscape* (1955). The following are examples drawn from the wide range of local studies: Maude E. Davies, *Life in an English Village* (1909), Alfred Williams, *Villages of the White Horse* (1913), Arthur H. Savory, *Grain and Chaff from an English Manor* (1920), Arthur G. Ruston and Denis Witney, *Hooton Pagnell : The Agricultural Evolution of a Yorkshire Village* (1934), W. G. Hoskins (ed.), *Studies in Leicestershire Agrarian History* (1949), W. G. Hoskins, *Devon* (1954), and Joan Thirsk, *English Peasant Farming : The Agrarian History of Lincolnshire from Tudor to Recent Times* (1957). The volumes of *The Victoria County Histories* are the

[1] *The Agricultural History Review*, vol. II (1954), p. 10.

most extensive source for local detail. From few of the books cited will the student obtain as vivid a picture of farming changes as that presented in the autobiographical reminiscences of A. G. Street, *Farmer's Glory* (1932). And from none will he savour so keenly the tang of the earth and feel so strongly the grip of its tyranny over most of those who lived by it, as from the trilogy of Flora Thompson, *Lark Rise to Candleford* (1945), or, if he is prepared to master the speech of the Mearns, as from two of the very few great novels written in Britain in the twentieth century, Lewis Grassic Gibbon (the pseudonym of James Leslie Mitchell), *Sunset Song* (1932) and *Cloud Howe* (1933). But, however extensive the bibliography and however diligent the reader, books will never be a substitute for boots.

Recently published books may usefully be supplemented by contemporary compilations and retrospective essays which conveniently permit speedy reference and afford a bird's-eye view. The heroic labours of G. R. Porter, *The Progress of the Nation* (1st edition, 3 vols., 1836, 2nd single volume edition, 1847, final revised edition by F. W. Hirst, 1912), assembles much useful agricultural information. The essays of William Johnston, *England As It Is* (2 vols., 1851), derive mainly from parliamentary papers, but, like the work of Porter, they bring a wide range of material into focus and supply a necessary corrective by treating the countryside as a segment of an industrialising society. The same attitude runs through the contribution of Andrew Steuart, " Agriculture in Britain at the Present Day ", to *Cambridge Essays* (1857). A " survey of fifty years of progress ", *The Reign of Queen Victoria*, ed. Thomas Humphry Ward (2 vols., 1887), contains the retrospective essay of Sir James Caird, " Agriculture " (vol. 2) ; the similar symposium, ed. James Samuelson, *The Civilisation of Our Day* (1896) is valuable for the essays of W. E. Bear, " The Land and the Cultivator " and of Richard Bannister, " The Food of the People ". R. E. Prothero wrote an account of " English Agriculture in the Reign of Queen Victoria " for the *Journal of the Royal Agricultural Society* in 1901 (vol. LXII). The statistical compilation of William Page, *Commerce and Industry* (vol. 2, 1919), prints selected returns running back, where available, to the Napoleonic Wars ; includes much statistical data relating to agriculture ; and is invaluable for easy reference. The diamond jubilee issue of *Agriculture*, the Journal of the Ministry of Agriculture (September, 1954), surveyed the years since 1894 in a series of short but authoritative articles.

The wealth of information concerning the practice and organisation of farming in different areas of the country at different times since the end of the eighteenth century, is remarkable. One such source is the observations of the many British and foreign travellers who trod paths first cut by Arthur Young and William Marshall. Their followers included many whose tours began and ended in Grub Street, but the books that follow, a small selection from a wide range, were all written from personal knowledge and observation and often provide a commentary upon many aspects of English society as well as upon its farming. In its three-volume manual on *Husbandry*, the Society for the Diffusion of Useful Knowledge, that steam engine of the moral world, included (vol. 3, 1840) a series of first-hand reports of select farms in several counties. An American from Boston, Henry Colman, came to Europe to report on its agriculture in 1843 and published his findings as *European Agriculture and Rural Economy from Personal Observation* (1st edition, 1844, 2nd edition, 2 vols., 1849). The title is misleading : all but two hundred pages of the second edition, over a thousand pages long, refer to Britain. (His continental tour was separately described in *The Agriculture and Rural Economy of France, Belgium, Holland, and Switzerland from Personal Observation* (1848).) His massive book has been wholly neglected when the histories of English farming have been put together.[1] Colman was clear-headed, lively in mind and style ; his point of view is fresh and unusual, and he assembles a mass of information not available in the justly more famous book of James Caird, *English Agriculture in 1850–51* (1852), also published in the United States and translated into French, German and Swedish. He was commissioned by *The Times* to undertake an enquiry into the " actual state of agriculture in the principal counties of England ". At this time, the French, who had repaid Arthur Young's interest in their affairs by an almost anthropological curiosity towards the English aristocracy and its land system, developed a keen enthusiasm for British farming practice. The *Société d'Agriculture, Sciences et Arts* of Meaux sent some of its members, who, like many others of the day, came over to visit the Great Exhibition, on a *voyage agronomique*. Their report, *Agriculture Anglaise : Situation Économique et Agricole ; Modes de Culture* (1852), may be read alongside that of Léonce de Lavergne, *The Rural Economy of England, Scotland, and Ireland* (1855). The *Memoir on the Agriculture of England and Wales*, produced by the Royal Agricultural Society

---

[1] The well-known book of F. Law Olmsted, *Walks and Talks of an American Farmer in England* (1852) is thin and limited to commonplace observations about Cheshire and Shropshire.

in 1878 for the International Agricultural Congress in Paris, serves as the epitaph for the halcyon days of high farming. Lavergne had confidently addressed " those especially who, like myself, after having tried other careers, have turned towards a country life, disgusted by the revolutions of the times. In the bosom of nature, which changes not, they will find what they seek,—occupation in undisturbed quiet, with an independence resulting from their labours. . . ."[1] Half a century later, Rider Haggard, turning from historical romances to present realities, surveyed, in *Rural England* (2 vols., 1st edition, 1903, 2nd edition, 1906), a countryside painfully aware that the bosom of nature could not always be relied upon by those seeking undisturbed quiet. *The Times* sent another agriculturist on his travels in 1910 ; the reports of Daniel Hall were published in book form as *A Pilgrimage of British Farm. ing* (1913). A new emissary from Printing House Square is long overdue. The inter-war years produced nothing to match the quality of the writing of the earlier *Times*' correspondents. The sketches of an Essex journalist, S. L. Bensusan, *Latter-Day Rural England, 1927* (1928) and the broadcast talks of J. A. Scott Watson, *Rural Britain To-Day and To-Morrow* (1934), though informative and knowledgeable, are slight.

Other sources which make less attractive reading are the County Reports of the old Board of Agriculture, the Prize Essays commissioned by the Royal Agricultural Society and published in its *Journal*, the evidence before the Royal Commissions appointed in 1879 and 1893, the Reports of the Land Enquiry and Land Committees of the Liberal Party, published in 1913 and 1925, and the National Farm Survey of 1941–1943. The County Reports, the Prize Essays and the evidence before the Richmond and Eversley Commissions are here listed county by county in order to indicate the range of material and to provide a means of easy, rapid and comparative reference.

Sir John Sinclair had intended to produce a Statistical Account of England on the model of his private enterprise survey of Scotland's parishes. Frustrated by the established Church which feared an enquiry into tithe, Sinclair and the Board compromised by initiating a series of County Reports published under the title : *General View of the Agriculture of* . . . The internal history of the Board and of its publications is authoritatively recounted by Rosalind Mitchison in " The Old Board of Agriculture, 1793–1822 " (*The English Historical Review*, vol. LXXIV, no. 290, 1959). The County Reports were first issued in quarto editions, printed manuscripts, as Sinclair described them, and

[1] p. vii.

circulated with a request to readers to return them to the Board with marginal comments; octavo editions then followed. The several editions and frequent changes in authorship are confusing. A complete list of the first quarto and first octavo editions is printed in the bibliography of W. F. Perkins, *British and Irish Writers on Agriculture* (1st edition 1929, 2nd edition 1932).[1] The following catalogue notes the first quarto and the last octavo editions.[2]

Bedfordshire	T. Stone (1794) : T. Batchelor (1808).
Berkshire	W. Pearce (1794) : W. Mavor (1813).
Buckinghamshire	William, James and Jacob Malcolm (1794) : St. J. Priest (1813).
Cambridgeshire	C. Vancouver (1794) : W. Gooch (1813).
Cheshire	T. Wedge (1794) : H. Holland (1813).
Cornwall	R. Fraser (1794) : G. B. Worgan (1815).
Cumberland	J. Bailey and G. Culley (1794) : J. Bailey and G. Culley (1805).
Derbyshire	T. Brown (1794) : J. Farey (3 vols., 1811–17).
Devonshire	R. Fraser (1794) : C. Vancouver (1813).
Dorset	J. Claridge (1793) : W. Stevenson (1815).
Durham	J. Granger (1794) : J. Bailey (1810).
Essex	Messrs. Griggs (1794) : C. Vancouver (1795) : A. Young (2 vols., 1807).
Gloucestershire	G. Turner (1794) : T. Rudge (1807).
Hampshire	Abraham and William Driver (1794) : C. Vancouver (1813).
Herefordshire	J. Clark (1794) : J. Duncumb (1805).
Hertfordshire	D. Walker (1795) : A. Young (1813).
Huntingdonshire	T. Stone (1793) : R. Parkinson (1813).
Kent	J. Boys (1794) : J. Boys (1813).
Lancashire	J. Holt (1794) : R. W. Dickson (1815).
Leicestershire	J. Monk (1794) : W. Pitt (1809).
Lincolnshire	T. Stone (1794) : A. Young (1808).
Middlesex	T. Baird (1793) : P. Foot (1794) : J. Middleton (1813).
Monmouthshire	J. Fox (1794) : C. Hassall (1812).

[1] Incomplete lists were printed in the books of J. R. McCulloch, *The Literature of Political Economy : A Classified Catalogue* (1845, reprinted 1938) and Ivy Pinchbeck, *Women Workers and the Industrial Revolution* (1930). The most useful bibliographical source for checking the various editions, once authorship is known, is the work of Mary S. Aslin, *Library Catalogue of Printed Books and Pamphlets on Agriculture Published Between 1471 and 1840* (Rothamsted Experimental Station, 1st edition 1925, 2nd edition 1940).

[2] A similar list was compiled by Sir Ernest Clarke, " The Board of Agriculture, 1793–1822 ", *J.R.A.S.E.*, 1898.

Norfolk	N. Kent (1794) : A. Young (1804) : N. Kent (1813).
Northamptonshire	J. Donaldson (1794) : W. Pitt (1813).
Northumberland	J. Bailey and G. Culley (1794) : J. Bailey and G. Culley (1805).
Nottinghamshire	R. Lowe (1794) : R. Lowe (1798).
Oxfordshire	R. Davis (1794) : A. Young (1809).
Rutland	J. Crutchley (1794) : R. Parkinson (1809).
Shropshire	J. Bishton (1794) : J. Plymley (1803).
Somerset	J. Billingsley (1794) : J. Billingsley (1798).
Staffordshire	W. Pitt (1794) : W. Pitt (1813).
Suffolk	A. Young (1794) : A. Young (1804).
Surrey	William, James and Jacob Malcolm (1794) : W. Stevenson (1813).
Sussex	Rev. A. Young (1793) : Rev. A. Young (1808).
Warwickshire	J. Wedge (1794) : A. Murray (1815).
Westmorland	A. Pringle (1794) : A. Pringle (1805).
Wiltshire	T. Davis (1794) : T. Davis (1813).
Worcestershire	W. T. Pomeroy (1794) : W. Pitt (1813).
Yorkshire	
East Riding	I. Leatham (1794) : H. E. Strickland (1812).
North Riding	J. Tuke (1794) : J. Tuke (1800).
West Riding	G. B. Rennie, R. Brown and J. Shirreff (1794): R. Brown (1799).
Wales	
North	G. Kay (1794) : W. Davies (1813).
South	

Brecknockshire	J. Clark (1794)	⎫
Cardiganshire	T. Lloyd and D. Turnour (1794)	⎪
Carmarthenshire	C. Hassall (1794)	⎬ W. Davies (2
Glamorgan	J. Fox (1796)	vols., 1815).
Pembrokeshire	C. Hassall (1794)	⎪
Radnorshire	J. Clark (1794)	⎭
Isle of Man	B. Quayle (1794).	
Jersey and Guernsey	T. Quayle (1815).	

These Reports must be used with great caution and only after careful enquiry into the qualifications of the reporters.[1] A criticism and

---

[1] A group study which also assessed the qualifications of the reporters would be invaluable. It should not be forgotten that similar Reports were prepared for Scotland and Ireland. They are listed in the bibliography of W. F. Perkins, *op. cit.*

summary of this massive, uncertain but inescapable source is available in the herculean undertaking of William Marshall, *A Review and Complete Abstract of the Reports to the Board of Agriculture on the Several Counties of England* (5 vols., 1817). But Marshall was disappointed and embittered ; his volumes must therefore be approached with the knowledge of the personal relationships and animosities clustering round the Board of Agriculture that may be obtained from the articles of Mrs. Rosalind Mitchison[1] and of G. E. Fussell, " My Impressions of Arthur Young" (*Agricultural History*, vol. 17, 1943), " My Impressions of William Marshall " (*ibid.*, vol. 23, 1949), and " My Impressions of Sir John Sinclair, Bt.: First President of the Board of Agriculture " (*ibid.*, vol. 25, 1951). The bibliography of G. E. Fussell, " English Agriculture: From Arthur Young to William Cobbett " (*The Economic History Review*, vol. VI, no. 2, 1936) places their labours in the contemporary setting.

The greatest days of the Royal Agricultural Society of England, recorded in the book of J. A. Scott Watson, *History of the Royal Agricultural Society of England, 1839–1939* (1939), span most of the years between the collapse of the old Board of Agriculture in 1822 and the establishment of the new in 1889 to take over the many agricultural functions and powers scattered among departments and *ad hoc* bodies. (The new Board's unimpressive history is unimpressively catalogued in the official history of Sir Francis Floud, *The Ministry of Agriculture and Fisheries* (1927).) During those early- and mid-Victorian years the Society acted persuasively as a central agency for English farmers and landlords, hortatory and supervisory ; and its Journal, under the great editorships of Philip Pusey[2] and H. M. Jenkins,[3] maintained the highest standards set by the best periodicals of the period, besides containing material of interest and importance to social as well as to agricultural historians. Before 1866, many of the contributions took the form of prize essays. Some of these reviewed, county by county, the recent progress and present state of agriculture and thus afford an invaluable comparison with the earlier County Reports. In the main they were written by farmers, land agents, parsons and land-owners, usually under the title " On the Farming of . . .". The prize winners received fifty sovereigns or a piece of plate of that value.

[1] *The English Historical Review, op. cit.*
[2] The obituary in the *D.N.B.* (vol. xlvii, 61) may be supplemented by the article of Paolo E. Coletta, " Philip Pusey, English Country Squire," *Agricultural History*, vol. 18, 1944.
[3] There is the essay of J. Chalmers Morton, " The Late Mr. H. M. Jenkins," *J.R.A.S.E.*, vol. XXIII (N.S.), 1887.

There is a brief discussion of the geographical and geological significance of the essays in the article of H. C. Darby, " Some Early Ideas on the Agricultural Regions of England " (*The Agricultural History Review*, vol. II, 1954).

The complete list is :

		Vol. no.	Year[1]
Bedfordshire	William Bennett	XVIII	1857
Berkshire	J. B. Spearing	XXI	1860
Buckinghamshire	Clare Sewell Read	XVI	1855
Cambridgeshire	Samuel Jonas	VII	1846
Cheshire	William Palin	V	1844
Cornwall	W. F. Karkeek	VI	1845
Cumberland	William Dickinson	XIII	1852
Derbyshire	John Jephson Rowley	XIV	1853
Devonshire	Henry Tanner	IX	1848
Dorset	Louis H. Ruegg	XV	1854
Durham	Thomas George Bell	XVII	1856
Essex	Robert Baker	V	1844
Gloucestershire	John Bravender	XI	1850
Hampshire	John Wilkinson	XXII	1861
Herefordshire	Thomas Rowlandson	XIV	1853
Hertfordshire	Henry Evershed	XXV	1864
Huntingdonshire	Gilbert Murray	IV (N.S.)	1868
Kent	George Buckland	VI	1845
Lancashire	William James Garnett	X	1849
Leicestershire	W. J. Moscrop	II (N.S.)	1866
Lincolnshire	J. A. Clarke	XII	1851
Middlesex	J. C. Clutterbuck	V (N.S.)	1869
Monmouthshire[2]	W. Fothergill	VI (N.S.)	1870
Norfolk[3]	Richard Noverre Bacon		
Northamptonshire	William Bearn	XIII	1852

---

[1] The system of numbering used by this journal causes much confusion and exasperation. The various series and their numbers are explained in the volume for 1901 (No. 62 of the entire series), p. 8. I have here followed the old numbers and (N.S.) therefore indicates the second series which began in 1865.

[2] This was not a prize essay.

[3] Mr. Bacon's desire to uphold " the skill and talent of Norfolk farmers " and his aim to maintain " with equal impartiality the true position of the OWNER, the just rights of the TENANT, and the welfare of the LABOURER " could not be accomplished in less than four hundred pages. The Society held them to be praiseworthy and prizeworthy, but declined to devote a whole issue of the Journal to their publication. They appeared as *The Report on the Agriculture of Norfolk*, . . . (1844) bearing " Veni, Vidi " on the title page.

		Vol. no.	Year
Northumberland	Thomas L. Colbeck	VIII	1847
Nottinghamshire	R. W. Corringham	VI	1845
Oxfordshire	Charles Sewell Read	XV	1854
Shropshire	Henry Tanner	XIX	1858
Somerset[1]	Thomas Dyke Acland jun.	XI	1850
Staffordshire	Henry Evershed	V (N.S.)	1869
Suffolk[2]	Hugh Raynbird	VIII	1847
Surrey	Henry Evershed	XIV	1853
Sussex	John Farncombe	XI	1850
Warwickshire	Henry Evershed	XVII	1856
Westmorland	Crayston Webster	IV (N.S.)	1868
Wiltshire	Edward Little	V	1844
Worcestershire	Clement Cadle	III (N.S.)	1867
Yorkshire :			
East Riding	George Legard	IX	1848
North Riding	M. M. Milburn	IX	1848
West Riding	John H. Charnock	IX	1848
Wales :			
North	Thomas Rowlandson	VII	1846
South	Clare Sewell Read	X	1849
Jersey, Guernsey,			
Alderney & Sark[3]	C. P. Le Cornu	XX	1859
Scilly Isles[3]	Laurence Scott and		
	Henry Rivington	VI (N.S.)	1870

The following articles were published at the same time as the prize essay series, and, though not part of it, they may usefully be added : John Grey, " A View of the Past and Present State of Agriculture in Northumberland " (vol. II, 1841) ; Philip Pusey, " On the Agricultural Improvements of Lincolnshire " (vol. IV, 1843) ; Barugh Almack, " On the Agriculture of Norfolk " (vol. V, 1844) ; Charles Stevenson, " On the Farming of East Lothian " (vol. XIV, 1853) ; John Parkinson, " On Improvements in Agriculture in the County of Nottingham since 1800 " (vol. XXII, 1861) ; and William Wright, " On the Improvements in the Farming of Yorkshire since the date of the last Reports in the Journal " (vol. XXII, 1861).

[1] Later expanded into a book as T. Dyke Acland and William Sturge, *The Farming of Somerset* (1851).
[2] Later expanded into a book as William and Hugh Raynbird, *On the Agriculture of Suffolk* (1849).
[3] These were not prize essays.

The Royal Agricultural Society's Prize Essays were more succinctly written than the old County Reports ; their authors worked to a plan and had the advantage of skilled editorship. Like the *Journal* in which they appeared, the Essays are an indispensable source.

The student of nineteenth- and twentieth-century economic and social history can call on little in the way of bibliographical aids. He must therefore be especially grateful for the devoted and magnificent enterprise of Professor and Mrs. Ford who have provided a *Select List of British Parliamentary Papers* (1953), *A Breviate of Parliamentary Papers, 1900–1916* (1957) and *A Breviate of Parliamentary Papers, 1917–1939* (1951). These volumes make it unnecessary here to list in detail the parliamentary papers relating to agriculture and allied subjects. Indeed, the virtues of the Ford *Breviates* are so great that their users are exposed to the temptation to read them as a substitute for the papers they so admirably summarise. The resistance of the indolent will be strengthened, and their labours eased, by a reading of the short, accompanying *Guide to Parliamentary Papers* (1955).

The evidence of the Assistant Commissioners before the Royal Commissions on the " Depressed Condition of the Agricultural Interests " and on " Agricultural Depression " is listed county by county as for the old Board of Agriculture Reports and the Royal Agricultural Society's Prize Essays. The alphabetical symbol in the column headed " *P.P.*" will enable the reader to extract the number of the parliamentary paper in which an Assistant Commissioner's report appears from the complete list here printed after the county table.

County or Area	Date	P.P.	Assistant Commissioner	Date	P.P.	Assistant Commissioner
Bedfordshire	1881	(a)	S. B. L. Druce	1895	(q)	R. H. Pringle
	1882	(c)				
Berkshire[1]	1881	(a)	W. C. Little	1895	(s)	Aubrey Spencer
	1882	(c)				
Buckinghamshire[2]	1881	(a)	S. B. L. Druce	1895	(l)	Aubrey Spencer
	1882	(c)				
Cambridgeshire	1881	(a)	S. B. L. Druce	1895	(r)	H. Wilson Fox
	1882	(c)				
Cheshire	1882	(d)	John Coleman			
Cornwall	1881	(a)	W. C. Little			
	1882	(b)				
Cumberland	1881	(a)	John Coleman	1895	(u)	H. Wilson Fox
Derbyshire	1881	(a)	S. B. L. Druce			
	1882	(c)				
Devonshire[3]	1881	(a)	W. C. Little	1895	(m)	Henry Rew
	1882	(b)				
Dorset	1881	(a)	W. C. Little	1895	(p)	Henry Rew
	1882	(b)				
Durham[4]	1881	(a)	John Coleman	1895	(n)	R. H. Pringle
Essex[5]	1881	(a)	S. B. L. Druce	1894	(h)	R. H. Pringle
	1882	(c)				
Gloucestershire[6]	1881	(a)	Andrew Doyle	1895	(s)	Aubrey Spencer
Hampshire[7]				1894	(f)	William Fream
Herefordshire	1881	(a)	Andrew Doyle			

[1] Spencer's report relates only to selected districts.
[2] Spencer's report relates only to the Vale of Aylesbury.
[3] Rew's report relates only to north Devon.
[4] Pringle's report relates only to south Durham.
[5] Pringle's report relates only to the districts around Ongar, Maldon, Braintree and Chelmsford.
[6] Spencer's report relates only to selected districts.
[7] Fream's report relates only to the Andover district.

County or Area	Date	P.P.	Assistant Commissioner	Date	P.P.	Assistant Commissioner
Hertfordshire	1881	(a)	S. B. L. Druce	1895	(l)	Aubrey Spencer
	1882	(c)				
Huntingdonshire	1881	(a)	S. B. L. Druce	1895	(q)	R. H. Pringle
	1882	(c)				
Kent[1]	1881	(a)	W. C. Little	1894	(f)	William Fream
Lancashire[2]	1882	(d)	John Coleman	1894	(e)	H. Wilson Fox
Leicestershire	1881	(a)	S. B. L. Druce			
	1882	(c)				
Lincolnshire[3]	1881	(a)	S. B. L. Druce	1894	(h)	R. H. Pringle
	1882	(c)		1895	(k)	H. Wilson Fox
Middlesex	1881	(a)	S. B. L. Druce			
	1882	(c)				
Monmouthshire	1881	(a)	Andrew Doyle			
Norfolk	1881	(a)	S. B. L. Druce	1895	(t)	Henry Rew
	1882	(c)				
Northamptonshire	1881	(a)	S. B. L. Druce	1895	(q)	R. H. Pringle
	1882	(b)				
Northumberland[4]	1882	(d)	John Coleman	1894	(e)	H. Wilson Fox
Oxfordshire[5]	1881	(a)	Andrew Doyle	1895	(s)	Aubrey Spencer
Rutland	1881	(a)	S. B. L. Druce			
	1882	(c)				
Shropshire	1881	(a)	Andrew Doyle			
Somerset[6]	1881	(a)	W. C. Little	1894	(q)	Jabez Turner
	1882	(b)		1895	(s)	Aubrey Spencer
Staffordshire	1881	(a)	Andrew Doyle			
Suffolk	1881	(a)	S. B. L. Druce	1895	(o)	H. Wilson Fox
	1882	(c)				
Surrey	1881	(a)	W. C. Little			
Sussex[7]	1881	(a)	W. C. Little	1895	(i)	Henry Rew

[1] Fream's report relates only to the Maidstone district.
[2] Fox's report relates only to the Garstang district.
[3] Pringle's report relates only to the Isle of Axholme ; that of Fox to the rest of Lincolnshire.
[4] Fox's report relates only to the Glendale district.
[5] Spencer's report relates only to selected districts.
[6] Turner's report relates only to the Frome district, that of Spencer to the Taunton district.
[7] Rew's report is on the *Poultry Rearing and Fattening Industry of the Heathfield District.* . . .

County or Area	Date	P.P.	Assistant Commissioner	Date	P.P.	Assistant Commissioner
Warwickshire[1]	1881	(a)	Andrew Doyle	1894	(q)	Jabez Turner
Westmorland	1881	(a)	John Coleman			
Wiltshire[2]	1881	(a)		1895	(j)	Henry Rew
				1895	(s)	Aubrey Spencer
Worcestershire	1881	(a)	Andrew Doyle			
Yorkshire[3]	1881	(a)	John Coleman	1895	(n)	R. H. Pringle

[1] Turner's report relates only to the Stratford-on-Avon district.
[2] Rew's report relates only to the Salisbury Plain district ; that of Spencer to other selected districts.
[3] Pringle's report relates only to selected districts in the East and North Ridings.

Numbering of the Parliamentary Papers cited above

1881	C.2778—II	xvi	(a)
	C.3375—I	xv	(b)
	C.3375—II	xv	(c)
	C.3375—V	xv	(d)

1894	C.7334	xvi Pt. I	(e)
	C.7365	xvi „	(f)
	C.7372	xvi „	(g)
	C.7374	xvi „	(h)

1895	C.7623	xvi	(i)
	C.7624	xvi	(j)
	C.7671	xvi	(k)
	C.7691	xvi	(l)
	C.7728	xvi	(m)
	C.7735	xvi	(n)
	C.7755	xvi	(o)
	C.7764	xvii	(p)
	C.7842	xvii	(q)
	C.7871	xvii	(r)
	C.7915	iii	(s)
	C.7915	xvii	(t)
	C.7915—I	xvii	(u)

A charge of tergiversation could never be upheld against Lloyd George in respect of his attitude to land. Throughout his career he always acted upon his conviction that " all down history nine-tenths of mankind have been grinding the corn for the remaining tenth, and been paid with the husks and bidden to thank God they had the husk ".[1] After he emerged from the Marconi affair, he determined to launch an agrarian programme which had the monopoly of the great landlords as its main target. To this end he set up the Land Enquiry Committee, with A. H. Dyke Acland as chairman, Roden Buxton as honorary secretary, and Seebohm Rowntree among its members, to collect information about the agricultural situation for the use of himself and his party. Its report, published in 1913 as *The Land : Volume I, Rural*,[2] deserves more careful attention than it has received. The research behind it earns a place in the history of social investigation as the first large-scale attempt to apply to rural problems the methods developed by Charles Booth in his study of *Life and Labour of the People in London* begun in 1886. Much data was collected by peripatetic investigators who obtained answers to carefully prepared questionnaires. The report surveys the wages, housing and condition of the agricultural labourer as well as many aspects of the cultivation of land, small holdings, land tenure and related topics. The Land Enquiry Committee's semi-official status and Lloyd George's declared objects (Charles Masterman recorded in his diary that " George is very pleased with the way his Land Campaign is going. [He said to Rufus Isaacs] : ' The Country is ripe for a Revolution ', and Rufus looked distinctly bored . . .")[3] provoked concern and indignation among landowners. The polemic of Charles Adeane and Edwin Savill, *The Land Retort* (1st edition February, 1914, 2nd edition March, 1914), was an immediate reaction : a more considered reply came from the Land Agents' Society which published *Facts About Land : A Reply to " The Land ", The Report of the Unofficial Land Enquiry Committee*, anonymously edited by R. E. Prothero,[4] in 1916. The Liberal Party was responsible for one further, useful survey, *The Land and The*

---

[1] Quoted Lucy Masterman, *C. F. G. Masterman* (1939), p. 150.
[2] Further volumes dealt with urban land and with urban and rural land in Scotland and Wales.
[3] Lucy Masterman, *op. cit.*, p. 261.
[4] *Whippingham to Westminster, op. cit.*, p. 207. " When the answers had all been collected I undertook to edit them anonymously . . . I can, therefore, say without risking the charge of self-advertisement, that no body of evidence on the condition of the agricultural industry in 1913, at once so detailed, so practical, and, allowing for the unconscious bias of land agents, so authoritative, is to be found elsewhere."

Nation (1925)[1] ; its tradition of enquiry into the problems of land then passed into the already cited work, organised in 1934 by Seebohm Rowntree and Viscount Astor. The needs of war-time policy and the refinements of survey techniques persuaded the Ministry of Agriculture to organise a national farm survey of England and Wales during the period 1941–1943. The survey fell into three parts : first, a farm survey record completed for every agricultural holding of five acres or above and providing detailed answers to questions about tenure, quality of management, availability of water and electricity, and the general condition of the farm and its buildings : second, the annual 4th June return supplemented by additional questions, and, thirdly, a scale plan of the holding. The summary report of the *National Farm Survey of England and Wales* (H.M.S.O., 1946) goes some way to support the claim of the Ministry of Agriculture to have compiled a " modern Domesday Book "[2] which disclosed, at one extreme, that 44 per cent of all holdings employed no labour whereas 20 per cent employed 70 per cent of the regular labour force, and, at the other, that 2 per cent of all holdings were infested with rats and mice. But there is one significant omission : no questions were asked about landownership, apart from the name and address of the owner. Thus, the survey revealed that 34 per cent of farmers owned their holdings,[3] but the rest is silence in the modern Domesday Book. The collection of information was carried out by the County War Agricultural Executive Committees in the course of farm inspections which were part of their war-time duties. Without doubt, the Survey is the most extensive ever made and its results the most objective yet obtained ; nevertheless they must be treated with some caution. They were partly based on a schedule containing many subjective questions which could be answered only by the exercise of skilled judgment by the staff employed. Inevitably, the criteria employed and the degree of skill available must have varied greatly from area to area, and there were grave temptations to take short cuts. It is, for example, within the personal knowledge of the present writer that all the schedules relating to the farms on one large estate were handed to the landlord's agent and completed by him in the estate office.

[1] A further volume, *Towns and the Land* (1925), was also published.
[2] p. 2.
[3] The United Nations Food and Agriculture Organisation's World Census of 1950 showed that 36 per cent of holdings in England and Wales were wholly owned by the occupier, 49 per cent were wholly rented and the remainder part-owned and part-rented.

No adequate, comprehensive account of the history of changing farming methods and practice since the end of the Napoleonic Wars has yet been written. The above (and other) sources have been used to reconstruct the development of farming in different counties, and there will be an increasing flow of books such as that of Joan Thirsk and Jean Imray (eds.), *Suffolk Farming in the Nineteenth Century* (1958). It is unlikely that they will add significantly to the conclusions of the continuing series of articles on this theme of G. E. Fussell. One series has appeared in *Economic Geography* (Worcester, Mass.) as " ' High Farming ' in Southwestern England, 1840–1880 " (vol. XXIV, 1948) ; " ' High Farming ' in the North of England, 1840–1880 " (vol. XXIV, 1948) ; " ' High Farming ' in the West Midland Counties, 1840–1880 " (vol. XXV, 1949) ; and " ' High Farming ' in the East Midlands and East Anglia, 1840–1880 " (vol. XXV, 1949). To these may be added " Home Counties Farming, 1840–1880 " (*Economic Journal*, LVII, no. 227, 1947) and " The Dawn of High Farming in England " (*Agricultural History*, vol. 22, 1948). A second series, under the general title, " Four Centuries of Farming in . . . 1500–1900 ", is widely scattered in diverse publications. " Cheshire " (*Transactions of the Historical Society of Lancashire and Cheshire*, CVI, 1955) ; " Derbyshire " (*Journal of the Derbyshire Archaeological and Natural History Society*, LXXI, 1951) ; " Devonshire " (*Transactions of the Devon Association*, XXXIII, 1951) ; " Dorset " (*Proceedings of the Dorsetshire Natural History and Archaeological Society*, 73, 1952) ; " Hampshire " (*Proceedings of the Hampshire Field Club*, XVII, Pt. 3, 1952) ; " Leicestershire " in *Studies in Leicestershire Agrarian History*, ed. W. G. Hoskins (1949) ; " Lincolnshire " (*Lincolnshire Archaeological and Archaeological Society Reports and Papers*, 4, Pt. 2, 1952) ; " Nottinghamshire " (*The Nottinghamshire Countryside*, nos. 2, 3, and 4, XVII, 1956–57) ; " Shropshire " (*Shropshire Archaeological Society*, Pt. I, LIV, 1951–52) ; and " Sussex " (*Sussex Archaeological Collection*, XC, 1952). To these may be added the pamphlet by G. E. Fussell, published by the Castle Museum, York, *Farming Systems from Elizabethan to Victorian Days in the North and East Ridings of Yorkshire* (1944).

The steady accumulation of detailed studies of farming practice helps to fulfil the agenda for the writing of agricultural history which Dr. W. G. Hoskins outlined to the newly-formed British Agricultural History Society in 1953. " We want . . . to study the actual farming practice of England more assiduously than has yet been done, and to correct what I believe to have been an over-emphasis on the legal and institutional side of agrarian history.   Historians have tended to study the *manor* rather than the village, the legal concept rather than the physical fact, and to be more interested in tenures and rents than actual farming : to give elaborate consideration to questions of land-ownership and land-occupation, and to give little consideration to land-use : to be interested, in short, in the details of the machinery and to forget what the machinery exists for."[1]   " We must be prepared to toil over minute details . . . there is no room for brilliant generalisations.   Let us leave those to the political historian.   We must get down to earth : to crops, animals, soils, buildings, implements."[2]   It may be that Dr. Hoskins provides a necessary corrective to the customary emphases in the writing of agrarian history in the pre-industrialist period.   But his directions, if followed by those primarily concerned with the history of agriculture in industrialised Britain, must lead directly into sterile antiquarianism.

Some sort of generalisation inevitably lurks behind all toil over minute details.   Too often it appears as the antiquarian's conviction that all facts are self-evidently important and of equal value.   But the historian's function is to assemble representative fact as a means to explain change.   Minute details become meaningful only in so far as they reflect and contribute to a conception of persistence, growth and change.   It is essential, as Dr. Hoskins insists, to remember that what matters is not only the machinery but also what the machinery exists for.   The implication that the machinery consists of a legal and institutional structure existing for the production of food, is fatal to an understanding of agrarian change in the nineteenth century.   For most of that period, British and Irish land sustained both an expanding urban population and the political power and social prestige of the class that owned it.   The conflict between these two functions of land was a major determinant of change in the Victorian period and its resolution helped to shape twentieth-century developments.   To give priority to minute inspections of rotations, implement sheds and middens necessarily leads to the pessimistic conclusion, gratifying to local historians, of Dr. Joan Thirsk's survey of " The Content and Sources of English Agrarian History after 1500 " that " in a sense . . . local studies have postponed immediate consideration of general development by showing

---

[1] W. G. Hoskins, " Regional Farming in England ", *The Agricultural History Review*, vol. II, 1954, p. 4.
[2] *ibid.*, p. 11.

that England has not one agrarian history but many . . .".[1] Much time and paper will be consumed before the social structure of every little local community has been defined and every rotation in every field in every parish in every county has been plotted, and, that labour accomplished, there will be the revelation that they varied in different places and at different times. This will be magnificent, but it will not be history. Local studies in the nineteenth century require a direction and purpose to be obtained only from some such general framework of interpretation as can be provided by the distinction between the food-producing and political functions of land.

Until the 1870's, British farmers were monopolists exploiting a market expanding both in numbers and in incomes. Like other producers, they suffered temporary difficulties and embarrassments stemming from inefficient, because precocious, financial and other sorts of institutions. The steadily growing volume of food imports supplemented but did not compete with home production, always the main source of food supply ; and the farmer's problem was never how to sell his output but how to produce enough to meet an insistent and shifting demand. In this setting, the treatment in *English Farming Past and Present* of the so-called agricultural depression after the Napoleonic Wars[2] and of the effects of the Corn Laws[3] is unsatisfactory. Lord Ernle's acceptance of the well conducted ululations arising from the Select Committees and Royal Commissions on agricultural distress must be qualified by the careful and sceptical studies of G. E. Fussell and M. Compton, " Agricultural Adjustments after the Napoleonic Wars " (*Economic History, A Supplement of The Economic Journal*, vol. III, no. 14, 1939) and of T. W. Fletcher, " The Great Depression in English Agriculture, 1873–1896 " (*The Economic History Review*) and " Lancashire Livestock Farming During the Great Depression " (*The Agricultural History Review*). [Both articles will be published in 1961.] His view of the Corn Laws must now be set against the revision of C. R. Fay, *The Corn Laws and Social England* (1932) and the several articles of G. Kitson Clark, for example, " The Repeal of the Corn Laws and the Politics of the

---

[1] *The Agricultural History Review* (vol. III, pt. II, 1955), p. 79. " For the eighteenth and nineteenth centuries," Dr. Thirsk writes (p. 77), " the evidence of agrarian change is to be found in Parliamentary enclosure awards, land tax assessments, rentals, tithe awards, and crop returns, and in officially printed population censuses, Parliamentary papers on the state of agriculture, agricultural statistics from 1866 onwards, and newspapers." Even for a short article, this seems an unduly narrow bibliography.

[2] 5th (Daniel Hall ed.) edition, ch. XV.

[3] *ibid.*, ch. XII.

Forties " (*The Economic History Review*, 2nd series, vol. IV, no. 1, 1951).

At the height of the Corn Law debate, an Edinburgh Reviewer surveyed the " Progress of Scientific Agriculture " : " If it be a difficult task for British Agriculture to fill with wholesome food the mouths of the present population of the island ", he asked, " how will it be able to fulfil this destiny sixty years hence, when, at the present rate of increase, the population will be doubled ? "[1] He went on to emphasise the importance of developments in field drainage and in knowledge of soil chemistry and manures as means of extending the techniques successful on light land in the eighteenth century to the heavy soils. Much of the relevant, technical literature is indicated and assessed in the bibliography of G. E. Fussell, " English Agriculture from Cobbett to Caird (1830–80) " (*The Economic History Review*, vol. XV, nos. 1 & 2, 1945). A number of important books has been published since Mr. Fussell's bibliography was compiled. The superb study of Charles A. Browne, *A Source Book of Agricultural Chemistry* (Chronica Botanica, Waltham, Mass., 1944) carries the story from ancient times to Liebig's removal from Giessen in 1852 ; the contribution of Rothamsted is traced in the book of Sir John Russell, *British Agricultural Research : Rothamsted* (1946) ; and the comparative study of L. F. Haber, *The Chemical Industry During the Nineteenth Century* (1958) fills in the industrial background. Surprisingly, what the editor of the translated manual of J. B. Bousingault, *Rural Economy in its relations with Chemistry, Physics and Meteorology* (2nd edition, 1845) described as " the capital, the all-important subject of Drainage, the master-engine of agricultural improvement", has not yet received the attention its importance warrants both as a means to increased productivity and as a source of capital expenditure. The administrative history of the public loans provided by Peel's government as a *quid pro quo* for the landed interest after the repeal of the Corn Laws, as well as of the private drainage companies, is usefully detailed in the article of J. Bailey Denton, " On Land Drainage and Improvement by Loans from Government or Public Companies " (*J.R.A.S.E.*, vol. IV, new series, 1868). Despite such assistance, the pace of improvement was slow. It can be assessed in the evidence and *Report* from the Select Committee of the House of Lords *On the Improvement of Land* (1873 VI, I).

The differentiation, specialisation and spread of farming techniques during the nineteenth century is succinctly discussed and illustrated in

[1] (Jan., 1845), p. 90.

an international context in the symposium (ed. Charles Singer, E. J. Holmyard, A. R. Hall, and Trevor I. Williams), *A History of Technology* : (vol. IV, *c. 1750–c. 1850*, 1958) ; Olga Beaumont and J. W. Y. Higgs, *Agriculture : Farm Implements*, and G. E. Fussell, *Agriculture*: *Techniques of Farming*, (vol. V, and, *c. 1850–c. 1900*, 1958) *Growth of Food Production*. The pre-tractor era of farm implements is also briefly, and very usefully, surveyed in the article of G. E. Fussell, " The Development of Farm Machinery in England " (*Engineering*, the 10th and 17th August, 1934) ; this journal also contains his account of the short but important period when steam was harnessed to agriculture, " Steam Cultivation in England " (the 30th July and the 13th August, 1943). The same author has treated these matters at greater length in *The Farmer's Tools, 1500–1900 : A History of British Farm Implements, Tools and Machinery before the Tractor Came* (1952). It is especially important in the history of farming to distinguish carefully between the invention of an implement or a machine and their generalised use which always depended on cheap, robust production. A history of the growth of a light engineering industry (that, for example, which grew up in the 1840's alongside the eastern counties railway) manufacturing reliable, standardised products at prices the ordinary farmer could pay, would be valuable. The story of the improvement of flocks and herds in the nineteenth century that followed the experiments of the great eighteenth-century improvers has been related by R. Trow-Smith in *A History of British Livestock Husbandry, 1700–1900* (1959). Changes in this as in other branches of farming were hastened and diffused by the growth of new agencies by which knowledge and, sometimes, error, could be disseminated. Of these, the Royal Agricultural Society was of dominating importance and influence with its *Journal* and Annual Show, but local societies, notably the Bath and West of England, founded in 1777 and revivified at the end of the forties by Thomas Dyke Acland, and the Yorkshire, founded in 1838, should not be neglected. The *Journals* of these two societies added to the increasing runs of agricultural periodicals and weekly journals briefly noted in the already cited book of J. A. Scott Watson and May Hobbs.[1] The results of applying new knowledge to the soil could not be measured nationally until provision had been made for the regular collection and publication of agricultural statistics. The article of J. T. Coppock, " The Statistical Assessment

---

[1] *Great Farmers*, ch. 10. The most important and useful agricultural journalism of the nineteenth century is to be found in the *Mark Lane Express*.

of British Agriculture " in *The Agricultural History Review* (vol. IV, pts. 1 and 2, 1956), traces the history behind the long-delayed Returns of 1866 and discusses their reliability and comparability. Local variations in weights and measures sometimes make the interpretation of statistical data a nightmarish procedure : some such handbook as the frequently reprinted *Agriculturist's Calculator* is often more useful than a calculating machine.

III

FARMERS, LABOURERS AND LANDLORDS

SCIENTIFIC and technical resources by which the productivity of English farming could be multiplied in the decades following the Hungry Forties[1] existed in abundance. Their utilisation was obstructed by three factors : first, by the conservatism and ignorance of most farmers, secondly, by the cheapness of agricultural labourers who were too thick on the ground, and, third, by the structure of tenurial relations.

Not least among the influences which made East Lothian farming the exemplar of "High Farming" was the educational system of Scotland. There, the attitude described by Sir Walter Scott, in *Tales of my Landlord*, early disappeared. " Your leddyship and the steward hae been pleased to propose that my son Cuddie suld work in the barn wi' a new-fangled machine for dighting the corn frae the chaff, thus impiously thwarting the will of Divine Providence, by raising the wind for your leddyship's own particular use by human art, instead of soliciting it by prayer, or waiting patiently for whatever dispensation of wind Providence was pleased to send upon the sheeling hill." Scots countryfolk were educated, and thereby acquired an open-mindedness (on matters unconnected with religion) that enabled them to welcome new ideas and to put them to profitable uses. In England, the rural middle class were poorly catered for by the provisions of a private enterprise educational system suited neither to their status nor to their pockets. " It is now fully admitted on all hands ", wrote R. Vallentine in the *Royal Agricultural Society's Journal*, " On Middle Class Education, having reference to the Improvement of the Education of those who depend upon the Cultivation of the Soil for their Support " (vol. II,

---

[1] It is suggested (W. H. Chaloner, *The Hungry Forties*, Historical Association Leaflet, 1957) that this phrase should be expunged from the literature, in part because the forties " were no hungrier than the 1830's or 1850's ". Even if this were true of England, the refusal of the title " Hungry " to a decade that witnessed the Irish Famine, mentioned by Dr. Chaloner only in order to be excluded from consideration, seems a curiously exaggerated form of parochialism.

N.S., 1866), " that the rural or National Schools are not suited to the requirements of farmers' sons. On the other hand, the leading public schools are too classical, and costly in both time and money. . . ." Discoursing on the same theme in the symposium, *Practice with Science : A Series of Agricultural Papers* (1867), the Rev. John Constable, Principal of the Royal Agricultural College, Cirencester, noted that " the majority of Scotch farmers are keenly alive to the value of scientific knowledge " whereas in England " the higher educators have not condescended to consider . . . the wants of those classes, the objects of whose education is to fit them for practical business results." " The dead languages educate our statesmen, warriors, and divines, as nothing else can educate them. . . . The requirements of the middle class may be met by other studies than those . . . Literary polish is no great object in middle class education ; [they are] to a certain extent alien from most money-making occupations."[1] Constable's plea that scientific education should be made accessible to English farmers was not quickly or extensively realised. The interesting curriculum at Cirencester is set out in the symposium, *Agricultural Education* (1864) and the outline history of agricultural education and of the advisory services is traced in a chapter (XXII), added to *English Farming Past and Present* by Sir Daniel Hall, whose biography by H. E. Dale, *Daniel Hall ! Pioneer in Scientific Agriculture* (1956) is also useful if read in conjunction with the essay of J. A. Hanley, " Agricultural Education in College and County " in the symposium, *Agriculture in the Twentieth Century* (1939).

Other factors, too, helped to promote conservative attitudes among farmers. Science without practice is as costly in farming as it is hazardous in medicine ; and English farmers had seen enough of the expensive novelties produced by the conspicuous expenditure of a leisured class to be cautious of innovation. Moreover, they worked to a calendar regulated by the rotation of the district. A. G. Street thus recalled his early memories of his father's farm worked on a four-course shift and running a Hampshire Down flock : ". . . it was the accepted practice of the district. Farms were laid out for it, and let on the understanding that the customary rotation would be followed. And once you were fairly into that system, it swept you with it, round and round, year after year, like a cog in a machine, whether you liked it or not . . . But would wheat pay? What were the prospects of the world's next harvest ? Don't be silly ! This land was due to come

[1] pp. 3–20.

into wheat, and wheat must be in by the end of November if possible, so you didn't worry over abstruse problems, but got down to the job."[1] Street's father was lucky : that rotation paid in his district in the early years of this century. But how many of his generation had spiralled with their customary rotations, round and round, year after year, into bankruptcy ? The inherited rotations were majestically though profitably insensitive to shifting market demands in the high, monopoly conditions before the deluge ; they rendered many arable farmers incapable of altering or adapting their practice ; and turned them into a race of bucolic robots. " The point I would again stress about this type of pre-war farming ", remarks Mr. Street, " is that one didn't consider whether the crop one was sowing would pay a profit over the cost of production or not. That never entered anyone's head."[2] Landlords' and farmers' subservience to unalterable rotations had been once a means to better farming ; in a competitive world, it became a major source of embarrassment.

In a famous phrase, Lord Ernle characterised the years 1852–63 as " the golden age of English Agriculture ". The human derelicts of an industrialising society who got subsistence by working on the land would not have recognised that description as an apt comment on their lives at any time during the nineteenth century. From 1795, when David Davies published *The Case of Labourers in Husbandry Stated and Considered* (2nd edition, abridged, 1828) to the appearance in 1925 of the moving indictment of J. W. Robertson Scott, *England's Green and Pleasant Land,*[3] the lot of most agricultural labourers was poverty, degradation and the deference enforced by their financial betters and spiritual leaders. No section of the population has suffered such continuous and consistent barbarity as that which larded English earth with its cheap sweat, and none has gained so much from the social emancipation of recent years. The transformation of social relationships and standards of living in the countryside is, indeed, one of the most heartening evidences of the growth of decency in English society.

The old Poor Law both saved the labourer in the midlands and south from starvation and distorted the intractable problem of rural indigence. In the view of the Commissioners of 1832, it gave him low wages, easy work and " also, strange as it may appear, what he values more, a sort

---

[1] *Farmer's Glory* (1932), p. 25.   [2] ibid.
[3] There was a 2nd edition (1931) and a 3rd (Penguin Books, 1947). The last edition usefully contains a " P.PS. After a Quarter of a Century ".

of independence. He need not bestir himself to seek work ; he need not study to please his master ; he need not put any restraint upon his temper ; he need not ask relief as a favour."[1] But the rat-catcher's remedies of 1834 were not based upon an adequate diagnosis of the problem, one aspect of which was clear enough to a Dorset labourer explaining to his parson in 1849 why he wished to emigrate. " I tell you what it is, Sir ; we are starving each other ; we be too thick in our place ; the best of us can't earn what will find us bread for our children and ourselves, let alone the clothing and the rent ; when we be gone 'twill be better for they we leave."[2] The agricultural labourer, who then stood to the soil that he tilled as the cotton operative to the machinery he tended, was too thick on the ground and too hard to shift either for his own comfort or for the rapid adoption of efficient, that is, labour-saving, methods of farming. His uneven distribution over the countryside from north to south was reflected in the history of internal migration as well as in the great variation in wages. Statistical data relating to the rural population are meticulously assembled and judiciously interpreted in the invaluable study of John Saville, *Rural Depopulation in England and Wales, 1851–1951* (1957), which also indicates, *inter alia*, the major sources of information concerning earnings and wages.[3] Mr. Saville's well grounded scepticism about their findings is in marked contrast with Sir John Clapham's confident treatment.[4] The enumeration of " allowances " is one of the most important of the many difficulties to be overcome in any reconstruction of the movement of agricultural earnings. The proportion of cash payments to " allowances ", the polite euphemism for truck in agriculture, has been persistently overestimated. In his well-known first *Report* to the Board of Trade *On the Wages and Earnings of Agricultural Labourers in the United Kingdom* (Cd. 346, 1900), Arthur Wilson Fox said of England that " generally speaking, there has been very little change either in the nature or quantity of these allowances during the (last) half-century ".[5]

This puzzling assertion must be qualified by the careful work of

[1] *Report (on) The Administration and Practical Operation of the Poor Laws* ((44) xxvii, 1834), p. 57. Lord Ernle's account of the Poor Law, ch. XV (Hall ed.) must be disregarded. The best approach to the Poor Law for the student of agricultural history is through the book of Karl Polanyi, *Origins of our Time* (1945).
[2] *The Letters of S.G.O.* (vol. 1), p. 169, cited above.
[3] p. 16, fn. 1, and the discussion, pp. 15–20.
[4] *An Economic History of Modern Britain* (vol. 2), pp. 284–96, and especially the graph of agricultural earnings, 1837–92, p. 286.
[5] p. 48.

W. Hasbach, *A History of the English Agricultural Labourer* (1908),[1] to which many other sources may be added. The Rev. Lord Sidney Godolphin Osborne, for example, indefatigable and striking correspondent to *The Times* on public, social and rural affairs for nearly half a century (Arnold White (ed.), *The Letters of S.G.O.*, 2 vols. (n.d.)), thus described the system in his native Dorset in the 1840's, as it applied to those working for the tenantry of the Prince of Wales. ". . . they are paid almost entirely on the truck system ; for a bushel of best wheat they pay 7*s.* the bushel, which is 56*s.* the quarter ; . . . for first tailings 6*s.* the bushel . . . it is charged at least 1*s.* a bushel too much. Now, Sir, I would have you bear in mind, that, having to pay 1*s.* rent, the labourer needing a bushel of grist a week, and having to pay 6*s.* for it, his wage being only 7*s.*, he is left wholly without money to buy anything else. In justice to the farmer, I will put the 1*s.* overcharge for the grist against the cider he gives in the field. But the farmer makes butter and cheese ; there are of these articles inferior qualities, for which there would be no market did not the labourer prove a customer. These, then, are taken by such men as may have children or wives at work on the farm in lieu of their wages . . . let me say I am speaking of large and apparently well-to-do farmers, renting, some of them, £400 a year." Osborne goes on to describe similar practices with respect to meat and mutton from diseased animals. " It may be said, are the men compelled to take these things ? They answer the question in various ways . . .' They just tell us they can't afford the money to pay us, and therefore axes of us to take the gristing and the cheese of them ; many of us must soon get in debt to them, and then you see, Sir, we must go on. As to the dead animals, they just says, " We can't be expected to take the whole loss ", and they then tells us as we should take some of it off their hands. There may be from twelve to a score of such sheep in the year on my master's farm. 'Tis the only meat we ever sees.' " And Osborne adds : " As to wages paid in money, I think you would be puzzled to find any one man who had drawn £2 in any one year ; by the help of clothing clubs, and an occasional job at draining out of the parish, they are alone enabled to get clothes."[2] This passage (there are many similar) has been quoted at length in order to illustrate the irrelevance of statistical computations as a means of judging the standard of living of agricultural workers in

---

[1] This book contains the only (but still inadequate) systematic account of the gang system. Its bibliography lists the relevant parliamentary papers.
[2] *The Letters of S.G.O., op. cit.*, vol. 1, pp. 16–18.

the nineteenth century. When C. W. Stubbs, vicar of Granborough, lectured[1] in his parish schoolroom on the labour question in 1872, he took from St. Paul the text : " To him that worketh, reward is reckoned not of grace, but of debt." That is the epitaph for Hodge as he emerged from the golden age of English agriculture.

The subsequent history of wages in agriculture may be quickly extracted from the book of A. L. Bowley, *Wages and Income in the United Kingdom since 1860* (1937) and from those already cited : the two Liberal Party enquiries of 1913 and 1925, and the survey, in 1938, of Astor and Rowntree. The benefits conferred upon workers in agriculture by Hitler's war are outlined in the official history of K. A. H. Murray, *Agriculture* (1955). Statistics of the changing occupational distribution of the population are summarised and explained in the book of D. C. Marsh, *The Changing Social Structure of England and Wales, 1871–1951* (1958).

Some observers were more optimistic in their assessments than Osborne or Stubbs. In a recent, widely read book, *The English Village* (1952), Mr. Victor Bonham-Carter insists that, " for a faithful portrait of English labouring life in mid-Victorian times, it is still best to turn "[2] to the profile of Wiltshire labourers sketched in a long letter to *The Times* by Richard Jefferies in 1872. " Since labour has become so expensive ", he wrote at a time when day labourers in work averaged eleven shillings a week in that county, " it has become a common remark among the farmers that the labourers will go to church in broadcloth and the masters in smock-frocks."[3] But, despite such opulence, " never once in all my observation have I heard a labouring man or woman make a grateful remark, and yet I can confidently say that there is no class of persons in England who receive so many attentions and benefits from their superiors as the agricultural labourers."[4] Some things are missing from this faithful portrait. One, rural housing, to which no reference is made in *English Farming Past and Present*, may help to explain the absence among rustics of that proper gratitude to their superiors which so dismayed Richard Jefferies. There is no comprehensive history of housing in industrial England ; when one comes to be put together, the blackest chapter, paradoxically, will be that devoted to the countryside. To the quality of brutish lives, mercifully shortened by preventible disease, huddled in over-

[1] Printed in *Village Politics : Addresses and Sermons on the Labour Question* (1878).
[2] p. 77.   [3] *ibid.*, p. 79.   [4] *ibid.*, p. 85.

crowded hovels set in the gentle English landscape, there is abundant and nauseating testimony. The inherited situation, described in the book of Edward Smith, *The Peasant's Home* (1876), was exacerbated by the Poor Law provisions for settlement and chargeability which, by giving landlords a powerful incentive to reduce the number of resident labourers, established the categories of close and open parishes. The consequences were described in the report of Dr. Hunter on the " Dwellings of the Poorer Labouring Classes in Town and Country ", printed in the *Seventh Annual Report* (1864) of the Medical Officer to the Committee of Council *on the State of the Public Health*. Hunter's investigation was pursued in every county and involved an examination of some five thousand houses. It showed " the insufficient quantity and miserable quality of the house-accommodation had by our agricultural labourers " and commented that " for many years past the state of the labourer in these respects has been deteriorating,—house-room being now greatly more difficult for him to find, and, when found, greatly less suitable to his needs, than perhaps for centuries has been the case. Especially within the last twenty or thirty years the evil has been in very rapid increase, and the household circumstances of the labourer are now in the highest degree deplorable."[1] Large proprietors reduced their liabilities to the poor-rate by depopulating their lands and " the nearest town or *open* village receives the evicted labourers. . . . In the open village, cottage-speculators buy scraps of land which they throng as densely as they can with the cheapest of all possible hovels. And into these wretched habitations . . . crowd the agricultural labourers of England." The *Report* is at pains to point out that such conditions were not universal. " There are some of the very largest land-properties in the country, where for generations there has been the tradition of a better treatment ; where at least no aggression has been made against the house-accommodation of the poor." Nevertheless, despite " these brighter but exceptional scenes, it is requisite in the interests of justice that attention should again be drawn to the overwhelming preponderance of facts which are a reproach to the civilisation of England. Lamentable indeed must be the case, when, notwithstanding all that is evident with regard to the quality of the present accommodation, it is the common conclusion of competent observers, that even the general badness of dwellings is an evil infinitely less urgent than their numerical

---

[1] The following quotations from the commentary on Dr. Hunter's report are taken from John Simon, *Public Health Reports* (2 vols., 1887), vol. II, pp. 182–9.

insufficiency. For years, the overcrowding of rural labourers' dwellings has been a matter of deep concern . . ."[1]

Over forty years later, it was still a matter of deep concern. The Select Committee on the Housing of the Working Classes Acts Amendment Bill had " abundant evidence before them as to the insufficiency of cottages in rural districts",[2] despite the Unions Chargeability Act of 1865 and the decline in the agricultural population. This sorry subject may be pursued in the very useful book of the Medical Officer of Health for Somerset, William G. Savage, *Rural Housing* (1915), and, for the inter-war years, *Rural Housing : Third Report of the Rural Housing Sub-Committee of the Central Housing Advisory Committee* (H.M.S.O., 1944), is valuable. In the last twenty years, new housing, publicly financed either directly or partly through tax concessions to owners of agricultural land, has become available for agricultural workers on an appreciable scale. Some country folk may doubt the truth of the verdict of Dr. Hoskins that " since the year 1914, every single change in the English landscape has either uglified it or destroyed its meaning, or both ".[3] Some meanings, it may be thought, have been over-long resistant to destruction.

These and other aspects of labouring life in the countryside have been examined in the book of G. E. Fussell, *The English Rural Labourer : His Home, Furniture, Clothing and Food from Tudor to Victorian Times* (1949). But what mattered in the long run was the painful struggle of cowed and battered human beings towards citizenship. The impulses behind that story may best be felt in the books of T. E. Kebbel, *The Agricultural Labourer* (1870), George C. T. Bartley, *The Seven Ages of a Village Pauper* (1874), F. G. Heath, *The English Peasantry* (1874) and *British Rural Life and Labour* (1911), F. E. Green, *The Tyranny of the Countryside* (1913), and J. W. Robertson Scott, *The Dying Peasant and the Future of His Sons* (1926). The precocious events at Tolpuddle in 1833, best explained in the book of H. L. Beales, *The Early English Socialists* (1933), inscribed a more permanent mark in history books than upon the lives of labourers. Their trade union history can be followed in the books of Ernest Selley, *Village Trade Unions in Two Centuries* (1919) and R. Groves, *Sharpen the Sickle* (1949) ;

[1] There is much information for the earlier period in the famous *Report . . . On an Inquiry into the Sanitary Condition of the Labouring Population of Great Britain* (1842, H.L., xxvi and xxvii) as well as in the literature cited in the bibliography of Ruth Glass, *Urban Sociology* (Current Sociology, Vol. IV, no. 4, 1955, UNESCO, Paris).

[2] 1906 (376) ix, 1, p. xi.

[3] W. G. Hoskins, *The Making of the English Landscape* (1955), p. 231.

and in the useful pamphlet of Rex C. Russell, *The " Revolt of the Field " in Lincolnshire* (1956). To these must be added two autobiographies of exceptional interest, *Joseph Arch : The Story of His Life, Told by Himself*, ed. the Countess of Warwick (1898) and Lord Snell, *Men, Movements and Myself* (1936).

The social costs of educationally deprived farmers and of cheap, inefficient labour were heavy burdens on English agriculture during the first three-quarters of the nineteenth century. In that period, too, the contractual and other relations between owners of land and the rural capitalists were ill-adapted to promote rapid technical innovation. George Hope, the greatest farmer of his generation, thus explained improving tenants' difficulties to the National Association for the Promotion of Social Science in 1863 in a paper " On the Conditions of Agricultural Success ".[1] " There are obstacles to success in farming peculiar to it, and I wish to bring under the notice of this Association some of the customary conditions and laws with regard to the holding of land by tenant farmers, which appear unjust in themselves, or which tend to prevent the temporary or permanent investment of capital in the soil, to the detriment and loss of owners, occupiers, agricultural labourers, and the general public. . . . In some districts of the country, there are large tracts of waste land which tenants are constantly improving and rendering arable . . . ; in others, many farms are wet and require draining, or are foul with root weeds, or filled with annuals which choke the crops, or then they are so low in manurial condition that they produce only half crops. None of these evils can be remedied without a large expenditure of capital, and most careful and thorough cultivation extended over a series of years. No man will sow unless he expects to reap, and I do not consider it is in human nature for an occupier of land to invest his capital in the soil, unless he has a guarantee that time will be afforded him to reap the fruits of his labour. And yet over a great part of England the land is occupied by tenants-at-will, liable at any moment to receive notice to quit at the caprice of the landlord, or, it may be, of his agent. I admit that instances may be found, even there, of . . . good and successful farming. But judging from what I have seen and heard, I consider farming, as a rule, to be inferior and much less spirited under such circumstances than it is in districts where leases prevail for nineteen or twenty-one years. On many large estates the family character of the owners and traditional usage are considered a species of security by the tenants, and practically

[1] *Transactions* (1863), pp. 776–82.

few changes are made amongst them ; often generation succeeds generation on the same farm, and that, too, at a very moderate rent. Still there is a want of enterprise amongst them ; they tread very much in the paths of their grandfathers, and the most is not made of the land. Furthermore, instances are not wanting where such tenants, who have ventured to improve their lands, have had their rents at once raised, while the sluggard continued to pay as formerly. Others, again, have received notice to quit for having refused to drink a toast, or it may have been from some paltry dispute about game, or the assertion of political or religious independence ; or, perhaps, to make way for some favourite in want of a farm. The agricultural papers frequently discuss very flagrant instances of such cases, and probably there are many we never hear of, but those made public are sufficient to damp the ardour for agricultural improvement over a wide circle."

Hope was drawing out the consequences of a situation that had been a recurring theme among commentators from the eighteenth century onwards and of the technical literature of agricultural improvement in the 1840's. In the new edition of *Hints to Gentlemen of Landed Property* (1793), containing *Supplementary Hints*, Nathaniel Kent emphasised " that leases are the first, the greatest, and most rational encouragement that can be given to agriculture . . . Let any impartial man take a view of two districts, where it is the custom to grant leases, and where it is not : in the former, he will generally find a respectable yeomanry, and a well cultivated country ; in the other, an indigent spiritless race, following a contracted system of husbandry, calculated to answer no permanent purpose of advantage to themselves or landlords. Yet, there are many gentlemen, who to have such people at immediate command, prefer the continuance of a slovenly, unproductive stile of husbandry, to neatness and fertility."[1] Discoursing on the same theme, James Donaldson, *Modern Agriculture ; or, the Present State of Husbandry in Great Britain* . . . (4 vols., 1795–96), regretted " that a prejudice against granting leases has rooted itself in the minds of a majority of English proprietors " who choose " to sacrifice their private interest in order to support their political consequence ".[2] The author of the anonymously published, *The Modern Land Steward ; in which the Duties and Functions of Stewardship are Considered and Explained. With their Several Relations to the Interests of the Landlord, Tenant and the Public* (1801), insisted that " to let out an estate on leases

---

[1] pp. 270–71.   [2] (vol. I), pp. 415–6 and 421–22.

of reasonable duration, is, beyond all question, of equal benefit to the lord, as to his tenants ; since the increased worth of the estate must arise from improvements, the completion of which require the expenditure of both much time and property, and which no tenant at will can risk ".[1] The magnificently exhaustive treatise of the great William Marshall, *On the Landed Property of England . . . Containing the Purchase, the Improvement, and the Management of Landed Estates* (1804), remarks the differences of landed from other forms of property. The landowner " is not in possession of the lands, only, but also of their inhabitants. For altho, in England, tenantry are not bought and sold, as other livestock ; yet, when an estate is under lease, tenants may be said to be as firmly rooted in the soil, as the oaks which grow upon it."[2] But annual holdings are, " to a tenant, most discouraging and, to improvements in agriculture, most unfriendly . . . While, to a proprietor, they are most convenient. He may be said to be in constant possession of his estate."[3] A Haddington farmer, Robert Brown, in his *Treatise on Rural Affairs* (2 vols., 1811), selected leases as a major factor promoting " the striking, and manifold improvements accomplished by British tenantry . . . The leasehold tenure has of late been discouraged in England, and the tenantry forced to trust the generosity of proprietors. According to our principles, a manifest obstruction to improvement is thereby created ; because the most implicit confidence does not furnish security equal to that conferred by a lease."[4] The first published work expressly upon the valuation of rents and tillages was that of a Yorkshire land-agent, J. S. Bayldon, *The Art of Valuing Rents and Tillages* (1st edition, 1823, 6th revised edition by John Donaldson, 1844), emphasised in all its editions " the relative superiority of holding by lease ".[5] Such observations were the commonplaces of all serious writing from the end of the eighteenth century to the repeal of the Corn Laws : they continued to parade as platitudes in the literature of farming for the next generation. The treatise of Robert E. Brown, *The Book of the Landed Estate Containing Directions for the Management and Development of the Resources of Landed Property* (1869), stressed the " great advantage " of a lease to a tenant, " and it is much more so to the landlord. Where a tenant has his farm from year to year, he has no security beyond the twelve months before him, and the consequence is, that he does not feel justified in expending his capital

---

[1] p. 72.   [2] p. 334.
[3] *ibid.*, p. 363.   [4] (Vol. 1), p. 131.
[5] (1st edition), p. 173, (6th edition), p. 10.

on the farm, as he knows quite well that he is liable at any time to receive a few months' notice to quit."[1]   Similarly, the compendium of G. A. Dean, *The Culture, Management, and Improvement of Landed Estates* (1872), instructed landlords that " no man of prudence, having capital commensurate with the size of his farm, would enter into possession, and farm highly without [security of tenure]."[2]

Particular urgency was given to these considerations in the 1840's, when, beneath the froth of the anti-Corn Law debate, there can be discerned in the technical literature an undercurrent of anxiety about the food supply for the next generation of the rapidly increasing population. " It cannot reasonably be expected ", wrote Joseph Russell (*A Treatise on Practical and Chemical Agriculture*, 1st edition, 1831, 2nd edition, 1840), " that a tenant will underdrain, raise fences and effect alterations and repairs in buildings, without a liberal allowance for materials, or a security for holding his farm for a term of years . . . Farms often lie in an impoverished state for ages, to the great injury of both landlord and tenant, for want of such covenants . . ."[3] In 1842, J. F. W. Johnston, Fellow of the Geological Society and Reader in Chemistry and Minerology in the University of Durham, asked " What Can be Done for English Agriculture ? " in *A Letter to the Marquis of Northampton*. His answer surveyed the isolation and educational backwardness of the agricultural population as well as the " mode of tenure by which the land is held. This point may appear at first sight to be very remotely connected with the application of scientific principles to the improvement of the soil . . . Scientific improvements generally demand a considerable outlay, which the yearly tenant is unwilling to sink upon a contingency."[4] Farmers and landowners, Johnston thought, " have clamoured for brass from [the] legislature, and have passed by the gold which nature, through the hands of science, willingly and plentifully offers."[5]

When David Low, Professor of Agriculture in the University of Edinburgh, looked *On Landed Property, and the Economy of Estates* in 1844, he found " that wherever tenancy at will is established, either the rents are comparatively low, or the income of the landlord is subjected to a perpetual drain . . . The country suffers by the less improved condition of the public territory. If we shall look at the finest parts of England, we might almost imagine that the purpose of agriculture was

---

[1] p. 64.   [2] (Pt. II), p. 51.
[3] p. 395.   [4] pp. 30–31.
[5] *ibid.*, p. 35.

to raise hay for horses, and not food for men. We find vast tracts of the finest land yielding wretched crops of hay . . . the greater part of the plains of England now is yielding not one-half of the quantity of human food which they [could] by suitable tillage. One of the causes of this state of affairs is the absence of proper tenures."[1] Joshua Trimmer, author of a book on *Practical Geology and Minerology* and, in 1847, of a pamphlet, *On the Improvement of Land as an Investment for Capital*, foresaw that drainage and science could produce a great increase in agricultural output and in the profitability of land. "But many landlords are as reluctant to bind themselves to pay a tenant for improvements, as they are to grant leases ; and the lawyers and auctioneers, to whom the management of landed property is usually confided . . . dislike trouble . . . and departure from established routine."[2] But the real obstacle lay elsewhere. "The foundation of any improvement", declared James Boydell in *A Treatise on Landed Property in its Geological, Agricultural, Chemical, Mechanical and Political Relations* (1849), "must be a new electoral system ; for as our House of Commons is constituted, there does not appear to be much probability of the adoption of measures [for improvement]."[3]

The long continued refusal of leases, like the obstinately delayed Ballot Act, can thus be understood only within the context of the character and structure of politics in this period. Traditionally, the influence of land, "the mountain backbone of the State, from which the streams flow down on either side",[4] has been underestimated. In part, as a result of the emphasis laid by Marx on the predominating priority of the class conflict between urban bourgeoisie and proletariat and, in part too, in the absence of adequate studies of the Victorian political system. Perspective may now be obtained from the recent meticulous investigations into the mechanics of politics of Norman Gash, *Politics in the Age of Peel* (1953) and of H. J. Hanham, *Elections and Party Management : Politics in the Time of Disraeli and Gladstone* (1959). But neither book is concerned, as is that of Asa Briggs, *The Age of Improvement* (1959), with the functions discharged by a political system designed to accord preponderating influence to landed property. No informed contemporary questioned its efficiency for this purpose. In his "Analysis of the House of Commons", in 1867, Bernard Cracroft traced the elements that "go to make up that truly and without

---

[1] p. 13.   [2] p. 29.   [3] p. 181.
[4] W. L. Newman, "The Land Laws," in the symposium, *Questions for a Reformed Parliament* (1867), p. 79.

exaggeration tremendous consent of power, often latent, often disguised, never absent, which constitutes the indirect representation of the aristocracy in the House of Commons ", where " the representatives of the landed class number about five-sevenths of the whole ". " The parliamentary frame ", Cracroft concluded, in a brilliant phrase, " is kneaded together almost out of one class ; it has the strength of a giant and the compactness of a dwarf."[1] So long as tenant farmers of more than £50 rental, enfranchised by the Chandos Clause in the Reform Act of 1832, had a part to play in such arrangements as open and subservient voters, so long did security of tenure remain an intrusion upon the rights of property repugnant to the majority of large English landlords for whom land represented not a source of food but the means of access to power.

In 1883, Frederick Pollock surveyed *The Land Laws* and thus defined the relation between landlord and tenant. " The farmer is legally bound to pay the full amount of his agreed rent, without regard to the goodness or badness of the season ; but in bad years it is a very common practice for the landlord to remit such a percentage of the rent as to leave the tenant answerable only for such as the farm seems fairly capable of paying under the circumstances. A great landlord who refused to follow this practice would be entirely within his legal rights, but would certainly be thought the worse of in the country. The landlord in return expects a certain amount of deference and compliance in various matters from his tenant. Not only does the farmer meet him half way on questions of shooting rights, and allow free passage to the hunt, but his political support of the landlord is not unfrequently reckoned on with as much confidence as the performance of the covenants and conditions of the tenancy itself. In the case of holdings from year to year it may be not unfairly said that being of the landlord's political party is often a tacit condition of the tenancy."[2] And the majority of English farmers were tenants holding at will from year to year. English farmers thus discharged their tenurial obligations. Admirers of aristocracy, then as now, sometimes assert that the farmer liked it that way, that his duty to vote Tory or Radical or Liberal with his landlord happened to coincide with his convictions; those with other preferences may assert otherwise. The fact was, as Mr. Hanham points out, that " even on the best-managed estates the 'free association' of landlord and tenant was based on coercion" and

---

[1] In the symposium, *Essays on Reform* (1867), pp. 173–4.
[2] pp. 150–51.

determined "the landlord-tenant alliance" which "dominated county politics ".[1]

This was the reality behind early- and mid-Victorian discussions of leases. It is not surprising that many tenants disliked them and that some refused their landlords' offer to convert annual into leasehold tenancies. The experience of leaseholders in the period of falling prices after 1815 discouraged many. Indeed, a study which will extend the illuminating article of R. A. C. Parker, " Coke of Norfolk and the Agrarian Revolution " (*The Economic History Review*, Second Series, vol. VIII, no. 2, 1955) into new areas and the following century, would be fruitful.[2] It might be developed on the lines suggested in the interesting lecture to the Royal Agricultural College, Cirencester, in 1864, of R. G. Welford, " On Leases ",[3] and assess the differing pressures upon farmers' and landowners' attitudes of price fluctuations and political considerations. Howsoever it was determined, David Low put his finger on the reluctance of many tenants to press for or to accept changes in their tenure. " Over the finest parts of England ", he wrote, " we find a tenantry nearly stationary in their habits and condition, opposed to innovations on established practices and educating their families as they themselves have been educated. Everywhere they will be found to prefer their tenancy at will, to all the advantages which a permanent tenure can afford, because they know that they pay a lower rent and can make it good by smaller exertions. The argument has again and again been used against the extension of leases, that the tenants themselves set no value upon them ; but to how different a conclusion ought the existence of such a feeling amongst the tenantry of a country conduct us. The fact itself shows that the absence of leases may render a tenantry ignorant of the means of employing their own capital to advantage, indisposed to the exertions which improvements demand, and better contented with an easy rent and dependent

---

[1] H. J. Hanham, op. cit., pp. 11–14. Mr. Hanham's use of the word " alliance " to describe this relationship is hardly in accordance with customary English usage.

Farmers did not always acquiesce quietly in this comfortable system : they became articulate in their Farmers' Clubs, e.g. *Reports of the Harleston Farmers' Club from its Establishment in 1838, to 1849* (1850), esp. pp. 151–4, and the report of a special meeting of the Midland Farmers' Club, R. H. Masfen, *On the Tenure of Land* (1864).

[2] There is one puzzling comment in Mr. Parker's article. He shows that " the main new feature of Coke's leases was the categorical prohibition of the growing of two white crops . . . in succession " (p. 163) and claims this as a " significant contribution " (p. 165) to the advance of farming technique in England. But what was or is reprehensible in two successive white crops?

[3] In the symposium, *Practice with Science : A Series of Agricultural Papers* (2 vols., 1867), pp. 219–91.

condition, than with the prospect of an independence to be earned by increased exertion."[1]

Twenty years later, H. S. Thompson, an M.P. and Yorkshire land-owner, writing on " Agricultural Progress and the Royal Agricultural Society ", succinctly summarised the landlord's point of view. He observed that " the comparative advantages of leases and agreements terminable at short notice, have been much discussed of late, though little progress has been made . . . in making converts from one side to the other . . . If . . . the case be argued on commercial principles alone, the advocates of leases have the best of the argument, and *doubtless in the long run this view of the matter will prevail ;* but there are important social questions connected with it, which have hitherto prevented the general adoption of leases, and will continue to exercise a powerful retarding influence.   Most landlords who let their land to tenants-at-will are aware that they could increase their rents by granting long leases ; but before realising this additional income tenants must be changed, farms consolidated, and farm-buildings augmented and improved.   Of these obstacles the least are those connected with providing the requisite capital.   Many landlords have funds at their disposal which they would be happy to invest on such undeniable security as their own land. . . .   One of the most general of these causes is the dislike on the part of landowners to dispossess deserving tenants . . between whom and their landlord there exist personal ties, which, if rudely severed, would be most inadequately replaced by a few shillings an acre additional rent.   Another equally influential cause is the repugnance of the proprietors to give up so much of the control over their estates as implied by a lease for any long term of years.   They let their land below its market value, for the sake of retaining the power of resuming possession at short notice, of any farm on which the tenant causes annoyance to his landlord, or sets a bad example to his neigh-bours.   The knowledge that such a power exists makes it rarely necessary to use it."[2]   Landowners were prepared, that is, to charge lower rents to the costs of maintaining political power.   As the history of the game laws demonstrates only too brutally, landlords never hesi-tated to coerce their tenantry to accept practices which increased the amenities of their estates ; among these, good farming did not figure importantly.

It was against this background that Philip Pusey attempted, in a

---

[1] *op. cit.,* p. 13.
[2] *J.R.A.S.E.,* vol. XXV, 1864, pp. 35–6. Ital. in original.

Bill of 1847, to circumvent the hostility to leases by giving tenants legal security for their capital expenditures on improvements to their farms by generalising, under statute, usages that had long prevailed in a few areas, notably Lincolnshire, by which, in practice, an incoming tenant paid compensation to the outgoer for tillage operations, straw, hay and dung left on the farm, and for growing underwood. The *Report* and *Evidence* of the *Select Committee on Agricultural Customs* (1847-8, 461 vii) to which Pusey's Bill was referred, is a useful source for attitudes and arrangements in different parts of the country. It must be supplemented by the information assembled in the important book of L. Kennedy and T. B. Grainger, *The Present State of the Tenancy of Land in Great Britain Showing the Principal Customs and Practices Between Incoming and Outgoing Tenants ; and the Most Usual Method under which Land is now Held in the Several Counties* (1828). This was discerningly discussed in the *Edinburgh Review*, " The Tenancy and Culture of Land in England " (vol. LIX, 1834), which noted " that rooted disinclination evinced by the generality of farmers to leave established practices ", refused to ascribe it, as many have done, " to the influence of tithes,[1] poor rates, or other public burdens ",[2] and pointed out that these pressed as, if not more, severely on farmers in Norfolk, Lincolnshire and Northumberland as on those in other districts. The entertaining compendium, packed with all sorts of fascinating information, such as the market price of advowsons, of Alfred Cox, *The Landlord and Tenant's Guide* (1853), contains a convenient and accurate summary of conditions of tenure, customs and the range of rents throughout the country. The cautious and limited survey of Clement Cadle, " The Farming Customs and Covenants of England ", is available in the *Journal of the Royal Agricultural Society* (vol. IV, N.S., 1868).

Pusey's efforts had no legislative result until Gladstone's Irish Land Act of 1870 and the achievement of secret voting in 1872 opened the way for Disraeli's concession of the first, permissive Agricultural Holdings Act of 1875. According to a survey printed in the *Journal*

---

[1] Richard Jones's Benthamite solution of the immediate tithe problem in the 1830's must be interpreted in the setting of the movements that produced the first, major interferences (canals and railways apart) with the rights of property in land in the nineteenth century ; the emancipation of the West Indian slaves and the establishment of the Ecclesiastical Commission. There is no adequate study of tithes. Apart from books already cited, the legislative history of the twenty-six-odd acts since that of 1836 is most quickly extracted from the *Report of the Royal Commission on Tithe Rentcharge in England and Wales* (Cmd. 5095, 1936), Pt. I, pp. 7-60. The establishment and early history of the Ecclesiastical Commission is admirably surveyed in the study of Olive J. Brose, *Church and Parliament* (1959).
[2] p. 391.

of the Farmer's Club in 1876, landlords, including the Crown and the Ecclesiastical Commissioners, hastened to contract out of its provisions. The first effective statutory interference with the contractual freedom of landlord and tenant came in 1883 and established a principle which was not significantly extended until the important Act of 1906 and the consolidating legislation of 1908. Tenants then secured freedom of cropping and compensation for unreasonable disturbance and for the depredations of game. The history and widening provisions of Agricultural Holdings legislation in the twentieth century may be traced in the various editions of standard legal texts listed in the bibliography of D. R. Denman, J. F. Q. Switzer and O. H. M. Sawyer, *Bibliography of Rural Land Economy and Landownership, 1900–1957* (1958).

The Holdings Act of 1883 came at a time when the direct link between property and political capacity had been severed. Although it marked a new stage in the contractual relations of landlords and tenants, its meaning and significance were not primarily legal or agricultural. Behind that act lay the halting but persistent urban middle-class attack on the entrenched power of an aristocracy sitting on land. The absence of a functional history of aristocracy is one of the astonishing gaps in nineteenth-century historiography ; and, lacking this essential background, agricultural history is in danger of becoming stuck in the groove of reproducing in ever greater detail the technical trivia of what happened on farms. Secondary writing has scarcely yet begun to tackle this part of the story, although the suggestive and gracefully learned books of O. F. Christie, *The Transition from Aristocracy* (1927) and *The Transition to Democracy* (1934) have cleared some of the ground. The book of E. Wingfield-Stratford, *The Squire and his Relations* (1956), concerned to " trace the unifying principle that makes it possible to comprehend so wide a diversity of persons within the compass of one brief monosyllable ",[1] promises more than it fulfils. " Squire " is never defined and the only observable principle employed is the inclusion of men (wives do not feature as " relations ") about whom anecdotes have survived. Less pretentious and more rewarding are the essays of Chester Kirby, *The English Country Gentleman* (n.d.).

There are two complementary approaches to this theme : the study of the aristocracy as a class and the investigation of the economy of estates. But details culled from estate papers will acquire meaning only if set in a purposive, general context. The best starting-point

[1] p. 2.

here for agricultural historians is the pioneering though regrettably unpublished thesis of Professor Ping-ti Ho, *Land and State in Great Britain, 1873–1910* (1951).[1] This admirable and suggestive research unravels the many strands which were caught up in John Stuart Mill's Land Tenure Reform Association of 1870, and examines the determinants of the rural and urban land policies of liberal governments in the early twentieth century. The important and detailed article of F. M. L. Thompson, " The Land Market in the Nineteenth Century " (*Oxford Economic Papers*, vol. IX, no. 3, 1957), estimates the level of land transfers and long-term movements in the quantity of land sold. Work on individual landowners and estates has produced a discussion[2] of aristocratic indebtedness which echoes with doubtful relevance the interests of sixteenth- and seventeenth-century historians, and articles relating[3] the political outlook of a great landowner to his ferruginous and other undertakings or extending in detail[4] the contemporary description of great estates, busily diversifying their economic activities, contained in the book of T. H. S. Escott, *England : Its People, Polity and Pursuits* (1st, 2 vol., edition, 1879, best revised editions those after 1885).[5] Undue concentration on the individual easily leads to curiously distorted judgments. " If one assumes," remarks Mr. Spring, " that the possession of 2,000–3,000 acres or more defined an aristocratic landowner, the class of landowners was formidable in number, roughly 2,000 in England alone."[6] This number, regarded as so many bundles of estate papers to be dredged, is doubtless formidably large ; seen through the eyes of late-Victorian critics as concentrated political power which could be easily and comfortably accommodated at one sitting in the Albert Hall, it was formidably small. And that was how they saw it.

The contemporary debate may usefully be approached through the books of Charles Seymour, *Electoral Reform in England and Wales* (New Haven, 1915), J. A. Thomas, *The House of Commons, 1832–1901 : A Study of its Economic and Functional Character*, and I. Jennings, *Party*

---

[1] Microfilm copies of this Ph.D. thesis may be obtained from the Library of Columbia University, New York.

[2] F. M. L. Thompson, " The End of a Great Estate ", and David Spring, " English Landownership in the Nineteenth Century : A Critical Note " (*The Economic History Review*, vols. VIII, no. 1, 1955, and IX, no. 3, 1957).

[3] David Spring, " Earl Fitzwilliam and the Corn Laws " (*The American Historical Review*, vol. LIX, no. 2, 1954).

[4] David Spring, " The English Landed Estate in the Age of Coal and Iron, 1830–1880 " (*The Journal of Economic History*, vol. XI, no. 1, 1951).

[5] Ch. III.

[6] David Spring, *op. cit.*, *The Economic History Review*, vol. IX, no. 3, 1957, p. 472.

Politics (vol. I, 1960); and the articles of W. L. Guttsman, "The Changing Structure of the British Political Elite, 1886–1935", and "Aristocracy and the Middle Class in the British Political Elite, 1886–1916" (*The British Journal of Sociology*, vols. II, no. 2, 1951, and V, no. 1, 1954).

In his last speech to constituents at Rochdale in 1864, Richard Cobden declared : "If I were five and twenty . . . I would have a League for free trade in Land just as we had a League for free trade in Corn."[1] That campaign ran back far beyond the Anti-Corn Law League, from which it gathered impetus ; found a focus in the Financial and Administrative Reform Associations ; and swelled into a vigorous attack on the monopoly of land which the Earl of Derby proposed to blunt by calling for a *Return of Owners of Land, 1873* (C. — 1097, 1875). Many laboured to correct the statistical ineptitude of the Local Government Board's four enormous volumes. The best compilation is that of John Bateman, *The Great Landowners of Great Britain and Ireland : A List of all Owners of Three Thousand Acres and Upwards, Worth £3,000 a Year* (best edition 4th revised, 1883). Thus revised, the New Domesday Book disclosed in full measure the remarkable achievement of the territorial aristocracy. Their artful use of the inscrutable and costly mysteries of conveyancers had prevented the alienation of estates and had promoted and sustained familial succession to a tenacious concentration of power and influence. The formal law of real property is a dark labyrinth through which untrained folk must stumble with such help as may be obtained from the standard text of Joshua Williams, *Principles of the Law of Real Property* (1st edition, 1845, 19th, 1901) and the modern works of A. D. Hargreaves, *Introduction to the Principles of Land Law* (3rd edition, 1952). The practical consequences of primogeniture, of settlement, of the discrimination between realty and personalty that accorded a privileged immunity to landowners, and of the maxim, written into English law by the Statute of Gloucester "quicquid plantatur solo, solo cedit," were exhaustively explored in an extensive polemical literature of very high quality. Outstanding were the books of Joseph Kay, *Free Trade in Land* (1879), Arthur Arnold, *Free Land* (1880), G. C. Brodrick, *English Land and English Landlords* (1881) and G. Shaw-Lefevre, *Agrarian Tenures* (1893). Behind such writing lay an interest in comparative systems of land tenure stimulated by experience in India and by the concern for

---

[1] John Bright and J. E. Thorold Rogers (eds.), *Speeches on Questions of Public Policy by Richard Cobden, M.P.* (2 vols., 3rd edition, 1908), vol. 2, p. 493.

Ireland which produced, *inter alia*, the very useful *Reports from Her Majesty's Representatives Respecting the Tenure of Land in the Several Countries of Europe* (C.—66, 1870). An economist, J. Shield Nicholson, provided in *Tenant's Gain not Landlord's Loss* (1883) a survey of the land question in the light of the principles of political economy which yielded the conclusion that " the presumption in favour of *laisser faire*, as the history of progressive societies clearly proves, tends to become stronger and not weaker as people . . . are inclined to imagine ".[1] Supporters of the existing system rarely wrote systematic defences : their views must be sought in parliamentary debates, in the reviews and in such books as that of Evelyn Cecil, *Primogeniture* (1895). The trend of serious, conservative opinion is best indicated in the study of J. A. R. Marriott, *The English Land System* (1914), written to underline Canning's belief that those who oppose improvement because it is innovation may have to submit to innovation which is not improvement. The contemporary flavour of the debate can be quickly savoured in the essays of J. A. Froude, " On the Uses of a Landed Gentry " and Arthur Arnold, " The Abuses of a Landed Gentry ".[2]

The campaign for free trade in land was a central issue in the politics of the decades that followed the repeal of the Corn Laws, and it has been unduly neglected in interpretations of the period. It necessarily forced questions of land taxation, long a bone of middle-class contention, into the open. There is no modern study. The best starting-point is the relevant section[3] in the last volume of the masterly *History of the English People in the Nineteenth Century* of E. Halévy. This may be supplemented briefly by the published lectures of J. S. Nicholson, *Rates and Taxes as Affecting Agriculture* (1905) or at length by the survey of T. P. Whittaker, *The Ownership, Tenure and Taxation of Land* (1914). The " dum-dum " duties introduced by the celebrated Harcourt budget of 1894 are firmly placed in their historical perspective in the unpublished thesis of LeRoy Dunn, " A History of Inheritance Taxation in England " (Ph.D., University of London, 1956).

No one has yet written an account of what the English at home have admired in foreign régimes during the last century and a half. Such a catalogue of sorry and curious naïvety would have to include reference

---

[1] p. 164.

[2] Froude's essay was reprinted in the many editions of *Short Studies on Great Subjects* (vol. III) and Arnold's in his volume *Social Politics* (1878).

[3] *The Rule of Democracy* (1952 edition), Bk. I, pp. 267–304.

to belief in the social benefits and political advantages of a general
diffusion of landed property which always survived the reports of travellers who, like Samuel Laing, saw " the whole body of small proprietors
[reduced] to the condition of a soldiery on leave ".[1] The pathetic
delusion that some sort of land settlement scheme contained the secret
cure for the ills of industrial society had great survival value. The
ideas of Spence, Ogilvie and Paine, easily accessible in the collection,
ed. M. Beer, *The Pioneers of Land Reform* (1920), flickered in the
labour movement throughout the first half of the nineteenth century.
Attempts to go back to the land were fortunately rare as such fiascos
as the Chartist Land Plan, described by Mrs. MacAskill in the symposium *Chartist Studies*, ed. Asa Briggs (1959), demonstrated. Similar
notions, suitably emasculated, attracted middle-class reformers in later
decades. They appear, for example, as home and oversea farm colonies
which were a feature of the scheme for salvation proposed by William
Booth, *In Darkest England and the Way Out* (1891). The most intriguing
case of an author who reconciled the irreconcilable in such a scheme is
Christopher Turnor. His *Land Problems and National Welfare* (1911)
carries an enthusiastic Introduction by Lord Milner, pays " a humble
tribute to Prince Kropotkin " and concludes by proposing that
the countryside should be utilised as a breeding pen for stocking
the Empire with sturdy emigrants. More sensible in conception was the
desire of *The Radical Programme* (1885) to secure " the permanent
improvement of our agricultural population " by giving " them an open
career on the land ". " Occupying ownership and peasant proprietary
established . . . by the aid of the state, acting through local authorities
seem to be the direction in which these objects "[2] can best be fulfilled.
The inevitably disappointing history of small holdings and the background of the important Small Holdings and Allotments Act, 1908, is
outlined in the admirable book of C. S. Orwin and W. F. Darke, *Back
to the Land* (1935). Their cool assessment of the results of this legislation may be compared with the hopes of one of its chief architects,
Jesse Collings, *Land Reform : Occupying Ownership, Peasant Proprietary
and Rural Education* (1906).

[1] *Observations on the Social and Political State of the European People in 1848
and 1849* (1850), p. 529.
[2] pp. 145–6.

IV

AGRICULTURE IN THE INDUSTRIAL SOCIETY

In the year of the Ballot Act, wheat stood at 57*s.* a quarter; when the scheme of elective local government in the countryside was completed in 1894 by the establishment of district and parish councils which finally rounded off the insidious work of the Reform Acts, it had fallen to 22*s.* 10*d.*, the lowest it reached in that century. The steep fall in the price of arable products is estimated by J. R. Bellerby, *Agriculture and Industry Relative Income* (1956), to have reduced farmers' average income by some 40 per cent. The calculation of H. A. Rhee, *The Rent of Agricultural Land in England and Wales* (1948), suggests that agricultural rents fell in the same period by some 25–30 per cent.[1] Mr. Rhee's estimate of the movement of rent between 1872 and 1941 is shown in the table below.

AVERAGE RENTS[2]

	Medium and Large Holdings		All Holdings	
	Without rough grazings	With rough grazings	Without rough grazings	With rough grazings
	s. *d.*	*s.* *d.*	*s.* *d.*	*s.* *d.*
1872–3	28 0	25 0	34 6	30 9
1935–9	21 6	19 0	26 6	24 0
1941	22 0	19 6	27 0	24 0

The debilitating effect of falling rents on that section of the landed interest dependent on arable farmland was exacerbated by other factors. A new monarch enforced a new definition of " Society " and facilitated the transition from aristocracy to plutocracy. The Liberal landslide

---

[1] This study, published by the Central Landowners' Association, contains a very useful bibliography of the writing concerning rent throughout the period.
[2] p. 5.

of 1905 opened Parliament to new men at a time when the old were finding it hard to make their incomes stretch to the maintenance of social and parliamentary appearances. Before the shadow of the Parliament Act fell across the statute book, a shrewd observer of English life noted that " the Capital and the Society of the new *régime* are to the survivor of the old English squirearchy what any other than its native element is to the denizen of the trout stream traversing his park . . . In the existing development (of the House of Commons), as in its social environment, the last of the long line of squires, with only a few thousands for revenue, knows there is no place for him."[1] That comment contained elements of exaggeration but it marked a new phase in the distribution of political power and in the decline of landed influence.

The perceptive pamphlet of James Caird, *High Farming Under Liberal Covenants the Best Substitute for Protection* (1849), went into five editions in four months. It noted that " the modes of management which have heretofore been confined to farms in the neighbourhood of large towns are now gradually extending to remoter parts of the country ; and the circle is every year widening within which dairy farming, the cultivation of vegetables for sale and for the production of butcher meat, with a more garden like management of the soil, are found the most profitable points to which a farmer can direct his labour ".[2] A generation later, in his *The Landed Interest and the Supply of Food* (1878) as well as in many public addresses, Caird was still emphasising the same point ; but, by then, its practical consequences were being drawn out in such books as those of R. S. Burn, *Suburban Farming* (1877) and D. Tallerman, *Agricultural Distress and Trade Depression* (1889). Tallerman's second edition appeared three years later under a title, *Farm Produce Realization*, that conceals his unusual freshness of outlook. But such is the backwardness of the agricultural history of this period that no study has yet related urbanisation to the growth of specialised farming in the manner of the pioneering article of F. J. Fisher, " The Development of the London Food Market, 1540–1640 " (*The Economic History Review*, vol. V, no. 2, 1935). Food distribution and marketing have received no systematic attention. Milk apart, who knows what happened to agricultural produce after it left the farm or arrived at a port ? And where are the studies that incorporate material in such books as that of George Dodd, *The Food of London : A Sketch of the Chief Varieties, Sources of Supply, Probable Quantities, Modes of Arrival, Processes of Manufacture, Suspected Adulteration, and Machinery of Distribution of the Food for a Community of Two Millions and a Half* (1856), and where the companion volumes to those of, say, Peter Mathias, *The Brewing Industry in England, 1700–1830* (1959) or J. T. Critchell and J. Raymond, *The Frozen Meat Trade* (1912) ?

The history of diet has been similarly neglected. The wide-ranging book of J. C. Drummond and Anne Wilbraham, *The Englishman's Food* (1st edition, 1939, 2nd revised edition with additional material by Dorothy Hollingsworth, 1957), stemmed, like much of the historical writing that has broken fresh ground in recent years, from the interests of a natural scientist. But it was exploratory, not definitive. Likewise, the sketchy survey of N. Curtis-Bennett, *The Food of the People : The History of Industrial Feeding* (1949), is useful because suggestive. Available information about poor people's food consumption in the nineteenth century is being put together by Dr. J. Burnett in a study arising from his as yet unpublished " The History of Food Adulteration in Great Britain in the Nineteenth Century " (Ph.D., University of London, 1958).[1] This thesis contains an extensive and valuable bibliography which lists the major sources, including the pioneering investigations in the early 1860's on the nourishment of distressed operatives and on the food of the poorer labouring classes conducted by Dr. Edward Smith and published as appendices to the fifth and sixth Reports of the Medical Officer to the Privy Council. The interests and anxieties reflected in such essays as those of Stephen Bourne, published under the title *Trade, Population and Food* (1880), reinforced the military considerations leading both to the appointment of the Royal Commission on the Supply of Food and Raw Material in Time of War[2] and to concern in the post-Boer War years for the quality and stamina of the industrial working class. The work of Gowland Hopkins and others created new methods and possibilities of measuring objectively the level and adequacy of diet. Thus, alongside the moving report of George Orwell, *The Road to Wigan Pier* (1937), may be placed the politically influential statistics of John Boyd Orr, *Food, Health and Income : A Survey of Adequacy of Diet in Relation to Income* (1936), which showed that nearly half the population was denied sufficient nutrients, and a considerable

---

[1] The most informative of the many centenary comments on the Act of 1860 was that of Dr. Burnett, " The Adulteration of Foods Act, 1860 " (*Food Manufacture*, November, 1960).

[2] *Report* (Cd. 2643, 1905) ; *Minutes of Evidence* (Cd. 2644–5, 1905).

proportion suffered from malnutrition because they had too little money and too little knowledge to buy enough of the right sorts of food. The social consequences and military implications of this situation in the hungry thirties,[1] documented in several surveys, were conveniently summarised in the book of F. Le Gros Clark and R. M. Titmuss, *Our Food Problem* (1939). The comparative diets of different social classes and of families of differing size in the more affluent fifties are examined with model clarity in the statistical compendium of A. M. Carr-Saunders, D. Caradog Jones and C. A. Moser, *A Survey of Social Conditions in England and Wales* (1958). The experience and machinery of food control and distribution during the Kaiser's and Hitler's wars is traced in the official histories of W. H. Beveridge, *British Food Control* (1928) and R. J. Hammond, *Food* (2 vols., 1951 and 1956). The short pamphlet, *How Britain Was Fed in War Time : Food Control 1939-1945* (H.M.S.O., 1946), provides an excellent summary.

The threat of war at the end of Victoria's reign brought many anxieties. In 1898, in the most famous Presidential Address ever delivered before the British Association, Sir William Crookes examined " the universal dearth to be expected " when pressure of population on wheat supply " will exhaust all the available store of nitrate of soda " within thirty years. He contrasted eager expenditure on armaments with the failure to " take necessary precautions to supply the very first and supremely important munition of war—food ". His book, *The Wheat Problem* (1898), was reprinted in 1917. He then noted the successful commercial fixation of atmospheric nitrogen by Haber and Bosch in Germany but transferred his pessimism to other predicted difficulties. Such Malthusian fears, the legend of an " agrarian revolution " in the eighteenth century and persistent sentimentality that regards agriculture as " a way of life " (whatever that may mean) have all helped to conceal the distinctiveness and range of technological advance during the last half-century. This is perhaps best symbolised by the tractor which came to meet the industry's basic requirement of a mobile source of power. Between 1931 and 1937, tractors first surpassed

---

[1] " . . . only the other day ", wrote Professor Ashton in 1951, " a magazine called *Womanfare* referred to the decade before the recent war as ' the hungry thirties '. A legend is growing up that the years 1930–39 were marked throughout by misery. In the next generation ' the hungry thirties ' may be common form." (F. A. Hayek (ed.), *Capitalism and the Historians*, 1954, p. 55). The phrase has been purposely used here in order both to suggest that inconvenient realities cannot be expunged from the record by describing them as " a legend " and to save those who share Professor Ashton's views the labour of searching files of women's magazines for writers to place in their pillory.

horses as a source of farm power in Britain, by 1939 they accounted for two-thirds and, by the end of the war, for 90 per cent.[1] The cumulative contribution of science and technology to food production is described in many of the books already cited.[2] Its utilisation, hindered by economic circumstances and promoted by war, is nowhere better described than in the official history of K. A. H. Murray, *Agriculture* (1955), which sets the details of war-time transformation in historical perspective.[3] The annual publication, *Britain: An Official Handbook*, prepared by the Central Office of Information, contains an admirably lucid and well-documented account of the structure and organisation of the industry.[4]

The fatal instability of English agriculture in the industrial society has been the dominating characteristic if its economic history. The monopoly conditions of the high farming period have been twice re-created by war, and the painful adjustments made by the generation of farmers before 1914 had to be endured again in the inter-war years after *The Agricultural Crisis, 1920–1923* described by R. R. Enfield in 1924. After 1945, the presuppositions of the Labour government, the effective organisation of the National Farmers' Union,[5] and the immediate food and dollar shortages all took the place of submarines as the determinants of policy. Since 1951, the belief of politicians that rural electors vote for an agricultural policy has given farmers ready access to public funds. This liberality is encouraged by the widespread notion that history will go on repeating itself, that English agriculture will again disintegrate if freely exposed to foreign competition. But the conditions of the past no longer exist at home or abroad. It may well be that the future of the most highly mechanised agriculture in the world no longer rests on a choice between protection or destruction but on price and other policies designed to put unproductive land and inefficient farmers out of business. Subsidies and fiscal policies are not,

[1] K. A. H. Murray, *op. cit.*, pp. 274–5.
[2] To these should be added the superb history of mycology, written for the untrained reader, of E. C. Large, *The Advance of the Fungi* (1940), and such a standard textbook as that of J. A. Scott Watson and J. A. More, *Agriculture: The Science and Practice of British Farming* (10th edition, 1956).
[3] The experience of the earlier war was recorded by T. H. Middleton, *Food Production in War* (1923).
[4] (1960 edition) pp. 265–89.
[5] Until the research of Dr. J. Finklestein is published, the only account of the N.F.U.'s history is that of L. Easterbrook, " Fifty Fighting Years " (*British Farmer*, 1st February, 1958). But the N.F.U., which must compete with the B.M.A. for the distinction of being counted the most successful pressure group of recent years, is already attracting the notice of political scientists, notably S. E. Finer, *Anonymous Empire* (1957). The history of the Central Chamber of Agriculture from 1865 to 1915 was recounted by A. H. H. Mathews, *Fifty Years of Agricultural Politics* (1915).

however, the only forms of protection. Primogeniture and entail had resulted in a divorce between the ownership and control of landed property long before this phenomenon was observed as a novel consequence of joint stock companies. What territorial families had self-helpfully achieved for their own security in the eighteenth and nineteenth centuries, statute gave to tenants in the twentieth. The processes by which tenant farmers acquired security of tenure within the institution of private property in land culminated in the Agriculture Act, 1947, and in the consolidating Holdings Act of 1948. The Minister of Agriculture then acquired exceptionally wide powers for enforcing good estate management and good husbandry upon landlords and tenants who became subject to compulsory purchase or to dispossession if they failed to respond satisfactorily to ministerial supervision. At that time, the new legislation opened two possibilities. On the one hand, it gave the Labour government many of the advantages and none of the administrative embarrassments of land nationalisation and hence the means to initiate a socialist reconstruction of the industry. On the other, improved estate management and better farming could have been exacted as a return for the community's willingness to guarantee the industry's profitability. In the event, the Labour government lacked the policy and determination to grasp either opportunity. None of the penal powers was effectively or extensively exercised, and post-war developments served only to complete the transformation of the once servile tenant into the spoilt darling of the legislature. The last half century's discrimination between the rights of landowners and those of other property owners was thus carried to an extreme conclusion by putting land tenure on a basis which, in practice, made a solvent sitting tenant irremovable, and kept rents substantially below their open market level. Subsequent political trends made adjustments in favour of the agricultural landlord inevitable, and the first instalment was made by the Agriculture Act, 1958. Present agricultural policies are thus a jumble of *ad hoc* expedients inherited from the past and administered, as Sir Daniel Hall once tartly remarked, by those " who have no idea that agriculture can be anything else than putting money in the pockets of weaker farmers so as to preserve farming as it is or was, not as it might be ".[1]

Foakes Jackson devoted one of his Lowell Lectures, *Social Life in England, 1750–1850* (1916), to " Sport and Rural England " because he

[1] *The Countryman* (Spring, 1943), p. 54 ; quoted by F. W. Bateson in his suggestive and cogent Fabian pamphlet, *Socialism and Farming* (1948).

thought " it is not possible to understand English life without study-ing "[1] a subject that he feared his Boston audience would regard as flippant. No apology was necessary. This aspect of country life has never been woven into the fabric of agricultural history. Indeed, no general study has yet been made of the consequences of urban indus-trialism for a sporting nation. Throughout the nineteenth century, sports dependent on the horse or the gun were the dominating activity of most landowners who organised their lives and the sittings of Parlia-ment according to the calendar of close and open seasons. The bibli-ography for such a study is enormous : all that can be attempted here is an indication of its resources.

The more important of *The Sports and Pursuits of the English*, which the Earl of Wilton praised in 1868 as bringing together " people of all grades, and thus for the time (breaking) down all class barriers " and pointing " to a reign of such universal freedom, such jostlings of high and low, such social mixtures, as could be found in no other country of the world ",[2] were not regulated by the principles of *laisser faire*. In 1820, Mr. Justice Best thus explained the lawfulness of spring guns loaded with live ammunition as deterrents to trespassers. " If you do not allow men of landed estates to preserve their game, you will not prevail on them to reside in the country. Their poor neighbours will thus lose their protection and kind offices ; and the government the support it derives from an independent, enlightened, and unpaid magistracy."[3] The history of the game laws and of the demand for free trade in game for middle-class tables is surveyed in the articles of Chester Kirby, " The English Game Law System " (*American Historical Review*, vol. XXXVIII, 1933) and " The Attack on the English Game Laws in the Forties " (*Journal of Modern History*, vol. IV, 1932). Astonishingly, no comprehensive account of this revealing element in aristocratic life has been written and there is no entry under " game laws " in the index of *English Farming Past and Present*.[4] Estate management, assisted by this formidable legal code, ensured that there was no shortage of game in the countryside. The Duke of Portland recalled the list of game killed by his friend, Lord de Grey, between 1867 and 1900. It included 562 deer, 27,686 hares, 29,858 rabbits,

---

[1] p. 302.   [2] p. 17.
[3] In the case of Ilott *v.* Wilkes, R. V. Barnewall and E. H. Alderson, *Reports of Cases Argued and Determined in the Court of King's Bench, in Hilary Term, 60th George III* (1820), p. 320.
[4] The contribution of Bernard Darwin, " Country Life and Sport ", to G. M. Young (ed.), *Early Victorian England, 1830–1865* (1934), is no more than a series of ill-assorted descriptive passages.

56,460 grouse, 97,759 partridges and 142,434 pheasants.[1] But it was the pigskin not the gun that produced the heroes and the literature. The book of George F. Underhill, *A Century of English Fox-Hunting* (1900) is a magnificently devoted introduction and guide to the world of Osbaldeston, Alken, " Nimrod ", " The Druid " and Surtees. No other bibliography of the nineteenth-century hunting field can compare with it. It may be supplemented by the Encyclopaedia of *Rural Sports or Complete Account (Historical, Practical and Descriptive) of Hunting, Shooting, Fishing, Racing, &c.* (1st edition 1840, many subsequent) of Delabere P. Blaine and the modern book of A. H. Higginson, *Two Centuries of Foxhunting* (1946). The centenary volume of R. N. Rose, *The Field: 1853–1953* (1953), provides a running commentary over the period.

As steam annihilated distance and carried aspiring English soap-boilers and ironmongers to grouse moors in Scotland and town dwellers to their expanding suburbs, so the claims of an urban society upon its land resources grew more urgent.   From the General Inclosure Act in 1845 to the end of the century, there was a battle to preserve public rights over commons and forests described in the informative book of G. Shaw-Lefevre, *English Commons and Forests* (1894), revised and reprinted as Lord Eversley, *Commons, Forests and Footpaths* (1910). The modern position is described in detail in the *Report* of the Royal Commission on Common Land (Cmd. 462, 1958).   In an important study of *Agriculture and Urban Growth* (1959), G. P. Wibberley has estimated that, by the year 2000, the area of crops and grass in England and Wales " is likely to be less by about 15 to 20 per cent of its 1900 acreage ".[2]   The policies of the future will have to assimilate the many claims upon land and must come to accord priority to the provision of amenities for an urban civilisation.   It is safe prediction that present arrangements (or the lack of them) for the acquisition and control of land for non-agricultural purposes must come up for drastic revision within the next generation.   Then old schemes of land nationalisation, gathered up in such books as those of C. S. Orwin and W. R. Peel, *The Tenure of Agricultural Land* (1925) or Daniel Hall, *Reconstruction and the Land* (1941) or C. S. Orwin, *Speed the Plough* (1942), will have to be looked at again in the course of fashioning a national land policy.

[1] *Men, Women and Things* (1937), pp. 228–9.   There can have been few quarters of the globe that did not echo to the crack of an English rifle in this period.   Lord de Grey, for example, boasted among his foreign bag, 2 rhinoceros, 11 tiger, 12 buffalo, 19 sambur and 97 pig.
[2] p. 59.

ENGLISH FARMING
PAST AND PRESENT

CHAPTER I

THE MANORIAL SYSTEM OF FARMING.

Virgin soils: traces of sites of early villages: "wild field-grass" husbandry; the permanent division of pasture from tillage; manors and trade-guilds; origin of manors; the thirteenth century manor and village; divisions of land according to differences of tenure; villages isolated and self-sufficing; importance of labour-rents in the economy of a manor; the cultivation of the demesne; the crops grown; the live-stock; miscellaneous produce; the manorial courts: the social grades among the villagers; the system of open-field farming; the arable land; the meadows; the hams; the pasture commons; the prevalence and permanence of the open-field system; the domestic industries of the village.

IMPROVEMENTS in the art and science of English agriculture were in its infancy dependent on the exhaustion of virgin soils. So long as land was abundant, and the people few or migratory, no rotation of crops was needed. Fresh land could be ploughed each year. It was only when numbers had increased and settlements became permanent, that farmers were driven to devise methods of cultivation which restored or maintained the fertility of their holdings.

The progress of farming is recorded in legal documents, in manorial accounts, in agricultural literature. But the story is also often preserved in the external aspect which the land, the villages, or the hedgerows bear in the twentieth century. Dry uplands, where the least labour told the most, were first occupied and cultivated; rich valleys, damp and filled with forest growth, remained uninhabited and untilled. In spite of difficulties of water-supply, light or sandy soils, or chalky highlands seem to have been the sites of the oldest villages. Patches of the lower slopes of downs were cleared of self-sown beech, and sheltered dips tilled for corn; the high ground behind was grazed by flocks and herds; the beech woods supplied mast for the swine. Salisbury Plain, a century ago, bore no sign of human life except the proverbial "thief or twain "—no contemporary mark of the hand of man but the gallows and their appendages. Yet here are to be found traces of numerous villages. Scored on the sides of the Wiltshire, Dorset, Hampshire, and Sussex downs, " Lynches," " Lynchets," or " Daisses,"—grass-grown terraces or benches,—still run horizontally, one above the other, along the slopes. The " elf-furrows " of Scotland seem to record a similar occupation of hill sites. Local tradition attributes their formation to spade husbandry. Marshall, in 1797, suggested, but only to reject, the operation of the plough. Fifty years later, Poulett Scrope adopted a similar suggestion : more recently Seebohm revived the same theory. Whatever explanation of the formation of these terraces may be correct, they indelibly indicate the sites of the earliest settlements, and the nature of the soil first selected for tillage. [1]

The most primitive form of agriculture is that known as " wild field-grass " husbandry. Joint occupation and joint tillage were probably its characteristics, as they afterwards were of tribal or village communities. The essential difference lies in this. In the open fields of the village, pasturage and tillage continue to be separated ; grass-land always remains meadow or pasture ; it is never broken up for tillage. Under the more primitive form of convertible husbandry, fresh tracts of grass were successively taken in, ploughed, and tilled for corn. As the soil became exhausted, they reverted to pasture. Such a practice may belong to some portions of the Celtic race, or to nomadic stages of civilisation. In 1804 Marshall thought that he could trace the " wild field-grass " system in a custom of the south-western counties. In some districts lords of the manor enjoyed rights of letting portions of the grass commons to be ploughed up, cultivated for corn, and after two years thrown back into pasture. Over the whole country, from the Tamar to the eastern border of Dorsetshire, he found that open commons, such as the wide expanse of Yarcombe and the hills above Bridport, which from time immemorial had never known the plough, were distinctly marked with the ridge and furrow. Other features of rural life, which a century ago were more peculiar to the south-west of England, suggest that arable tillage by village communities, if it ever prevailed in this district, was soon exchanged for a system of convertible husbandry better suited to a damp climate. The cultivated land is divided into little patches by the high Devonshire earthwork, or hedge ; the

[1] Lynchets are to be found in many parts of the country other than on the chalk downs. The method and date of their formation is still in dispute. See *The Countryman*, Jan. 1936.

large open-fields of the parish can rarely be traced ; fewer of the inhabitants are collected into villages, more are scattered in single houses or tiny hamlets. Cornwall and parts of Devonshire, like Brittany, are a country of hedges, and of a Celtic race.

This " wild field-grass " husbandry was displaced in most parts of England by the permanent separation of arable from pasture land. The change indicates an advance towards a more settled state of society, but not necessarily an advance in agricultural practice. The fixed division of tillage and grass may have been introduced into this country by a people accustomed, like the Romans or the Anglo-Saxons, to a drier and less variable climate. If so, it was on this alien system that the agricultural organisation of the mediaeval manor was based. On it also were founded the essential features of those village communities which at one time tilled two-thirds of the cultivated soil of England, survived the criticism of Fitzherbert in the sixteenth century, outlived the onslaught of Arthur Young in the eighteenth century, clung to the land in spite of thousands of enclosure Acts, were carried to the New World by the Pilgrim Fathers, and linger to this day in, for instance, the Nottinghamshire village of Lexington, now Laxton,[1] where half the land of the parish is tilled by an agricultural association of partners.

In the early stages of history, the law itself was powerless to protect individual independence or to safeguard individual rights. Agriculture, like other industries, was therefore organised on principles of graduated dependence and collective responsibility. Mediaeval manors, in fact, resembled trade guilds, and it would be difficult to frame an organisation which, given the weakness of law and the infancy of agriculture, was better calculated to effect the object of mutual help and protection. Communities grouped together in villages were less liable to attack than detached farmhouses and buildings ; common methods of farming facilitated that continuous cultivation which otherwise might have been interrupted by the frequent absence of the able-bodied men on military expeditions ; the observance of common rules of management may have hindered improvement, but, if strictly enforced, it also prevented deterioration. Thus the system was suitable to the times and their conditions.

The origin of the legal relation of manors to village communities lies outside the scope of the present enquiry. It concerns tenures

[1] See Orwin. *Laxton.* Oxford Univ. Press, 1935.

rather than systems of cultivation. Two theories explain the rights of manorial lords and rights of common exercised over manorial lands. The legal theory, in its crudest form, is that the lord of the manor is the absolute owner of the soil of his manor, and that rights acquired over any part of it by freeholders and tenants are acquired against him, and originate in his grant or sufferance. The historical theory, stated baldly, is that self-governing, independent communities of freemen originally owned the land in common, and were gradually reduced to dependence by one of their members, or by a conqueror, who became the lord of the soil. There seems to be no doubt that individual ownership belongs to an earlier stage of civilisation than communal ownership. But if the second theory is correct, the legal position of the lord of the manor represents a series of encroachments, which transformed the Mark of freemen into the Mark of bondmen, and changed the rights of the villagers over the wastes of the district into customary rights of user over the lord's soil. Questions of the origin and antiquity of manors, and the extent to which they prevailed before the Norman Conquest, have been to a great degree reopened by recent studies. Seebohm, for example, practically supported the legal view by historical argument. He traced the feudal manor to the Roman *villa*, with the lord's estate as the centre round which clustered cultivators, who tilled the soil under servile or semi-servile conditions. This system, according to his view, was taken over by the Anglo-Saxon invaders, and the agrarian results of the Teutonic occupation may be summed up in the transfer of the Roman *villa*, with its servile labourers, to the conquerors. As a complete explanation of social development the legal theory, in spite of this historical support, seems inadequate. But whether the early stages of village communities reveal a movement from serfdom or originated in freedom, whether their relations to manors represent encroachments by the lord or advances by the serf, whether the rights of agrarian associations underlay, or were acquired against, the manorial rights of the feudal baron—whether, in other words, the land-law of the noble became the land-law of the people, or the reverse—is here immaterial. Roughly and generally speaking, the immediate lordship of the land farmed by a village community, including the wastes and commons, was, after the Norman Conquest, vested in the lord of the manor, subject to regulated rights enjoyed by its members.

On a manorial estate, at the beginning of the thirteenth century, only the church, the manor-house, and perhaps the mill, rose out conspicuously. There were no detached, isolated farm-houses ; but the remaining buildings of the village, grouped together in a sort of street, were the homes of the peasantry, who occupied and cultivated the greater part of the land. At some little distance from the village stood the manor hall or grange, with its out-buildings, garden, and fishpond, surrounded by clay-built walls with thatched tops. The style and extent of the buildings depended on whether the house was the permanent or occasional residence of the lord ; they also varied with the importance of the manor, and the wealth of its owner. The house itself was built either of timber and clay, or of stone, for brickmaking was still a forgotten art. It often consisted of a single hall, plastered inside, open to the roof, and earth-floored, which served as court of justice, dining-room, and bedchamber. At one end of the central room was a stable ; at the other a chamber, kitchen, or larder. Below one part of the ground floor was a cellar ; above another part was, perhaps, a " solar," or parlour, approached by an outside staircase. If the manor was sufficiently important, there were probably added a detached building for the farm servants, and a chamber for the bailiff. The outbuildings consisted of bake-house, stables, dairy, cattle and poultry houses, granary, and dove-cote. Some of the oldest specimens of domestic architecture are granaries, like Hazelton or Calcot in Gloucestershire, or the dove-cotes which still in country districts mark the former sites of manor-houses. Repairs of the walls and buildings of the manor-house were among the labour services of the tenantry, who dug, tempered, and daubed the clay, cut and carted the timber, and gathered the straw or reeds for thatching. Where technical skill was needed they were aided by craftsmen, who either held land in reward for their special services, or, on the smaller manors, were hired for the occasion.

Tufts of trees, conspicuous in the hedgeless expanse of arable land by which they were surrounded, marked the sites of villages, as they still do in the high table-land of the Pays de Caux. Under their shelter clustered the homes of the peasantry, clay-walled, open-roofed, earth-floored, chimneyless sheds, covered in with straw or reeds or heather, and consisting of a single room. Here, divided by a hurdle or wattle partition, lived, not only the human inhabitants, but their cows, pigs, and poultry. Close by were the tofts and crofts of the open-field farmers, each with its miniature hay-rick and straw-stack ; and the cottages and curtilages of the cottagers, " fencèd al aboute with stikkes." Here were the scanty gardens in which grew the vegetables, few but essential to the health of a population which lived almost entirely on salted meat and fish—often half-cured and half-putrid. These homesteads were in early times the only property held by members of the township in exclusive separate occupation. They were also, at first, the only permanent enclosures on the commonable land. But, as agriculture advanced, pasture paddocks (" gerstuns " or " garstons ") for rearing stock, calves, or fattening beasts, or for the working oxen, which could not endure his " warke to labour all daye, and then to be put to the commons or before the herdsman," were enclosed in the immediate neighbourhood of the village. In these enclosures, or " happy garstons " as they were called at Aston Boges, were held the village merrymakings, the rush-bearings, the May games, the summerings at St. John's Eve, the public breakfasts, and the distribution of bread and ale in Rogation week.

The land comprised in a thirteenth century manor was generally divided into four main portions, and, speaking generally, was cultivated on co-operative principles ; the demesne or " board " land, reserved for the lord's personal use, surrounding the manorhouse, and forming the smaller portion of the whole ; the free land occupied by freemen holding by military service, or by some form of fixed rent in money or in kind ; the unfree land, occupied by various classes of bondmen, holding by produce-rents and labour services which varied with the custom of the manor ; the common pastures and untilled wastes on which the tenants of the manor and the occupiers of certain cottages, in virtue of their holdings, fed their live stock. This right of pasture must be clearly distinguished from those rights which, at certain seasons of the year, were exercised by the associated partners over the cultivated arable and meadow lands of the village farm.[1] Thus the lord's demesne, using the word in its narrower sense, might be kept in hand, or let on lease to free or unfree tenants, or thrown

[1] By "village farm" is meant the land in the village which was occupied by an association of partners, who were bound by the same rules of cultivation, held intermixed strips of arable land over which at certain seasons the whole body exercised common rights, annually received allotted portions of meadow for hay, and enjoyed, in virtue of their arable holdings, the right to turn out stock on the common pasture. This open-field system of farming is described pp. 23-27.

into the village farm, or dealt with as to portions in each of these three ways. But whether the land was treated as a compact whole, like a modern home-farm, or whether the landlord, as a shareholder in the village association, allowed it to be cut up into strips and intermixed with other holdings, the demesne was mainly cultivated by the labour services of the unfree peasantry. The rest of the land of the manor, forming the larger portion of the cultivated area, was farmed by village partners, whose rent chiefly consisted in the labour, more or less definite in amount, which they were obliged to perform on the lord's demesne.

In this method of cultivating a manorial estate there are many contrasts with the modern system. The three-fold division of the agricultural interests into landlord, tenant farmer, and wage-earning labourer was practically unknown. Landowner and tenant-labourer owned, occupied, and cultivated the soil, and the gradual relaxation of the labourer's tenure of the land, and the inter-position of the tenant farmer between the two existing classes, sum up the early social history of English farming. In the thirteenth century, muscles were more essential to the prosperity of the land-lord than money rents. The cultivators of the soil grew their produce, not for sale, but for their own consumption. Each manor or village was isolated and self-sufficing. Only in the neighbour-hood of towns was there any market for the produce of the farm. Few manufactured articles were bought. Salt, tar, iron (bought in four-pound bars), mill-stones, steel for tipping the edges of imple-ments, canvas for the sails of the wind-mill, cloths for use in the dairy, in the malthouse, or in the grange, together with the dresses of the inhabitants of the hall, and a few vessels of brass, copper, or earthenware, satisfied the simple needs of the rural population. Hands were therefore more required than money on manorial estates. If the manor was well stocked with labour, the land paid ; when the stock of labour shrank, the profits dwindled. It was in order to retain a sufficient supply of labour on the land that bond-men were restrained from leaving the manor to assume the tonsure of the clerk or the flat cap of the apprentice, to become soldiers or to work outside the manor. Even their marriages were carefully controlled by licences. It was, again, in order to exact and super-vise the due performance of labour services that the lord of the manor maintained his large official staff—his seneschal, if he owned several manors, his steward, his bailiff, and the various foremen of the labourers, such as the reeve, the hayward, the head-reaper, and the granger. But with the thirteenth century begins the practice of keeping estate accounts, in which the amount and cash values of the labour services are entered. Thus the uncertainty of villein-tenure was modified, and the means were prepared for commuting obligations to work into their money equivalents. Already the causes were operating which hastened the process, and changed agriculture from a self-sufficing industry into a commercial system of farming for profit. Population was increasing; trade was growing; urban classes, divorced from rural pursuits, were forming; means of communication were improving; money taxes took the place of personal services; the standard of living rose; coin was needed, not only to meet the demands of the government, but to buy the luxuries of more civilised life.

The obligations of the peasantry to cultivate the demesne varied, not only with local customs, but with the seasons. Their most important services were the autumnal, Lenten, and summer ploughings on the three fields, into which the arable land of the demesne was generally divided. The crops grown were, as winter seeds, wheat and rye, and, as spring seeds, oats, barley, beans, peas, or vetches. In smaller quantities, flax, hemp, and saffron were locally raised in separate plots. Roots, clover and artificial grasses were still unknown. Rotations of crops, as they are now understood, were therefore impossible. The soil was rested by fallowing the one-half, or the one-third, of the arable land required by the two or the three course system. Red rivet, or a lost white variety, was then recommended for wheat-sowing on light land, red or white pollard for heavy soils, "gray" wheat for clays. But on the tenants' land, rye was the chief grain crop. It is the hardiest, grows on the poorest soils, makes the toughest straw. Rye was then the bread-stuff of the English peasantry, as it still is in Northern Europe. The flour of wheat and rye were often mixed together, and bread made in this form was called "maslin."[1] It retained its moisture longer than pure wheaten bread, and, as Fynes Moryson

[1] Lat. *mixtilio*; "mestilon," anon. author of *Hosebonderie* (thirteenth century); "miscellin," Harrison (sixteenth century); "massledine," Henry Best (1641); "mashelson," Yorkshire (1797). In *The Compleat Farmer* (1760) it is called "maislen"; but the writer says that it is "ill husbandry to grow wheat and rye together." Fitzherbert (1523) recommends rye and wheat to be sown together as the surest crop to grow and good for the husbandman's household. But he does not believe in the slowness of rye in ripening.

says in his *Itinerary* (1617), was used by labourers because it " abode longer in the stomach and was not so soon digested with their labour." Wheat and rye were sometimes sown together. But as rye was slower to ripen, the better practice was to sow it alone and earlier, lest, as Tusser (1557) writes, " rye tarry wheat, till it sheds as it stands." The mixed cultivation was, however, recommended as a cure for mildew, and for this reason prevailed in Yorkshire in 1797. Barley was the drink-corn, as rye was the bread-corn, of the Middle Ages. It was of two kinds. The head with two rows of grain seems to have been used exclusively for brewing ; the coarser four-rowed head, known as " drage," was used partly for brewing, partly for feeding pigs and poultry. Barley and oats were often sown together. In the North, oats were extensively cultivated ; but they were grey-awned, thin, and poor. In the Midlands and South of England they were comparatively rare on tenants' land.

The fallows were three times ploughed in preparation for wheat and rye. The seed began to be sown after Lammas Day (August 12),[1] and at latest was completed by Hallowmas (November 1). For oats, beans, and peas, the land was ploughed and the seed sown between the Feast of Purification (February 2) and Easter. Oats were said to be best sown in " the dust of March." " On St. Valentine's Day cast beans in clay. But on St. Chad sowe good or bad." That is to say, the time for sowing beans was between February 14 and March 2. Barley came last. The land was ploughed and sown between Hoke-tide (the third Tuesday after Easter) and Pentecost. The ploughings were performed, and the teams supplied and driven, partly by the servants of the demesne, partly by the tenants. Sometimes ploughmen seem to have been hired. The harrowings were similarly provided for, and the harrow, often a hawthorn tree, weighted on its upper side with logs, was supplied from the lord's waste. Here also harrowers seem to have been sometimes specially hired. In this case they possibly provided their own home-constructed implements with sharp points or teeth like the modern type of harrow. When the fallows were first broken up, as was then the practice, in March, or when the land was prepared for barley, the ground was often so hard that the clods had to be subsequently broken. For this

---

[1] The Julian calendar was in force. To make the dates correspond with those of the present Gregorian calendar, twelve days have to be added.

purpose the ploughman, holding the principal hale of the plough in his left hand, carried in his right a "clotting beetle," or "maul," such as that which is depicted in the Cotton MSS. A "Dover-court beetle" was a necessary tool in the days of Tusser; and Plot, whose *Natural History of Oxfordshire* appeared in the seventeenth century, recommends its use after the land was harrowed.

The amount of wheat, rye, beans, and peas usually sown to the acre was only two bushels; and of oats and, strangely enough, of barley, four bushels. The yield of wheat rarely exceeded five-fold, or ten bushels to the acre; that of the leguminous crops ranged from three- to six-fold, or from six to twelve bushels to the acre; that of oats and barley varied from three- to four-fold, or from twelve to sixteen bushels to the acre. Considerable care was exercised in the choice and change of the seed-corn, which was often one of the produce-rents of the tenants. On the Berkeley Estates (1321) the seed was changed every second or third year; the upland corn being sown in the vale, and *vice versa*. Wheat rarely followed a spring grain crop. If it did, it may be supposed that it received the greater part of the manure mixed with earth, which the tenants carted from the demesne yard, and spread on the manor farm. From the point of view of manuring the land, the right of folding was a valuable privilege. Tenants, unless they purchased a licence to fold their sheep on the land they occupied, were often obliged to feed and fold their flocks on the lord's land for fallow or in his own fold. Sometimes the herbage of the lord's land for fallow was sold to a sheep-master to be depastured on the land. Lime was used on heavy clays, or to destroy moss. The value of marl in improving the texture of sandy soils and some kinds of clays was appreciated. On the Berkeley Estates it was first used in the fortieth year of Henry III. But the cost was excessive. "Marl," says Fitzherbert,[1] "is an excellent manure, and . . . exceeding chargeable." Sea sand was used near the coast; soot and even street refuse were employed on home farms. Drainage, except in the form of ridging the surface of wet soils, was rarely practised. Sometimes, as Palladius recommends (Book VI. st. 6), shallow trenches filled with gravel, stones, or hollow alder stems, and turfed over, were cut, and, on the manors belonging to

---

[1] Fitzherbert's *Book of Husbandry*, book i. c. 20 (ed. 1598). For agricultural literature, see Chronological List in Appendix I.

the Collegiate body of St. Paul's Cathedral, it was one of the labour services to clean out the ditches. But the science of deep drainage made little progress before the nineteenth century. Beans were often dibbed ; but all other seed was sown broadcast. The actual labour of sowing was probably performed by the lord's bailiff, or the hayward, with his own hand, as, at the beginning of the last century, all seed was sown by the farmer himself. The hoeing and the weeding of the crops were among the labour services of the tenants. In cleaning land the maxim was ancient :

"Who weeds in May
Throws all away,"

and the crops were generally weeded in June or the first few days of July. Walter of Henley [1] (thirteenth century) gives St. John's Day (June 24) as the earliest date for cleaning the land. " If," he says, " you cut thistles fifteen days or eight before St. John's Day, for each one will come two or three." On a Suffolk manor, in the fourteenth century, sixty " sarclers," or weeders, were employed in one day, armed, if the weather was dry, with a hook or forked stick, and, in wet weather, with nippers.

The meadows of the demesne were mown, and the hay made, carted, and put on the manorial ricks, by the labour services of the tenants. They also reaped, bound, gathered, loaded, carted, and stacked the corn crops in the lord's grange. They also threshed the corn, and winnowed it, unless, as was sometimes the case, the duty of winnowing fell to the dairywoman, or " Daye." If any corn was sent for sale to the markets, it was carried there by the labour services of the tenants, in their carts drawn by their teams. Harvestings in the Middle Ages were picturesque scenes of bustle and of merriment among the thousands to whom they meant the return of plenty. On 250 acres at Hawstead in Suffolk, towards the close of the fourteenth century, were grown wheat, oats, barley, peas, and " bolymong," a mixture of tares and oats. The grain crops were cut and housed in two days. On the first day appeared thirty tenants to perform their " bedrepes," and 244 reapers ; on the second day, the thirty tenants and 239 reapers, pitchers, and stackers. Many of this assembly were the smaller peasantry on the manor ; the rest were the lord's farm servants, together with

[1] *Walter of Henley's Husbandry, together with an anonymous Husbandry, Seneschaucie, and Robert Grosseteste's Rules*, ed. E. Lamond, 1890. For agricultural literature, see Chronological List in Appendix I.

wandering bands of " cockers " or harvesters, who had already begun to travel the country at harvest time. A cook, brewer, and baker were hired to supply dinner at nine and supper at five. Reapers were organised in bands, or "setts," of five. The anonymous author of *Hosebonderie* [1] (thirteenth century) calculates that each band could reap and bind two acres a day. Barley and oats, as well as peas and beans, were generally mown ; rye and wheat were reaped. But the reaping, as in Roman times, seems to have consisted of two operations : the first was to cut the ears, the second to remove part of the straw for thatching, or to be used as forage for cattle, as litter for strewing the sheep-house, folds, and yards, or as bedding for men. Often the value of the straw of thin short corn hardly paid for the expense of removal, and the stubble was either grazed or burned on the ground, or ploughed in.

The most important crops of the farm were the corn crops of wheat, rye, and barley, which were raised for human food and drink. Their consumption, especially if the lord of the manor lived on the estate, was enormous. Domestic households were considerable, and often only the bailiff was paid money wages. Rations were also allowed to tenants when performing many of their services. Though the manual and team work of the tenants provided most of the labour of the farm, the lord also employed a large permanent staff of agricultural servants, most of whom were occupied in the care of live-stock. Such were the horseman or waggoner, oxherd or ploughman, cowherd, shepherd, swineherd, warrener, and keepers of hawks and dogs, whose wages were mostly paid in kind. There were, besides, other servants in husbandry, hired for special occasions, whose food and drink formed a large portion of their payment. The granary was, therefore, rarely so full that any surplus remained for sale. For such ready-money as he needed, the lord looked mainly to the produce of his live-stock. For their consumption were grown the remaining crops—the hay, beans, peas, and oats ; though oats were not only used for human food, but in some districts were brewed into inferior beer.

Horse-farms appear in some estate accounts ; but they probably supplied the " great horse " used for military purposes. On an ordinary farm the horses used for farm-work were mostly home-bred, and were divided into cart-horses, and—under the names of stotts, " affers," or " avers "—plough-horses. Colts, not needed

---

[1] *Hosebonderie* in *Walter of Henley's Husbandry*, ed. E. Lamond, 1890.

to keep up the supply, were sold. Plough-teams were seldom made up of horses only ; if horses were used at all, they were mixed with oxen. But, as a rule, oxen were preferred to horses. Though horses worked more quickly, when the ploughman allowed them to do so,—they pulled less steadily, and sudden strains severely tested the primitive plough-gear. On hard ground they did less work, and only when the land was stony had they any advantage, Economical reasons further explain the preference for oxen. From St. Luke's Day (October 18) to April, both horses and oxen were kept in the stalls. During these twenty-five weeks neither could graze, and Walter of Henley calculates that the winter-keep of a horse cost four times that of an ox. Horses needed more attendance ; they required to be rubbed, curried, and dressed. Oxen were less liable to sickness than horses. The harness of the ox, mainly home-made from materials supplied on the estate, was cheaper to provide and repair. Shod only on the forefeet, the shoeing of the ox cost less than that of the horse. When either horse or ox was past work, the profit of the one lay in his hide ; of the other, not only in his hide, but the larder : the ox was " mannes meat when dead, while the horse is carrion." Great care was taken both of horses and of oxen. In *Seneschaucie* [1] (thirteenth century) the duties both of the waggoner and oxherd are carefully defined ; each was expected to sleep every night with his charges.

Cattle were seldom fatted even for the tables of the rich ; oxen were valued for their power of draught : cows for their milk. It may, indeed, be said that fresh butcher's meat was rarely eaten, and that, if it was, it was almost universally grass-fed. No winter-keep or feeding stuff was available ; not even carrots or parsnips were known. The commons, generally unstinted, carried as much stock as could keep skin and bone together in the winter, and the lord could not only turn out on them his own sheep and cattle, but license strangers for money payments to do the same. Even if the commons were stinted, the margin was too bare to mean abundance. The best pastures were either in the lord's own hands, and were saved by him at the expense of the commons, or were let out to individuals in separate occupations. Even among these superior feeding-grounds, there were few enclosures which would fatten a bullock. At the wane of the summer, the cattle had the

---

[1] In *Walter of Henley's Husbandry*, ed. 1900.

aftermath of the hay meadows, and the stubble and haulm of the arable lands. During this season they were at their best. They only survived the winter months in a state of semi-starvation on hay, straw, and tree-loppings. It was, therefore, the practice at the end of June to draft the aged cows, worn-out oxen, and toothless sheep, or " crones," prepare them as far as possible for the butcher, slaughter them in the autumn, and either eat them fresh or throw them into the powdering tub to be salted for winter consumption. " For Easter at Martilmas (November 11) hange up a biefe " is the advice of Tusser.

The dairy produce was a greater source of money revenue, though the home consumption of cheese must have been very large. But the management was necessarily controlled, like the management of the stock, by the winter scarcity. The yield of a cow during the twenty-four weeks from the middle of April to Michaelmas was estimated at four-fifths of her total annual yield. Six to ten ewes gave as much milk as one cow ; but the best practice was to cease milking ewes at Lammas Day (August 12). Cheese-making formed an important part of the dairywoman's duties, and the purchase of the cloths and utensils used in its manufacture are a serious item in estate accounts. Cheese seems generally to have been made of skim-milk, though superior varieties were doubtless found on the lord's table. Most of the butter made in the summer months was either sold, or salted and preserved in pots and barrels for winter use.[1] The butter-milk was either drunk, made into curds, or more rarely used to fatten pigs. The curds were eaten with wine or ale ; the whey, under the name of " whig," made a cool and wholesome summer drink. During the winter months, milk fetched three times its summer price, and was generally sold. For this, among other reasons, calves were timed to fall before autumn. In the scarce months of winter, the price obtained for milk during eight weeks was supposed to be worth more than the calf. Small open-field farmers must usually have sold their calves as soon as possible. The same practice prevailed on the demesne. The total

---

[1] Rogers, noticing that butter was sold by the gallon, seems to have concluded that it was melted (*Six Centuries of Work and Wages*, ed. 1890, pp. 94-5). But it would seem from the thirteenth century writings of Walter of Henley and the anonymous author of *Hosebonderie*, that two pottles of butter made 1 gallon of 7 lbs., 2 gallons made 1 stone ; and 14 stone 1 wey. Whatever inference may be founded on the use of a liquid measure, it is discounted by the use of the pottle and the stone.

number of live-stock, including horses but not including sheep, sold from the manor of Forncett in thirteen years, between 1272 and 1306, was 152.[1] Out of this total 99 were calves. The cows of the demesne were under the care of a cowherd, who was required to sleep every night with his charges in the sheds.

Sheep were the sheet anchor of farming. But it was not for their mutton, or for their milk, or even for their skins, that they were chiefly valued. Already the mediaeval agriculturist took his seat on the wool-sack. As a marketable commodity, both at home and abroad, English long wool always commanded a price. It was less perishable than corn, and more easily transported even on the worst of roads. To the Flemish weavers it was indispensable, for Spanish wool could not be used alone, and the supply from Saxony was not as yet developed. The washing and shearing of sheep were among the labour services of the tenantry. Certain districts, especially Shropshire, Leominster, and the Cotswolds, were from very early times famous for the excellence of their wool. So far as its quality depended on breed rather than on soil, some care, as evidenced by the higher prices paid for rams, was taken to improve the flocks. From Martinmas to Easter sheep were kept in houses, or in moveable folds of wooden hurdles, thatched at the sides and tops. During these months they were fed on coarse hay or peas-haulm, mixed with wheaten or oaten straw. For the rest of the year they browsed on the land for fallows, in woodland pastures, or on the sheep commons. But in the autumn they were not allowed to go on the ground, till the sun had purified the land from the " gelly or matty rime," which was supposed to engender scab. So also they were driven from the damp, low-lying grounds lest they should eat the white water-snails which our ancestors suspected of breeding the rot. These two diseases made sheep-farming, in spite of its profits, a risky venture. The scab does not seem to have attacked sheep before the latter end of the thirteenth century ; but, from that time forward, the tar-box was essential to every shepherd. The rot is carefully treated by Walter of Henley, if he is the real author of the passage interpolated in the Bodleian manuscript of his work.[2] The writer discusses the

---

[1] *The Economic Development of a Norfolk Manor* (1086-1565). By Frances Gardiner Davenport, pp. 33-35.

[2] *Walter of Henley*, 1890, ed. E. Lamond. The passage is given on pages 37-8, and its genuineness is disputed in the Introduction, p. xxii.

symptoms of the disease. White veins under the eyelids, wool that can be easily pulled away from the ribs, a skin that will not redden when rubbed, are signs of unsoundness. Another sign is when the November hoar-frost melts rapidly on the fleece, for the animal is then suffering from an unnatural heat. The losses of the flockmasters from the " murrain," to use the generic term for diseases employed by mediaeval writers, were so severe as to create another danger. The minute instructions against fraud given to the official staff show that shepherds not infrequently produced the skin, and explained the disappearance of the carcase by death from disease. " Let no sheep," says the author of *Seneschaucie*, " be flayn before it be seen and known for what fault it died." The value of the flock made the shepherd one of the most important of farm servants. He was required to be a patient man, " not over-hasty," never to be absent without leave at "fairs, markets, wrestling-matches, wakes, or in the tavern," and always to sleep in the fold together with his dog. Later writers insist on the value of lameness in the shepherd, as a lame man was unlikely to over-drive his sheep.

Swine were the almost universal live-stock of rich and poor. As consumers of refuse and scavengers of the village, they would, on sanitary grounds, have repaid their keepers. But mediaeval pigs profited their owners much, and cost them little. It was a Glouces-tershire saying :

> " A swine doth sooner than a cowe
> Bring an ox to the plough."

In other words, a pig was more profitable than a cow. For the greater part of the year pigs were expected to pick up their own living. When the wastes and woodlands of a manor were extensive, they were, except during three months of the year, self-supporting. They developed the qualities necessary for taking care of themselves. The ordinary pigs of the Middle Ages were long, flat-sided, coarse-boned, lop-eared, omnivorous animals, whose agility was more valuable than their early maturity. Growth and flesh were the work of time : so also were thickened skin, developed muscles, and increased weight of bone. The styes were often built in the woods, whence the pigs were only brought to feed on the arable land after the crops were cleared, or, at times of exceptional frost, to subsist on the leavings of the threshing-floor. During most months of the year they ranged the woods for roots, wild pears, wild plums, crab apples, sloes, haws, beech-mast, and acorns. Only when the sows were farrowing, or when animals were being prepared for the rich man's table, were they specially fed. Pigs were fatted on inferior corn, especially coarse barley, peas, beans, skim- and butter-milk, or brewers' grains which were readily obtainable when nearly every household brewed its own barley beer. The amount consumed varied with the purpose intended to be served. The boar was fatted for the feast on ten times the grain bestowed in finishing ordinary animals for conversion into salted pork or smoke-dried bacon. Walter of Henley implies that some attention was given to breed, as he recommends the use of well-bred boars. But the only quality on which he insists is that the animal should be able to dig, or, in other words, support itself. Modern ideas of purchasing corn for fattening purposes, or of converting into pork or bacon farm-produce for which no ready market was available, scarcely entered into the heads of mediaeval farmers. On the contrary, they tell us that, if pigs were entirely dependent on the crops of the arable land, they could not be kept at a profit, when the wages of the swineherd, the cost of the grain consumed, and the damage done to growing crops had been taken into account. Some trade was, however, carried on in stores. This is proved by the records of Forncett manor (*A Norfolk Manor*, 1086-1565), which show that, in years when no pigs were kept, stores were bought and fatted for the larder.

The poultry yard was under the care of the dairywoman, who sometimes seems to have had the poultry to farm at so much a head. Ducks are not mentioned in any of the mediaeval treatises on farming, though they appear in the Berkeley accounts in 1321 : guineafowl and turkeys were unknown. But the number of geese and fowls, and, on important estates, of peacocks and swans, was large, and it was swollen by the produce-rents which were often paid in poultry and eggs. The author of *Hosebonderie* gives minute instructions as to the produce for which the dairywoman ought to account. " Each goose ought to have five goslings a year : " each hen was to answer for 115 eggs and seven chickens, " three of which ought to be made capons, and, if there be too many hen chickens, let them be changed for cocks while they are young, so that each hen may answer for three capons and four hens a year. And for five geese you must have one gander, and for five hens one cock." Besides the poultry yard, the dove-cote or pigeon-house was a source of profit to the lord and of loss to the tenant. Prodigious numbers of pigeons were kept ; not only were they eaten, but their dung was prized as the most valuable of all manures. The privilege of keeping a pigeon-house was confined to manorial lords and jealously guarded, and every manor had its dove-cote. The story of the French Revolution shows how bitterly the peasants resented the plunder of their hard-earned crops by the lord's pigeons. Doubtless many a British peasant in mediaeval times was stirred to the same hostility by the same nuisance.

To the produce of the crops and the live-stock of the demesne must be added game, rabbits from the "conygarth" or warren, cider from the apples, oil from the nuts, honey and wax from the bee-hives, and sometimes grapes from the vineyards. Bee-keeping was an important feature of agricultural industry. The ancient proverb says : "He that hath sheep, swine, and bees, sleep he, wake he, he may thrive." Honey, besides being the only sugar, was invaluable in the still-room, and in the arts of the apothecary, physician, and "chirurgeon." It was an ingredient in mead and metheglyn. It was used in embalming, in medicines, and in such decoctions as mulse water, oenomel, honey water, rodomel, or quintessence. Wax was not only necessary for the candles of the wealthy, but, like honey, was largely used in mediaeval medicine. Mixed with violets, it was a salve : it was also one of the ingredients of "playsters, oyntementes, suppositories, and such like." In some districts of England, vineyards formed part of the equipment of manors ; one was made by Lord Berkeley towards the close of the reign of Edward III., and his biographer suggests that he learned the "husbandry . . . whilst hee was prisoner in ffrance or a Traveller in Spaine." Few great monasteries were without vineyards, which are mentioned thirty-eight times in Domesday Book. It is not necessary to explain the disappearance of the vine by a change of climate. Wine was then often sweetened with honey and flavoured with blackberries and spices. Unless it came from abroad, it was rarely drunk in its pure state. It would, therefore, be unsafe to found any theory of climatic change upon the production of a liquid which, in its natural state, may frequently have resembled vinegar.

Besides the produce of the live-stock and crops of his demesne, the lord of the manor had other sources of revenue. There were the fixed money or produce rents for their land paid by free tenants and bondmen, and the money payments which were sometimes accepted in lieu of labour services. Sales of timber and underwood, of turf, of herbage, licences to fold on the tenant's land, or licences to turn pigs into the lord's woods for beechmast or acorns, brought in varying sums of money. The mill at which the tenants ground their corn was his property. Whether the miller was his servant, or farmed the receipts, a considerable proportion of the tolls went into the landlord's purse, though the cost of repairs and upkeep diminished the net profits. On some manors the oven in which the bread was baked was also the property of the lord of the manor. The fees and fines levied and settled by the manorial courts in the course of a year were surprisingly large ; besides their administrative work, they were at once the guardians and the interpreters of the customs of the manor. The range of business administered in these courts, to which the tenants, both free and bond, were summoned as jurors, therefore embraced the domestic and financial affairs of the manor. Here were paid the fees for permission to reside outside the manor, to send children to school, to enter minor orders, to apprentice a son to a trade, or to marry a daughter. Here too were imposed the fines for slovenly work at harvest, for selling cattle without the lord's leave, for appropriating commons and wastes, for moving a neighbour's landmark, for neglecting to repair a cottage, for failing to discharge labour dues. Here too were fixed the contributions of the tenantry in money or labour towards the maintenance of the by-roads within the manor, and the fines for neglect of the duty to keep their surfaces in repair, to provide for their proper drainage, and to remove obstructions. Here also crime was punished ; offenders against life or property, as well as poachers, were mulcted ; wrangling scolds and tavern-hunters were presented ; idlers were deprived of their holdings, and, as a last resort, expelled from the manor. Here too were fixed and levied the necessary contributions for the repair of the stocks, the pillory, the ducking-stool, and the pound. Here the miller would be fined for mixing rubbish with his flour, the baker for selling short weight, the brewer who adulterated his beer, the ale-wife and tavern-keeper who used false measures or mixed the drink they sold with peony seed, salt or garlick, the carrier for failing to deliver goods, the householder who harboured a stranger without a licence. Here also were received and entered the fees of tenants for admission to their holdings, and the payment of fines by sons who succeeded their fathers. Here, finally, on the sworn evidence of a body of jurors chosen from the tenants, were drawn up the surveys of the manor which recorded the exact condition of the estate—the total acreage of the demesne, and of each of the arable fields, of the meadows, the several pastures and the pasturage, and their annual values ; the state of the woods and the coppices, how much could be cut, and what they were worth yearly ; the acreage of the commons and the stock which they would carry ; the number of the live-stock of various kinds ; the holdings of the free tenants, and their rents or services ; the holdings of the villeins, bordars, and cottagers, their services and money equivalents ; the profits of fisheries, mills, and incidental manorial rights ; the number of tenants who had finally commuted their services for fixed payments in cash, of those who, at the discretion of the lord, either rendered labour services or paid the money values, and of those who still discharged their personal obligations by actual work.

The remainder of the cultivated land of the manor was occupied by tenants who paid rents in the form of military or labour services, or money, or produce. Their farm practices, crops, and live-stock were the same as those of the demesne, though their difficulties in combating winter scarcity were greater. Free tenants, whose tenure was military service, or who had commuted the personal obligations for quit-rents, may sometimes have held land, like modern farmers, in their exclusive occupation for individual cultivation. But the area of free land was comparatively small, and, as often as not, it was thrown into the village farm, occupied and cultivated in common by an agrarian association of co-partners, free and unfree.

The varieties of tenure were great. So also were the varieties of social condition, and of the obligations by which the grades of those social conditions were governed. The distinctions between freemen and bondmen and between freehold and bond tenure had been, in the eye of the law, broad and deep. But custom had gradually intervened, and, with endless variety of practice, mitigated the severity of legal theory. At law the bondman's position was subject to the lord's caprice. Unlike the freeman, he was tied to the manor ; he could not leave it without licence from the lord, and payment of a fine. His services were uncertain in amount, and could be increased at the lord's pleasure. He paid a fine to marry his daughter, to send his son to school, to make him a priest or an apprentice. His lands and his goods and chattels might be seized by his lord, and when he died, his holding was given to whom the lord willed : his heir bought a licence to inherit even his moveables, and paid a fine when he was admitted to his father's tenancy. In the thirteenth century, some at least of these conditions had been modified. The bondman's services had become fixed ; he could buy and sell, hold property, and dispose of his possessions by will. In theory he might still be at the mercy of the lord's will : but custom had so regulated the exercise of that will that it could no longer be capricious.

Speaking broadly,[1] the mass of the occupiers of land were, in the eye of the law, unfree—bondmen who rented the shares in the land which they cultivated for themselves by labour services on the lord's demesne. It was the amount and certainty of their services which determined the rank of the unfree. Sometimes the service was for the autumn only, or for autumn and spring work, whether on specified days or at particular periods ; sometimes of team work, sometimes of manual labour, sometimes of both ; sometimes of week-work throughout the year, and either of one, two, or three days in each week. All their spare time was spent on their own holdings. Of this semi-servile class the villeins formed the aristocracy. The villein was neither a servant in husbandry nor a labourer for wages. He occupied land, and, like Chaucer's ploughman, had " catel " of his own. He was a partner in the village association, holding land of various amounts. In theory the size of his holding was based on the number of oxen which, in discharge of his share of the joint liability, he could contribute to the manorial plough-team.[2] A " hide " of land, which Professor

---

[1] Students of Professor Maitland's invaluable works will recognise the danger of broad and general statements, to all of which there are exceptions and modifications.

[2] The hide, or "carucate" of Domesday Book, or "ploughland," which averages 120 acres, is sometimes said to have been as much land as a team of 8 oxen could plough in a year of 44 weeks of working days. But Walter of Henley, who is the authority for this statement, only tries to show that the area should be 160, or even 180, acres ; he does not say that it actually was of this larger size. It does not seem likely that a fiscal unit varied with the nature of the soil, the weight of the plough, the condition of the team, the configuration of the land, and the temperament of the ploughman. It seems more probable that the hide or carucate was the definite area of 120 acres. Therefore a quarter of a carucate (30 acres) was the Domesday " virgate," which, under the name of " broad ox-gang," " husband-land," " farm-hold " or " farm " in the North, " yardland " in the Midlands, " full land " in Cambridgeshire, and " living " or " whole place " in Dorsetshire, formed the typical arable

Maitland considers to have been " the land of a household," was treated as the area which a team of eight oxen could plough in a working year. Its extent may have varied. But, if the size was 120 acres, then each hide consisted of four portions of 30 acres, called "virgates," or 8 portions of 15 acres, called "bovates." Thus the eighth part of the hide, or "bovate," was the land of one ox ; the fourth part of the hide, or "virgate," was the land of two oxen ; and the whole hide was the land of the complete team of eight oxen. It was on this basis that the tenemental land, in theory, and sometimes in practice, was divided. The typical holding of the villein was regulated by his capacity to furnish one or two oxen to the team. In other words, it was the "virgate" or "yardland" of 30 acres, though one-ox holdings or "bovates" of 15 acres, and even half-ox holdings, were frequent.

Villeins of the higher grade were generally distinguished from inferior orders of the semi-servile classes of the peasantry by the size of their holdings in the village farm, by the certainty of their agricultural services on the demesne, and by the obligation to do team-work rather than manual labour. The smaller the holding, the vaguer the labour obligations, the more manual the work,— the lower was the grade of the villein. Besides the villeins there were other orders of bondmen—such as the rural handicraftsmen who were specially provided with land, and the bordars and cottars, who rented particular cottages and garden ground, which often carried with them from two to five acres of arable land, together with common rights. The two latter classes, besides their obligatory manual services, probably eked out their subsistence either as hired labourers on the demesne or by supplying the labour for which their wealthier neighbours were responsible. At the bottom of the social ladder were the serfs, to whom strict law assigned no rights, though there were many varieties in their grades and position. Their chief badge of serfdom was the indeterminate character of their services—the obligation to labour in the manner, at the time, and for the wage, if any, which the lord directed. But

holding of the common-field farmer. It was in fact as much as two oxen could plough in the working year. There were, however, also " one-ox men," whose holdings of 15 acres were an eighth of a carucate, and were called in Domesday Book " bovates," and at later stages " narrow oxgangs," or " half places." Smaller holdings consisting of half bovates, like the " farthing holds " of Dorsetshire, " fardels " of Somersetshire, or " farrundells " of Gloucestershire, were by no means uncommon, and in practice there was no fixed area for the arable holdings of open-field farmers.

the serf might occupy land, own cattle, and labour for himself. Thus, out of these various classes, free and unfree, sprang small landowners, tenant farmers, copyholders,[1] and wage-earning labourers.

Round the village, or " town," in which were gathered the homesteads of the inhabitants, lay the open arable fields, which were cultivated in common by the associated partners. Here were grown the crops which Shakespeare enumerates. These were the lands " of Ceres " :

" —thy rich leas
" Of wheat, rye, barley, vetches, oats, and peas."[2]

Here, at harvest time, the yellow of the corn crops alternated with the dark and light greens of beans or peas and the brown of the bare fallows. This cultivated area, which included the driest and soundest of the land, was hedgeless, open, and unenclosed, divided by turf-grown balks into fields—two, three, or, rarely, four in number. If the former, one field lay fallow, while the other was under tillage for corn, or beans, or peas. This dual system still prevailed near Gloucester in the nineteenth century, and existed at Stogursey in Somersetshire in 1879. But from the Norman Conquest onward the three-field system was the most prevalent. Down to the middle of the reign of George III. the arable land received the unvarying triennial succession of wheat or rye, of spring crops such as barley, oats, beans, or peas, and of fallow.

---

[1] The term " copyholder " belongs to a later date. In the thirteenth century, practically all land held in villeinage, or in bondage, was held " according to the custom of the manor " (secundum consuetudinem manerii). The title was the sworn testimony of those who knew the custom. Land was said to be held not only " according to the custom of the manor," but " at the will of the lord " (ad voluntatem domini). By the thirteenth century, however, the will of the lord was no longer arbitrary, but could only be exercised according to manorial custom.

Towards the end of the fourteenth century, another expression was added. Tenants were said to hold land " according to the custom of the manor by a copy of the entry on the court roll " (per copiam rotuli curiœ). Probably it had been found that, owing to the increased mobility of the rural population, oral testimony was not always available. Hence it became the practice to enter the incidents of the tenure of customary land on the rolls of the manorial court, and tenants were called copyholders, because the copy of the entry was the evidence of their title. The words " at the will of the lord " were still retained, and it has been suggested (The End of Villainage in England, by T. W. Page, American Economic Association, May, 1900, pp. 84-5) that the use of the words indicated an increased power on the part of the lord to abolish or alter the custom.

[2] Tempest. Act iv. Sc. i. 60-1.

During these seven centuries a more scientific rotation was in some districts adopted. Thus at Aston Boges, in Oxfordshire, a fourth course was interposed. But, speaking generally, open-field husbandry rather retrograded than advanced, as the discipline of manorial officials relaxed.

Each of the three arable fields was subdivided into a number of shots, furlongs, or flats, separated from one another by unploughed bush-grown turf balks, varying in width from two to sixteen feet.[1] These flats were in turn cut up into parallel acre, half-acre, or quarter-acre, strips coinciding with the arrangement of a ploughed field into ridges and furrows. If the strips[2] were acre strips, they were a furlong in length (220 yards) and 4 rods (22 yards) in breadth. Ploughmen still measure the acre in the same way as the open-field strip. Theoretically each flat was square, with sides of 40 poles, containing 10 acres ; in practice every variety of shape and admeasurement was found. But, though the pole from which the acre was raised varied from the $13\frac{1}{2}$ feet of Hampshire to the 24 feet of Cheshire, two sides of the flats always ran parallel. Thus each of the three arable fields resembled several sheets of paper, cut into various shapes, stitched together like patch-work, and ruled with margins and lines. The separate sheets are the flats ; the margins are the headlands running down the flats at right angles to, and across the ends of, the parallel strips which are represented by the spaces between the lines. The lines themselves are the " balks " of unploughed turf, by which the strips were divided from each other. The strips appear under different names. For instance, in Scotland and Northumberland they were called " rigs " ; in Lincolnshire " selions " ; in Nottinghamshire " lands " ; in Dorsetshire " lawns " ; in North Wales " loons " ; in Westmorland " dales," and their occupiers " dalesmen " ; in Cambridgeshire " balks " ; in Somersetshire " raps " ; in Sussex " pauls " ; elsewhere in southern counties " stitches." When the strips were stunted by encountering some obstacle, such as a road or river, they were called " butts." [3] Stray odd corners which did not fit in with the parallel arrangement of the flats were " crustæ," [4] that

---

[1] The balks appear under a variety of names, such as "rames," "reins," "walls," "meres," "lynches," "lantchetts," "landshares," "launchers," or "edges."

[2] The strips rarely seem to possess areas that are fractions or multiples of the statutory acre ; they may be related to the local customary acre but often do not suggest any common unit.

[3] As in Newington Butts.

[4] *Registry of Worcester Priory* (Camden Society), 1865, p. 18A.

is, pieces broken off, "pightels," "gores,"[1] "fothers,"[2] and "pykes," because, as Fitzherbert explains, they were "often brode in the one ende and a sharpe pyke in the other ende."

The arable fields were fenced against the live-stock from seed-time to harvest, and the intermixed strips were cultivated for the separate use of individuals, subject to the compulsory rotation by which each of the three fields was cropped. On Lammas Day separate user ended, and common rights recommenced; hence fields occupied in this manner were, and are, called Lammas Lands or "half-year lands." After harvest the hayward removed the fences, and the live-stock of the community wandered over the fields before the common herdsman, shepherd, or swineherd. The herdsman, in the reign of Henry VIII., received 8d. a year for every head of cattle entrusted to his care, and the swineherd 4d. for every head of swine. When sheep were folded on the cultivated land, each farmer provided, during the winter months, his own fold and fodder for his flock. Richard Hooker, while he held the country living of Drayton Beauchamp in Buckinghamshire, was found by two of his former pupils, "like humble and innocent Abel, tending his small allotment of sheep in a common field." That no occupier might find all his land fallow in the same year, every one had strips in each of the three arable fields. If the holding of the open-field farmer consisted of thirty acres, there would thus be ten acres in each field. In other words, he would have ten acres under wheat and rye, ten acres under spring crops, and ten acres fallow. The same care was taken to make the divisions equal in agricultural value, so that each man might have his fair proportion of the best and worst land. To divide equally the good and bad, well and ill situated soil, the bundle of strips allotted in each of the three fields did not lie together, but was intermixed and scattered.

In the lowest part of the land—if possible along a stream—lay the "ings," "carrs," "leazes," or meadows, annually cut up into lots or doles, and put up for hay. These doles were fenced off to be mown for the separate use of individuals either from Candlemas (February 2), or, more usually, from St. Gregory's Day (March 12)

---

[1] As in Kensington Gore.
[2] Cf. Chaucer (Prologue, 530):

"A ploughman was his brother,
That hadde y-lad of dong ful many a fother,"

where the word is generally taken to mean a load.

to Midsummer Day ; from July to February, or later, they were open, common pasturage. Sometimes the plots, which varied in size from a half-acre downwards, went with the arable holdings, so that the same man annually received the same portion of meadow. Sometimes the plots were balloted for every year. Each lot was distinguished by a name, such as the cross, crane's foot, or peel, *i.e.* baker's shovel, which will often explain puzzling field-names. Corresponding marks were thrown into a hat or bag and drawn by a boy. This balloting continued up to the last century in Somersetshire, and still continues at Yarnton in Oxfordshire.[1] After the hay had been cut and carried, the meadows reverted to common occupation, and were grazed indiscriminately by the live-stock of the village, till they were again fenced off, allotted, and put up for hay.

On the outskirts of the arable fields nearest to the village lay one or more " hams " or stinted pastures, in which a regulated number of live-stock might graze, and therefore supplying superior feed. Brandersham, Smithsham, Wontnersham, Herdsham, Constable's Field, Dog Whipper's Land, Barber's Furlong, Tinker's Field, Sexton's Mead, suggest that sometimes special allotments were made to those who practised trades of such general utility as the stock-brander, the blacksmith, the mole-catcher, the cowherd, the constable, the barber, the tinker, and the sexton. The dog-whipper's usefulness is less obvious ; but possibly he was employed to prevent the live-stock from being harried by dogs. Even the spiritual wants of the village were sometimes supplied in the same way. Parson's Close and Parson's Acre are not uncommon. It is significant that no schoolmasters seem to have been provided for by allotments of land.

Besides the open arable fields, the meadows, and the stinted hams, there were the common pastures, fringed by the untilled wastes which were left in their native wildness. These wastes provided fern and heather for litter, bedding, or thatching ; small wood for hurdles ; tree-loppings for winter browse of live-stock ; fuzre and turves for fuel ; larger timber for fencing, implements, and building ; mast, acorns, and other food for the swine. Most of these smaller rights were made the subject of fixed annual payments to the manorial lord ; but the right of cutting fuel was generally attached to the occupation, not only of arable land, but

[1] As described by R. H. Gretton in *The Economic Journal* for March, 1912.

of cottages. The most important part of these lands were the common pastures, which were often the only grass that arable farmers could command for their live-stock. They therefore formed an integral and essential part of the village farm. No rights were exercised upon them by the general public. On the contrary, the commons were most jealously guarded by the privileged commoners against the intrusion or encroachments of strangers. The agistment of strange cattle or sheep was strictly prohibited : commoners who turned out more stock than their proper share were " presented " at the manorial courts and fined ; cottages erected on the commons were condemned to be pulled down ; the area within which swine might feed was carefully limited, and the swine were to be ringed.[1] Those who enjoyed the grazing rights were the occupiers of arable land, whose powers of turning out stock were, in theory, proportioned to the size of their arable holdings, and the occupiers of certain cottages, which commanded higher rents in consequence of the privilege. It was on these commons that the cattle and sheep of the village were fed. Every morning the cattle were collected, probably by the sound of a horn, and driven to the commons by the village herdsman along drift ways, which were enclosed on either side by moveable or permanent fences to keep the animals from straying on to the arable land. In the evening they were driven back, each animal returning to its own shelter, as the herd passed up the village street. Similarly, the sheep were driven by the village shepherd to the commons by day, and folded at night on the wheat fallows. Sheep were the manure carriers, and were prized as much for their folding quality as for their fleeces. In some districts they were kept almost entirely for their agricultural value to the arable land. Until the winter they were penned in the common fold on the fallows or the stubbles. After the fallows had been ploughed, and before the crops on the other fields were cleared, they had only the commons. During winter each commoner was obliged to find hay for his sheep and his own fold, the common shepherd penning and folding them so as gradually to cover the whole area.

The open-field system, thus briefly sketched with its arable, meadow, and permanent pasture land, prevailed at some time or

---

[1] The Regulations for " Common Rights at Cottenham and Stretham " are printed by Dr. Cunningham in the *Camden Miscellany*, vol. xii. (1910), pp. 173-296.

other throughout England, except perhaps in the south-west. The following description of the crofters' holdings in Skye in 1750 might have been written, with but few alterations, of half the cultivated area of England in the eighteenth century : " A certain number of tacksmen formed a copartnery and held a tract of land, or township, for which they paid tribute to the chief, and each member was jointly and severally responsible. The grazing was in common. All the arable land was divided into ridges, assigned annually by lot among the partners. Each might have a dozen or more of these small ridges, and no two contiguous except by accident ; the object being to give each. partner a portion of the better and inferior land. The copartner appears to have had cotters under him, for whose work he paid." The prevalence of the system may still be traced with more or less distinctness in rural England. The counties in which it was most firmly established are counties of villages, not of scattered farmsteads and hamlets. Turf balks and lynches record the time when " every rood of ground maintained its man." Irregular and regular fences, narrow lanes and wide highways, crooked and straight roads, respectively suggest the piecemeal or the wholesale enclosure of common fields. The waving ridges on thousands of acres of ancient pasture still represent the swerve of the cumbrous village plough with its team of eight oxen. The age of the hedgerow timber sometimes tells the date of the change. The pages appropriated to hedges by agricultural writers of the eighteenth century indicate the era of the abolition of open fields, and the minuteness of their instructions proves that the art of making hedges was still in its infancy. The scattered lands of ordinary farms, compared with the compact " court," " hall," or " manor " farm, recall the fact that the lord's demesne was once the only permanent enclosure. The crowding together of the rural population in villages betrays the agrarian partnership, as detached farmsteads and isolated labourers' dwellings indicate the system by which it has been supplanted. *

Accurate comparison between the conditions of the rural population in the thirteenth and twentieth centuries seems impossible. Calculations based on the prices of commodities, involving, as they must, the translation of the purchasing power of mediaeval money into its modern equivalent, are necessarily guess-work. They are also to a great extent irrelevant, for few of the necessaries of life were ever bought by the cultivators of the soil, and whether the corn that they raised was fetching 3s. or 6s. the quarter in a distant market made little difference to the inhabitants of villages. They grew it for their own consumption. Owing to difficulties of communication, every village raised its own bread-supply. Hence a great extent of land, which from a farming point of view formed an excessive proportion of the total area, was tilled for corn, however unsuited it might be for arable cultivation. As facilities of transport increased, this necessity became less and less paramount. Land best adapted to pasture no longer required to be ploughed, but might be put to the use for which it was naturally fitted. Improvements in means of communication were thus among the changes which helped to extinguish village farms. But for the time, and so long as the open-field system prevailed, farming continued to be in the main a self-sufficing industry. Except for the payment of rent, little coin was needed or used in rural districts. Parishes till the middle of the eighteenth century remained what they were in the thirteenth century—isolated and self-supporting. The inhabitants had little need of communication even with their neighbours, still less with the outside world. The fields and the live-stock provided their necessary food and clothing. Whatever wood was required for building, fencing, and fuel was supplied from the wastes. Each village had its mill, and nearly every house had its oven and brewing kettle. Women spun and wove wool into coarse cloth, and hemp or nettles into linen ; men tanned their own leather. The rough tools required for cultivation of the soil, and the rude household utensils needed for the comforts of daily life, were made at home. In the long winter evenings, farmers, their sons, and their servants carved the wooden spoons, the platters, and the beechen bowls. They fitted and riveted the bottoms to the horn mugs, or closed, in coarse fashion, the leaks in the leathern jugs. They plaited the osiers and reeds into baskets and into " weeles " for catching fish ; they fixed handles to the scythes, rakes, and other tools ; cut the flails from holly or thorn, and fastened them with thongs to the staves ; shaped the teeth for rakes and harrows from ash or willow, and hardened them in the fire ; cut out the wooden shovels for casting the corn in the granary ; fashioned ox-yokes and bows, forks, racks, and rack-staves ; twisted willows into scythe-cradles, or into traces and other harness gear. Travelling carpenters, smiths, and tinkers visited detached farmhouses and smaller villages, at rare intervals, to perform those parts of the work which needed their professional skill. But every village of any size found employment for such trades as those of the smith and the carpenter, and the frequency with which "Smiths Ham" appears among field names suggests the value which the inhabitants attached to the forge and the anvil. Meanwhile the women plaited straw or reeds for neck-collars, stitched and stuffed sheepskin bags for cart-saddles, peeled rushes for wicks and made candles. Thread was often made from nettles. Spinning-wheels, distaffs, and needles were never idle. Homemade cloth and linen supplied all wants. Flaxen linen for boardcloths, sheets, shirts or smocks, and towels, as the napkins were called, on which, before the introduction of forks, the hands were wiped, was only found in wealthy houses and on special occasions. Hemp, in ordinary households, supplied the same necessary articles, and others, such as candle-wicks, in coarser form. Shoethread, halters, stirrup-thongs, girths, bridles, and ropes were woven from the "carle" hemp ; the finer kind, or "fimble" hemp, supplied the coarse linen for domestic use, and "hempen homespun"[1] passed into a proverb for a countryman. Nettles were also extensively used in the manufacture of linen ; sheets and tablecloths made from nettles were to be found in many homes at the end of the eighteenth century. The formation of words like spinster, webster, lyster, shepster, maltster, brewster, and baxter indicated that the occupations were feminine, and show that women spun, wove, dyed, and cut out the cloth, as well as malted the barley, brewed the ale, and baked the bread for the family.

[1] *Midsummer-Night's Dream*, Act iii. Sc. 1.

---

* Though the question requires further investigation there is evidence that the open field farming with its system of strips was far from being universal in England. This form of tenure was associated only with certain of the Teutonic tribes who settled in the east and midlands of England at an early date, in many cases probably before the Roman conquest. With other tribes, *e.g.* the Jutes in the south-eastern counties, settlement appears always to have been on separate holdings (see Joliffe. *Pre-feudal England: The Jutes.* Oxford, 1933). In many parts of the country the evidences of "stripping" are few, nor can enclosures be traced. The manorial system was, therefore, something superimposed upon various forms of tenure.

CHAPTER II

THE BREAK-UP OF THE MANOR. 1300-1485.

Great landlords as farmers : horrors of winter scarcity : gradual decay of
the manorial system and the increased struggle for life : aspects of the
change : common rights over cultivated and uncultivated land : tendency
towards separate occupation : substitution of labour-rents for money-
rents; the Black Death; Labour legislation, and its effect; Manor of Castle
Combe and Berkeley Estates ; new relations of landlords and tenants
substituted for old relations of feudal lords and dependents ; tenant-
farmers and free labourers ; leases and larger farms ; increase of separate
occupations : William Paston and Hugh Latimer ; wage-earning labourers ;
voluntary surrender of holdings ; freedom of movement and of contract.

CHANGES in farming practices are always slow ; without ocular
demonstration of their superiority, and without experience of
increased profits, new methods are rarely adopted.   In the Middle
Ages agriculture was a self-supporting industry rather than a
profit-making business.   The immediate neighbourhood of large
towns created markets for the surplus produce that remained after
satisfying the needs of the cultivators of the soil.   But remoter
villages contained neither buyers of produce nor pioneers of improvements.   Edward I. was a gardener, and Edward II. a farmer,
horse-breeder, and thatcher.   These royal tastes may have set the
fashion.   Here and there great lay landowners, as well as great
ecclesiastics, actively interested themselves in farming progress.
Thomas, first Lord Berkeley, who held the family estates from 1281
to 1321, encouraged his tenants to improve their land by marling,
or by taking earth from the green highways of the manors.   Another
famous farmer was his grandson, the third Lord (1326-61).   Feudal
barons are rarely represented as fumbling in the recesses of their
armour for samples of corn.   But " few or noe great faires or
marketts were in those parts, whereat this lord was not himself, as
at Wells, Gloucester, Winchcomb, Tetbury, and others ; where also
hee new bought or changed the severall grains that sowed his

arrable lands."[1] These mediaeval prototypes of "Farmer George," of "Turnip" Townshend, or of Coke of Norfolk were rare. Few of the baronial aristocracy verified the truth of the maxim that "the master's foot fats the soil." The strenuous idleness or the military ardour of youthful lords was generally absorbed in field sports and martial exercises—in tilting at the ring, in hawking, hunting the buck, or lying out for nights together to net the fox. Grown to man's estate, they congregated for a month at a time at "tylts, turnaments, or other hastiludes," or exchanged the mimicry of war for its realities in France, or on the borders of Scotland and Wales. Most of the lay barons rebelled against the minute and continuous labour of farming, and this contempt for bucolic life may be illustrated from heraldry. Its emblems are drawn from sport, war, mythology, or religion. Products and implements of husbandry are despised, unless, like the "garb" or sheaf of the Washbournes, the scythe of the Sneyds, or the hay-wains of the Hays, they had been ennobled by martial use.

Few landowners, except the wealthiest, had as yet built permanent residences on their distant estates. Content with temporary accommodation, they travelled with their households and retinues from manor to manor, and from farmhouse to farmhouse, in order to consume on the spot the produce of their fields and live-stock. It was the practice of the first Lord Berkeley to go "in progress from one of his Manor and farmehouses to an other scarce two miles a sunder, making his stay at each of them . . . and soe backe to his standinge houses where his wife and family remayned . . . sometymes at Berkeley Castle, at Wotton, at Bradley, at Awre, at Portbury, And usually in Lent at Wike by Arlingham, for his better and neerer provision of Fish." His example was followed by his successors. But in the frequent absences of manorial lords on military service at home or abroad, their wives played important parts in rural life. Joan, wife of the first Lord Berkeley, "at no tyme of her 42 yeares mariage ever travelled ten miles from the mansion houses of her husband in the Countyes of Gloucester and Somersett, much lesse humered herselfe with the vaine delights of London and other Cities." She spent much of her time in supervising her "dairy affairs," passing from farmhouse to farmhouse, taking account of the smallest details. The family tradition

---

[1] *The Lives of the Berkeleys*, by John Smyth of Nibley, ed. Maclean (1883), vol. i. p. 300

lingered long. The same housewifely courses were followed by the widowed Lady Berkeley, who administered the estates during her son's minority in the reigns of Henry VIII. and Edward VI., and died in 1564. At all her country houses she "would betimes in Winter and Somer mornings make her walkes to visit her stable, barnes, dayhouse, pultry, swinetroughs, and the like." Her daughter-in-law's tastes were different. She was a sportswoman, delighting in buck-hunting, skilled with the cross-bow, an expert archer, devoted to hawking, commonly keeping "a cast or two of merlins, which sometimes she mewed in her own chamber, which falconry cost her husband each yeare one or two gownes and kirtles spoiled by their mutings." Well might the elder lady "sweare, by God's blessed sacrament, this gay girle will begger my son Henry!"

Great ecclesiastics made their progresses from manor to manor like the lay barons, and for the same reason. But in many instances monks were resident landowners, and by them were initiated most of the improvements which were made in the practices of mediaeval farming. They studied agriculture in the light of the writings of Cato, Varro, and Columella : the quaintly rhymed English version of Palladius was probably the work of an inmate of a religious house at Colchester ; the *Rules* for the management of a landed estate are reputed to be the work of one of the greatest of thirteenth century churchmen, Robert Grosteste, Bishop of Lincoln ; Walter of Henley is said to have been a Dominican, and manuscripts of his work, either in the original Norman French or translated into English or Latin, found a place in many monastic libraries. Throughout the Middle Ages, both in England and France, it was mainly the influence of the monks which built roads and bridges, improved live-stock, drained marshes, cleared forests, reclaimed wastes, and brought barren land into cultivation.

Large improvements in the mediaeval methods of arable farming were impossible until farmers commanded the increased resources of more modern times. There was little to mitigate, either for men or beasts, the horrors of winter scarcity. Nothing is more characteristic of the infancy of farming than the violence of its alternations. On land which was inadequately manured, and on which neither field-turnips nor clovers were known till centuries later, there could be no middle course between the exhaustion of continuous cropping and the rest-cure of barrenness. The fallow was *un véritable Dimanche accordé à la terre*. As with the land, so with its products. Famine trod hard on the heels of feasting. It was not only that prices rose and fell with extraordinary rapidity ; but both for men and beasts the absolute scarcity of winter always succeeded the relative plenty of autumn. Except in monastic granges no great quantities of grain were stored, and mediaeval legislators eyed corn-dealers with the same hostility with which modern engineers of wheat corners are regarded by their victims. The husbandman's golden rule must have been often forgotten— that at Candlemas half the fodder and all the corn must be untouched. Even the most prudent housekeepers found it difficult always to remember the proverbial wisdom of eating within the tether, or sparing at the brink instead of the bottom. Many, like Panurge, eat their corn in the blade. Equally violent were the alternations in the employment afforded by mediaeval farming. Weeks of feverish activity passed suddenly into months of comparative indolence. Winter was in fact a season to be dreaded alike by the husbandman and his cattle, and it is not without good cause that the joyousness of spring is the key-note of early English poetry.

Under the conditions which prevailed in the fourteenth and fifteenth centuries, little advance in farming practices could be expected. During the greater part of the period, therefore, the history of agriculture centres round those economic, social, and political changes which shaped its future progress. Under the pressure of these influences the structure of feudal society was undermined. The social mould, in which the mediaeval world had been cast, crumbled to powder under a series of transformations, which, though they worked without combination or regularity, proved to be, from the latter half of the fourteenth century onwards, collectively and uniformly irresistible. From within, as well as from without, the manor as an organisation for regulating rural labour and administering local affairs was breaking up. As money grew more plentiful, it became more and more universally the basis on which services were regulated. Commerce, as it expanded, created new markets for the sale of the produce of the soil. Parliament assumed new duties ; the Royal Courts of Justice extended their jurisdiction ; and, as a consequence, manorial courts lost some of their importance in matters of local self-government. Land was beginning to be regarded as a source of income, not of military power. As landowning became a business and farming a trade, agricultural progress demanded less personal dependence, a freer hand, a larger scope for individual enterprise. The foundations of feudalism were thus shaken, though the Hundred Years' War maintained its superstructure intact. It is this contrast between reality and appearance which gives an air of hollowness and artificiality to the splendour of the reign of Edward III.

The break-up of the manorial system accompanied the transition from an age of graduated mutual dependence towards an age of greater individual independence. It meant the removal of restrictions to personal freedom, the encouragement of individual enterprise, the establishment of the principle of competition in determining both money rents and money wages. From another point of view the results were not entirely advantageous. Against the older system it might be urged that it created a lack of opportunity which caused local stagnation. In its favour might be pleaded that it maintained a certain level of equality among the households in village communities, presided over by the lord of the manor. Now, however, the struggle for life becomes intensified ; the strong go to the front, the weak to the wall ; for one man who rises in the social scale, five sink. Here, one prospers, laying field to field, adding herd to herd and flock to flock. Here, others sell their live-stock, yield their strips of land to their more enterprising neighbour, and become dependent upon him for employment and wages. From the fourteenth century onwards the agricultural problem of holding the balance even between the economic gain and social loss of agricultural progress has puzzled the wisest of legislators.

The manorial organisation of labour suffered no sudden or universal collapse, due to any improvements in the methods or alteration in the aims of farming. It rather underwent a gradual and local decay which originated in economic, social, and political causes, and proceeded most rapidly in the neighbourhood of trading centres or sea-ports. It would be inaccurate to attempt to divide this process into successive stages, because they always overlapped, were generally simultaneous, and were often almost complete on one manor before they had begun on another. But from one point of view, the movement increased the number of holdings which were separately occupied ; in another aspect, it exchanged labour services for their cash values, and altered the relations between feudal lords and their retainers into those of employer and employed, and of the letter and the hirer of land ; in another, it applied principles of competition to money rents and money wages ; in another, it encouraged enterprising tenants to recognise that the best results of farming could only be obtained on compact holdings, large enough for the employment of money as well as of labour. The tendency towards the separate occupation and individual management of land had already begun, though it was most marked on the new land which was brought into cultivation. On the ancient arable land it was checked by the rights of common which were enjoyed, not only over the waste, but over the open arable fields. In their origin these rights were arable and attached only to arable land. Each occupier of an arable holding was entitled to graze on the common pastures the horses and oxen required for his tillage operations, and to feed the sheep needed for manuring his cultivated land. Without this right the associated partners in the common venture of farming would have had no means of supporting their beasts after the crops were sown. Common rights of pasture were therefore integral portions of, and essential adjuncts to, the ancient tillage system.[1] No rights of common of pasture could be claimed by the general public. The only persons by whom they could be acquired and enjoyed were the occupiers of arable holdings. It was as occupiers of portions of the tilled land, which was in fact or in theory attached to their homes, that cottagers claimed and exercised grazing privileges. On most manors three distinct kinds of common rights existed. The first kind is, in this connection, unimportant, though its creation marks an improvement in agricultural practices and a step towards the break-up of the early open-field system. It arose when the partners in a village farm agreed, with the sanction of the lord of the manor, to set aside a portion of their joint arable holdings for pasture, to be used in common in a " stinted " or regulated manner. " There is commonly," says Fitzherbert,[2] " a common close taken in out of the common fields by tenants of the same towne, in which close every man is stinted and set to a cer-taintie how many beasts he shall have in common." The second class of common of pasture consisted of rights enjoyed by the partners of the agrarian association over the whole *cultivated* area of the village farm, both over the arable portion that lay fallow in rotation, and over all the other arable lands and meadows, after the crops had been cleared and before the land was again sown or

---

[1] See chapter i. pp. 23-27.

[2] *The Boke of Surveyeng and Improvementes* (1523), ed. 1539, chap. ix.

put up for hay. The third kind of common of pasture consisted of rights over that part of the manor which was neither arable nor meadow,—the outlying portions, which were left in their natural condition,—the pastures, moors, wastes, woods, and heaths, *which had never been tilled.* These rights were attached to the arable holdings of manorial tenants, and to the occupation of particular cottages on the manor, and, when the strictness of the ancient system relaxed, might also be acquired by neighbours and strangers who neither lived nor held land within the manor. " In these commons," says Fitzherbert,[1] " the lord should not be stinted because the whole common is his own."

Rights of common of pasture over cultivated or commonable land, under the second heading, were enjoyed by the partners in the village farm, were exercised in virtue of their arable holdings, were limited to the extent of the farm, and could only be extinguished by the agreement of the co-partners. But if the lord of the manor, as a partner in the farm, had allowed portions of his demesne to be intermixed with the strips of his tenants, he could withdraw those portions at will, even though their withdrawal diminished the commonable area of cultivated land. With this exception, land subject to these rights of common could not be freed by any individual tenant, unless the main body of his farming partners assented.

Rights of common of pasture over the untilled land, under the third heading, were at first confined to the occupiers of arable holdings on the manor. In process of time, however, they were less narrowly limited. They could not be enjoyed by a landless public ; but they might be exercised by persons living both within and without the manor. In the case of persons living within the manor, the enjoyment of common rights belonged to the occupation of arable holdings or of particular cottages to which arable land had been or was attached. In the case of persons living outside the manor, rights might be acquired by neighbours and strangers, either by direct grant from the lord of the manor, or, through his sufferance, by long usage. As a general rule, the number of livestock which each manorial tenant or freeholder could pasture on the wastes was fixed, or capable of being fixed, in proportion to his holding. Vaguer rights were acquired by neighbours and strangers, and it was in these cases mainly that the lord's right of enclosure was successfully resisted. At common law it seems that, against

[1] *Surveying,* chap. iv.

his own customary tenants, the lord of the manor could always enclose the wastes at pleasure. Whether before 1236 he had the same power at common law against the free tenants of the manor is disputed. Be this as it may, the Statute of Merton [1] in that year empowered the lord of the manor to enclose against his free tenants, provided enough pasture was left to satisfy his previous grants of rights of common. Fifty years later, the Statute of Westminster [2] (1285) extended the lord's right of enclosure to the case of those neighbours and strangers who had acquired grazing rights, subject to the same condition of sufficiency of pasture. Practically the existence of rights of common of pasture only prevented enclosures when the rights were enjoyed by the associated body of tenants over one another's cultivated and commonable land, or when general rights, vaguely expressed, had been acquired by strangers or neighbours over the untilled wastes of the lord of the manor. Unless a custom to the contrary could be established, an enclosure of untilled waste by the lord of the manor would be upheld in the law courts, provided that the number of live-stock which could be turned out by the commoners was certain or capable of being ascertained, and that enough pasture was left to satisfy the grazing rights.

As early as the end of the twelfth century, landlords had begun to withdraw their demesne lands from the village farm, to consolidate, enclose, and cultivate them in separate ownership. They had also pared the outskirts of their woods and chases, reclaimed and enclosed these " assart " lands, as they were called, and either added them to their demesnes or let them in several occupations. They had also begun to encourage partners in village farms [3] to agree among themselves, to extinguish their mutual rights of common over the cultivated land which they occupied, to consolidate their holdings by exchange, and to till them as separate farms. The pace at which these enclosures proceeded, and the extent to which they were carried, varied with each county and almost with each manor. But by the end of the fifteenth century, though the great bulk of the village farms remained untouched, the area of land over which manorial tenants enjoyed rights of common was considerably diminished, partly by the action of lords of the manors, partly by that of the tenants themselves. Portions of the

---

[1] 20 Hen. III. c. 4   [2] 13 Ed. I. c. 46.
[3] See chapter i. p. 6, note 1, and pp. 23-27.

untilled waste had been enclosed, reduced to cultivation, and let in separate farms to rent-paying leaseholders, and to copyholders, who were admitted to their tenancies in the Court Baron and entered as tenants on the court roll. "Many of the lordes," says Fitzherbert, "have enclosed a great part of their waste grounds, and straightened their tenants of their commons within." So also, by withdrawing those parts of the cultivated demesne which lay in the village fields, and letting them in small compact holdings, they had reduced the area of cultivated land over which common of pasture was enjoyed. Fitzherbert notes that "the mooste part of the lordes have enclosed their demeyn landes and medows, and kepe them in severaltie, so that theyr tenauntes have no comyn with them therein." Finally, the tenants themselves followed the example of their landlords. Wherever the custom of the manor permitted the practice, tenants and partners in the village farms accepted "licenses to enclose part of their arable land, and to take in new intakes or closes out of the commons," or agreed with their fellow-commoners to extinguish, temporarily or permanently, their mutual rights to graze each other's arable and meadow lands after the crops had been cleared.

At first the holdings, whether separate or associated, were, as has been previously described, rented by labour services or produce-rents. But from the latter half of the thirteenth century onwards a change had been taking place. Landowners, who were themselves exchanging their personal services for cash equivalents, needed money not only to make the purchases required by an advancing standard of living, but to satisfy the demands of the royal tax-collectors. In their land they found a new source of income. They still kept their demesnes in hand ; but they preferred to cultivate these home farms by the contract services of hired men, whether servants in husbandry or day labourers, instead of relying on the compulsory labour of tenants, which it was difficult and expensive to supervise. They were, therefore, willing to commute for money payments the team dues, and, to a less extent, the manual dues, by which much of the manorial land was rented— whether in the whole or in part, whether temporarily or permanently. Those who owed the personal services were on their side eager to pay the cash equivalents. The money payments freed them from labour obligations which necessarily interfered with their own agricultural operations, and enabled them to devote themselves, continuously and exclusively, to the cultivation of their own holdings. Their places on the demesne land were taken by wage-earning farm-servants or hired labourers, recruited from the landless sons of tenants, or from cottagers who either had no holding at all or not enough to supply them with the necessaries of life. Thus there were hired farm-servants and day-labourers cultivating the demesne land for money wages; tenants paying money rents only for their holdings; others who still paid their whole rent in produce or in labour; others whose labour services had been partially commuted for money payments, either for a period or permanently.

The local and gradual break-up of the manorial organisation of agricultural labour was accelerated by the Black Death (1348-9). Entering England through the port of Weymouth in August, 1348, the plague spread to the north before it died out in the autumn of the following year. It had been preceded by several years of dearth and pestilence, and it was succeeded by four outbreaks of similar disease before the end of the century. During its ravages it destroyed from one-third to one-half of the population. Lords of manors suffered both as owners of land and as employers of labour. Whole families were swept away, and large quantities of land were thrown on the hands of landlords by the deaths of free-holders and customary tenants without heirs or descendants. Numbers of bondmen took advantage of the general confusion, threw up their holdings, escaped into the towns, or joined the ranks of free labourers. Their derelict holdings increased the mass of untenanted land, and their flight diminished the amount of resident labour available for the cultivation of the home farm. Those tenants who remained on the manor found in the landlord's diffi-culty their opportunity of demanding increased wages, of commuting labour services for money payments,[1] of enlarging the size of their

---

[1] Before the Black Death, on 81 manors, the services of tenants supplied the necessary farm labour on the demesnes in the following proportions : on 44, the whole ; on 22, the half ; on 9, an inconsiderable portion ; on 6, all labour services were commuted. After the Black Death (1371-80), on 126 manors, the proportions were as follows : on 22, the whole ; on 25, the half ; on 39, an inconsiderable portion ; on 40, all labour services were com-muted. *The End of Villainage in England*, by T. W. Page (Publications of the American Economic Association, May, 1900), pp. 44-46, 59-65.

Miss Davenport (*The Economic Development of a Norfolk Manor*, 1906, pp. 52, 58) says that, out of 3219 services charged on the lands of Forncett Manor in 1376, only 195 were available in 1406.

holdings, of establishing the principle of competitive rents. The "Great Death" in fact produced the natural results. There was a fall in rents and a rise in wages, because the supply of land exceeded the demand, and the demand for labour was greater than the supply.

Legislation came to the aid of landowners by endeavouring to maintain the supply of labour and to regulate the rise both of wages and of prices. The statutes clearly illustrate the difficulties of landlords and consumers. The crisis was so abnormal that unusual action seemed justifiable. In the plague years of 1348-9 agricultural labour was so scarce that panic wages were asked and paid. A similar rise in prices took place simultaneously. So exorbitant did the demands both of labourers and producers appear, following as they did on a previous rise in both wages and prices, that a royal proclamation was issued in 1349. It ordered all men and women, "bond or free,"—unless living on their own resources, tilling their own land, employed in merchandise, or exercising some craft,— to work on the land where they lived at the rate of wages current in 1346. Those who gave or took higher wages were fined treble or double the sums so given or received. The claim of lords of manors to the services of their own men was acknowledged. But their claim was no longer exclusive ; they were not to employ more labour than they absolutely required. The king's proclamation was not universally obeyed. Employers had either to lose their crops or yield to "the proud and covetuous desires" of the men. They were indeed placed in a difficulty. On the one hand, men could not be hired under threepence to perform the same services which had been recently commuted for a half-penny. On the other hand, the strike was well-aimed and well-timed. It hit the most vulnerable points. The classes of agricultural labourers against whom the proclamation was specially directed were ser- vants in husbandry, mowers, reapers, and harvesters. Servants in husbandry, boarding at the home-farm or the houses of the larger tenants, were the ploughmen, carters, cowherds, shepherds, milk- maids, and swineherds, who had the care of the live-stock. They, like the harvesters, were indispensable. If the crops were not harvested when ripe, they spoiled ; if the live-stock were neglected, they died. To solve the difficulty Parliament itself intervened. The provisions of the proclamation were supplemented by the first Statute of Labourers (1349, 23 Ed. III.), and expanded by a series of Acts extending over the next 150 years.[1] The stocks, imprisonment, outlawry, and branding were the punishments of those who refused to work, or absented themselves without licence from the hundred where they lived. Every boy or girl, who had served in husbandry up to the age of twelve " at the plough or cart," was bound to " abide at the same labour." Justices, either of the Peace or under a special commission, were sworn to enforce the Acts, and to fix the rates of wages at which labourers could be compelled to serve.

How far this legislation attained its immediate ends it is difficult to say ; the repeated re-enactment of labour laws, the petitions of employers, and the preambles to successive statutes may seem to suggest that it failed. On the other hand, there is abundant evidence [2] that the law was rigorously enforced, and this would naturally be inferred from the fact that its administration was entrusted to officials who were directly interested in compelling obedience to its provisions. The rise both in wages and prices was great. But the statutes undoubtedly prevented either from reaching famine height. Whether they were completely successful or not, they embittered the relations between employers and employed, and so prepared the ground for the Peasants' Rising of 1381. Confronted by a discontented peasantry, burdened with large tracts of land which threatened to pass out of cultivation, hampered by the scarcity and dearness of labour, landlords turned in new directions for relief. Here and there, where the climate favoured the expedient, they reduced their labour-bills by laying down tracts of arable land to pasture. Elsewhere the demesnes were let off in separate farms at money rents. Often, in order to secure tenants, the land was let on the " stock and land " system, similar to that of the *métayer*, the landlord finding the stock and implements. Sometimes the entire manor was leased to one or

---

[1] *E.g.* 1360-1 (34 Ed. III. cc. 10, 11) ; 1368 (42 Ed. III. c. 6) ; 1377 (1 Ric. II. c. 6) ; 1385 (8 Ric. II. c. 2) ; 1388 (12 Ric. II. cc. 3-9) ; 1402 (4 Hen. IV. c. 14) ; 1405 (7 Hen. IV. c. 17) ; 1423 (2 Hen. VI. c. 18) ; 1427 (6 Hen. VI. c. 3) ; 1429 (8 Hen. VI. c. 8) ; 1444 (23 Hen. VI. c. 12) ; 1495 (11 Hen. VII. c. 22) ; 1496 (12 Hen. VII. c. 3) ; 1514 (6 Hen. VIII. c. 3) ; 1563 (5 Eliz. c. 4).

[2] Miss Putnam's *Enforcement of the Statutes of Labourers* (1908), (Columbia University : Studies in History, Economics and Public Law, vol. xxxii.) is an exhaustive commentary on the administration of the law from 1349 to 1360

more tenants, who paid a fixed annual rent for the whole, and then sub-let portions of the land.[1]

Two examples of this gradual transformation of the manorial system may be quoted. In the first instance—that of Castle Combe [2] in Wiltshire—the neighbourhood of a clothmaking industry may have made the process of change exceptionally rapid, even for the south of England. At the Domesday Survey the manor contained 1200 acres under the plough. Of this arable land, 480 acres were in the lord's demesne, cultivated by 13 serfs and the team and manual labour of the manorial tenants. The remainder of the arable area (720 acres) was occupied by 5 villeins, 7 bordars, and 5 cottagers. There was a wood of a mile and half in length by three quarters of a mile in breadth. There were also three water mills. The whole population consisted of bondmen : none were, in the eye of the law, free. In 1340 the tenemental land had increased to nearly 1000 acres. There were ten freemen, holding between them 247 acres of arable land. Of these freemen, one of the three millers held an estate of inheritance to himself and his heirs, at a fixed quit-rent, subject to a heriot and attendance at the manorial courts. The nine remaining freemen, among whom were the other two millers, held their land at will at fixed money rents and similar services. The rest of the inhabitants were still bondmen. Fifteen customary tenants occupied for the term of two lives 540 arable acres, in holdings of from 60 to 30 acres, partly by money rents, partly by labour services. Eleven others held 15 acres each (165 acres) for two lives, paying their rent only by labour on the demesne ; but in addition nine of them also held crofts, for which they paid annual money rents. All these classes, in virtue of their holdings, were protected against caprices of the lord's will by manorial customs. Many of them remained bondmen in status, but the condition of their tenure was raised. Eight " Monday-men " held cottages and crofts or curtilages by labour services only. These thirty-four bondmen, at the will of

[1] Thus the land of the manor of Hawsted in Suffolk was let in 1410 by Sir William Clopton to Walter Bone, Sir William reserving the manor-house and the fines and other legal rights of a manor (*History of Hawsted*, pp. 193-5).

[2] *History of the Manor, etc. of Castle Combe*, by G. Poulett Scrope (1852). The areas are calculated on the assumption that the local " carucate " contained 120 acres. Whatever the actual acreage may have been, the proportions remain the same.

the lord, could buy themselves out of their labour obligations on payment of the cash values which are entered against their services in the steward's book. In this event substitutes were provided in the twelve cottagers, who paid a fixed money rent for their cottages. Immediately after the " Great Death " the final stage is reached. In 1352 the demesne was cut up into separate farms, and let on money rents. Labour services were therefore no longer needed, and were either merged in the copyhold rents or allowed to die out.

The second instance, that of the vast estates of the Berkeleys, covers a wider area. The policy adopted by the family in the management of their manors in Gloucestershire, Somersetshire, Essex, and elsewhere, was in one important respect consistent from 1189 to 1417. Throughout the whole period, successive lords aimed at increasing their enclosures. They began to withdraw those portions of the demesne which lay in " common fields, here one acre or ridge, and there an other, one man's intermixt with an other," to consolidate them, free them from common, and enclose. By exchange with free tenants, other lands were thrown together and similarly treated. The skirts of woods and chases were taken in hand, and hundreds of acres of " assart " land were enclosed. Sometimes these enclosures were made by agreement ; sometimes without. Maurice de Berkeley (1243-81) had within his manor of Hame " a wood called whitclive wood, adioinynge whereunto were his Tenants' arrable and pasture grounds and likewise of divers freeholders. This hee fancieth to reduce into a parke ; hee treateth with freeholder and tenant for buyinge or exchanginge of such of their lands lyeing neere the said wood as hee fancied : In which wood, also, many others had comon of pasture for theire cattle all tymes of the yeare, (for noe woods or grounds, in effect, till the Eve of this age, were inclosed or held in severaltie :) with theis also hee treatieth for releases of their comon : After some labor spent, and not prevailinge to such effect as hee aymed at : hee remembered (as it seemeth) the Adage, *multa non laudantur nisi prius peracta* : many actions are not praisworthy till they bee done : Hee therefore on a sodaine resolutely incloseth soe much of each man's land unto his sayd wood as hee desired : maketh it a parke, placeth keepers, and storeth it with Deere, And called it, as to this day it is, Whitclyve parke. They seeing what was done, and this lord offeringe compositions and exchanges as before, most of them soone agreed, when there was noe remedy. . . . Those few that remayned obstinate fell after upon his sonne with suites, to theire small confort and less gaines."[1]

For the first 140 years of the period (1189-1417) the lords of Berkeley steadily pursued the plan of converting customary tenancies and tenancies of newly enclosed lands into freeholds of inheritance at fixed quit-rents which represented the rack-rents then current. They seem to have feared that in future years the income of their land would fall rather than rise. Robert de Berkeley began the policy (1189-1220) ; it was continued by his successor, Maurice ; it culminated in the time of Thomas, first Lord Berkeley (1281-1321), who himself created 800 of these freeholds, many of which still remained when John Smyth wrote the history of the family in 1628. This family policy was, however, completely reversed by his grandson Thomas, third Lord Berkeley (1326-61). Many hundreds of the freeholds created by his predecessors were repurchased, and let at rack-rents. His example was, for the next half century, actively followed by his successors. But for this reversal of the family policy, Smyth calculates that three-quarters of the Berkeley Estates would have been freeholds of inheritance, paying fixed quit-rents of fourpence or sixpence an acre for land which in 1628 was worth twelve shillings.

At no time during the period (1189-1415) was any large proportion of the demesne lands divided and let on lease. The Berkeleys themselves farmed on a gigantic scale through their bailiffs and their reeves. Thus the third lord (1326-61) kept in his own hands the demesnes of upwards of 75 manors, stocking them with his own oxen, cows, sheep, and swine. On no manor did the flock of sheep number less than 300 ; on some it reached 1500. At Beverston in Gloucestershire, in the seventh year of Edward III., he sheared 5775 sheep. From these manors his supplies were drawn to feed each day at his " standing-house " 300 persons and 100 horses. Thence came every year geese, ducks, peacocks, capons, hens and chickens,—200 of each kind, many thousands of eggs and 1000 pigeons, coming from a single manor,—stores of honey, wax, and nuts, an " uncredible " number of oxen, bullocks, calves, sheep and lambs, and vast quantities of wheat, rye, barley, oats, pease, beans, apples, and pears. All was accounted for with minute detail by the stewards, reeves, and bailiffs. Their accounts for

[1] *Lives of the Berkeleys,* vol. i. pp. 140-1.

the manors and for the household show what amount of corn remained in the granary from the previous year; how much was each year reaped and winnowed, sold at markets, shipped to sea; how much was consumed in the lord's house, in his stable, in his kennels, in the poultry yard, or in the falcons' mews; how much was malted; how much was given to the poor, to friars and other religious orders by way of yearly allowances.

The policy of repurchasing freeholds and of increasing enclosures was pursued by the fourth lord (1361-68) and by his son (1368-1417). But from 1385 onwards the practice of farming the demesne lands through the reeves was abandoned. "Then," says Smyth, "began the times to alter, and hee with them (much occasioned by the insurrection of Wat Tyler and generally of all the Comons in the land,) And then instead of manureing his demesnes in each manor with his own servants, oxen, kine, sheep, swine, poultry and the like, under the oversight of the Reeves of the manors. . . . This lord began to joyst and tack in other mens cattle into his pasture grounds by the week, month, and quarter: And to sell his meadow grounds by the acre; and so between wind and water (as it were) continued part in tillage, and part let out and joysted as aforesaid for the rest of that kings raigne. And after, in the time of Henry the fourth, let out by the year stil more and more by the acre as hee found chapmen and price to his likeing." [1] The landlord was ceasing to be a patriarchal farmer and becoming only a rent-receiver. The process went on with increasing rapidity. By the end of the reign of Edward IV. the greater part of the manors and demesnes had been let to tenants, either on rack-rents or at lesser rents with the reservation of a fine. The day-works due from the old customary tenants, in proportion to their holdings of yard-lands and "far-rundells," together with their produce rents, were commuted into money equivalents and added to the new rents.

The story of the Manor of Castle Combe and of the estates of the Berkeleys holds true, with many variations, of England generally. Everywhere the cultivation of demesnes by the labour services of manorial tenants was gradually abandoned, and the older system replaced by separate farms, let for money rents to individual occupiers. The change proceeded more rapidly in the south and south-west than in the north and east. But as the fifteenth century

[1] *Lives of the Berkeleys*, vol. ii. pp. 5-6.

neared its close the relations between owners, occupiers, and cultivators of land had, in many parts of England, assumed a more
modern aspect. There was a large increase in the number of freeholders, and of leaseholding or copyholding farmers renting land in
individual occupation ; there was also an increase in the number
of free labourers whose only capital was their labour. The complete
abolition of villeinage had been demanded by the people in the
rising of 1381, and one of the principal objects of the rioters had
been the destruction of the rolls of the manor courts, which were
the evidence not only of their titles but of their disabilities. Possibly
they may have hoped that, if the court rolls were destroyed, they
would be left in undisturbed possession of their holdings. Possibly
they may have expected to escape the payment of the vexatious
fines and licences incidental to the tenure, and there is some suggestion
that landlords were endeavouring to recoup themselves for the loss
of income, which the commutation of labour services and the
decrease of the manorial population had produced, by the stricter
exaction of payments. Eighty years later the class of villeins,
which once had included the great mass of the rural population,
was fast disappearing. The more prosperous members of the class
had retained their hold on the land, whether on the demesnes, the
assart lands, or the village farms. Some had become freeholders ;
others rented their holdings at fixed money rents on leases for a
term of years or for lives ; others, whose rights were derived from
ancient customs, were admitted as copyholders for lives and possibly
of inheritance on the court roll of the manor. The uncertainty of
villein tenure was gone, and its brand of personal servitude could
not long continue when the old relation of feudal lord and dependent
was exchanged for that of landlord and tenant or of employer and
employed, and was expressed in cash instead of personal services.
Even landless bondmen had for the most part gained their personal
freedom. Some purchased freedom by money payments ; on
some the influence of the Church, or the pricking of conscience
conferred it by a deathbed emancipation ; the legal presumption
of natural liberty and the decisions of the law courts bestowed it
on others. Here a bondman escaped from the manor and was
lost sight of ; here a man took refuge in a town ; another accepted
the tawny livery of the Berkeleys or of some other great lord ; a
fourth received the tonsure, or took service in a monastery, as a
lay brother ; a fifth made freedom the condition on which he would

take up land. In numerous cases the services were lost from neglect, because they ceased to be profitable when landlords abandoned farming and became only rent-receivers. In all these ways the ranks of freemen and free labourers were recruited. The numbers of villeins dwindled fast. But the tenure survived the Tudor period. Its abolition was demanded in the eastern counties during Kett's rebellion (1549), and all men who had not been legally emancipated lived throughout the reign of Elizabeth in peril that its incidents might be revived against them. Even the old personal services still lingered. Till the end of the eighteenth century, labour dues as part of the rent of land were enforced in the north-west of England. Half the county of Cumberland was still unenclosed in 1794. " By far the greatest part of this county was held under lords of manors, by that species of vassalage, called *customary tenure*; subject to the payment of fines and heriots, on alienation, death of the lord, or death of tenant, and the payment of certain annual rents, and performance of various services, called *Boon-days*, such as getting and leading the lord's peats, plowing and harrowing his land, reaping his corn, hay-making, carrying letters, etc., etc., whenever summoned by the lord." [1]

The fifteenth century lies midway between two recognised periods of distress among the rural population. Agriculturally, its history is almost a blank. The silence has been interpreted in different ways. Some writers have considered it as a time of progress ; others have read it as the reverse. There is evidence that the principal sufferers by the dynastic and aristocratic struggle of the Roses were the nobility and the soldiers, that country districts were not laid waste, and that villages and their populations were neither destroyed nor harried. If so, rural life may have advanced peacefully, profiting by the absorption of landowners in more exciting pursuits than the administration of their estates. When once the struggle was ended, a new world began to piece itself together. Accepting the spirit of the coming age, agriculture reorganised itself on a money basis, and two classes emerge into prominence—capitalist tenant-farmers and free but landless labourers. Both had been slowly forming during the first three quarters of the century : both were equally essential to the changed conditions of farming. The tenant-farmer had risen in the social

[1] *General View of the Agriculture of the County of Cumberland*, by John Bailey and George Culley (1794), p. 11.

scale ; the labourer, if the possession of land alone measured his position in society, had fallen. Mediaeval organisations of trade were undergoing a similar transformation. Guilds, like village farms, had maintained a certain equality of wealth and position among the master craftsmen, and apprentices and journeymen not only looked to become masters themselves, but shared in the advantages of membership of the organised crafts. At the close of the fifteenth century, the wealthier liveried masters began, like capitalist tenant-farmers, to form a higher rank within the guild, and to control and administer its policy. Below them in the scale a new class was coming into existence. Independent journeymen were increasing in number—hired artisans who derived no benefits from the guilds, enjoyed no prospect of becoming master-craftsmen, and depended for their livelihood, like the free labourer divorced from the soil, on employment and wages. For the rising classes, the fifteenth century may have been a period of prosperity ; for the classes which were in some respects falling, it was probably a time of adversity. Only thus can the rose-coloured descriptions of writers like Sir John Fortescue be reconciled with the darker accounts which might be put together from other sources. It is not in the gay holiday scenes of a Chaucer, but in the grimly realistic pictures of a Langland that the features of rural life are most truly painted.

Leaseholders and copyholders in separate occupation of farms had increased rapidly in number as well as in importance. Their ranks were swollen by the tenants of the reclaimed wastes, by those among whom the demesne was now divided, and by holders of the " stock and land " leases who had saved sufficient capital to stand on their own feet ; by men of capacity and enterprise, who realised the superior advantages of a separate holding, however small ; by hundreds of the old customary tenants, who found that the rents for which their personal services had been commuted were higher than the competitive money rents which land could command when the supply was excessive. The terms for which leases ran grew longer. They advanced from a year to five years, then to seven years, then to ten years, then to twenty-one, then to lives, and often to fee farm. The increasingly prolonged term illustrates the greater confidence in the stability of the government. It also indicates, on the part of the farmer, a growing sense of the legal security which leases afforded ; on the part of landowners, the wish to retain as long as possible their responsible tenants ; and, among the more far-sighted of the tenantry, a desire to rid themselves of the imperfect ownership which customary tenure implied. Finally, farms were increasing in size. The word " farm " was itself changing its meaning from the stipulated rent to the area of land out of which the payment issued. In this transition another meaning of the word was lost. In many parts of England at that time, and in the north of England down to the last century, a farm meant that definite area of land which afforded a living to the occupier and his family.[1] By the end of the fifteenth century it had acquired its modern sense of an indefinite area of land occupied by one tenant at one rent. Complaints of the practice of throwing together a number of men's " livings " into one holding in one man's occupation begin to be frequent, and are directed against the absorption of the small arable holdings of from ten to thirty acres. They occur in sermons, in Petitions to the King, in doggerel verse. The letter of the Vicar of Quinton in Gloucestershire, written to the President of Magdalen College, Oxford,[2] at the close of the fifteenth century. breathes the spirit of the twentieth century. Magdalen College owned an estate in the parish of Quinton, and the president hesitated whether the College should let the land as one farm, or, as we should now say, let it in small holdings. The vicar appeals on behalf of his parishioners. " Aftur my sympull reson," he writes, " it is mor meritory to support and succur a comynte [community] then one mane, yowre tenan[ts] rathere then a stronge man, the pore and the innocent for [instead of] a gentylman or a gentylman's man."

Whatever may have resulted from the vicar's appeal, circumstances generally favoured the multiplication of separate holdings and their increase to a size which rendered the employment of money as well as of labour remunerative. Practical agriculturists, like Fitzherbert, urge every man to " change fields with his neighbour, so that he may lay his lands together," keep more live-stock, improve the soil by their " compostynge," and rest his corn land when it becomes impoverished. The long wars with France were over ; the civil strife between York and Lancaster was ended ; the central government under Henry VII. was firmly established ; trade was beginning to expand ; population, arrested in its increase since the death of Edward I., was once more growing. On the

---

[1] *The Ancient Farms of Northumberland*, by F. W. Dendy (1893), pp. 11-19.
[2] *England in the Fifteenth Century*, by the Rev. W. Denton (1888), p. 318

other hand, land had depreciated in value ; rents had declined ; farming had deteriorated ; useful practices had been discontinued ; cattle were dwindling in size and weight ; the common pastures had become infected with " murrain " ; the arable area of open-fields had grown less productive, and without manure its fertility could not be restored. Land was cheap to buy and cheap to rent. Enterprising purchasers and farmers could make it pay, if they realised the advantages of separate occupation, of employing money on the land, of reviving obsolete practices like marling, and, in certain climates, of adopting a convertible husbandry that adapted itself to fluctuating needs better than the open-field system, which rigidly regulated the cultivation of the soil and permanently separated arable land from pasture. The one obstacle to the success of the new tenant-farmer was the scarcity and dearness of labour. But sheep-grazing cut down labour bills, while legislation checked the natural rise of wages, and barred the outlet into towns against agricultural labourers and their sons. Even a high rate of wages often proved nominal rather than real, for, under the Statutes of Labourers, farmers had the option of paying their men in corn at the statutory price of 6s. 8d. a quarter when corn fell below that price, or in money when the price of corn approached or exceeded the statutory figure.

Two contemporary pictures have been painted of the lives of tenant-farmers, who were fathers of famous sons—one at the opening, the other at the close, of the fifteenth century. Each picture seems to be more or less typical of the farming class at the periods to which they belong. Clement Paston, at the beginning of the century, lived at the village of Paston, near Mundesley in Norfolk.[1] " He was," says an anonymous writer who was no friend to the family, " a good plain husband(man), and lived upon his land that he had in Paston, and kept thereon a plough all times in the year, and sometimes in barlysell two ploughs. The saide Clement yede (went) at one plough both winter and summer, and he rode to mill on the bare horseback with his corn under him, and brought home meal again under him, and also drove his cart with divers corns to Wynterton to sell as a good husband(man) ought to do." He had at the most 100 or 120 acres of land, some of it copyhold, and a " little poor water-mill." He married a bond-woman. Their son William, who was kept at school, often on

---

[1] *Paston Letters*, ed. Gairdner, Introduction, vol. i. pp. 28-30.

borrowed money, became a distinguished lawyer, a sergeant-at-law, in 1429 a Judge of the Common Pleas, and the founder of the Paston family. At the close of the same century, Hugh Latimer the father of the Bishop of Worcester, was a farmer in Leicestershire. Preaching before Edward VI.,[1] the son describes his father's circumstances. The elder Latimer rented some 200 acres of arable land with rights of common of pasture, employed half a dozen men on his farm besides women servants, ran 100 sheep, milked 30 cows, owned oxen for ploughing, and a horse for riding or for the king's service. He portioned his daughters with £50 or £60 apiece ; and, besides teaching his son to " lay his body in the bow," sent him to school and college. He was hospitable to his neighbours and charitable to the needy. And this he did out of the profits of his farm.

For wage-earning landless labourers, the last 130 years of the period from 1200 to 1485 were probably, in some respects, unprosperous. They now were exposed to the fluctuations, not only of the price of necessaries, but of the labour market. Yet agricultural change had not affected them wholly for the worse. The bright side was the bondman's passage towards personal freedom ; the darkest feature was his divorce from the soil. To some extent his severance from the land was the means and the price of his personal emancipation.

.The surrender of the hold on the land was, at this period, mainly due to voluntary action by the villeins themselves ; it was not caused by clearances for sheep farming. A landlord had no desire to lose them either as tenants or as labourers. Their flight threw more land on his hands, and at the same time increased the scarcity of labour for its cultivation. But villeins, whose holdings were small, had little inducement to retain them, and much to gain by escape. The sentimental objection to the tenure had been deepened and embittered by the teaching of wandering friars and " poor preachers." Freedom meant the rise out of a condition, the degradation of which they had begun to feel with a new acuteness. It meant also new possibilities. Beyond the limits of their own manor, they might, as freemen, acquire other holdings, or join the ranks of free labourers, or settle behind a city wall and practise some handicraft. After the " Black Death " the prospect of employment in towns was good. Hands were at a premium. The great

---

[1] *Sermons* (Parker Society), p. 101.

scarcity of labour is proved by the fact that the severity of the Labour Statutes was relaxed in the case of immigrants into London, and, temporarily, into Norwich. That the chances of town life were in themselves sufficient inducements for flight from the manor is shown by the willingness of villeins to surrender their holdings, and purchase licences to live within the walls of cities. But very often another cause must have made the voluntary severance from the land a Hobson's choice. The yield of arable land on open-field farms was so small that farming scarcely provided necessaries. Throughout the closing years of the fifteenth century, successive outbreaks of murrain had killed numbers of cattle and sheep, swept off geese and poultry, and even destroyed the bees. If the results of similar outbreaks in the sixteenth century justify the conclusion, it may be supposed that it was the live-stock of open-field farmers which suffered most. Without stock small holders or cottagers found common rights valueless, and their few acres of arable land rather a burden than a profit. To such men the voluntary surrender of holdings, with or without flight, might well seem the choice of a lesser evil. For a time they may have prospered as labourers for hire. But when the conversion of tillage to pasture had begun, their daily employment and their harvest earnings were in peril. In such conditions it must have been useless, if not impossible, to enforce residence within the limits of the manor.

The possibility that the manor itself might not provide work for its inhabitants was recognised in the labour legislation of the period. Indirectly the Labour Statutes, though manifestly not passed in the interest of labourers, aided their progress towards freedom of movement and of contract. They broke down the exclusive right which lords of the manor claimed over the personal services of their manorial dependents. Hitherto no one could employ a villein from another manor without the risk that this superior claim might be asserted. Under the king's proclamation of 1349, the lord's right is recognised, preferentially, but not exclusively. He has the first claim, not the only claim, to the services. He may not employ more labour than he absolutely needs. When his requirements are satisfied, his villein may, and on demand must, work for other employers. In the statutes themselves the same principle is carried further. Servants in husbandry are bound to appear, tools in hand, in market towns to be publicly hired, as, five centuries later in many parts of England, they frequented the local statute fairs, or mops—cowmen with the hair of cows twisted in their button-holes, or carters and ploughmen with whip-cord in their hats. Thus the very legislation which was designed to maintain the supply of rural labour and check migration into towns, introduces that principle of freedom of movement which is essential to the modern relations of employer and employed. In another respect, also, the Labour Statutes loosened the dependence of bondmen on their manorial lords. The jurisdiction of the king's law courts was extended till it invaded the sacred precincts of the manor court, and settled disputes between the lord and his villeins. Wages even were no longer to be fixed as between a bondman and his feudal lord ; they were to be controlled by Justices of the Peace acting as the king's agents. It is not suggested that the fifteenth century labourer benefited by a change which virtually transferred the right of fixing wages to an association of employers. But the transfer of authority was a not unimportant step towards the complete collapse of the manorial organisation, and towards free competition as the true basis of money wages.

CHAPTER III

FARMING FOR PROFIT: PASTURE AND SHEEP-GRAZING. 1485-1558

The passing of the Middle Ages : enclosures in the sixteenth and eighteenth centuries compared ; the commercial impulse and its results ; conversion of tillage to pasture : enclosures and depopulation : legislation against enclosures ; literary attack on enclosures ; the practical defence of enclosures : larger farms in separate occupation : loss of employment ; enclosures equitably arranged, or enforced by tyranny ; legal powers of landowners ; open-field farmers not the chief sufferers by enclosures ; scarcity of employment and rise in prices ; the new problem of poverty : the ranks of vagrants ; the Elizabethan fraternity of vagabonds.

OUT of wars at home and abroad, and pestilences destructive both to man and beast, emerged one great agricultural change which by 1485 was practically completed. Feudal landowners, instead of pursuing the patriarchal system of farming their own demesnes by the labour services of their dependents, had become receivers of rent. Home-farms and " assart " or reclaimed lands were cultivated, not by lords of the manor through bailiffs and labour-rents, but by freeholders, leaseholders, copyholders, and hired labourers. Further changes were close at hand. With the dawn of the Tudor period began the general movement which gradually transformed England into a mercantile country. The amount of money in actual use was increasing ; men possessed more capital, could borrow it more easily, and lay it out to greater advantage. Commerce permeated national life. Feudalism was dead or dying, and trade was climbing to its throne. The Middle Ages were passing into modern times.

On the agricultural side, the spirit of trading competition gave fresh impulse to an old movement which, in spite of a storm of protest, continued in activity throughout the Tudor period, and, after a century and a half of silent progress, became once more the centre of literary controversy before it triumphed at the close of the reign of George III. That movement is described as enclosure, and it is generally treated as necessarily destructive to the old village farms. But the word includes various processes, some of which rather strengthened than weakened the open-field system. Some enclosures, such as closes for stock-feeding, intakes from the common for arable purposes, even the not uncommon practice of fencing portions of the open-fields for several occupation, whether temporarily or permanently, were really efforts to adapt village farms to changing needs. Another form of enclosure was the cultivation of new land obtained by clearing forests, approving portions of wastes, or draining fens. Here also village farms were not directly affected. Indirectly, indeed, these new enclosures produced a considerable effect. Much of the reclaimed land was tilled for corn ; thus the ancient arable soil was relieved from the former necessity of bearing grain crops, and might not improbably be put to the use for which it was best adapted. A third process was the direct enclosure of open-fields and pasture commons. This form generally appeared in the neighbourhood of towns, where the demand for animal food and dairy produce was greatest and labour found a ready market, or in counties where some manufacturing industry prevailed and small grass holdings made a less exacting claim on the time of the handicraftsmen than tillage. But whatever form the enclosure took, the general drift of the movement was towards individual occupation of land. It was therefore always, and particularly in the sixteenth and eighteenth centuries, directly opposed to the open-field system of farming in common.

At both periods that special form of enclosure was prominent which meant the break-up of the mediaeval agrarian partnerships and the substitution of private enterprise for the collective efforts of village associations. But in details the earlier and the later movements were strongly contrasted. In the sixteenth century, the change was opposed and partially arrested by legislation ; in the eighteenth century, it received from Parliament encouragement and support. Under Henry VIII., it was mainly inspired by commercial advantage ; under George III., it was alleged to be enforced by necessity. In the sixteenth century some of the grass-land was undoubtedly used for grazing beasts. But it was mainly to supply the growing wool trade that Tudor husbandmen substituted pasture for tillage, sheep for corn. They took their seats on the woolsack, and maidens of all degrees were spinsters. Hanoverian farmers reversed the process ; they valued sheep for their mutton instead of their fleeces, and concentrated their energies on the production of bread and meat for the teeming populations of manufacturing cities. Dearth of bread was in Tudor times the most effective cry against enclosures ; under George III. it was the unanswerable plea for their extension. At the opening of the sixteenth century, enclosure did not always mean improved farming ; the conversion of arable land into inferior sheep-walk was rather retrogression than progress. At the close of the eighteenth century, it at least meant the opportunity for advance and for the introduction of better practices. To some extent, indeed, the different developments of the two movements measure the improvements in the methods and the increase in the resources of Hanoverian farmers. The Tudor husbandman might devote himself exclusively to the one or the other of the two branches of farming ; but he had not mastered the secret of their union. If he changed from tillage to pasture, he did so completely. He could not, like his successor, combine the two, and by the introduction of new crops, at once grow more corn and carry more stock.

Agriculturally, the period which opens with the Battle of Bosworth and ends with the early years of Elizabeth is one of transition towards the modern spirit and forms of land cultivation. Like all transition periods, it is full of suffering for those who were least able to adapt themselves to altered conditions. The ruin of noble families by the Wars of the Roses, the lavish expenditure which Henry VIII. made fashionable, the rise in prices, and the difficulty of raising rents, compelled many " unthrifty gentlemen " to sell their estates. The break-up of landed properties and their passage into new hands favoured the introduction of the commercial impulse. The landholders whose " unreasonable covetousness " is most loudly condemned were mainly speculators in land, men who had made money in business, had capital to invest, could afford the expense of enclosures, and were determined to make their estates pay. Such were " the Merchant Adventurers, Clothmakers, Goldsmiths, Butchers, Tanners, and other Artificers," [1]—" the merchants of London " who "bie fermes out of the handes of worshypfull gentlemen, honeste yeomen, and pore laborynge husbandes."[2]   Translated

---

[1] Petition to Henry VIII. (1514), quoted by F. J. Furnivall in *Ballads from MSS.*, p. 101 (Publications of the Ballad Society, vol. i.).

[2] Thomas Lever's *Sermons* (1550) ; Arber's *Reprints*, p. 29.

into the language of to-day, the old landlords had been satisfied
to draw from their estates certain advantages and a low percentage
of profit ; the new men required at the least a four per cent. return
in money on their investments. Feudal barons had partly valued
their land for the number of men-at-arms it furnished to their
banners ; Tudor landowners appraised its worth by the amount of
rent it paid into their coffers. Mediaeval husbandmen had been
content to extract from the soil the food which they needed for
themselves and their families. Tudor farmers despised self-
sufficing agriculture ; they aspired to be sellers and not consumers
only, to raise from their land profits as well as food. As trade
expanded, and towns grew, and English wool made its way into
continental cities, or was woven into cloth by English weavers,
new markets were created for agricultural produce. Fresh in-
centives stimulated individual enterprise, and both landlords and
tenants learned to look on the land they respectively owned or
cultivated as a commercial asset.

Among the results of this conquest of agriculture by the new
spirit of commercial competition three may be noticed—firstly,
the clearer recognition of the advantages of farms held in individual
occupation, large enough to make the employment of capital
remunerative ; secondly, the substitution of pasture for tillage,
of sheep for corn, of wool for meat ; thirdly, the attack upon the
old agrarian partnerships in which lords of the manor, parsons,
freeholders, leaseholding farmers, copyholders, and cottagers had
hitherto associated to supply the wants of each village. Legislation failed to prevent a movement which harmonised and syn-
chronised with the progressive development of the nation on
commercial lines. But in its earlier stages, the consequences to
the rural population were serious. Many tenants lost their holdings, many wage-earning labourers their employment, when landlords " turned graziers," and farmers cut down their labour-bills
by converting tillage into pasture. It is impossible to doubt the
reality of the distress. From 1487 onwards, literature, pamphlets,
doggerel ballads, sermons, liturgies, petitions, preambles to statutes,
Commissions of Enquiry, Acts of Parliament, bear witness to a
considerable depopulation of country districts. In the numerous
insurrections, which marked the sixteenth century and the early
years of the reign of James I., rural distress undoubtedly contributed its share. But zealous advocates of Roman Catholicism

found it useful to ally agrarian discontent with religious reaction, and men like Protector Somerset thought it politic to attribute anti-Protestant risings entirely to agricultural causes.

There was no novelty in the withdrawal of demesne lands from the open-field farm and their partition into individual occupations ; or in fencing off portions of the home-farm and of the reclaimed " assart " lands as separate plots ; or in the appropriation of parts of the commonable waste for private use ; or in the encouragement given to partners in the village association to throw their scattered strips together into one compact holding. Each of these processes had been for many years in progress ; each had necessitated enclosures ; none had required the decay of farm-houses and cottages, loss of employment, eviction of tenants, or rural depopulation. But from the Tudor enclosing movement these consequences did necessarily result, because its objects were the promotion of sheep-farming, the conversion of tillage into pasture, the consolidation and enlargement of grass holdings. If farmers had not yet at their disposal the means of realising the full truth of the maxim that " the foot of the sheep turns sand into gold," the new commercial aristocracy were quick to see that money was to be made, or at least to be saved, by the growth of wool. It is true that down to 1540 the prices of wool remained low ; but some at least of the grass was taken up by the graziers, and the saving in labour effected by pasture farming was great. Sheep could not be herded with success on open commons, still less on the arable lands of village farms, and small holdings were incompatible with large flocks. It was these new elements which upset the calculations of agriculturists like Fitzherbert (1523), or Cardinal Pole [1] in Starkey's Dialogue (1536), or Tusser (1557), or Standish (1611), who hoped that the economic advantages of enclosure might be secured without the social loss which the conversion of large tracts of arable land into wide pasture farms inflicted on the rural population.

If evidence which is rarely impartial may be implicitly trusted, considerable tracts of cultivated land were converted into wildernesses, traversed only by shepherds and their dogs ; roofless granges and half-ruined churches alone marked the sites of former hamlets ; the " deserted village " was a reality of the sixteenth

[1] In the Dialogue between Cardinal Pole and Thomas Lupset, Pole defends enclosures for pasture on the plea that cattle, as well as corn, were necessary for human food (*England in the Reign of Henry VIII.*, ed. J. M. Cowper, E.E.T.S., extra series xxxii. 1878).

century. Already anxious for the maintenance of the national supply of corn, men began to be alarmed at another result of the movement which became increasingly prominent. John Rous [1] (1411-91), chantry priest of Guy's Cliffe and Warwickshire antiquary, was the first to protest against the decay of population caused in the midland counties by enclosures for pasture farming. To this rural exodus the attention of Parliament had been called by the Lord Chancellor in the first year of Richard III. (1484). Francis Bacon, writing of the opening years of the reign of Henry VII., says : [2] " Inclosures at that time began to be more frequent, whereby arable land, which could not be manured without people and families, was turned into pasture which was easily rid by a few herdsmen ; and tenances for years, lives, and at will, whereupon much of the yeomanry lived, were turned into demesnes. This bred a decay of people." So formidable did the danger begin to appear, that in 1489 two Acts of Parliament were passed for its prevention. The first Act was local, dealing with the effects of enclosures in the Isle of Wight from the point of view of national defence ; the second is general, directed " against the pulling down of tounes " (i.e. townships or villages). These Acts were the precursors of many others throughout the sixteenth century,[3] forbidding the conversion of arable land into pasture, ordering newly laid pasture to be restored to tillage, directing enclosures to be thrown down, requiring decayed houses to be rebuilt, limiting the number of sheep and of farms which could legally be held by one man, and imposing severe penalties for disobedience to the new provisions.

No favour was shown by Parliament to enclosers, except perhaps in the case of deer-parks. On the contrary, strenuous efforts were repeatedly made to stop the process of enclosure. Nor was the Government satisfied with passing laws and imposing penalties. Wolsey personally interested himself in enforcing obedience to the laws against the decay of houses and farm-buildings and against

[1] *Historia Regum Angliae*, ed. 1745, pp. 116-24. But Thomas Hearne was not always a reliable editor.

[2] *History of King Henry the Seventh (Works,* ed. Spedding, vol. vi. pp. 93-4).

[3] *E.g.* 1489 (4 Hen. VII. cc. 16, 19) ; 1514 (6 Hen. VIII. c. 5) ; 1515 (7 Hen. VIII. c. 1) ; 1533-4 (25 Hen. VIII. c. 13) ; 1535-6 (27 Hen. VIII. c. 22) ; 1551-2 (5 and 6 Ed. VI. c. 5) ; 1555 (2 and 3 Phil. and Mary, c. 2) ; 1562-3 (5 Eliz. c. 2) ; 1593 (35 Eliz. c. 7, repealing part of 5 Eliz. c. 2) ; 1597-8 (39 Eliz. c. 1) ; 1601 (43 Eliz. c. 9) ; in 1624 the enclosure laws were repealed.

the conversion of arable land to pasture. Active steps were taken to see that buildings were restored and enclosures and ditches levelled. In default, heavy penalties were exacted. A Commission was appointed in 1517,[1] which enquired into all cases where farm-houses had been destroyed since 1485, or where ploughs had been put down by the increase of pasture farming. Similar enquiries were held in 1548, 1566, and 1607. No doubt these strenuous efforts checked the movement. But they failed to stop it altogether. In this respect they succeeded no better in encouraging tillage than the quaint pedantry of the law, which gave arable land precedence over other land, or conferred on beasts of the plough privileges that were denied to other animals. The new legislation seems to have been satisfied, or evaded, without serious difficulty; partly, because compositions for breaches of its provisions might be paid or exemptions purchased; partly, no doubt, because the administration of the law was often entrusted to those who were interested in making it a dead letter. The destruction of farm buildings was forbidden ; but it was easy to keep within the statute by retaining a single room for the shepherd or the milk-maid ; a solitary furrow driven across newly laid pasture satisfied the law that it should be restored to tillage ; the number of sheep to be owned by one man was limited, but the ownership of flocks might be fathered on sons or servants. Down to the middle of the reign of Elizabeth the enclosing and grazing movement continued. At subsequent intervals it renewed its special activity throughout the seventeenth century, when dairying began to claim a larger share of the attention of farmers. It was restrained or encouraged rather by natural causes than by legislation. Fluctuations in the prices of wool or corn, the increased profits of improved methods of arable farming, and the restoration of the fertility of the ancient tilled land, which was brought back to the plough after an enforced rest from excessive cropping, gradually restored the preponderance of tillage over pasture.

The grievances of the rural population are to be gathered not only from legislation, proclamations, petitions, articles of complaint, the Returns of Commissioners, or the records of the law courts. They are also written large in More's *Utopia*, and in much of the ephemeral literature of the sixteenth and seventeenth centuries. The cry of the people is heard, often in exaggerated tones, in the sermons of

[1] *The Domesday of Inclosures* (1517-8), by I. S. Leadam, 2 vols. 1897.

popular preachers like Tyndale, Becon, and Latimer, in the pamphlets of such writers as Simon Fish, Henry Brinklow, or Philip
Stubbes, or in the rhymes of versifiers like " Sir " William Forrest,
Robert Crowley, and Thomas Bastard, or in such anonymous
ballads as " Nowe-a-dayes " : [1]

> " The townes go down, the land decayes ;
> Off cornefeyldes, playne layes (grass-land) ;
> Gret men makithe now a dayes
> A shepecott in the church.
>
> . . . . . .
>
> Commons to close and kepe ;
> Poor folk for bred to cry and wepe ;
> Towns pulled downe to pastur shepe ;
> This ys the new gyse ! "

Throughout the burden is the same—enclosure of commons, conversion of plough-land into pasture, sheep-farming, excessive rents,
exorbitant fines, consolidation of small holdings into large farms,
decay of houses and farm-buildings, formation of deer-parks, and.
more rarely, enclosure of open-field arable farms. Here are to be
found fierce denunciations of the " caterpillars of the common-
weal," [2] who " join lordship to lordship, manor to manor, farm to
farm, land to land, pasture to pasture," and gather many thousands
of acres of ground " together within one pale or hedge " ; or of the
unchristian landlords, who " rack and stretch out the rents of
their lands," taking " unreasonable fines," " setting their pore
tenants so straitely uppon the tenter hookes as no man can lyve
on them " ; [3] or of the insatiable " cormorants " who " let two or
three tenantries unto one man," " take in their commons " till not
so much as a garden ground is safe, and make " parks or pastures
of whole parishes " ; [4] or of the " unreasonable covitous persones
whiche doth encroche daily many ffermes more than they can be
able to occupye or mainteyne with tilth for corne as hath been used
in tymes past, forasmoche as divers of them hath obteyned and
encroched into their handes, X, XII, XIV, or XVI fermes in oon
mannes hand attons " ; [5] or of the " ambicious suttletie " of those

---

[1] " Nowe-a-dayes," *Ballads from MSS.*, ed. F. J. Furnivall (Publications of
the Ballad Society, vol. i. p. 97, 1868).

[2] Thomas Becon, *Jewel of Joy* (Parker Society, *Becon's Works*, p. 432).

[3] Philip Stubbes, *Anatomy of Abuses* (1583), (New Shakespeare Society.
p. 116).

[4] William Tyndale, *Doctrinal Treatises* (Parker Society, p. 201).

[5] *Petition to Henry VIII.*, quoted in *Ballads from MSS.*, vol. i. p. 101.

who make "one fearme of two or three," and even sometimes " bringe VI to one " ; or of the greed of " step-lords," like the " rich franklings," [1] who

> " Occupyinge a dosen men's lyvyngis
> Take all in their owne hondes alone."

Nor do the innocent causes of much of the trouble escape attack ; sheep " that were wont to be so myke and tame, and so smal eaters, now, as I heare saie, be become so greate devowerers, and so wylde, that they eate up and swallow down the very men themselfes," [2] drive " husbandry " out of the country, and thrust " Christian labourers " off the land.

> " Sheepe have eate up our medows and our downes,
> Our corne, our wood, whole villages and townes ;
> Yea, they have eate up many wealthy men,
> Besides widowes and orphane childeren ;
> Besides our statutes and our Iron Lawes,
> Which they have swallowed down into their maws :—
> Till now I thought the proverbe did but jest,
> Which said a blacke sheepe was a biting beast." [3]

Enclosers were condemned by preachers as " guilty before God of the sin in the text—' they have sold the righteous for silver and the poor for a pair of shoes.' " A playwright like Massinger did not draw entirely on his imagination, but expressed the feeling of the day when he painted his portrait of a Sir Giles Overreach, insensible to pity for his victims and justly called :

> " Extortioner, Tyrant, Cormorant, or Intruder
> On my poor neighbour's right, or grand Incloser
> Of what was common to my private use." [4]

In the passion for sheep and hedges, which changed " merrie England " into " sighing or sorrowful England," men saw the fulfilment of the prophecy " Horne and Thorne shall make England forlorne." [5] Superstitions enforced the popular judgment, and legend doomed " emparkers," like Sir John Townley, to haunt the solitudes they had created, uttering bitter cries of unavailing remorse.

---

[1] " Rede me and be nott Wrothe." By William Roy (1527), Arber's *Reprints*, 28.

[2] More's *Utopia*, bk. i. (Ralph Robynson's Translation), ed. Lupton, p. 51.

[3] Bastard's *Chrestoleros* (1598), bk. iv. Epigram 20.

[4] *A New Way to pay Old Debts*, Act. iv. Sc. 1.

[5] Francis Trigge, *Humble Petition of Two Sisters : the Church and the Commonwealth* (1604).

It was easy for popular preachers and pamphleteers to excite popular passion against the "greedy gulls" and "insatiable cormorants," who advocated and practised enclosures, and to denounce the agricultural tendencies of Tudor times as solely guided by selfish greed. But there are practical and broader sides to the question. When once land was regarded as an important asset in the wealth of the nation, national interests demanded that it should be utilised to the greatest possible advantage. Without enclosures, the soil could not be used for the purposes to which it was best adapted, or its resources fully developed. If money was to be made out of land, or if its full productive power was to be realised, it was individual enterprise alone that could make or realise either. Under the open-field system one man's idleness might cripple the industry of twenty : only on enclosed farms, separately occupied, could men secure the full fruit of their enterprise. This fact had slowly revealed itself during the last two centuries. To exchange intermixed lands, to consolidate compact holdings, and fence them off in separate occupation, had long been the aim both of landlords and tenant-farmers. Few practical men would have disputed the truth of Fuller's statement : ' The poor man who is monarch of but one enclosed acre will receive more profit from it than from his share of many acres in common with others."

Tudor agriculturists went further in their zeal for farming progress. They saw that a small enclosed plot of 15 acres could be used with less advantage than a large enclosure of 150 acres which enabled the tenant to invest money in the land, carry more stock, provide his cattle with more winter food, and, if the climate permitted, adopt convertible husbandry. This was recognised both by landowners and farmers of the progressive school, and the increased size even of arable farms continues to be a feature in sixteenth century changes. For successful sheep-farming, a large stretch of land, held in individual occupation, was still more essential. From this point of view the untilled common wastes were unprofitable. Whether land was enclosed for tillage or as sheep runs, its productiveness was increased by enclosure. Finally, the natural fertility of arable land on open unenclosed farms was becoming exhausted. The system was one of taking much from the land and putting little back. The soil, lightly ploughed, seldom manured, often foul, was in some districts worn-out. From 1349 to 1485, that is, from the Black Death to the Battle of Bosworth, its yield had declined ; its farming had deteriorated. Fitzherbert, writing in the first quarter of the sixteenth century, notes that useful agricultural practices had in many parts become obsolete, that crops were smaller, and methods of husbandry more slovenly. The fall in rentals had been general. But it was on demesne lands, or on enclosed farms, that the fall in rents had been least. These were the lands which were in the best condition, because on them most manure had been expended. Open-field farmers commanded little or no manure for their arable land, and were practically dependent on sheep for fertilising the soil. Yet in winter, animals, reduced to the lowest possible number, barely survived on straw and tree-loppings. The miserable condition of live-stock on open-field farms and commons exposed the sheep to the scab and the rot, and the cattle to the murrain. It was no uncommon spectacle to see the head of an ox impaled on a stake by the highway, as a warning that the township was infected.

Agriculturists might with good reason plead that the changes which they advocated were justified, if not necessitated, by the progress of farming. They hoped that even open-field farmers might themselves recognise the advantages of enclosure, and would agree to consolidate their intermixed holdings and extinguish their reciprocal rights of common. Fitzherbert in his *Book of Husbandry* argues strongly in favour of enclosures, and especially insists on their advantages in keeping live-stock, which, he says, thrive best and cost least on enclosed land. If a farmer has only a twenty years' lease of his land, it will pay him to go to the expense of fencing off his land in separate parcels with hedges and ditches. Common-field farmers have to pay 2d. a quarter for each head of cattle, and 1d. a quarter for each head of swine, under the care of the common herdsman and swineherd. If they wish to thrive, each must keep a shepherd of his own. The hire of the herdsman and the swineherd, together with the wages and board of the shepherd, and the cost of hurdles and stakes put together, double the rent. If a farmer encloses, he may have to pay three times this annual cost in one year ; but he has no further expense. " Than hathe he euery fyelde in seueraltie : and by the assente of the lordes and the tenauntes euery neyghbour may exchaunge landes with other. And than shall his farme be twyse so good in proffite to the tenaunte as it was before, and as muche lande kepte

in tyllage, and than shall not the ryche man ouer-eate the poore
man with his cattell, and the fourth parte of haye and strawe shall
serue his cattell better iu a pasture than foure tymes so muche
will dooe in a house, and less attendaunce, and better the cattell
shall lyke, and it is the chiefest sauegarde for corne bothe daye
and nyght that may be." To the same effect wrote Tusser in the
comparison between " champion " (or open-field) " and severall "
(or enclosed) in his *Five Hundreth Good Pointes of Good Husbandrie*
(1573).

> " More profit is quieter found,
> (Where pastures in severall bee);
> Of one seelie aker of ground
> Than champion maketh of three.
>
> . . . . .
>
> The t'one is commended for grain,
> Yet bread made of beanes they doo eate;
> The t'other for one loafe have twaine
> Of mastlin, of rie, or of wheate."

But the agriculturists did not anticipate that one shepherd, with
his dog, his crook, shears, and tar-box, might take the place of
many ploughmen. They had not reckoned on the strength of the
new commercial spirit, and of the impulse which it gave to large
grazing farms. The area of land actually returned as enclosed and
converted to pasture was relatively small. It has been oalculated
that, during a period of nearly two centuries,—that is, from 1455
to 1637,—the total acreage enclosed and converted did not exceed
750,000 acres, and that the total number of persons thrown out of
work was not greater than 35,000.[1] At the present day, four
million acres of arable land may in fifteen years be converted into
pasture without calling the serious attention of a single statesman

[1] Mr. Gay's estimate of the total area affected between the years 1455 and
1607 is 516,673 acres ("Inclosures in England " in *Quarterly Journal of
Economics*, vol. xvii. pp. 576-97). He admits that this is probably an under-
estimate. The figure given in the text is the calculation made by the Rev.
A. H. Johnson in *The Disappearance of the Small Landowner* (1909), pp. 48, 58.
   On the other hand, a contemporary writer (*Certayne causes gathered
together*, Four Supplications, E.E.T.S. extra series xiii.,pp. 101-2) estimates that
at that time (1551) 50,000 ploughs had been put down, and that each plough
not only maintained six persons, but provided food in addition for 7½ persons.
In other words, upwards of 650,000 persons lost their means of support. This
is an obvious exaggeration.
   More than two-thirds of the area affected lay in the Midland counties
(" in umbelico regni," as Rous writes), and especially in Northamptonshire,
Oxfordshire, Bucks, Warwick, Berkshire, Leicestershire, Bedfordshire, and
Huntingdonshire. The northern and southern counties were almost untouched.
In the west, Gloucestershire, and in the east, Norfolk, were the only districts
seriously affected.

to the consequent loss of employment and rural depopulation. But small though the acreage may have been, it was considerable in proportion to the cultivated area, and the suffering was undeniably great. The distress was aggravated by the disbanding of the great retinues which had been maintained in feudal households, and by the consequent disturbance of the labour market. It was still more intensified by the suppression of the monasteries (1536-42). Not only were a very large number of dependents deprived of their livelihood, but enclosures on the old ecclesiastical estates were carried out with peculiar harshness. The new owners among whom the monastic lands were distributed, bound by no sentimental tie to the existing tenants, claimed that the royal grant annulled all titles derived from the previous owners, entered on their possessions as though they were vacant of leaseholders or copyholders, and enclosed the land for sheep-runs. The doggerel ballad, " Vox Populi, Vox Dei " (1549),[1] laments the consequences of the change of ownership :

> " We have shut away all cloisters,
> But still we keep extortioners :
> We have taken their lands for their abuse,
> But we have converted them to a worse use."

Voluntary agreements for the valuation and commutation of rights of common were often entered into between tenants and landowners, and bargains were struck on equitable terms. Instances like that given in the following extract from Kennet's *Parochial Antiquities*[2] might be indefinitely multiplied : " The said Edmund Rede, Esquire granted and confirmed to Thomas Billyngdon one close in Adyngrave, in consideration whereof the said Thomas Billyngdon quitted and resigned his right to the free pasturage of four oxen to feed with the cattle of the said Edmund Rede and all right to any common in the said pasture or inlandys of the said Edmund." Here in 1437 was the principle of commutation of rights of common accepted and enforced by private contract. In other cases a semblance of agreement may have been secured by threats. But justice was not always perverted in the interests of landlords. Attempted acts of oppression were frequently checked by the courts of law. As an instance may be quoted the proposed enclosure of the common-fields at Welcombe, near Stratford-on-,

---

[1] *Ballads from MSS.* ll. 538-41. The spelling is modernized. (Publication of the Ballad Society, vol. i. p. 139.)

[2] Vol. ii. 324.

Avon.[1] The example is the more interesting because it reveals one of the rare appearances of William Shakespeare in public life. In 1614 William Combe, of Stratford-on-Avon, the Crown tenant of the "College," wished to withdraw his arable land from the open-field farm of Welcombe, enclose it, and lay it down to pasture. He also wished to enclose so much of the ancient greensward or pasture as his rights of pasturage represented. To his scheme he had obtained the consent of Lord Chancellor Ellesmere, as representative of the Crown, and the active co-operation of the Chancellor's steward. Shakespeare, however, was in a position to be a formidable opponent, for he not only owned land adjoining, but also held the unexpired term of a lease of half the tithes of the open-fields. But a deed, dated October 28 1614, secured him from any loss of tithe through the conversion of tillage into pasture, and his consent to the enclosure was obtained. Combe had now only to deal with the Corporation of Stratford, who offered a strenuous resistance. Strong language did not move them ; in the Corporation MS. the witnesses are duly noted who heard him call them " Purtan knaves," " doggs and curres." Tempting offers were refused, though Combe proposed to compensate them in more than the value of the tithe, to undertake the perpetual repair of the highways passing over the land, and to increase the value of the rights of freeholders and tenants by waiving part of his claim to turn out sheep and cattle on the commons. Then Combe took matters into his own hands, and prepared to enclose his land by surrounding it with a ditch. This brought the dispute to a crisis. Not apparently without the knowledge of the Town Clerk, the townspeople filled in the ditch. A breach of the peace seemed imminent. The matter was, there-fore, referred to the law-courts, and at Warwick Assizes, on March 27, 1615, Lord Chief Justice Coke made an order that " noe inclosure shalbe made within the parish of Stratforde." The Dingles, which formed part of the common-fields of Welcombe, remain uninclosed to this day.

Instances of the tyrannical use of power could also be quoted. The Tudor age was rough, and might was sometimes right. Sir Thomas More in his *Utopia* (1516) paints this side to the picture, when he speaks of " husbandmen . . . thrust owte of their owne, or els by coveyne and fraude or by vyolent oppression they be put

---

[1] *Shakespeare and the Enclosure of Common Fields at Welcombe*, edited by C. M. Ingleby (1885).

besydes it, or by wronges and injuries they be so weried that they
be compelled to sell all." If a small freeholder or copyholder proved
obstinate, the proceedings of Sir Giles Overreach, in *A New Way
to Pay Old Debts* (Act ii. Sc. 1), may illustrate the methods by which
a Naboth's vineyard, even when it belonged to a manorial lord,
might be appropriated by a wealthy capitalist :

> " I'll therefore buy some cottage near his manor,
> Which done, I'll make my men break ope his fences,
> Ride o'er his standing corn, and in the night
> Set fire on his barns, or break his cattle's legs,
> These trespasses draw on suits, and suits expenses
> Which I can spare, but will soon beggar him.
> When I have harried him thus two or three year,
> Though he sue *in formâ pauperis*, in spite
> Of all his thrift and care he'll grow behindhand.
>
> . . . .
>
> Then, with the favour of my man at law,
> I will pretend some title : want will force him
> To put it to arbitrement. Then if he sell
> For half the value he shall have ready money,
> And I possess his land."

Considerations of mutual advantage, equitable bargains, fair purchase, superior force, legal chicanery, threats and bullying, were
all at work to hasten the change to the individual occupation of
land, and the consolidation of separate holdings. If copyholders
or commoners appealed to the law-courts, matters, no doubt, sometimes ended as they were friended. " Handy-dandy " was in the
Middle Ages a proverbial expression for the covert bribe offered by
a suitor, and the occasional perversion of justice is enshrined in the
Latin jingle : *Jus sine jure datur, si nummus in aure loquatur.*

Illegal evictions are not included among the grievances alleged by
the leaders in any of the risings of the peasantry which marked
the Tudor period. Their absence from these lists justify the conclusion that open illegality was at least rare. But the law itself
gave landowners abundant opportunities of regaining possession of
the land. Leaseholders for a term of years or for lives had no legal
claim to a renewal of their leases, when the term of years had
expired or the last life had dropped. Rents might then be raised
to an exorbitant sum or extravagant fines exacted, and, unless the
tenant was prepared to pay the increased charge, he must surrender
his holding. Cottagers or squatters on the waste could rarely
show any legal claim to the occupation of land, and the tenancy of
a cottage to which rights of common attached could be practically
determined by enhancing the rent. Copyholders were, in all

probability, almost equally insecure in their holdings. So long as they were in possession, the court roll was evidence of the incidents of their tenure. But the law was still vague as to rights of succession to copyholds. It may be doubted whether copyholds of inheritance were yet known, and it is reasonably certain that the normal copyhold was for a term of years or for lives. At the expiration of the term of years or of the last life, normal copyholders were at the mercy of the lord. Even if copyholds of inheritance were recognised by lawyers in the sixteenth century, they were still insecure. Their titles must often have been incapable of legal proof ; they might be forfeited by some real or technical breach of custom ; their renewal was subject to the payment of fines on admittance, which might, where no manorial custom fixed the sum, be arbitrary in amount. It was not till the close of the eighteenth century that the law fixed the limits of a reasonable fine, and, if the fines were arbitrary, the landlord had a weapon with which even copyholds of inheritance, as understood by modern lawyers, might be determined. It is impossible to doubt that exorbitant rents and excessive fines, of which the peasant leaders, preachers, and pamphleteers so bitterly complain, were sometimes used to dispossess leaseholders and copyholders. The powers were legal ; but their exercise often worked injustice. Yet it should be remembered, on the other side, that the raising of rents or the enhancing of fines, whenever the opportunity occurred, were the only means of adjusting the landlord's income to the great rise in the prices of agricultural produce. In the *Compendious or Briefe Examination* [1] the Knight puts the landlord's case. " In all my life time," he says, "I looke not that the thirde part of my lande shall come to my dispocition that I may enhaunce the rent of the same, but it shalbe in mens holdinges either by lease or by copie graunted before my time. . . . We cannot rayse all our wares as youe maye yours." Rents, based on the commutation of labour services at a fixed annual sum in the fourteenth century did not represent the annual value of the land in 1550. Nor were fines for renewal or on

---

[1] *The Compendious or briefe Examination of certayne ordinary complaints of divers of our countrymen in these our dayes* was printed in 1581, and the authorship is attributed to " W. S. Gentleman." But Miss Lamond discovered, edited, and published (1893) an edition from a MS. probably written in 1549. She gives reasons for assigning its authorship to John Hales. " W. S." may have been William Stafford (1554-1612) ; but that he was not the writer appears to have been conclusively proved.

admittance always excessive. Roger Wilbraham,[1] of Delamere in Cheshire, about the middle of the seventeenth century, left behind him instructions for his heir : " It will be expected of my heir that he deale no worse with tenants than I have done. And for his directions I have set down ye yearly values according to which I deale and wold have him to deale with the tenants. My rule in leasing is to take for a fine from ancient tenants : 8 years' value for 3 lives, 5 years' value to add 2 lives to 1, 2 years' value to add 1 life to 2, 1 year's value to change a life, or more if there is any great disparity in years betwixt the lives." When, therefore, rents were raised or fines enhanced, the landlord was not always trying to dispossess his tenant. As often as not, he was claiming his proper share of the tenant's " unearned increment."

Against these weapons of the law the cultivators of the old home-farms and of the assart lands were practically defenceless. It is therefore natural to suppose that they were the principal sufferers by the enclosing movement. In their case enclosures did not of necessity involve any breach of the old or new law. Even the provisions of the Tudor legislation were not infringed, unless the land, thus cleared of its cultivators, was so used as to throw any number of holdings together into the hands of one man, to " decay " farm-buildings or houses, to convert tillage into pasture, and so put down ploughs, or to carry an illegal number of sheep. But open-field farmers were in a stronger position. The common rights, which each partner in the association enjoyed over the whole cultivated area of the village-farm, could only be extinguished by agreement, real or enforced, among the commoners. Nor was this consent the only obstacle to enclosure which the system presented. The intermixture of the strips is recognised as a protection against enclosure by the ablest of the sixteenth century writers on the subject. In the *Compendious or Briefe Examination* both the Doctor and the Husbandman agree as to the difficulties which these two features of the open-field system threw in the way of any general enclosure. The same points are insisted upon by eighteenth century writers. It is not, of course, asserted that the difficulties of enclosing open-field farms were insuperable. Ever since the thirteenth century, village farms had been broken up, both by large landowners and comparatively small freeholders.

But, before the enclosure acts of the eighteenth century, it was a slow and piecemeal process, by which the principal landlord, or some freeholder who was a partner in the farm, gradually consolidated in his own hands the whole or a part of the commonable cultivated land, enclosed it, and freed it from common rights. No doubt the enclosure of uncultivated wastes injured the tenants of village farms, because it restricted the area of rough pasture grazed by their live-stock. Enclosures of this kind, carried out without leaving a sufficiency of common pasture, were the chief grievance of the peasantry in Kett's rebellion in Norfolk. In this connection the re-enactment by Edward VI. of the statutes of Merton and Westminster,[1] is significant. But the meaning is obscure. It may have been intended to increase the amount of tillage by bringing new land under the plough in exchange for that which had been laid down to grass. Except through the attack upon their pasture commons, it is reasonable to conclude that open-field farmers escaped the storm of sixteenth century enclosures more lightly than the less protected cultivators of demesnes and " assart " lands. This seems to have been the case. Bitter complaints were made against the enclosure of open-fields. But the outcry was practically confined to the corn-growing counties of the Midlands, which throughout the whole period were seething with discontent and insurrection. Yet even here, with the exception of Leicestershire, the enclosing movement cannot have, to any great extent, succeeded, since these are the very counties which, in the eighteenth century, still contained the largest proportion of " champion " or open land.

Advanced free-traders might agree with Raleigh that England, like Holland, could be wholly supplied with grain from abroad without troubling the people with tillage. Others of a less theoretical turn of mind looked no further than the immediate distress which the abandonment of tillage produced. If the enclosing movement had been accompanied by a large extension of arable farming, the market for agricultural labour might have been so enlarged as substantially to relieve agrarian distress. But the extension of pasture and the substitution of a shepherd and his dog for the ploughmen and their teams only increased the scarcity of employment. Tenant-farmers lost their leaseholds ; copyholders were dispossessed of their holdings ; squatters and cottagers,

---

[1] **3 and 4 Edward VI. c. 3.** (See p. 38.)

who had eked out their harvest earnings by the produce of the
live-stock which they maintained on the commons, were ruined ;
servants in husbandry and labourers for weekly wages were thrown
out of work. The high prices of necessaries, combined with the loss
of commons, the ravages of the murrain, and a succession of dry
summers, had driven many small cultivators over the narrow
border-line which separated them from starvation. Rents rose
exorbitantly till, for farmers at rack-rent, existence became a
misery. There was an ominous growth of middlemen, "lease-
mongers, who take groundes by lease to the entente to lette them
againe for double and tripple the rente,"[1] and battened on the
land-hunger of the people. Legislators were bewildered by currency
questions, and violent changes in the standard purity of the gold
and silver coinage aggravated the distress by raising or lowering
prices. As gold and silver poured into the Old World from America,
prices rose throughout Europe. The rise was in England attributed
to every cause other than the cheapening of the precious metals.
While from one or the other of these causes the purchasing power
of wages rapidly diminished, their nominal value remained stationary, and labourers were forced to accept the statutory rates.

It was on those agriculturists who were unwilling or unable to
adapt themselves to the times that the blow fell with the greatest
severity. The Husbandman in the *Compendious Examination* knew
several of his neighbours who had " turned ether part or all theire
arable grounde into pasture, and therby have wexed verie Rich
men." These were the men of whom Harrison and Sir Thomas
Smith speak as " coming to such wealth that they are able and
do daily buy lands of unthrifty gentlemen and make . . . their
sons gentlemen." But the Husbandman himself, having " enclosed
litle or nothinge of my grownd, could never be able to make up
my lorde's rent, weare it not for a litle brede of neate, shepe, swine,
gese and hens." Hence it is that, while Latimer laments the
degradation of small yeomen who, like his father, had farms of
" three to four pounds a year at the uttermost," Harrison describes
the rise of substantial farmers and of the middle classes, and their
improved standard of living. The distribution of wealth was
becoming more and more unequal ; the problem of poverty was
acquiring a new significance. In the growing struggle for existence

---

[1] Robert Crowley's *Way to Wealth* (1550). See also his *Epigrams* " of
Leasemongars " and " of Rent raysers."

it was possible for men, who were neither infirm nor idle, to lose
their footing. Voluntary almsgiving was tried and proved inadequate. Gradually and cautiously the legislators of the reign of
Elizabeth were forced to apply the principle of compulsory provision for the relief of the necessitous.[1] Previous legislation, in
dealing with the impotent poor, had outlined the systems of local
liability and of settlement which were adopted in the later poor-
laws ; but it had been mainly concerned with the suppression of
those persons who were styled idle rogues and vagabonds. The
object explains, though to modern ideas it cannot justify, the
harshness of the law. Able-bodied men and women, who were
willing to work but had lost their livelihood, were unknown to the
legislators who had sketched the first poor-laws for the relief of
the impotent poor and the punishment of sturdy beggars (*validi
mendicantes*). Our ancestors did not discriminate closely between
the different sources of poverty. To them, as is stated in the
preamble to the statute of Henry VIII.,[2] " ydlenes " was the
" mother and rote of all vyces." The " great and excessive
nombres " of idle rogues and vagabonds were a crying evil. To
this class belonged the men who committed " contynuall theftes,
murders, and other haynous offences, which displeased God,
damaged the King's subjects, and disturbed the common weal of
the realm." Apart from the committal of serious crime, the mass
of idle vagrants was in country districts a nuisance and a danger.
The kidnapping of children was not uncommon. Housewives were
robbed of their linen, and their pots and pans, or terrified by threats
of violence into parting with their money. Horses were stolen
from their paddocks, or, still more easily, from the open-field balks
on which they were tethered ; pigs were taken from their styes,
chickens and eggs from the henroosts. Men and women, as they
returned from markets, were waylaid by sturdy ruffians. Shops,
booths, and stalls were pilfered of their contents. Tippling-houses
were converted into receivers' dens for stolen goods. The comparative leniency of the laws of Henry VII. had failed ; therefore
the evil must be stamped out with a severity which was not only
unsentimental but ferocious.

Here the interesting point is whether the ranks of idle rogues

[1] See Appendix II., *The Poor Laws*, 1601-1834.
[2] 1530-1 (22 Henry VIII. c. 12) supplemented in 1536 by 27 Henry VIII.
c. 25.

were to any large extent swollen by agriculturists, driven to want and desperation by the loss of their holdings. The sturdy beggars, against whom Richard II. had legislated, had not the excuse of want of employment. They consisted, partly of disbanded soldiers who had so long followed the trade of war that they knew no other ; partly of men who had suffered that general moral deterioration which often resulted from great catastrophes like the successive visitations of the " Black Death." In the fifteenth century, the close of the French war and of the Wars of the Roses again recruited the ranks of idle poverty and crime. To them were added, at a later date, the disbanded retinues of great nobles, " the great flock or train," to quote More's *Utopia*, " of idle and loitering serving-men, which never learned any craft whereby to get their living." Finally, the suppression of the monasteries displaced and threw upon the world a large number of dependents, many of whom, from inclination or necessity, joined the army of sturdy beggars. Disbanded soldiers, discharged serving-men, and dismissed dependents of monastic institutions account for a formidable total of unemployed labour, without the addition of clothiers out of work or displaced agriculturists. But the evidence of More's *Utopia* cannot be ignored. The passage is familiar [1] in which he speaks of the husbandmen " thrust owte of their owne " by enclosures ; compelled to " trudge out of their knouen and accustomed howses " ; driven to a forced sale of their " housholde stuffe " and " constrayned to sell it for a thyng of nought." " And when they have, wanderynge about, sone spent that, what can they els do but steale, and then justelye, God wote, be hanged, or els go about a beggyng? And yet then also they be cast in prison as vagaboundes, because they go about and worke not ; whom no man will set a worke, though they never so willingly offer them selfes therto." More's eloquent appeal may have produced effect. In the year after the publication of *Utopia*, the first and most important Commission was issued (1517-19) to enquire into the progress and results of enclosures in the twenty-four counties principally affected. The Returns of the Commissioners in Chancery are admittedly imperfect. But they justify the conclusion [2] that More's picture, though true

[1] *Utopia*, bk. i., ed. Lupton, pp. 53-4.

[2] Hypothetical tables based on these returns have been constructed by Mr. Gay, showing that the total number of persons displaced by enclosures during the period 1485-1517 did not much exceed 6931. See Johnson's *Disappearance of the Small Landowner*, p. 58.

in particular instances, is as a general description of rural conditions too highly coloured. Dispossessed agriculturists undoubtedly contributed some proportion of the class which the Government grouped under the heading of idle rogues. Contemporary writers imply that the proportion was large : modern research, based on contemporary enquiries and returns, suggests that it was relatively small. The evidence seems insufficient for a decision. In coping with a real evil, the Government attempted no classification. The innocent suffered with the guilty, and men and women, whether many or few, who had lost their means of livelihood and were willing to work, were the victims of severe punishment designed for the class of professional vagabonds.

Something is known of the degrees, practices, and jargon of the Elizabethan fraternity of vagabonds. Awdelay and Harman [1] describe the " Abraham man," or " poor Tom," bare-legged and bare-armed, pretending madness; the "Upright man " with his staff, and the " Ruffler " with his weapon ; the " Fraters," Pedlars, and Tinkards ; the " priggars of Prauncers," or horse-stealers, in their leather jerkins ; the " Counterfet Cranke," feigning the falling sickness, with a piece of white soap in his mouth which made him foam like a boar : the " Palliards," with their patched cloaks, and self-inflicted sores or wounds ; and many others of the twenty-three varieties, male and female, of the professional beggar. But even Harman seldom enquired into their previous life. Some, like the " Ruffler," had either " serued in the warres or bene a seruinge man " ; others, like the " Uprights," have been " serueing men, artificers, and laboryng men traded up to husbandry." The " wild Roge " was a " begger by enheritance—his Grandfather was a begger, his father was one, and he must nedes be one by good reason." Few allusions can be gleaned from Shakespeare's writings to the agricultural changes which were taking place around him. But when we pass from the movement itself to some of the results which it helped to produce, his references are many and clear. The mass of " vagrom men " was a real social danger which exercised the wits of wiser men than Dogberry.[2]

---

[1] *The Fraternity of Vacabondes*, by John Awdeley (1561) and *A Caveat or Warening for Commen Curseters*, by Thomas Harman (1567-8).

[2] Many of the types of beggars appear in Shakespeare's pages. There is Harman's " Ruffler," " the worthiest of this unruly rablement " :

    ". . . fit to bandy with thy lawless sons
     To ruffle in the commonwealth of Rome."

            (*Tit. Andr.* Act i. Sc. 1, ll. 312-3.)

There is the " pedlar," the aristocracy of the profession, a clever plausible rascal like Autolycus. " The droncken tyncker " is represented by Christopher Sly—" by birth a pedlar . . . by present profession a tinker "—drunk on the heath, and in debt for ale to Marian Hacket (*Tam. Sh. Ind.* ii. ll. 19-22). There is the " prygger " or " prygman," who " haunts wakes, fairs, and bear-baitings " (*Wint. Tale*, Act iv. Sc. 2, l. 109). There is Awdeley's " choplogyke," who gives " XX wordes for one," to whom Capulet likens his daughter Juliet (*Rom. and Jul.* Act iii. Sc. 5, l. 150). There is Harman's " Rogue," or " Wild Rogue," in the " rogue forlorn," who shares the hovel and the straw with King Lear and the swine (*Lear*, Act iv. Sc. 7, l. 39). Edgar, disguised as a madman and calling himself ' poor Tom " (*Lear*, Act iii. Sc. 4, l. 57), is Awdeley's " Abraham man," who " nameth himselfe ' poore Tom.'. . Whipped from tithing to tithing," he had only received the punishment to which an Elizabethan statute (39 Eliz. c. 4) sentenced " all fencers, bearwards, common players, and minstrels ; all jugglers, tinkers and petty chapmen," and other vagrants who were adjudged to be rogues, vagabonds, and sturdy beggars.

CHAPTER IV

THE REIGN OF ELIZABETH

Paternal despotism : restoration of the purity of the coinage ; a definite commercial policy : revival of the wool trade : new era of prosperity among landed gentry and occupiers of land : a time of adversity for small landowners and wage-earning labourers : Statute of Apprentices ; hiring fairs ; growth of agricultural literature : Fitzherbert and Tusser : their picture of Tudor farming : defects of the open-field system : experience of the value of enclosures ; improvement in farming : Barnaby Googe ; Sir Hugh Plat : progress in the art of gardening.

THE reign of Elizabeth marks a definite stage in English history. The mediaeval organisation of society, together with its trade guilds and manorial system of farming, had broken down. Out of the confusion order might be evolved by a paternal despotism. The Queen's advisers, with strong practical sagacity, set themselves to the task. They sate loosely to theories and rode no principles to death. But so firmly did they lay their foundations, that parts of their structure lasted until the nineteenth century. National control displaced local control. The central power gathered strength: it directed the economic interests of the nation ; it regulated industrial relations ; through its legislation and administration it fostered the development of national resources.

The restoration of the standard purity and weight of the coinage was resolutely taken in hand. Its debasement had been the cause of much of the economic distress in previous reigns ; credit was ruined, and the treasury bankrupt. The debased, sweated, and clipped silver coinage was called in, and new coins were issued. As silver flowed into the country from the New World, the amount of money in circulation increased. More capital was available in a handy form, and, when legitimate interest ceased to be confused with usury, more people could borrow it on reasonable terms. The way was thus paved for a new era of commercial prosperity.

In mediaeval times the whole external trade of the country had been in the hands of foreigners. Elizabeth followed and developed the commercial policy of England, which first assumed a deliberate continuous shape under Henry VII. Foreign traders were discouraged, and English merchants favoured. The Hanseatic League lost the last of its privileges ; the Venetian fleet came to England less and less frequently, and at last ceased altogether to fly its flag in the Channel. The import of manufactured goods was checked. The export of raw material and of English sheep was narrowly restricted, though long wool, as the staple of a great trade, was still sent abroad freely. The Government realised to the full all the abuses of patents and monopolies ; but they did not hesitate to grant both privileges in order to stimulate native enterprise. Companies were formed with exclusive rights of trading in particular countries. The oldest and most powerful of these Companies, the Merchant Adventurers, obtained a royal charter in 1564. The Muscovite, Levant or Turkey, Eastland or Baltic, and Guinea or African Companies were formed to push English trade in foreign parts. In 1600 the East India Company was chartered. The mercantile marine was encouraged by fishery laws, which gave English fishermen a monopoly in the sale of fish. Men who argued that abstinence from meat at certain seasons was good for the soul's health risked the stake or the rack ; but, for the sake of multiplying seamen, the Government did not hesitate to ordain fast-days on which only fish was to be eaten.[1] To foster the home manufacture of cloth, it was made a penal offence for any person over the age of six not to wear on Sundays and holy days a cap made of English cloth. Stimulated by such methods, trade throve apace, and English goods were carried in English-built ships, owned by Englishmen, and manned by English seamen. While foreign merchants were discouraged, foreign craftsmen, especially religious refugees from France or Flanders, were welcomed as settlers, bringing with them their skill in manufacturing paper, lace, silk, parchments, light woollens, hosiery, fustians, satins, thread, needles, and in other arts and industries.

The English wool trade was restored to more than its former

---

[1] The rule of eating fish twice a week was continued from Catholic times ; but a third day was added by Elizabeth from motives of " civile policy." " Accounting the Lent Season, and all fasting daies in the yeare, together with Wednesday and Friday and Saturday, you shall see that the one halfe of the yeare is ordeined to eate fish in " (Cogan, *Haven of Helthe*, ed. 1612, p. 138).

prosperity. On it had long depended the commercial prosperity of the country. John Cole, " the rich clothier of Reading " at the end of the thirteenth century, was as famous as his fellow-craftsman, John Winchcomb, the warlike " Jack of Newbury," became in the days of Henry VIII. Wool was the chief source of the wealth of traders and of the revenues of the Crown. It controlled the foreign policy of England, supplied the sinews of our wars, built and adorned our churches and private houses. The foreign trade consisted partly in raw material, partly in semi-manufactured exports such as worsted yarns, partly in wholly manufactured broad-cloth. As the manufacture of worsted and cloth goods developed in this country, the demand and consumption rapidly increased at home. According to the purpose for which it was to be used, wool was divided into long and short. In England, long wool was employed mainly for worsted fabrics, but also to give strength and firmness to cloth. Abroad, it was eagerly bought in its raw state for both purposes. In long wool, or combing-wool, England had practically a monopoly of the markets, and to it the export trade of raw material was almost exclusively confined. Short wool, on the other hand, was used for broad-cloth. In its raw state it had a formidable rival abroad in the fleeces of the Spanish merino. Only in the manufactured state did it compete with Flemish and French fabrics on the Continent, and often found itself unable, owing to the excellence of merino wool and the skill of foreign weavers, to maintain its hold on the home market. Wool-staplers were the middlemen. They bought the wool from the breeder, sorted it according to its quality, and sold it to the manufacturer. Dyer,[1] two centuries later, describes their work :

> " Nimbly, with habitual speed,
> They sever lock from lock, and long and short,
> And soft, and rigid, pile in several heaps.
> This the dusk hatter asks ; another shines,
> Tempting the clothier ; that the hosier seeks ;
> The long bright lock is apt for airy stuffs :
>
> .   .   .   .
>
> If any wool, peculiar to our isle,
> Is given by nature, 'tis the comber's lock,
> The soft, the snow-white, and the long-grown flake."

In the long-wooled class Cotswold wool held the supremacy, with Cirencester as its centre, though the " lustres " of Lincolnshire always commanded their price. Among short-wools, Ryeland had

---

[1] *The Fleece* (1757), bk. ii ; ll. 83-88 and 445-47.

the pre-eminence, with Leominster as the centre of its trade. " Lemster ore " was the equivalent of the " golden fleece " of the ancients, and poets compared the wool for its fineness to the web of the silk-worm, and for its softness to the cheek of a maiden.

During the Tudor period, a change was passing over the wool trade, which may have influenced the labour troubles of the period as well as the policy of land-holders. As enclosures multiplied, sheep were better fed, and the fleece increased in weight and length, though it lost something of the fineness of its quality. In other words, the wool was less adapted for the manufacture of broadcloth. The old pastures were also wearing out. During long and cold winters, if the sheep is half-starved, the fleece may retain its fineness, but it loses in strength. There also was a deterioration in the quality of short wool. How far these considerations may have influenced pasture-farming is necessarily uncertain. But it is at least a coincidence that, in spite of the increase in the number of sheep, there was, in the early years of the Tudor period, considerable distress in the clothing trade. As the reign of Elizabeth advanced, the great development of home manufactures provided a remedy. The newly established Merchant Companies opened up fresh markets abroad for English cloth. At the same time France and the Low Countries, distracted by civil or religious wars, ceased for the moment to be our rivals in the trade. English broadcloths were exported abroad in increasing quantities. The suspension of continental manufactures checked the exportation of English long wool. But again the religious troubles of the Continent relieved the situation. Foreign refugees settled in England, bringing with them secrets in the manufacture of worsted, light woollen stuffs, and hosiery, for all of which English wool was specially adapted.

Thus England was once more growing prosperous, and farming shared in the general prosperity. As the reign advanced, agricultural produce rose rapidly in price. The rise no longer depended on those fluctuations in the purity of the coinage, which had been so frequent that no man knew the real value of the coin in which he was paid. For a time the influx of silver had cheapened the precious metals, diminished their purchasing power, and so created dearness. But the great expansion of trade gradually absorbed the new supply of silver. The later rise in agricultural prices was due to the relative scarcity of produce, which was caused by the increased consumption consequent on revived prosperity, by a higher standard of living, and by a growing population. The necessary spur of profit was thus applied to farming energies. Leaseholders for a long term or for lives, and copyholders at fixed quit-rents had their golden opportunity, and many of them used it to become wealthy.

Of the general prosperity of the landowning and land-renting portion of the rural community, there is sufficient evidence. Every man, says Harrison,[1] turned builder, " pulled downe the old house and set up a new after his owne devise." In ten years more oak was used for building than had been used in the previous hundred. Country manor-houses were built not of timber, but of brick or stone, and they were furnished with "great provision of tapistrie, Turkie work, pewter, brasse, fine linen and . . . costlie cupbords of plate." Ordinary diet had become less simple. " White-meats," —milk, butter, eggs, and cheese,—were despised by the wealthy, who preferred butcher's-meat, fish, and a " diversitie of wild and tame foules." The usual fare of the country gentleman was abundant, if not profuse. The dinner which Justice Shallow ordered for Falstaff might be quoted as an illustration. But more direct evidence may be produced. Harrison says that the everyday dinner of a country gentleman was " foure, five, or six dishes, when they have but small resort." Gervase Markham in his *English Housewife* gives directions for a " great feast," and for " a more humble feast, or an ordinary proportion which any good man may keep in his family, for the entertainment of his true and worthy friend." The " humble feast " includes " sixteen dishes of meat that are of substance and not emptie, or for shew." To these " sixteen full dishes," he adds " sallets, fricases, quelque choses, and devised paste, as many dishes more, which make the full service no lesse then two-and-thirtie dishes." In dress, also, the country gentry were growing more expensive, imitating the " diversities of jagges and changes of colours " of the Frenchman. Already, too, as Bishop Hall has described in his *Satires*, they were in the habit of deserting their country-houses for the gaiety of towns, and the " unthankful swallow " " built her circled nest " in

> " The towered chimneys which should be
> The windpipes of good hospitalitie."

Of the yeomen, who included not only farming owners, but

[1] Harrison, *Description of England* (1577), bk. ii. cc. vi. xii. xxii.

lessees for lives and copyholders, Harrison says that they " commonlie live wealthilie, keepe good houses, and travell to get riches." [1] Their houses were furnished with " costlie furniture," and they had " learned also to garnish their cupbords with plate, their joined beds with tapistrie and silke hangings, and their tables with carpets and fine naperie." Though rents had risen and were still rising, " yet will the farmer thinke his gaines verie small toward the end of his terme if he have not six or seven yeares rent lieing by him, therewith to purchase a new lease, beside a faire garnish of pewter on his cupbord, three or foure featherbeds, so manie coverlids and carpets of tapistrie, a silver salt, a bowle for wine, and a dozzen of spoones to furnish up the sute." Old men noted these changes in luxurious habits—" the multitude of chimnies latelie erected," " the great amendment of lodging," and " the exchange of vessel as of treene platters into pewter and wodden spoones into silver or tin." Writing of the Cheshire yeomen in 1621, William Webb says : [2] " In building and furniture of their houses, till of late years, they used the old manner of the Saxons ; for they had their fire in the midst of the house against a hob of clay, and their oxen also under the same roof ; but within these forty years it is altogether altered, so that they have built chimnies, and furnished other parts of their houses accordingly. . . . Touching their housekeeping it is bountiful and comparable with any shire in the realm. And that is to be seen at their weddings and burials, but chiefly at their wakes, which they yearly hold . . . for this is to be understood that they lay out seldom any money for any provision but have it of their own, as beef, mutton, veal, pork, capons, hens, wild fowl, and fish. They bake their own bread and brew their own drink. To conclude, I know divers men, who are but farmers, that in their housekeeping may compare with a lord or a baron in some countries beyond the seas. Yea, although I named a higher degree, I were able to justify it." In the Isle of Wight, Sir John Oglander [3] compares the state of the country at the close of Elizabeth's reign with that at the outbreak of the Civil War. At the former period he says that " Money wase as plentiful in yeomens purses as nowe in ye beste of ye genterye, and all ye genterye full of monyes and owt of debt."

---

[1] *Description*, bk. ii. ch. v.

[2] Quoted in King's *Vale Royal* (1778), vol. i. pp. 30, 31.

[3] *Oglander Memoirs* (1595-1648), p. 55.

The small copyholder's house is described by Bishop Hall as being :

> " Of one bay's breadth, God wot, a silly cote
> Whose thatched spars are furred with sluttish soote
> A whole inch thick, shining like blackmoor's brows
> Through smoke that downe the headlesse barrel blows,
> At his bed's feete feeden his stalled teame,
> His swine beneath, his pullen o'er the beame."

The outside walls were made of timber uprights and cross-beams, forming raftered panels which were thickly daubed with clay. But the fare which the small copyholder enjoyed was at least as plentiful as that of landless labourers in modern times. In one of the Elizabethan pastoral poems a noble huntsman finds shelter under a shepherd's roof. The food, even if something is allowed for Arcadian licence, was good, though, in the language of the day, it consisted mainly of " white meat." The guest was supplied with the best his host could provide :

> " Browne bread, whig, bacon, curds, and milke,
> Were set him on the borde."

Fresh butcher's meat was rarely seen on the table. Of the " Martylmas beef," hung from the rafters and smoked, Andrew Borde [1] thought little. If, he says, a man have a piece hanging by his side and another in his belly, the piece which hangs by his side does him more good, especially if it is rainy weather. Bacon, souse, and brawn were the peasant's meat. " Potage," Borde elsewhere writes, " is not so moch used in all Crystendom as it is used in England." It was part of the staple diet of the peasant, whether made of the liquor in which meat had been boiled, thickened with oatmeal, and flavoured with chopped herbs and salt, or made from beans or pease. Oatmeal, porridge, and " fyrmente," made of milk and wheat, were largely used. His bread was generally made of wheat and rye, often mixed, as Best states,[2] with pease—a peck of pease to a bushel of rye, or two pecks of pease to the same quantity of rye and wheat. Even " horse-bread," as Borde calls it,[3] made of pease and beans, was better than the mixture of acorns which Harrison says [4] was eaten in times of dearth. Yet the husbandman had his feastings, such as " bridales, purifications of women and such od meetings, where it is incredible to tell what meat is consumed and spent."

---

[1] Andrew Borde's *Dyetary* (1542), ch. xvi. [2] *Farming Book*, p. 104.
[3] Borde's *Dyetary*, ch. xi. [4] *Description*, ch. vi.

The prosperity of the rural community was not universal. For many of the smaller gentry, and for day-labourers for hire, times were hard. Landowners, whose income was more or less stationary, suffered from the rise in prices, accompanied, as it was, by a higher standard of luxury. When leases fell in, or lives were renewed, or copyholders were admitted, rents might be increased or fines enhanced. But in an extravagant age, when country gentlemen began to be attracted to London, such opportunities, if the tenants belonged to a healthy stock, might come too rarely or too late. Many owners were compelled to sell their estates. Land was often in the market. Thus two opposing tendencies characterised the sixteenth and seventeenth centuries. The division of church lands among grantees who already owned estates strengthened the landed aristocracy, while continual sales democratised the ownership of land. It is said that only 330 families can trace their titles to land beyond the dissolution of the monasteries. In the two centuries that followed, few of the gentry retained their hold on their estates, unless they were enriched by wealthy marriages, by trade, or by the practice of the law. The buyers generally belonged to the rising middle classes. Harrison, in his *Description of England*,[1] says that yeomen, " for the most part farmers to gentlemen," by attention to their business " do come to great welth in somuch that manie of them are able and doo buie the lands of unthriftie gentlemen." Fynes Moryson, in his *Itinerary* [2] (1617) notes that the English " doe . . daily sell their patrimonies, and the buyers (except-ing Lawyers) are for the most part Citizens and vulgar Men." Sir Simon Degge [3] (1669), a learned lawyer, declares that in Stafford-shire, during the past sixty years, half the land had passed into the possession of new men. He attributes this change of ownership, partly to divine punishment for the sacrilege of those who were grantees of ecclesiastical property, partly to the extravagance of the country gentry who now took pleasure in spending their estates in London. He makes these comments on Erdeswick's *Survey of Staffordshire*, drawn up between 1593 and 1603, and goes on to say that there were then in the county only " three citizen owners "

---

[1] Bk. ii. ch. v. The *Description* was published in 1577. The same passage occurs in Sir Thomas Smith's *De Republica Anglorum*, bk. i. ch. xxiii. published in 1583.

[2] Part III. bk. iii. ch. iii.

[3] Degge's Letter is printed as a supplement to Erdeswick's *Survey of Staffordshire* in the edition of 1717.

of land, and that now, in 1669, there were three Barons, four Baronets, and twenty calling themselves Esquires who had bought estates with money made in trade. Similar is the evidence of the compiler of *Angliae Notitia* [1] (1669). "The English," he says, "especially the *Gentry* are so much given to *Prodigality* and *Slothfulness* that Estates are oftner spent and sold than in any other Countrey . . . whereby it comes to passe that *Cooks, Vintners, Innkeepers,* and such mean Fellows, enrich themselves and begger and insult over the *Gentry* . . . not only those but *Taylors, Dancing Masters* and such *Trifling Fellows* arrive to that Riches and Pride, as to ride in their *Coaches,* keep their Summer Houses, to be served in Plate, etc. an insolence insupportable in other well-govern'd Nations."

Another class, that of labourers, suffered from the dearness of agricultural produce, because their wages were fixed by law, and only by slow degrees followed the upward tendency of prices. In some respects the worst evils of the period 1485-1558 were passing away, or were modified by the expansion of trade. Enclosures still continued. Acts of Parliament [2] were still passed against the decaying of towns and against the substitution of pasture for tillage, and one of the most vehement of protests against enclosures, was made by Francis Trigge,[3] in 1604. But land was now more frequently enclosed for arable farming, and there was consequently less displacement of labour. The great extension of gardens attached to country houses provided new occupations. Industries like spinning, weaving, and rope-making, which were previously confined to particular towns by the craft-organisations of guilds, spread into rural districts, and employed villagers in supplying not merely their domestic wants but the needs of manufacturers. Agriculturally, a change was taking place in the labourer's condition. For the cultivation of the soil, farmers, except in the North and East, looked less to servants in husbandry and more to the day-labourers, whose wages assumed a new importance in the assessments of the Justices of the Peace. As the prices of agricultural produce rose, and as, here and there, the improvement of roads brought new markets within the reach of farmers, it was cheaper

---

[1] In the 1692 edition of *Angliae Notitia* the words "Prodigality, Sports, and Pastimes" are substituted for "Prodigality and Slothfulness."

[2] 1562-3, 5 Eliz. c. 2 ; 1597-8, 39 Eliz. cc. 1 and 2 ; 1601, 43 Eliz. c. 9.

[3] *The Humble Petition of Two Sisters : the Church and Common-wealth.*

to pay wages to hired labourers than to board agricultural servants, especially if, as Tusser says, they required roast meat on Sundays and Thursdays. Free labour, sometimes, but not invariably, still associated with the occupation of land, was becoming in the southern and midland counties the chief agent in cultivating the soil. Where enclosures were fewest, the largest number of labourers supplemented their wages by the profits of their land, their rights of common, and their goose-runs. Where enclosures were most extensive, those labourers were most numerous who were dependent only on their labour-power. Apparently there was difficulty in lodging this increasing class of landless labourers, and an attempt was made to use existing cottages as tenement houses. The Government endeavoured to check these tendencies by legislation.[1] Not more than one family was allowed to occupy each cottage, and to every cottage four acres of land were to be attached.

But the most important attempt to regulate the labour-market was the Statute of Apprentices (1563).[2] This industrial code "touching divers orders for artificers, labourers, servants of husbandry, and apprentices" deals with labour in the towns as well as in the country. It was framed, partly as a consolidating Act, partly because, as the Preamble states, the allowances limited in previous legislation had, owing to the advance in prices, become too small. It was passed in the hope that its administration would "banish idleness, advance husbandry, and yield unto the hired person both in the time of scarcity and in the time of plenty a convenient proportion of wages." It proceeds on the old lines that men could be compelled to work. But it contemplates a minimum wage at the rates current in the district, establishes a working day for summer and winter, and endeavours to provide for technical instruction by a system of apprenticeship. Any person between the age of twelve and sixty, not excepted by the Statute, could be compelled to labour in husbandry. All engagements, except those for piecework, were to be for one year. Masters unduly dismissing servants were fined. Servants unduly leaving masters were imprisoned. No servant could leave the locality where he was last employed without a certificate of lawful departure. Hours of labour were twelve hours in the summer and during daylight in winter. Wages were to be annually fixed by the Justices of the Peace, after considering the

[1] 1589, 31 Eliz. c. 7. [2] 1562-3, 5 Eliz. c. 4.

circumstances, in consultation with "such grave and discreet persons as they shall think meet." No higher wages than those settled under the assessment were to be given, or received, under severe penalties. At harvest time, artificers and persons "meet to labour" might be compelled to serve at the mowing or "inning" of hay and corn. Persons over twelve and under eighteen might be taken as apprentices in husbandry and compelled to serve till the age of twenty-one. By agreement the age might be extended to twenty-four.

Under the provisions of this Statute agricultural labourers and servants were engaged annually. Shortly before Martinmas, the chief constable of the division sent out notices that he would sit at a certain town or village on a given day, and required the petty constables to attend with lists of the masters and servants in their districts. At the appointed place and time the chief constable met his subordinates and the masters : the servants also assembled, all "cladde," as Henry Best describes them,[1] "in their best apparrell," in the market square, the churchyard, or some other public place. The chief constable took the lists, called each master in turn according to the entries, and asked him whether he was willing to set such and such a servant at liberty. If the master replied in the negative, the constable stated what were the wages fixed by the Justices, received a penny fee from the master, and bound the servant for a second term of a year. If the answer was in the affirmative, the constable received from the servant a fee of twopence, and gave him his certificate of lawful departure. Meanwhile masters who wished to hire labourers, whether men or women, walked about among the assembled crowd in order to choose likely-looking servants. When a master had made his choice, his first enquiry was whether the man was at liberty. If the servant had his ticket, the master took him aside, and asked where he was born, where he was last employed, and what he could do. Best once heard the answer :

> " I can sowe,
> I can mowe,
> And I can stacke,
> And I can doe
> My master too,
> When my master turnes his backe."

If the last employer was present at the sitting, he was sought out,

---

[1] *Rural Economy in Yorkshire in 1641, being the Farming and account Books of Henry Best* (Surtees Society, vol. xxxiii. 1857), pp. 132-6.

and asked whether the man-servant was " true and trustie . . . gentle and quiett . . . addicted to company-keepinge or noe," or whether the woman-servant was a good milker, not " of a sluggish and sleepie disposition for dainger of fire." Then followed the bargaining for wages. Sometimes the servant asked for a " gods-penny " on striking the bargain, " or an old suite, a payre of breeches, an olde hatte, or a payre of shoes ; and mayde servants to have an apron, smocke, or both." Sometimes it was a condition to have so many sheep wintered and summered with the master's flock, and to have the twopence which was paid for the certificate refunded before handing over the ticket to the new master. Once hired, the servant could not leave the master, nor the master dismiss the servant, without a quarter's warning. In Yorkshire a servant liked to come to a new place on Tuesday or Thursday. Monday was counted an unlucky day, and the proverb ran :

> " Monday flitte
> Never sitte."

Farming annals are comparatively silent as to the conditions in which day-labourers for hire lived in the reign of Elizabeth. But in one respect, as has been said, they undoubtedly shared the general prosperity. Though their wages remained low, and only fitfully rose as the purchasing power of money declined, they were more secure of employment. In the increased demand for labour resulting from improved methods of agriculture lay their best hopes for the future. It is probable that the decay and ultimate dissolution of the monasteries had for the time inflicted a heavy blow on the development of agriculture as an art. To English farming in the early centuries the monks were what capitalist land-lords became in the eighteenth century. They were the most scientific farmers of the day : they had access to the practical learning of the ancients ; their intercourse with their brethren abroad gave them opportunities of benefiting by foreign experience which were denied to their lay contemporaries. Already, however, there were signs that their places as pioneers would be occupied. Throughout Europe agricultural literature was commencing, and writers were at work urging upon farmers the improved methods which enclosure revealed to them. In Italy Tarello and the translators of Crescentius, in the Low Countries Heresbach, in France Charles Estienne and Bernard Palissy, in England Fitzherbert and Tusser, wrote upon farming. It was not long before the gentry began to pay attention to agriculture. As Michel de l'Hôpital solaced his exile with a farm at Étampes, so Sir Richard Weston in the reign of Charles I., and Townshend in that of George II., occupied their leisure in farming, and in their retirement conferred greater benefits on the well-being of England than they had ever done by their political activities.

Up to the sixteenth century Walter of Henley's farming treatise had held the field. Now it was superseded. In 1523 appeared the *Boke of Husbandrye*, " compyled," as Berthelet says in his edition of 1534, "sometyme by mayster FitzHerbarde, of Charytie and good zele that he bare to the weale of this moost noble realme, whiche he dydde not in his youthe, but after he had exercysed husbandry with greate experyence XL yeres." In the same year was also printed, by the same author, the *Boke of Surveyinge and Improvements*. The *Book of Husbandry* is a minutely practical work on farming, written by a man familiar with the Peak of Derbyshire and by a horsebreeder on a large scale who possessed " 60 mares or more." The *Book of Surveying* is a treatise on the relations of landlord and tenant and on the best methods of developing an estate. Only an experienced farmer could have written the first ; the second required no greater acquaintance with law than might be acquired by a shrewd landowner in the administration of an estate. The authorship of the two books has been claimed for Anthony Fitz-herbert, who was knighted in 1521-2 on becoming a Justice of the Common Pleas, and also for his elder brother John Fitzherbert.[1] It is difficult to credit the Judge—immersed in judicial and political duties, and absorbed in the composition of legal works—with the practical knowledge of farming displayed in the *Book of Husbandry*.

[1] The dispute as to the authorship of the Books of *Husbandry* and *Surveying* is ancient. Professor Skeat (*Introduction to the Book of Husbandry*, English Dialect Society, 1882), and Mr. Rigg (*Dictionary of National Biography*) champion Sir Anthony : the Rev. Reginald Fitzherbert (*English Historical Review*, April, 1897), Sir Ernest Clarke, whose knowledge of agricultural bibliography is unrivalled (*Transactions of Bibliog. Soc.* 1896, p. 160), and Mr. Gay (*Quarterly Journal of Economics,* 1904) support the elder brother, John. The Catalogue of the British Museum now attributes the authorship of both books to John Fitzherbert. Berthelet, who printed the edition of 1534, speaks of the author, in the passage quoted in the text, as though he were dead. This would be true of John Fitzherbert, who died in 1531, but not of Sir Anthony, who lived till 1538. The "XL yeres " experience, from which the author wrote, could not be claimed by Sir Anthony in 1523 ; it might well have belonged to John, who was his elder brother. It is known that John Fitzherbert was for four years a student at the Inns of Court, where he might have laid the foundation of his legal knowledge.

It is much less difficult to imagine that John Fitzherbert should combine minute experience of agricultural details with a sufficient knowledge of law to write the *Book of Surveying*. At any rate, the *Book of Husbandry* became, and for more than half a century remained, a standard work on English farming.

Thirty-four years later appeared Thomas Tusser's *Hundreth Good Pointes of Husbandrie* (1557). The work was afterwards expanded into *Five Hundreth Pointes of Good Husbandrie, united to as many Good Pointes of Huswifery* (1573). Like Fitzherbert, Tusser was a champion of enclosures, and his evidence is the more valuable because he was not only an Essex man, a Suffolk and a Norfolk farmer, but began to write when the agitation against enclosures in the eastern counties was at its height. His own life proved the difficulty of combining practice with science, or farming with poetry. " He spread his bread," says Fuller, " with all sorts of butter, yet none would ever stick thereon." He was successively "a musician, schoolmaster, serving-man, husbandman, grazier, poet—more skilful in all than thriving in his vocation." To the present generation he is little more than a name. But his doggerel poems are a rich storehouse of proverbial wisdom, and of information respecting the rural life, domestic economy, and agricultural practices of our Elizabethan ancestors. His work was repeatedly reprinted. It is also often quoted by subsequent writers, as, for example, by Henry Best in his *Farming Book* (1641), by Walter Blith in his *English Improver Improved* (1649), and by Worlidge in the *Systema Agriculturae* (1668-9). The practical parts of the poem were edited in 1710 by David Hillman under the title of *Tusser Redivivus*, with a commentary which continually contrasts Elizabethan practices with those of farmers in the reign of Queen Anne. When Lord Molesworth in 1723 proposed the foundation of agricultural schools, he advised that Tusser's " Five hundred points of good husbandry " should be " taught to the boys to read, to copy and get by heart."

From the pages of Fitzherbert and Tusser may be gathered a picture of Tudor agriculture at the time when Elizabeth came to the throne. But even in this literature, which probably represents the most progressive theory and practice of farming, it is difficult to trace any important change, still less any distinct advance on thirteenth century methods. Here and there, on the contrary, there are signs that farmers had gone backwards instead of forwards. Agricultural implements remained unaltered. Ploughs were still the same heavy, cumbrous instruments, though several varieties are mentioned as adapted to the different soils of the country. But Fitzherbert was familiar with the same device for regulating the depth and breadth of furrows, which was one of the most notable improvements in the eighteenth century ploughs. Oxen were still preferred to horses for ploughing purposes by both Fitzherbert and Tusser. Iron was more used in the construction of ploughs ; both share and coulter were more generally of iron, and the latter was well steeled. Iron also entered more largely into the building of waggons. Instead of the broad wheels made entirely of wood, Fitzherbert recommends narrower wheels, bound with iron, as more lasting and lighter in the draught. So long as artificial grasses and roots were unknown, the farmer's year necessarily remained the same—its calendar of seasonable operations regulated by the recurrence of saints' days and festivals, and controlled by a belief in planetary influences as unscientific as that of Old Moore or Zadkiel. Since the Middle Ages, the only addition to agricultural resources had been hops, introduced into the eastern counties from Flanders at the end of the fifteenth century. The date 1524, which is usually given for their introduction, is too late ; so also is the rhyme, of which there are several variations :

> " Hops, reformation, bays, and beer,
>   Came into England all in one year."

Hops were apparently unknown in 1523 to Fitzherbert in Derbyshire ; but in 1552 they were sufficiently important to be made the subject of special legislation by Edward VI. In Tusser's day they were extensively cultivated in Suffolk. On enclosed land their cultivation rapidly increased. Harrison (1577) questions whether any better are to be found than those grown in England. Reginald Scot, himself a man of Kent, published his *Perfite Platforme of a Hoppe Garden* in 1574, with minute instructions for the growing, picking, drying and packing of hops. The book was reprinted in 1575, and again in 1576. It was still the standard work in 1651. In Hartlib's *Legacie* it is called " an excellent Treatise, to the which little or nothing hath been added, though the best part of an hundred years are since past."

Fitzherbert starts his *Book of Husbandry* with the month of January. But Tusser begins his farmer's year at Michaelmas as the usual date of entry. Both writers note that an open-field farmer entered by custom on his fallows on the preceding Lady-Day, in order that he might get or keep them in good heart for his autumn sowing. As the Julian Calendar was still in force, the dates are twelve days earlier than they would be under the present Gregorian Calendar. Even with this difference, few farmers of to-day would accept Tusser's advice to sow oats and barley in January ; they would be more likely to agree with Fitzherbert that the beginning of March is soon enough. All wheat and rye were sown in the autumn,—from August onwards,—and the heaviest grain was selected for seed by means of the casting shovel. Neither of the writers speak of spring wheat, possibly because the preparation for it would not fit in with the rigid rules of open-field farming ; but both mention other varieties in the three corn crops. Fitzherbert thinks that red wheat, sprot barley, and red oats are the best, and peck wheat, bere barley, and rough oats the worst varieties. Mixed crops were popular, such as dredge, or barley and oats ; bolymong, or oats, pease, and vetches ; and wheat and rye. As to the mixed sowing of wheat and rye, the authors differ. Probably their respective experiences in Derbyshire and Suffolk diverged. Fitzherbert advises that wheat and rye should be sown together, as the blend makes the safest crop and the best for the husbandman's household ; but he recommends that white wheat be chosen because it is the quickest to arrive at maturity.[1] He was therefore no believer in the slowness of rye to ripen. Tusser, on the other hand, condemns the practice of sowing the two corns together because of the slow maturity of rye as compared with the relative rapidity of wheat. If they are to be blended, he says, let it be done by the miller. The seed was to be covered in as soon as possible. On the time-honoured question whether rooks are greater malefactors than benefactors,—whether they prefer grubs and worms to grain,— neither writer has any doubt. Both give their verdict against the bird, in the spirit of the legislation of their day.[2] As soon as the corn is in, says Fitzherbert, it should be harrowed, or " croues, doues, and other Byrdes wyll eate and beare away the cornes." Tusser advises that girls should be armed with slings, and boys with bows, " to scare away pigeon, the rook, and the crow." Both writers urge the preparation of a fine tilth for barley,—in rural

[1] Henry Best, writing a century later (1641), preferred " Kentish wheate . . . or that which (hereabouts) is called Dodde-reade " (*Farming Book*, p. 45).

[2] *E.g.* 24 Hen. VIII. c. 10 ; 8 Eliz. c. 15.

phrase " as fine as an ant-hill,"—and advise that it should be rolled. Tusser recommends that wheat should also be rolled, if the land is sufficiently dry. For seeding, Fitzherbert adopts the mediaeval rule of two bushels of wheat and rye to the acre. All seeds were scattered broadcast by the hand from the hopper. Neither writer mentions the dibbing of beans, though that useful practice had been introduced by thirteenth century farmers. For barley, oats, and " codware," Fitzherbert recommends a thicker seeding than was practised in mediaeval farming. The best yield per acre is obtained from moderate or thin sowing. But it has been suggested that Elizabethan farmers more often allowed their land to become foul, and that crops were more thickly sown in the hope of saving them from being smothered. The suggestion is perhaps confirmed by the space which Fitzherbert devotes to weeds, and by his careful description of the most noxious plants. At harvest, wheat and rye were generally cut with the sickle, and barley and oats were mown with the scythe. Fitzherbert advises that corn ricks should be built on scaffolds and not on the ground. In the eighteenth century the advice was still given and still unheeded.

In their treatment of drainage and manure, neither author makes any advance on mediaeval practice. To prevent excessive wetness, both advise a water-furrow to be drawn across the ridges on the lowest part of the land ; but neither describes the shallow drains, filled with stones, and covered in with turf, which were familiar to farmers in the Middle Ages. Mole-heaps, if carefully spread, are not an unmixed evil. But when Tusser champions the mole as a useful drainer of wet pastures, it is evident that the science of draining was yet unborn. In choice of manure, neither writer appears to command the resources of his ancestors. The want of fertilising agencies was then, and may even now prove to be, one of the obstacles to small holdings. At the present day the small cultivator can, if he has money enough, buy chemical manures, and, unlike his Elizabethan ancestor, he no longer uses his straw or the dung of his cattle as fuel. But when chemical manures were unknown, it was imperatively necessary to employ all natural fertilisers. Fitzherbert does indeed deplore the disappearance of the practice of marling.[1] But Tusser does not mention the value

[1] Arthur Standish, writing in 1611, says that straw and dung were used as fuel (*The Commons Complaint*, p. 2), and Markham (*Enrichment of the Weald of Kent*) shows the antiquity of the practice of marling by saying that trees of 200 or 300 years old may be seen in " innumerable " spent marl-pits.

of marl, lime, chalk, soot, or town refuse, all of which were used in
the Middle Ages, and it is doubtful whether mediaeval farmers
followed his practice of rotting straw in pits filled with water, or of
carting manure on to the land and leaving it in heaps for a month
before it was spread or ploughed in.   One new practice, and that
a miserable one, is recommended.   It is suggested that buck-wheat
should be sown and ploughed in, in order to enrich the soil.

Both Tusser and Fitzherbert advise that on open-field land the
sheep should be folded from May to early in September.   But
Fitzherbert believed that folding fostered the scab.   Among the
practical advantages of enclosures which he urges is the opportunity
that they afforded to farmers of dispensing with the common fold,
saving the fees to the common shepherd and the cost of hurdles and
stakes, and keeping their flocks in better health.   June was the
month for shearing.   Fitzherbert recommends that sheep should
be carefully washed before they were shorn, " the which shall be
to the owner greate profyte " in the sale of his wool.   Probably
the modern farmer has found that his unwashed wool at a greater
weight but a lower price is worth as much as his washed wool at
less weight and a higher price.   Fitzherbert considers sheep to be
" the most profitable cattle that any man can have."   But, until
the introduction of turnips, the true value of sheep on arable land
could not be realised.   Hence the two branches of farming, which
are now combined with advantage to both the sheep farmer and
the corn-grower, were entirely dissevered.   Until clover, artificial
grasses, turnips, swedes, mangolds took their place among the
ordinary crops for which arable land was cultivated, no farmer
experienced the full truth of the saying that the foot of the sheep
turns sand into gold.   The practice of milking ewes still continued.
Fitzherbert condemns it ; but Tusser, though he notices the injurious results, weakens the effect of his warning by promising that
five ewes will give as much milk as one cow.   Neither Fitzherbert
nor Tusser has anything to say on the improvement of breeds of
cattle for the special purposes that they serve.   The " general
utility " animal was still their ideal.   Yet the root of the matter is
in Fitzherbert, when he says that a man cannot thrive by corn
unless he have live-stock, and that the man who tries to keep live-
stock without corn is either " a buyer, a borrower, or a beggar."
If once the difficulty of winter keep could be solved, here was the
secret of mixed husbandry realised, and the truth of the maxim

verified that a full bullock yard makes a full stack-yard. On horses and horse-dealing Fitzherbert is full of shrewdness. He defines the horse-master, the " corser " and the " horse leche." " And whan these three be mette," he dryly observes, " if yeh adde a poty-carye to make the fourthe, ye myghte have suche foure, that it were harde to truste the best of them."

The times at which Fitzherbert and Tusser respectively wrote give special interest to their championship of enclosures. As has been already noticed, both wrote when the agitation against the progress of the movement was at its height, and Tusser was familiar with the eastern counties at the moment of Kett's insurrection in Norfolk. As practical farmers both writers insist on the evils of the open-field system ; but it fell within the province of neither to criticise the tyrannical proceedings by which those evils were often remedied. They rather dwell on the superior yield of enclosed lands,[1] and on the obstacles to successful farming presented by open-fields—the perpetual disputes, the damage to crops, the waste of land by the multitude of drift-ways, the cost of swineherds, cowherds, and shepherds who were employed as human fences to the corn and meadows. Incidentally also they reveal many practical difficulties of the open-field farmer in ploughing and draining. During the winter months, he was obliged to bring his live-stock in sooner, keep them longer, and feed them at greater cost, than his neighbour on enclosed land. For winter keep, when his hay and straw were running out, he had nothing to rely on but " browse " or tree-loppings. In rearing live-stock he was heavily handicapped. Unless he had pasture of his own, he was forced to time his lambs to fall towards the middle of March. Hence the proverb :

> " At St. Luke's day (Oct. 18, Greg. Cal.)
> Let tup have play."

Thus he risked losing lambs because the common shepherd had too much on his hands at once ; his lambs lost a month on the meadow before it was put up for hay ; and the owner missed the profits of an early sale at Helenmas (May 21) and had to sell, if he sold at all, at the same time as all other open-field farmers. The same restrictions hampered him in rearing calves. He could not afford to keep the cow and calf in the winter ; therefore he was obliged to time the calf to come after Candlemas.

These and other disadvantages convinced practical agriculturists

---

[1] See ch. iii. pp. 65-66.

of the inferiority of the open-field system. Experience was in favour of enclosures. Fitzherbert points to the prosperity of Essex as an example of the advantage of enclosures. The author of the *Compendious or Briefe Examination* says that " the countries where most enclosures be are most wealthie, as Essex, Kent, Devenshire." So also Tusser compares " champion " (open) counties, like Norfolk and Cambridgeshire, with " enclosed " counties, like Essex and Suffolk and says that the latter have

> " More plenty of mutton and biefe,
> Corne, butter, and cheese of the best,
> More wealth anywhere, to be briefe,
> More people, more handsome and prest. . . ."

The proverbial expression " Suffolk stiles " seems to point to the early extinction of open-fields. Norden in his *Essex Described* [1] (1594) calls the county the " Englishe Goshen, the fattest of the Lande ; comparable to Palestina, that flowed with milke and hunnye." So " manie and sweete " were the " commodeties " of Essex, that they compensated for the "moste cruell quarterne fever" which he caught among its low-lying lands. Every practical argument that could be pleaded against open-field farms in the days of Henry VIII. or Elizabeth might be urged against the system with treble force from the end of the eighteenth century onwards, when farming had grown more scientific, when new crops had been introduced, when drainage had been reduced to a science, and when, under the pressure of a rapidly increasing population, farms were becoming factories of bread and meat.

Enclosures undoubtedly assisted farming progress. Before the end of the reign the effect of the movement, combined with increased facilities of communication, is distinctly visible. Under the spur which individual occupation and better markets gave to enterprise, " the soil," as Harrison says, " had growne to be more fruitful, and the countryman more painful, more careful, and more skilful for recompense of gain." Increased attention was paid to manuring. In Cornwall, farmers rode many miles for sand and brought it home on horseback ; sea-weed was extensively used in South Wales ; in Sussex, lime was fetched from a distance at heavy expense ; in Hertfordshire, the sweepings of the streets were bought up for use on the land. The yield of corn per acre was rising. On the well-tilled and dressed acre, we are told that wheat now averaged

twenty bushels, and that barley sometimes rose to thirty-two
bushels, and oats and beans to forty bushels. The improvement of
pastures is shown in the increased size and weight of live-stock.
The average dead weight of sheep and cattle in 1500 probably did
not exceed 28 lbs. and 320 lbs. respectively. At the beginning of
the seventeenth century the dead weight of the oxen and sheep
supplied to the Prince of Wales's household was no doubt exceptional ; but the difference is considerable. " An ox should weigh
600 lbs. the four quarters . . . a mutton should weigh 46 lbs. or 44
lbs." A new incentive to improvement in arable farming and stock-
rearing was supplied by the lower price of wool, consequent partly
on over-production, partly on deterioration in quality. This
deterioration was in some cases the result of enclosures. The wool
was sacrificed to the mutton, and the demand for butcher's meat
was not yet sufficient to make the sacrifice profitable. When
English wool first came into the Flemish market, it was distinguished
for its fineness, and sold at a higher rate than its Spanish rival. It
was indispensable for the foreign weaver. The best fleeces were
those of the Ryeland or Herefordshire sheep, for which Leominster
was the principal market. In the days of Skelton, Elynour Rum-
mynge, ale-wife of Leatherhead, had no enviable reputation ; but
when her customers made a payment in kind, she was a shrewd
judge of its value :

> " Some fill their pot full
> Of good Lemster wool."

Drayton's Dowsabel had a " skin as soft as Lemster wool." Rabe-
lais makes Panurge cheapen the flock of Ding-dong ; and when the
latter descants upon the fineness of their wool, the English translator
(Motteux, 1717) compares them to the quality of " Lemynster
wool." From the preamble to a statute of the reign of James I.
(4 Jac. I. c. 2.) it would seem that Ryeland flocks were cotted all
the year. The second price was fetched by Cotswold wool. The
sheep that are kept on downs, heaths and commons produce the
finest, though not the heaviest, fleeces. It was the experience of
Virgil :

> " Si tibi lanicium curae, . . . fuge pabula laeta."

In the same sense wrote Dyer :

> " On spacious airy downs, and gentle hills,
> With grass and thyme o'erspread, and clover wild,
> The fairest flocks rejoice ! "

As the commons and wastes of England began to be extensively enclosed, the quality of the fleece deteriorated. Heavier animals— better suited to fat enclosed pastures, and producing coarser wool— were introduced. English wool lost its pre-eminence abroad ; and, though still commanding high prices, was no longer indispensable for foreign weavers. The loss was to a great extent counterbalanced by increased consumption at home. But, at the time, the decrease in value was at least as influential in checking the conversion of arable land to pasture as were Acts of Parliament.

Open-field farms were not as yet such obstacles to agricultural progress as they became after the discovery of new resources and new rotations of crops which could only be utilised to full advantage on enclosed lands. But already these new sources of wealth were in sight. The great difficulties in the way of mediaeval and Tudor farmers were want of winter keep and lack of means to maintain or restore the fertility of exhausted soils. In the agricultural literature of Elizabeth the remedy for both is dimly suggested.

In 1577 appeared *Foure Bookes of Husbandry*,[1] to which Barnaby Googe, a better poet than Tusser, gave his name. The work was a translation of Heresbach, with 16 additional pages by the translator. Googe mentions Fitzherbert or Tusser as writers worthy to be ranked with " Varro, Columella, and Palladius of Rome " ; advises agriculturists to read " Maister Reynolde Scot's booke of Hoppe Gardens " ; and quotes an imposing list of " Aucthors and Hus-bandes whose aucthorities and observations are used in this book." By this reference he does not necessarily mean that all the men whose names he mentions had written books on farming, but rather that he had consulted those who were reputed to be most skilful in its practice. In other words, there were already agriculturists, like " Capt. Byngham," " John Somer," " Richard Decryng," " Henry Denys," or " William Pratte," whose methods were an object lesson to their less advanced neighbours. Googe's book has been despised because it was " made in Germany." But in this fact lies its chief value. The farming of the Low Countries was better than the farming of England, and Googe gives English agri-culturists the benefit of foreign experience. He is the first writer to mention a reaping machine—" a lowe kinde of carre with a couple of wheeles and the frunt armed with sharpe syckles, whiche,

[1] *Foure Bookes of Husbandry*, collected by M. Conradus Heresbachius ... Newely Englished and increased by Barnabe Googe Esquire, London, 1577.

forced by the beaste through the corne, did cut down al before it." He insists on the extreme importance of manure, and the value of marl, chalk, and ashes. But he does not consider that farmers can thrive by manure alone. On the contrary, he thinks that " the best doung for ground is the Maister's foot, and the best provender for the house the Maister's eye." He also gives a caution against the persistent use of chalk, because, in the end, it " brings the grounde to be starke nought, whereby the common people have a speache, that grounde enriched with chalke makes a riche father and a beggerly sonne." He mentions the use of rape in the Principality of Cleves, a valuable suggestion whether for green-manuring, for the oil in its seeds, or for use as fodder for sheep. He commends " Trefoil or Burgundian grass," which he believes to be of Moorish origin and Spanish introduction, for " there can be no better fodder devised for cattell." He says that turnips have been found in the Low Countries to be good for live-stock, and that, if sown at Midsummer, they will be ready for winter food. In English gardens turnips were already known. They appear under the name of " turnepez " among " Rotys for a gardyn " in a fifteenth century book of cookery recipes ; Andrew Borde [1] (1542) recommends them " boyled and eaten with flesshe " ; William Turner, the herbalist, mentions that " the great round rape called a turnepe groweth in very great plenty in all Germany and more about London then in any other place of England " : Tusser classes them among " roots to boil and to butter " ; but Googe, though only as a translator, was the first writer to suggest that field cultivation of turnips which revolutionised English farming.

Another Elizabethan writer makes the first attempt to combine science with practice. Sir Hugh Plat was an ingenious inventor, and, as Sir Richard Weston calls him, " the most curious man of his time." He devotes the second part of his *Jewell House of Art and Nature* (1594) to the scientific manuring of arable and pasture land. Manure presents itself to his poetic mind as a Goddess with a Cornucopia in her hand. If land, he says, is perpetually cropped, the earth is robbed of her vegetative salt, and ceases to bear. The object, therefore, of the wise husbandman must be to restore this essential element of fertility. His list of manurial substances is long. He recommends not only farm-yard dung, but marl, lime, street refuse, the subsoil of ponds and " watrie bottomes," salt,

[1] *Dyetary,* ch. xix.

ashes from the burning of stubble, weeds, and bracken ; the hair of beasts, malt dust, soap-ashes, putrified pilchards, garbage of fish, blood offal and the entrails of animals. He warns farmers of the difficulty in discovering the right proportion of marl to lay on different sorts of soil. He condemns the waste of the richest properties of farm-yard manure, and recommends the use of covers to all pits used for its accumulation. He himself used a barn roof at his farm at St. Albans, which moved up and down on upright supports, so that the muck-heap could be raised, yet always remain under cover. In his *Arte of setting of Corne* (1600) he advocates dibbing as superior to broadcast sowing. He traces the origin of the practice to the accident of a silly wench, who deposited some seeds of wheat in holes intended for carrots. He goes so far as to say that, by dibbing, the average yield of wheat per acre would be raised from 4 quarters to 15 quarters !

The growth of an agricultural literature, as well as Googe's list of notable authorities, suggest that landowners were beginning to interest themselves in corn and cattle. Probably their taste for farming was encouraged by the fashionable love for horticulture. In the fourteenth and fifteenth centuries both had declined : in the Tudor age both revived. The garden was the precursor of the home-farm. In the reign of Elizabeth, gardening became one of the pursuits and pleasures of English country life. The art was loved by Bacon ; it was patronised by Burghley and Walsingham ; it gathered round it a rich literature ; it claimed the services of explorers and builders of Empire like Sir Walter Raleigh. Tudor architects used pleasure gardens to carry on and support the lines of their main buildings, and even repeated the patterns of their mural decorations in the geometrical " Knots " of their flower borders ; but they banished kitchen gardens out of sight. The cultivation of vegetables made less progress than that of flowers and fruits. This useful side of horticulture, like farming, was as yet comparatively neglected by the Tudor gentry. But an advance was made. The first step was to recover lost ground. In order to flatter Elizabeth, Harrison probably exaggerated the disuse of vegetables before the accession of her father. He over-states his case when he says that garden-produce, which before was treated as fit for hogs and savage beasts, now supplied not only food for the " poore commons " but " daintie dishes at the tables of delicate merchants, gentlemen, and the nobilitie." It was doubtless true that the art of gardening, like that of farming, had declined during the period which preceded Tudor times. Yet in the decadent fifteenth century, rape, carrots, parsnips, turnips, cabbages, leeks, onions, garlic, as well as numerous " Herbes for Potage," and " Herbes for a salade " appeared in a book on gardens,[1] or in the recipes of cookery books. On the other hand, it is said that, in the reign of Henry VIII., Queen Catherine was provided with salads from Flanders, because none could be furnished at home, and that onions and cabbages, known in the reign of Henry III. and praised by Piers Plowman, were in the first part of the fifteenth century imported from the Low Countries. Now, however, in the reign of Henry VIII. and onwards, gardening, as Fuller says, began to creep out of Holland into England. In Shakespeare's day, it may be remembered that potatoes[2] as yet only " rained from the sky " and that Anne Page would rather

> " be set quick i' the earth,
> And bowled to death with turnips,"

than marry the wrong man. Sandwich became famous for its carrots, and in the neighbourhood of Fulham, and along the Suffolk coast, gardens were laid out in which vegetables were extensively cultivated. In rich men's gardens potatoes found a place after 1585, though for some years to come, they were regarded, and sold, as luxuries. Here then were accumulating new sources of future advance in farming. Yet progress must have been slow. Robert Child, writing anonymously on the " Deficiencies " of agriculture in 1651,[3] says : " Some old men in *Surrey*, where it (the *Art of Gardening*) flourisheth very much at present, report, That they knew the first *Gardiners* that came into those parts, to plant *Cabages*, *Colleflcwers*, and to sowe *Turneps*, *Carrets*, and *Parsnips*, and to sowe *Raith* [early] *Pease*, all of which at that time were great rarities, we having few, or none in *England*, but what came from *Holland* and *Flaunders*." He goes on to say that he could name " places, both in the *North* and *West* of *England*, where the name of *Gardening* and *Howing* is scarcely knowne, in which places a few *Gardiners* might have saved the lives of many poor people, who have starved these dear years."

---

[1] *The Feate of Gardeninge*, by Mayster Ion Gardener, printed in *Archaeologia*, vol. liv., with a glossary by Mrs. Evelyn Cecil.

[2] *Merry Wives of Windsor*, Act. v. Sc. 5 and Act iii. Sc. 4.

[3] Hartlib's *Legacie* (1651), pp. 11-12.

CHAPTER V

FROM JAMES I. TO THE RESTORATION (1603-1660)

FARMING UNDER THE FIRST STEWARTS AND THE COMMONWEALTH

Promise of agricultural progress checked by the Civil War: agricultural writers and their suggestions : Sir Richard Weston on turnips and clover : conservatism of English farmers ; their dislike to book-farming not un reasonable : unexhausted improvements discussed ; Walter Blith on drainage : attempts to drain the fens in the eastern counties ; the resistance of the fenmen : new views on commons : Winstanley's claims : enclosures advocated as a step towards agricultural improvement.

THE beginning of the seventeenth century promised to usher in a new era of agricultural prosperity. During the first four decades of the period prospects steadily brightened. No general improvement in farming practices had been possible until a considerable area of land had been enclosed in one or other of the various forms which enclosures might assume. Under the Tudor sovereigns—in the midst of much agrarian suffering and discontent—this indispensable work had been begun, and it continued throughout the seventeenth century. Estates were consolidated ; small farms were thrown together ; open village farms in considerable numbers gave place to compact and separate freeholds or tenancies ; agrarian partnerships, in which it was no man's interest to be energetic, made way, here and there, for that individual occupation which offered the strongest incentive to enterprise. Thus opportunities were afforded for the introduction of new crops, the application of land to its best use, and the adoption of improved methods. Dairying was extended in the vales of the West and South West ; corn and meat found better and dearer markets ; under the spur of increased profits arable farming again prospered, and the conversion of tillage to pasture was arrested. New materials for agricultural wealth were accumulating ; turnips, already grown in English gardens, were recommended for field cultivation ; twenty years later, potatoes were suggested as a farming crop ; the value of clover and other artificial grasses had been recognised, and urged upon English farmers. Methods became less barbarous. An Act of Parliament was passed " agaynst plowynge by the taile," and the custom of " pulling off the wool yearly from living sheep " was declared illegal. Drainage was discussed with a sense and sagacity which were not rivalled till the nineteenth century. Increased care was given to manuring ; new fertilising agencies were suggested ; the merits of Peruvian guano were explained by G. de la Vega at Lisbon in 1602 ; the use of valuable substances, known to our ancestors but discontinued, was revived. Attention was paid to the improvement of agricultural implements. Patents were taken out for draining machines (Burrell, 1628) ; for new manures (1636) ; for improved courses of husbandry (Chiver, 1637 and 1640) ; for ploughs (Hamilton, 1623 ; Brouncker, 1627 ; Parham, 1634) ; for instruments for mechanical sowing (Ramsey, 1634, and Plattes, 1639). On all sides new energies seemed to be aroused.

Much of the land had changed hands during the preceding century, and the infusion of new blood into the ownership of the soil introduced a more enterprising and business-like spirit into farming. The increased wealth of landowners showed itself in the erection of Jacobean mansions ; farmer owners, tenant-farmers for lives or long terms of years, copyholders at fixed quit-rents, made money. Only the agricultural labourer still suffered. His wages rose more slowly than the prices of the necessaries of life ; his hold on the land was relaxing ; his dependence upon his labour-power became more complete. He was more secure of employment ; but in this respect alone was his lot altered for the better.

The promise of improvement was checked by the outbreak of the Civil War. Excepting those who were directly engaged in the struggle, men seemed to follow their ordinary business and their accustomed pursuits. The story that a crowd of country gentlemen followed the hounds across Marston Moor between the two armies drawn up in hostile array, may not be true ; but it illustrates the temper of a large proportion of the inhabitants. It was the prevailing sense of insecurity, rather than the actual absorption of the whole population in the war, that caused the promise of agricultural progress to perish in the bud. In more settled times under the Commonwealth, farming prospects again brightened. But practical progress was once more suspended by the social changes and political uncertainties of the last half of the seventeenth century. Agriculture languished, if it did not actually decline. It is a significant fact that between 1640 and 1670 not more than six patents were taken out for agricultural improvements. Country gentlemen ceased to interest themselves in farming pursuits. " Our gentry," notes Pepys, " are grown ignorant in everything of good husbandry." Without their initiative progress was almost impossible. Open-field farmers could not change their field-customs without the consent of the whole body of partners. Farmers in individual occupation of their holdings had not, as a general rule, the enterprise, the education, the capital, or the security of tenure, to conduct experiments or adopt improvements.

But the period was one of active preparation. A crowd of agricultural writers followed in the train of Fitzherbert, Tusser, and Googe. Leonard Mascall in his *Booke of Cattell* (1591) had instructed husbandmen in the more skilful " government " of horses, oxen, cattle, and sheep. Gervase Markham wrote on every variety of agricultural subjects, multiplying his treatises under different titles with a rapidity which gained for him the distinction of being the " first English hackwriter," and proved that books on farming found a sale.[1] Horses were made the subject of special treatment. Blundeville's *Fower chiefyst offices belonging to Horsemanshippe* (1565-6) was followed by such books as Markham's *Discourse on Horsemanshippe* (1593) and *How to Chuse, Ride, Trayne,*

[1] As an agricultural writer, Markham's reputation was doubtful, in spite of the many editions which were published of his works. In Hartlib's *Legacie* (1651) R. Child in his " Large Letter " had spoken of the want of a complete book on English husbandry. On this a critic had remarked " England hath a perfect systeme of Husbandry, viz. Markham." The author replies (*Legacie*, 3rd edition, 1655) : " He speaketh more of Markham than ever I heard before, or as yet have seen. In general he is accounted little more than a Translator, unless about Cattle, and yet I cannot but in that question his skill. . . . The works which I have seen of his are, first, the great book translated out of French " (*The Country Farm*, 1616, a revision of Surflet's translation of the *Maison Rustique*, with additions from foreign writers), " which whether well or ill done, I will not declare ; but I am sure our Husbandmen in *England* profit little by it. Secondly I have seen five several bookes bound up together, two or three of which he acknowledgeth to be anothers, as *The Improvement of the Wild of Kent*, also his *Houswifery* he acknowledgeth to have had from a Countess, also part of his *Farewell* is borrowed, and what he owneth, if I have seen all, are very short in many particulars. . . . Yea, if I understand any thing, he setteth down many gross untruths, which every Countryman will contradict." He quotes instances, and concludes " he hath done well in divers things, and is to be commended for his industry."

and Dyet both Hunting and Running Horses (1599), by Grymes's
Honest and Plaine Dealing Farrier (1636), and by John Crawshey's
Countryman's Instructor (1636). Then, as now, horsedealing was a
trial of the sharpest wits, blunted by the fewest scruples. Crawshey,
who describes himself as a " plaine Yorkshire man," warns his
readers against being deceived when buying horses in the market,
" for many men will protest and sweare that they are sound when
they know the contrary, onely for their private gaine." Where so
much is strange in farming matters, it is refreshing to find familiar
features. The proper treatment of woodlands was discussed by
Standish (1611). Rowland Vaughan (1610), struck by the sight of
a streamlet issuing from a mole-heap in a bank, discussed new
methods of irrigation, or " the summer and winter drowning " of
meadows and pasture. Even the smaller profits of farming received
attention. Numerous books were published on orchards, and on
gardens, in which were now accumulating such future stores of
agricultural riches as turnips, carrots, and potatoes. Mascall in
1581 had written on the " husbandlye Ordring of Poultry " ; Sir
Hugh Plat had instructed housewives in the art of fattening fowls
for the table ; and John Partridge published a treatise on the same
subjects, in which he gives recipes for keeping their natural foes at
bay. The following may be recommended to Hunt Secretaries, who
are impoverished by demands on their poultry funds. " Rub your
poultry," says Partridge, " with the juice of Rue or Herbe grasse
and the wesels shall do them no hurt ; if they eate the lungs or
lights of a Foxe, the Foxes shall not eate them." Nor were bees
neglected. Thomas Hill (1568), and Edmund Southerne (1593) had
written on the " right ordering " of bees. But Charles Butler's
Feminine Monarchie (1609), and John Levett's *Orderinge of Bees*
(1634) became the standard authorities on the subject. Both books
were known to Robert Child, author of the *Large Letter* on the
deficiencies of English husbandry, published by Hartlib in 1651.[1]
He says that Butler " hath written so exactly, and upon his owne
experience " that little remained to be added. Henry Best (1641),[2]
however, preferred Levett to any other writer on bee-keeping.
" Hee is the best," he thinks, " that ever writte of this subjeckt."
During the same period men like Gabriel Plattes or Sir Richard

[1] Hartlib's *Legacie*, p. 64.   Robert Child in the 1651 edition speaks of Levett
as " Leveret."

[2] *Farming Book*, p. 68

Weston were suggesting new agricultural methods, or introducing new crops which were destined to change the face of English farming. Plattes (1638), who seems to have been of Flemish origin, urged that corn should be steeped before sowing, and not sown broadcast but set in regular rows. To those who adopted the suggestion of the " corn setter," he promised a yield of a hundred-fold, and he invented a drill to facilitate and cheapen the process. Plattes was on the verge of a great improvement. But men who looked for no larger return than six-fold or eight-fold on the grain sown, regarded his promise as the dream of a visionary who had not travelled beyond the sound of Bow Bells. Unfortunately, the career of Plattes confirmed the contempt with which practical farmers were ready to regard the theories of agricultural writers. Like Tusser, he failed in farming. As Tusser died (1580) in the debtor's prison of the Poultry Compter, so Plattes is said to have died starving and shirtless in the streets of London.[1]

Sir Richard Weston could at least lay claim to thirty years experience in the successful improvement of his estates at Sutton in Surrey " by Fire and Water." He had enriched his heathy land by the process of paring and burning, " which wee call Devonshiring " ; he had also adopted Vaughan's suggestion of irrigation, and proved its value on his own meadows. But the important change with which Weston's name will always be associated is the introduction of a new rotation of crops, founded on the field cultivation of roots and clover. As Brillat-Savarin valued a new dish above a new star, so Arthur Young regards Weston as " a greater benefactor than Newton." He did indeed offer bread and meat to millions. Whether Weston had visited Flanders before 1644 is uncertain. His attempt to make the Wey navigable by means of locks suggests that he was acquainted with the foreign system of canals. On the other hand, his treatise on agriculture implies that he paid his first visit to the country in that year as a refugee. A Royalist and a Catholic, Weston, at the outbreak of the Civil War, was driven into exile, and his estates were sequestrated. He took refuge in Flanders. There he studied the Flemish methods of agriculture, especially their use of flax, clover, and turnips. For the field cultivation of clover he advises that heathy ground should be pared, burned, limed, and well ploughed and harrowed ; that the seed should be sown in April, or the end of March, at the rate

[1] Hartlib's *Legacie* (3rd edition, 1655), p. 183.

of ten pounds of seed to the acre ; that, once sown, the crop should
be left for five years. The results of his observations, embodied in
his *Discours of the Husbandrie used in Brabant and Flanders*, were
written in 1645 and left to his sons as a "*Legacie.*" The subsequent history of the "Legacie" is curious. Circulated in manuscript, an imperfect copy fell into the hands of Samuel Hartlib,
who piratically published it in 1650, with an unctuous dedication
"to the Right Honorable the Council of State." In the following
year Hartlib seems to have learned the name of the author and to
have obtained possession of a more perfect copy. He therefore
wrote two letters to Weston, asking him to correct and enlarge his
"Discourse." Receiving no answer, he republished the treatise in
1651. Eighteen years later, the *Discours* was again appropriated—
this time by Gabriel Reeve, who, in 1670, reprinted it under the
title of *Directions left by a Gentleman to his Sons for the Improvement of Barren and Heathy Land in England and Wales.*

Roots, clover, and artificial grasses subsequently revolutionised
English farming ; but it was more than a century before their use
became at all general. Other crops were pressed by agricultural
writers upon the attention of farmers—such as flax, hemp, hops,
woad and madder for dyes, saffron, liquorice, rape, and coleseed.
A more important suggestion was the field cultivation of potatoes
which hitherto had been treated as exotics, rarely found except in
the gardens of the rich. In 1664 John Forster [1] urged farmers to
grow them in their fields. He distinguishes "Irish Potatoes" from
Spanish, Canadian, or Virginian varieties, points to their success in
Ireland, notices their introduction into Wales and the North of
England, and recommends their trial in other parts of the country.
It was not till the Napoleonic wars that the advice was taken to
any general extent. None of these crops, it may be observed,
could be introduced on an open-field farm, unless the whole body
of agrarian partners agreed to alter their field customs.

Another noteworthy book is the *Legacie* (1651), which passes
under the name of Samuel Hartlib, who has gained undeserved credit
by his piracy of Weston's work. By birth a Pole, Hartlib had come
to England in 1628. By his *Reformation of Schooles* (1642), translated from Comenius, he forced himself on the notice of Milton, who
in 1644 curtly addressed to him his Tract *Of Education.* From
Weston's *Discours*, Hartlib stole the title of the *Legacie* (1651), com-

----
[1] *England's Happiness Increased*, etc., by John Forster *Gent.* 1664.

posed of letters from various writers on the defects of English agriculture, and their remedies. Five-sixths of the *Legacie* are taken up with "A Large Letter . . . written to M. Samuel Hartlib," signed (1655), by R. Child. It throws a clear light on some of the conditions of English farming in the middle of the seventeenth century.

In the "Large Letter" the cumbrousness of the English ploughs, carts, and waggons is noticed. Clumsy implements and bad practices were said to exist side by side with obvious improvements, which yet found no imitators. Some Kentish farmers used "4, 6, yea 12 horses and oxen" in their ploughs, and in Ireland farmers fastened their horses by the tails. Yet in Norfolk the practice was to plough with two horses only, while in Kent itself, a certain Colonel Blunt of Gravesend ploughed with one horse, and an ingenious yeoman had invented a double-furrow plough. Men who perplexed their brains about perpetual motion would, says the writer, have used their ingenuity to more effect if they had tried to improve the implements of agriculture. Cattle-breeding, except "in *Lancashire* and some few *Northern Counties*" was not studied ; no attempt was made to improve the best breeds for milking or for fattening. Dairying needed attention ; butter might be "better *sented* and *tasted*" ; our cheeses were inferior to those of Italy, France, or Holland.[1] Various remedies against the prevalence of smut and mildew in wheat are suggested, including lime, change of seed, early sowing, and the use of bearded wheat. Flax and hemp were unduly neglected, though both might be grown, it is suggested, with profit to agriculturists, and to the great increase of employment ; as a remedy against this persistent neglect, the author advocates compulsory legislation, to force farmers, "even like brutes, to understand their own good." Twenty-one natural substances are recommended as manures, the value of which had been proved by experience. Among them are chalk, marl, lime, farm-yard dung, if it is not too much exposed to the sun and rain ; "snaggreet," or soil full of small shells taken out of rivers, and much used in Surrey ; owse, from marshy ditches or foreshores ; seaweed ; sea-sand, as used in Cornwall ; "folding of sheepe after

---

[1] This also had been the opinion of Googe, who places the Parmesan cheese of Italy first. Then follow, in order of merit, the cheeses of Holland, Normandy, and lastly, of England. Among English cheeses the best came from Cheshire, Shropshire, Banbury, Suffolk, and Essex. "The very worste" is "the Kentish cheese."

the Flaunders manner, (viz.) under a covert, in which earth is
strawed about 6 inches thick "; ashes, soot, pigeons' dung,
malt-dust, blood, shavings of horn, woollen rags as used in
Hertfordshire, Oxfordshire, and Kent. It need scarcely be pointed
out that for none of these fertilisers was the agriculturist indebted
to chemistry, and that no attempt was as yet made to restore to
the soil the special properties of which it is impoverished by particular crops. To meadows and pasture no attention was paid ;
mole-heaps and ant-hills were not spread and levelled ; in laying
down land to graze, little care was taken to sow the best and sweetest
grasses. Clover, sainfoin, and lucerne were generally ignored.
The practice of " soiling," that is, of cutting clover green as fodder
for cattle, is, however, commended. Large tracts of land were
allowed to lie waste, so " that there are more *waste* lands in *England*
than in all Europe besides, considering the quantity of land."
Among the waste lands he includes " dry heathy commons." " I
know," he adds, " that poore people will cry out against me because
I call these waste lands : but it's no matter."

The destruction of woods for fuel is condemned. For this consumption the glass furnaces of the South, the salt " wiches " of
Cheshire, and, above all, the iron-works of Surrey, Sussex, and
other counties, were responsible. The writer probably alludes to
" Dud " Dudley's experiments, when he expresses the hope that
the difficulties of using " sea-coal " for the smelting of iron might
be overcome so as to save our timber. Experiments were not
sufficiently tried, and a " *Colledge of Experiments*," already recommended by Gabriel Plattes, is once more suggested. Men do not
know where to go if they want advice, or to obtain reliable seeds
and plants. Some means was needed of bringing home to other
husbandmen a knowledge of the improvements made by their more
skilful brethren. Another deficiency in English husbandry was its
insular repugnance to foreign methods and new-fangled crops.
Men objected that the new seeds " will not grow here with us, for
our forefathers never used them. To these I reply and ask them,
how they know ? have they tryed ? Idlenesse never wants an
excuse ; and why might not our forefathers upon the same ground
have held their hands in their pockets, and have said, that *Wheat*,
and *Barley*, would not have grown amongst us ? " The same complaint, it may be added, is made by Walter Blith in *The English
Improver Improved* : " The fourth and last abuse is a calumniating

and depraving every new Invention ; of this most culpable are your mouldy old leavened husbandmen, who themselves and their forefathers have been accustomed to such a course of husbandry as they will practise, and no other ; their resolution is so fixed, no issues or events whatsoever shall change them. If their neighbour hath as much corn of one Acre as they of two upon the same land, or if another plow the same land for strength and nature with two horses and one man as well as he, and have as good corn, as he hath been used with four horses and two men yet so he will continue. Or if an Improvement be discovered to him and all his neighbours, hee'l oppose it and degrade it. What forsooth saith he, who taught you more wit than your forefathers ? " Seventeenth century farmers did not lack descendants in later generations. It took a heavy hammer and many blows to drive a nail through heart of oak.

It would be unjust to lay on agriculturists the whole blame for neglect of improvements. Much deserves to rest on the agricultural writers themselves. Their promises were often exaggerated beyond the bounds of belief ; mixed with some useful suggestions were others which were either ridiculous or of doubtful value. Men actually and practically engaged in cultivating the soil were, therefore, justified in some distrust of book-farmers. Turnips were undoubtedly an invaluable addition to agricultural resources. But it was an exaggeration to say with Adolphus Speed [1] that they were the only food for cattle, swine, and poultry, sovereign for conditioning " Hunting dogs," an admirable ingredient for bread, affording " two very good crops " each year, supplying " very good Syder " and " exceeding good Oyl." Nor was confidence in Speed's advice on other topics likely to be inspired by his promise that land, rented at £200 a year, might be made to realise a net annual profit of £2000 by keeping rabbits. Similarly the remedy which is suggested in Hartlib's *Legacie* (3rd edition, 1655) " against the Rot, and other diseases in Sheep and Horses " is enough to cast suspicion on the whole book : " Take Serpents or (which is better) Vipers," advises the writer, " cut their heads and tayls off and dry the rest to powder. Mingle this powder with salt, and give a few grains of it so mingled now and then to your Horses and Sheep." Other suggested remedies are, at least, more easy of application. " The colicke or pain in the belly (in oxen) is put away in the beholding of geese in the water, specially duckes." If a horse sickens from

[1] *Adam out of Eden* (1659).

some mysterious ailment, " a piece of fern-root placed under his tongue will make him immediately voyde, upward and downward, whatsoever is in his body, and presently amende." Again, neither silkworms nor vineyards, though both are favourites with the Stewart theorists, commend themselves strongly as a safe livelihood to practical men who farmed under an English climate. Nor was it possible to take seriously the proposed introduction of " *Black Foxes, Muske-cats, Sables, Martines*," etc., suggested by Robert Child, the author of the principal tract in the *Legacie*, as an addition to the agricultural wealth of the country. He adds to his list " the Elephant, the greatest, wisest, and longest-lived of all beasts . . . very serviceable for carriage (15 men usually riding on his backe together)." It would have added variety to English rural life to see the partners in a village farm conveyed to their holdings on the back of a co-operative elephant, and dropping off as they arrived at their respective strips. But it is doubtful whether they would have found their four-footed omnibus " not chargeable to keepe." Literary and experimental agriculturists naturally gained a reputation similar to that of quack medicine vendors. In practice they often failed. Like ancient alchemists, they starved in the midst of their golden dreams. Tusser, teaching thrift, never throve. Gabriel Plattes, the corn setter, died for want of bread. Donaldson, the author of the first Scottish agricultural treatise, admits that he took to writing books because he could not succeed on the land. Even Arthur Young failed twice in farm management before he began his invaluable tours.

In the " Large Letter " on the defects of English farming, and their remedies, from which quotations have been already made, Child also notices the amount of land that lay waste from want of drainage. This was one of the crying needs of agriculture. Without extensive drainage, the introduction of new crops and improved practices was impossible. With the hour comes the man. The necessity and methods of drainage were ably discussed by Walter Blith. Writing as " a lover of Ingenuity," he published his *English Improver* in 1649. His treatise, interlarded with biblical quotations, was the first which dealt with draining. As the Puritans of the day sought Scriptural authority for their political constitution, so the Puritan farmer justifies his advocacy of drainage by references to the Bible. " Can the rush," he asks with Bildad, " grow without mire or the flagg without water ? " In other ways also Blith's work is significant of the era of the Civil War. He himself beat his ploughshare into a sword, became a captain in the Roundhead army, dedicated his second edition under the title of *The English Improver Improved* (1652) " to the Right Honourable the Lord Generall Cromwell," adorns it with a portrait of himself arrayed in full military costume, and adds the legend ' *Vive La Re Publick*.'

Among the remedies which Blith suggests for the defects of English farming, he urges the employment of more capital ; enclosures, with due regard to cottiers and labourers ; the abolition of " slavish customs " ; the removal of water-mills ; the extinction of " vermine " ; the recognition of tenant-right. It is an indication of agricultural progress that the question of tenants' improvements should be thus forcing itself to the front. Sir Richard Weston in his *Discours* called attention to the Flemish custom, unknown to him in England, of " *taking a Farm upon Improvement*." In Flanders leases for twenty-one years were taken on condition that " whatsoever four indifferent persons (whereof two to bee chosen by the one, and two by the other) should judg the Farm to bee improved at the end of his Leas, the *Owner* was to paie so much in value to the Tenant for his *i*mproving it." In the Preface to his *Legacie*, Hartlib had imitated Weston in urging the adoption of this custom in England. Blith, who also quotes the Flemish lease with approval, points out the injustice of the English law and the hindrance to all improvements which it created. " If," he says, " a Tenant be at never so great paines or cost for the improvement of his Land, he doth thereby but occasion a greater Rack upon himself, or else invests his Land-Lord into his cost and labour *gratis*, or at best lies at his Land-Lord's mercy for requitall ; which occasions a neglect of all good Husbandry. . . . Now this I humble conceive may be removed, if there were a Law Inacted, by which every Land-Lord should be obliged, either to give him reasonable allowance for his clear Improvement, or else suffer him or his to enjoy it so much longer as till he hath had a proportionable requitall." The question had not yet become acute ; but, with the insecurity of tenure which then prevailed, it was not surprising that tenant-farmers were averse to improvements. Their experience was embodied in the proverbial saying current in Berkshire :

> " He that havocs may sit
> He that improves must flit."

The same experience inspired the popular saying prevalent in the Lowlands of Scotland. Donaldson, in his *Husbandry Anatomised* (1697) says that, when a tenant improves his land, "the Landlord obligeth him either to augment his Rent, or remove, insomuch that it's become a Proverb (and I think none more true), *Bouch and Sit, Improve and Flit.*"

In treating of drainage, Blith deals not only with surface water but the constant action of springs and stagnant bottom water. He urges that no man should attempt to lay out his drains by the eye alone, but by the aid of "a true exact Water Levell," an instrument which he carefully describes and depicts. No drain, he said, could touch the "cold spewing moyst water that feeds the flagg and rush," unless it was "a yard or four feet deep," provided with proper outfalls. The drains were to be filled with elder boughs or with stones, and turfed over. He insists that they should be cut straight, not, as open-field farmers were compelled to cut them— for want of space or from the opposition of their neighbours—with turns and angles. His views are sound and advanced on general schemes of drainage, which, for "the commonwealth's advantage" should, he suggests, be enforced by compulsory powers upon landowners.

When Blith wrote, the condition of the fens had become a matter of national importance. It was now that the great work of draining and reclaiming the drowned district had been for the first time seriously undertaken on a scale commensurate with the magnitude of the task. It is singular that foreigners should have taught the English how to deal, not only with land, but with water. As farmers, the Low Countries were far in advance of England, and from them came the most valuable improvements in agricultural methods, as well as the most useful additions to agricultural resources. Dutchmen drained our fens ; irrigation, warping, canals were all foreign importations. The irrigation of meadows, which M. de Girardin described as a sound insurance against drought, is said to have been first practised in England in modern times by the notorious "Horatio Pallavazene," of Babraham . . . "who robbed the Pope to lend the Queen." Warping was brought from Italy to the Isle of Axholme in the eighteenth century, and by its means the deposits at the estuary of the Humber were converted into "polders." The Dutch and Flemings had mastered the secret of locks and canals long before any attempt was made to render

English rivers navigable, or available for water-carriage in inland districts. The great French " Canal du Midi " was completed in 1681, nearly a century before the example was followed in England. In this connection it may be also noticed that a colony of Walloon emigrants, settled at Thorney towards the middle of the seventeenth century, introduced into the district the practice of paring and burning the coarse tussocks of grass, and the paring plough was long known as the French plough.

Robert Child in his " Large Letter " on the most notable deficiencies of English agriculture, printed in Hartlib's *Legacie* (1651), suggests that the drainage of marshes was not begun till the reign of Elizabeth. " In Qu. Elizabeth's dayes," he writes, " *Ingenuities, Curiosities*, and *Good Husbandry* began to take place, and then *Salt Marshes* began to be fenced from the Seas." In this he is mistaken. Some progress had been made at an earlier date. A number of Acts were passed in the reign of Henry VIII. for the reclamation of marshes and fens by undertakers, who were usually rewarded with half the reclaimed land. Thus Wapping Marsh was reclaimed by Cornelius Vanderdelf in 1544, and the embankment of Plumstead and Greenwich Marshes was begun in the same reign. Isolated marshes had been drained in the eastern counties during the reign of Elizabeth. Norden (*Surveyor's Dialogue*, 1607) says : " much of the Fennes is made lately firme ground, by the skill of one Captaine Lovell, and by M. William Englebert, an excellent Ingenor." But it was not till the reign of Charles I. that any serious attempt was made to deal with the Great Level of the Fens, which extended into the six counties of Cambridge, Lincoln, Hunting-don, Northampton, Suffolk, and Norfolk.

Seventy miles in length, and varying in breadth from ten to thirty miles, the fens comprised an area of nearly 700,000 acres. Now a richly fertile, highly cultivated district, it was, in the seven-teenth century, a wilderness of bogs, pools, and reed-shoals—a vast morass, from which, here and there, emerged a few islands of solid earth. Here dwelt an amphibious population, travelling in punts, walking on stilts, and living mainly by fishing, cutting willows, keeping geese, and wild-fowling. " H. C." who, in 1629, urged upon the Government the *Drayning of Fennes*, paints an unattractive picture of the country : " The Aer Nebulous, Grosse, and full of rotten Harres ; the Water putred and muddy, yea, full of loath-some Vermine ; the Earth spuing, unfast, and boggie ; the Fire noysome turfe and hassocks ; such are the inconveniences of the Drownings." Eight great principal rivers—the Great Ouse, the Cam, the Nene, the Welland, the Glen, the Milden-hall or Lark, the Brandon or Lesser Ouse, and the Stoke or Wissey [1]—carry the upland waters through this wide stretch of flat country towards the sea. Whenever the rains fell, the rivers rose above their banks, and, especially if the wind was blowing from the east or south, flooded the country for miles around. It was only in the map that they reached the ocean at all. Two causes principally contributed to make the country a brackish swamp. The outfalls of the rivers had become silted up so that their mouths were choked by many feet of alluvial deposit.[2] Twice every day the tides rushed up the channels for a considerable distance, forcing back the fresh water, and converting the whole country into one vast shallow bay. Efforts had been made by the Romans to reclaim these flat levels, and their " causey " is still in existence. In the palmy days of the great monasteries of Crowland, Thorney, Ely, and Ramsey, isolated districts were occupied, and highly cultivated. William of Malmesbury, writing in the reign of Henry II. (1143), describes the district round Thorney as " a very Paradise in pleasure and delight ; it resembles heaven itself—it abounds in lofty trees, neither is any waste place in it ; for in some parts there are apple trees, in other vines which either spread upon the ground or run along poles." Such a description applies only to the islands on which the great monasteries were situated. The rest of the country had become, at some unknown period of history, an unproductive bog, affording little benefit to the realm other than fish and fowl, and "overmuch harbour to a rude and almost barbarous sort of lazy and beggarly people."

No important effort was made to reclaim the district till the time of John Morton, Bishop of Ely, afterwards a Cardinal and Archbishop of Canterbury, in the reign of Henry VII. As a grower of strawberries he is enshrined in literature ; but in the history of

---

[1] Sir Jonas Moore, *History of the Great Level of the Fennes* (1685), p. 9 ; Wells, *History of the Bedford Level* (1830), vol. i. p. 6 ; Vermuyden, *Discourse touching the Drayning of the Great Fennes* (1642), p. 4.

[2] Andrewes Burrell, in his *Briefe Relation Discovering Plainely the True Causes why the Great Levell of Fenes . . . have been drowned* (1642), says that, when working for the Earl of Bedford in 1635 in deepening " Wisbeach River," he " discovered a stony bottome upon which there was found lying at severall distances seven boates, which for many yeares had laine buried eight foot under the bottome of the river."

farming his principal achievement was probably suggested by his residence in Flanders from 1483 to 1485 as a political refugee. A cut, forty feet wide and four feet deep, running from Peterborough to Wisbech, still bears the name of "Morton's Leam" and still plays an important part in the drainage of the country. Other local efforts were made, which proved for the most part ineffective. In spite of individual enterprise, the general condition of the district was so deplorable that it attracted the attention of the Government. The fens were surveyed, Commissioners and Courts of Sewers appointed, and an Act (1601) was passed for the drainage of the Great Level. In 1606, under a local Act, a portion of the Isle of Ely was reclaimed, the undertakers receiving two-thirds of the land thus recovered from the water. In 1626 the drainage of Hatfield Chase, Ditchmarsh, and all the lands through which crept the Idle, the Aire, and the Don, was commenced by Cornelius Vermuyden. Three years later, the greater task was attempted of draining that portion of the fens which was afterwards known as the Bedford Level. In 1630 the local gentry who formed the Commissioners of Sewers, contracted with Vermuyden (now Sir Cornelius) to execute the work, and the fourth Earl of Bedford headed the undertaking. The work began vigorously enough. In 1637 the Commissioners of Sewers certified its completion ; but the winter rains flooded the country ; the Earl of Bedford was at the end of his resources ; he had spent £100,000, and was in danger of losing it all. The certificate of completion was reversed. Charles I. intervened ; fresh arrangements were made for the allotment of the recovered land ; a new Company of Adventurers was formed ; Vermuyden still directed the operations, although his skill was attacked by Andrewes Burrell in his *Briefe Relation* (1642). Vermuyden in his defence (*Discourse*, 1642) pleaded that the only purpose of the first Agreement was to make the land "summer ground." The new venture was more ambitious. Though the work was partially suspended during the Civil War, it proceeded under the Commonwealth. In 1649 the fifth Earl of Bedford joined the undertaking, and, four years later, the drainage was finished. New channels and drains were made to carry off the surface water ; existing drains were scoured and straightened ; banks were raised to restrain the rivers within their beds ; new outfalls into the sea, provided with sluices, were made, and old ones deepened and widened ; numerous dams were erected to keep out the sea. In 1652 Sir C.

Vermuyden reported the completion of the work to the Council, saying that " wheat and other grains, besides unnumerable quantities of sheep, cattle, and other stock were raised, where never had been any before." The Bedford Level was the largest work undertaken. It was also the most complete, though even here for a time there were failures. Other marshes were attacked by improvers, with more or less success. From various causes, however, the water often regained its hold on the country. In some cases the work was only partially finished ; in others, it was so inadequately executed by persons whom Blith calls " mountebank engineers, idle practitioners, and slothful impatient slubberers," that it broke down under the rainfall of the first wet season ; in others, the wind-mills, which were used to raise the water of the interior districts to the levels of the main rivers, could not cope with a flood ; in others, the works were destroyed by the fenmen, and were not really restored till the beginning of the nineteenth century.

The marshes were to fenmen what wastes and commons were to dwellers on their verge. Catching pike and plucking geese were more attractive than feeding bullocks or shearing sheep. Any change from desultory industries to the settled labour of agriculture was in itself distasteful to the commoners, and little, if any, com-pensation was made for their rights or claims to pasture, turf-cutting, fishing, or fowling. All over the fen districts there were, on the one side, outbursts of popular indignation, and, on the other, complaints of the " riotous letts and disturbances of lewd persons." The commoners were called to arms by some Tyrtaeus of the fens, whose doggerel verses have been preserved by Dugdale in his *History of Imbanking and Draining* :

Come, brethren of the water, and let us all assemble,
To treat upon this matter which makes us quake and tremble,
For we shall rue it, if't be true, the Fens be undertaken,
And where we feed in fen and reed, they'll feed both beef and bacon.

*   *   *   *   *   *   *   *   *   *   *   *   *

Behold the great design, which they do now determine,
Will make our bodies pine, a prey to crows and vermin ;
For they do mean all fens to drain and waters overmaster ;
All will be dry and we must die, 'cause Essex calves want pasture.

*   *   *   *   *   *   *   *   *   *   *   *   *

The feathered fowls have wings to fly to other nations,
But we have no such things to help our transportations ;
We must give place (oh grievous case !) to hornéd beasts and cattle,
Except that we can all agree to drive them out by battle.

Wherefore let us entreat our ancient water nurses
To show their power so great as t'help to drain their purses,
And send us good old Captain Flood to lead us out to battle,
Then Twopenny Jack with skales on's back will drive out all the cattle.

The Civil Wars gave the fenmen their opportunity. Vermuyden seems to have been personally unpopular : he was a Zealander ; most of his workmen were foreigners ; the adventurers who settled on the lands which they had reclaimed were French or Dutch Protestants. The commoners, moving swiftly and silently in their boats, broke down the embankments, fired the mills, filled up the drains, levelled the enclosures, turned their cattle into the standing corn. They attacked the workmen, threw some of them into the river, held them under the water with poles, and burnt their tools. The perpetrators of the outrages worked so secretly that they could rarely be identified. Sometimes their action was bold and open. In the neighbourhood of Hatfield Chase, near the Isle of Axholme, every day for seven weeks, gangs of commoners, armed with muskets, drew up the flood-gates so as to let in the flowing tide, and at every ebb shut the sluices, threatening that they "would stay till the whole level was well drowned, and the inhabitants forced to swim away like ducks." Even the religion of the French and Dutch Protestants was not respected. From Epworth in 1656 comes their petition that the fenmen had made their church a slaughter-house and a burying-place for carrion. Major-General Whalley was entrusted by Parliament with the task of protecting the adventurers. But agitators like Lilburne and Noddel were at work among his soldiers, and the commoners showed no respect for the authority of Parliament. "They could make as good a Parliament themselves ; it was a Parliament of clouts." In some cases the resistance of the fenmen secured them further concessions ; in others they succeeded in destroying the works of the undertakers. It was not till after 1714 that the riots caused by the reclamation ceased to disturb the peace of the country. By that time the object was partially achieved, and many of the swamps and marshes of the fen districts were restored to the ague-shivering, fever-stricken inhabitants in their primitive unproductiveness.

The struggle for the reclamation of the waste-lands of the water-drowned fens is another aspect of the older land-battle between enclosers and commoners. Men like Robert Child in his *Large Letter* in Hartlib's *Legacie*, or Walter Blith, championed reclamation for the same reasons that they advocated enclosures. The former, writing in 1650 before the drainage was complete, speaks of " that great *Fen* of Lincolneshire, Cambridge, Hungtingdon, consisting, as I am Informed, of 380,000 Acres, which is now almost recovered." " Very great, therefore," he continues, " is the improvement of draining of lands, and our negligence very great, that they have been waste so long, and as yet so continue in divers places : for the improving of a Kingdome is better than the conquering a new one." Blith, writing three years later (*English Improver Improved,* 1652), speaks of the work as finished. " As to the Drayning, or laying dry the Fenns," he says, " those profitable works, the Common-wealths glory, let not Curs Snarl, nor dogs bark thereat, the unparralleld advantages of the World." But when these and other writers of the period dealt with enclosures, they treated the subject from a new point of view. As a matter of farming, their arguments were sound. But economic gain might involve social and moral loss, and the Stewart writers on agriculture tried to reconcile the two aspects of the question. In the interests of agricultural progress, they are practically unanimous in their advocacy of individual as opposed to common occupation of arable land. But in the case of commons of pasture, they vigorously defended the claims of the commoners, both tenant-farmers and cottagers. More advanced members of the Republican party went beyond the recognition of pasture rights, and claimed the common, not for the open-field farmers to whose arable holdings it was historically attached, but for the general public—irrespective of claims arising from neighbourhood or from the tillage of adjacent land. On the practical assertion of such claims a curious side-light is thrown by the proceedings of Jerrard Winstanley in 1649.

Winstanley and his friends sought to establish a society having all things in common. With this object they settled on the common lands of St. George's Hill, near Walton-on-Thames, and began to plough, cultivate, and enclose the land. Lord Fairfax's soldiers burned their huts, and turned them off. Winstanley, in the jargon of the day, identified the struggle, in which his personal profits were staked, with the prophetic Armageddon " between the Lamb of Righteousness . . . and the Dragon of Unrighteousness." Needless to say, he found himself a champion of the former. He sets forth his claims in a pamphlet addressed to the General as *A Letter to the Lord Fairfax and his Council of War :* . . . *Proving it an undeniable Equity That the Common People ought to dig, plow, plant, and dwell upon the Commons, without hiring them or paying Rent to any.* He was, he says, opposed by none save " one or two covetous freeholders that would have all the commons to themselves, and that would uphold the Norman tyranny over us, which by the victory that you have got over the Norman successor is plucked up by the roots and therefore ought to be cast away." In other words, the effect of the Civil War and of the defeat of Charles I., as interpreted by his school of thought, was to establish the rights of the people to " have the land freed from the entanglement of lords, lords of manors, and landlords, which are our taskmasters," " to enter on their inheritance," and "dig, plow, plant and dwell upon the Commons " without rent, and improve them " for a public treasury and livelihood." Instead of the existing law, the rule was to be established of " First come, first served." For this appropriation and improvement of the commons the inspiration of the " Lamb of Righteousness " was claimed, so long as the new possessors were Winstanley and his communistic society ; but the same processes were the direct suggestion of the " Dragon of Unrighteousness," if the work was carried out by the adjacent owners and cultivators of the soil. In either case, whoever was the encloser, the general public gained no advantage ; the pasture commons were ploughed, enclosed, and appropriated to individuals.

The episode is significant. Probably Winstanley had, and has, sympathisers. But the views of those practical agriculturists, who were interested in the enclosure and tillage of open-fields and commons in order to accelerate farming progress, were less revolutionary. Had they been carried into effect, much social loss might have been averted. From the purely commercial side, their arguments in favour of converting open-field land into separate holdings and of enclosing the commons and wastes were overwhelming. There need be no depopulation, for tillage would be increased. If the rights of commoners were respected, the social drawbacks to the change might be removed. The whole question was assuming a new form. The improvements in arable farming suggested by Stewart and Commonwealth writers minimised the social loss caused by enclosures, at the same time that they magnified the economic waste of the open-field system.

Tudor farmers had treated arable and pasture farming as two distinct branches, which could not be combined. On open-field land, though some live-stock was maintained by means of commons, the energies of farmers were almost exclusively concentrated on corn. On enclosed land, corn might be comparatively relegated to the background, and the farmer's mainstays were meat, dairy produce, and, if a flock-master, wool. So long as this rigid distinction was maintained, enclosures often meant depopulation and a dwindling wheat-area. Experience was crystallised into the proverb " No balks, no corn." It is true that, towards the end of Elizabeth's reign, the advantages which enclosures gave to the enterprise of the arable farmer were realised, and land began to be fenced off, not for pasture only, but also for tillage. But the economic case for enclosures was enormously strengthened, when the real pivots of mixed husbandry were discovered, and when Stewart agriculturists found that neither turnips, nor clover, nor artificial grasses, nor potatoes, nor drainage, were possible on open-fields which were held in common for half the year. Yet the experience of the previous two hundred years had created a mass of well-founded prejudice, which fought stubbornly against any extension of the practice of enclosing land. It is for this reason, probably, that the best writers of the Stewart and Commonwealth period labour hard to prove that enclosures of open-fields and commons, whatever their past history had been, necessitated neither depopulation nor decay of tillage, and might even promote not only economic but social gain.

In his *Book of Surveying* (1523) Fitzherbert had written on the way " to make a township that is worth 20 marks a year worth £20 a year." His plan was to discover, first, how many acres of arable land each man occupied in the open-fields, how much meadow, and what proportion of common pasture were attached to his holding ; and secondly, by means of exchange, to consolidate these lands, lay them together, and enclose them in several occupation. Every man should have " one little croft or close next to his house." In the *Briefe Examination* (1549) the Doctor, who represents the author's views, only condemns those enclosures of land which were made for the conversion of tillage into pasture, or " without recompence of them that have right to comen therein." It was on this principle that in 1545 the Royal wastes of Hounslow Heath were enclosed under the award of Commissioners, who set out a portion of the heath to each inhabitant ; either as copyholds, or on leases for terms of years.

Sentiments like these became the commonplaces of Stewart and Commonwealth writers. The demand for "three acres and a cow" can show an origin of respectable antiquity. Gabriel Plattes (1639) [1] pleads that all parties would gain by enclosures, landowners by increased rents, clergymen by improved tithes, the poor by increased employment. "I could wish," he adds, "that in every Parish where Commons are inclosed, a corner might be laid to the poore mens houses, that every one might keep a Cow or for the maintenance of his familie two." The wish of the Stewart writer had been expressed by a Tudor predecessor a century earlier. Thomas Becon [2] in 1540 had suggested that landlords should attach to every cottage enough " land to keep a cow or two." Walter Blith [3] argues vigorously in favour of enclosures, and quotes with approval the whole of Tusser's poem comparing " champion " (open) and " severall " land. Of open-field farmers he says " live they do indeed, very many in a mean, low condition, with hunger and care. Better do those in Bridewell. And for the best of them, they live as uncomfortably, moyling and toyling and drudging. What they get they spend." But in all enclosures he expressly makes the condition that all interests should be provided for—not only those of the landlord, but those also of the " Minister to the People," the " Freeholder Farmer or Tenant," and the " Poor Labourer or Cottier." All these, he says, would gain by the process. He takes the last first : " Look what right or Interest he hath in Common, I'll first allot out his proportion into severall with the better rather than with the worse, a Proportion out of everyman's inheritance." At the same time he condemns " depopulating Inclosure . . . such as former oppressive times by the will and power of some cruell Lord either through his greatness or purchased favour at Court, or in the Common Courts of *England*, by his purse and power could do anything, inclose, depopulate, destroy, ruine all Tillage, and convert all to pasture without any other Improvement at all . . . which hath brought men to conceive, that because men did depopulate by Enclosure, therefore it is now impossible to enclose without Depopulation."

To the same effect as Blith writes Robert Child in the " Large

---

[1] *A Discovery of Infinite Treasure, Hidden since the World's Beginning*, by G(abriel) P(lattes), 1639.

[2] *The Jewel of Joy*, 1540.

[3] *English Improver Improved*, ed. 1653.

Letter" in Hartlib's *Legacie* (1651). He regards wastes and commons as defects in English husbandry, and in defence of his position asks eight questions, which he does not attempt to answer, preferring to leave " the determination for wiser heads."

1. " Whether or no these lands might not be improved very much by the *Husbandry of Flaunders* (viz.) by sowing *Flax, Turneps*, great *Clover-Grasse*, if that *Manure* be made by *folding Sheepe* after the *Flaunders* way, to keepe it in heart ?

2. " Whether the *Rottennesse* and *Scabbinesse* of *Sheepe, Murrein* of Cattel, *Diseases* of horses, and in general all diseases of Cattel do not especially proceed from *Commons* ?

3. " If the rich men, who are able to keepe great stockes are not great gainers by them ?

4. " Whether Commons do not rather make poore, by causing idlenesse than maintaine them : and such poor, who are trained up rather for the Gallowes or beggery, than for the Commonwealths service ?

5. " How it cometh to passe, that there are fewest poore, where there are fewest *Commons*, as in *Kent*, where there is scarce 6 *Commons* in the *County* of a considerable greatnesse ? [1]

6. " How many do they see enriched by the *Commons* : and if their Cattel be not usually swept away by the *Rot*, or starved in some hard winters ?

7. " If that poore men might not imploy 2 Acres enclosed to more advantage, than twice as much in a Common ?

" And lastly, if that all Commons were *enclosed*, and part given to the Inhabitants, and part rented out, for a stock to set the poore on worke in every County."

Blith not only quoted Tusser in support of his opinion, but adds that " all that ever I yet saw or read " held the same opinion. " Tis true I have met with one or two small Pieces, as M. Spriggs, and another whose name I remember not, that write against depopulating Inclosure, with whom I freely joyn and approve." It is probable that he alludes to Henry Halhead's *Inclosure Thrown Open* etc. (1650), to which Joshua Sprigge of Banbury contributed a

---

[1] Tusser held the same opinion that poverty and commons go together. In his comparison between " Champion Country and Severall " he writes :

" T'one barefoot and ragged doth go
And ready in winter to starve ;
When t'other you see not do so,
But hath that is needful to serve."

Preface. The tract is an appeal against enclosures, mainly based on past history. It probably belongs to a group of pamphlets dealing with the Midland counties, where the enclosing movement seems to have been active. Halhead describes how would-be enclosers begin by upsetting the field customs by which the cultivation of the land was regulated ; how they tell the people that they will be three times as well off, that enclosure stops strife and contention, " nourisheth Wood in hedges," and keeps sheep from rotting. If they cannot prevail by these promises, they begin a suit at law, and make the resisters dance attendance at the law-courts for months and even years. Then they pull out their purses, and offer to buy them out. If this fails, on goes the suit till a decree against the open-field partners is granted in Chancery. The description bears the stamp of accuracy. But, logically, neither the old methods of enclosing nor the results of the conversion of tillage into pasture really met the case put forward by the new advocates of enclosure as an instrument both of social and agricultural progress.

The case for enclosures of open-field farms and commons is vigorously stated in three tracts ; one by S. Taylor (1652) ; [1] another by Adam Moore (1653) ; [2] the third by Joseph Lee (1656).[3] Their arguments are mainly based on the wretched conditions of the commons, the poor farming of open-field land, and the social and agricultural gain which, as Lee's practical experience had shown, resulted from individual occupation. None of the three authors alludes to the recent discoveries of roots, clover, and grasses, or to the improved methods of drainage, on which Blith and others so strongly relied. Of Taylor nothing is known, except that his tract shows him to have been a vehement assailant of ale-houses. Moore tells us that he was a Somersetshire man. The Rev. Joseph Lee was a Leicestershire " Minister of the Gospel " at Cotesbach, who had been violently attacked by his professional brethren for the

---

[1] *Common Good : or the Improvement of Commons, Forests, and Chases by Inclosure*, by S. T(aylor), 1652.

[2] *Bread for the Poor . . . Promised by Enclosure of the Wastes and Common Grounds of England*, by Adam Moore, Gent. 1653.

[3] Εὐταξία τοῦ ᾿Αγροῦ; *or a Vindication of a Regulated Enclosure*, etc., by Joseph Lee, Minister of the Gospel, 1656.

The contrary view to that taken by Lee was stated by John Moore, Minister of Knaptoft in Leicestershire, whose tract *The Crying Sin of England of not Caring for the Poor*, etc., was published in 1653. Moore's *Scripture Word against Inclosure* (1656) was an answer to Lee.

share he had himself taken in the enclosure of Catthorp Common. In his Epistle to the Reader he explains why he preferred to reply to these attacks by a tract and not by a sermon : " I am very sensible that if our pulpits had sounded more of the things of Christ, and lesse of the things of the world, it had been better with us then it is this day." Part of the tract consists of hard text-fighting ; but its value lies in the facts which he quotes from his own experience.

Enclosures, in the opinion of the three writers, are not only " lawfull " but " laudable. ' They injure none, but profit all. Lee considers that five classes ought to be considered, landlords, ministers, the poor, cottagers, and tenant-farmers. Moore omits the ministers, but asserts the claims of the remaining four classes. All three writers agree that a certain portion of the Commons ought to be set aside for the poor, and the rest proportionately divided. This, says Lee, was the principle adopted at Catthorp. If this were done, there need be no depopulation. In proof Lee mentions a number of parishes in Leicestershire, where the land had been enclosed without any decay of population, houses, or tillage. Neither would it lead to any diminution of useful employment. The same number of maid-servants would be employed ; and though there might be fewer lads, they would be more useful citizens if set to some trade. On the industrial gain thus derived from enclosures the three authors are also agreed. They in effect answer the fourth question asked by Robert Child in the " Large Letter " in Hartlib's *Legacie* with an unhesitating affirmative. At the beginning of the century, Norden (1607) had drawn attention to the character of the squatters who settled on the edges of wastes and commons. He describes them as " people given to little or no labour, living very hardly with oaten bread and sour whey and goat's milk, dwelling far from any church or chapel, . . . as ignorant of God or of any civil course of life as the very savages among the infidels, in a manner which is lamentable and fit to be reformed by the lord of the manor." Fifty years later, according to the three authors, commons were blots on the social life of the nation. Children, says Taylor, are " brought up Lazying upon a Common to attend one Cow and a few sheep," and " being nursed up in idleness in their youth they become indisposed for labor, and then begging is their portion or Theevery their Trade. . . . The two great Nurseries of Idlenesse and Beggery etc. are Ale-houses and

Commons." Taylor says that " people are nowhere more penurious than such as border on commons." " This poverty." he explains, " is due to God's displeasure at the idleness of the Borderers," or commoners. They have no settled industry. They look to the profits of a horse or cow, if they can keep one ; if not, they can at least " compass a goose or a swine." If they have no live-stock at all, they are " sure of furze, fern, bushes, or cowdung, for fuel to keep them warm in winter." They can beguile, writes Moore, the " silly Woodcock and his feathered fellows by tricks and traps of their own painful framing," and so gain money enough to keep them till they have to work again. Sometimes they earn a few shillings by guarding the flocks and herds of others. But, if a sheep or a cow is missing, the " chuck-fists " will not pay them their wage, but suspect them of theft, and proceed against them by law. The Commons are, in fact, " Nurseries of Thieves and Horse-stealers." Lee is of the same opinion that commons fostered idleness. Perhaps, he admits, " 3 or 4 shepherd boyes " by enclosures " will be necessitated to lay aside that idle employment ; . . . destructive to the soules of those Lads, in that, poor creatures, they are brought up by this means without either civill or religious education." When they should have been at school or at church, they were " playing at nine-holes under a bush," while their cattle make a prey on their neighbour's corn, and " they themselves are made a prey to Satan." Other moral gains are alleged that by enclosure an end is put to occasions for litigation and strife between common-field farmers, or for quarrels between herdsmen, and that there are fewer opportunities for pilferings of land and of corn, or for the destruction of a neighbour's crops by turning in horses and cattle under pretence that they have broken loose from their tethers.

It is not true, in the opinion of the three writers, that enclosures necessarily destroy tillage. On the contrary, the cheapness and abundance of corn are due to the opportunities that enclosures afford for breaking up worn-out pastures which yield double the quantities produced on common fields. Nor is it true that enclosers are under a curse so that the land passes out of their families. Instances to the contrary are adduced from Leicestershire, and that cannot be a special curse on enclosures which is a fate common to all other landholders. Enclosure may diminish the number of horses ; but one horse well kept is worth three so " jaded and tyred as are the horses of common-field farmers." Nor is it any tyranny for the majority to enforce enclosure where the whole body of partners are not unanimous. At Catthorp one man with common for seven sheep stood out. The rest overruled him ; but he lost nothing. All that the other commoners did was to enclose their portions of the common away from him. That the agricultural gain is great, scarcely admits of a doubt. On open-fields the corn-land is worn out. It can only be induced to bear at all by constant ploughings and liberal manurings, which absorb all profits in labour and charges. Even then there is often little more than a bare return of seed, poor in quality—" small humble-Bee-Ears with little grains." The pease land is no better ; it may provide enough for seed and keep of the horses ; but it yields no clear profit. The live-stock that are reared on the commons are dwarfed and under-sized ; they are driven long distances to and fro, so that they have neither rest nor quiet. Colts, raised on the commons, by cold and famine come to no good. " Cattle, nurtured there, grow to such brockish and starved stature " that, living, they grieve the owner's eye, and, dead, deceive the Commonwealth. Sheep do better ; but they even are so pinched that they make little profit. One sheep in an enclosure is worth two on a common. There are five rots in the open-fields to one rot in enclosed land. The commons are over-stocked. They are, says Moore, " Pest-houses of disease for cattle. Hither come the Poor, the Blinde, Lame, Tired, Scabbed, Mangie, Rotten, Murrainous." No order is kept ; but milch cows, young beasts, sheep, horses, swine—often unringed—and geese are turned out together. Furze and heath are encouraged by com-moners, because they keep cattle and sheep alive in hard winter when fodder is scarce ; but the same space covered with grass would be more useful. That which is every man's is no man's, and no one tries to better the commons. When it is everybody's interest to improve the pasture, it is nobody's business to do the work.

The whole subject of enclosures had yet to be fought out. From the point of view of production, the change was desirable ; no pressure of population as yet made it necessary. Commons were essential to the existence of those open-field farms, which advocates of agricultural improvement recognised as an obstacle to pro-gress ; but new methods and new resources had as yet hardly advanced beyond theories. Neither the argument from increased productiveness, nor the appeal for progress, had gained their full force. Yet the altered tone of agricultural writers is significant. It was almost as incontestably in favour of enclosure as the tone of Elizabethan writers had been opposed to the process. Generalisation from handfuls of particular instances is always easy. A large tract of country might have been improved and enclosed with the approval of all parties. But there were the widest differences between commons, or between commons and moors, wastes, and bogs. Moore himself reserves his bitterest condemnation for what he calls "marish," as opposed to "uplandish," commons. Stress might be laid on the moral influences of common land either way, and self-interest or bias is always prone to conceal itself under the mask of moral motives. The same rights might encourage industry and thrift, or idleness and crime. It was doubtless illogical to argue that enclosures must always depopulate, whether the change was effected with or without regard to the claims of cottagers and small commoners, or for the purpose of increasing the area either of tillage or of pasture. Yet those who had suffered from enclosures were not unjustified in the conclusion that history would repeat itself. Whichever way the question was ultimately decided, it could not fail to affect the condition of the rural population for better or for worse, and to affect it profoundly. Unfortunately the decision was made, in the eighteenth and nineteenth centuries, under an economic pressure which completely overrode the social considerations that should have controlled and modified the process of enclosure.

CHAPTER VI

THE LATER STEWARTS AND THE REVOLUTION
1660-1700

Worlidge's *Systema Agriculturae* (1669) : improvements suggested by agricultural writers ; tyranny of custom ; contempt for book-farming ; slow progress in farming skill ; general standard low ; horses, cattle, sheep, and pigs in the seventeenth century ; want of leaders ; growing influence of landowners ; the finance of the Restoration, and the abolition of military tenures ; legislation to promote agriculture ; Gregory King on the *State and Condition* of England and Wales in 1696 : the distribution of population and wealth.

THE practical improvements, which had been suggested by " Rustick Authors " in the first sixty years of the seventeenth century, were collected by John Worlidge in his *Systema Agriculturae* (1669). Five editions of this " first systematic treatise on farming " show that it was for some time regarded as a standard authority. Free from the extravagant promises of his predecessors, Worlidge summarises their most useful recommendations. Inordinate space is still allotted to such topics as trees, orchards, " garden tyllage," bees, and silkworms, which occupy 106 pages out of a total number of 217. On the side of stock-breeding and stock-rearing his book remains especially defective. For information on this subject he merely refers readers in a general way to other writers. Three pages only are devoted to the section " Of Beasts," in which the special qualities required for the different uses of horses, cattle, and sheep are wholly ignored ; only in the case of dogs does Worlidge appear to recognise the variety of purposes for which animals are bred.

Even the most practical work on farming which was published in the seventeenth century is ill-balanced and defective. Yet it is remarkable how many of the triumphs of nineteenth century farming were anticipated by these early writers, a century and a half before the improvements were generally adopted. Already the germs of a proper rotation of crops had been implanted, and a few advanced husbandmen, familiar with the methods of the Low Countries, had realised that, in roots and clovers, they commanded the means, not only of keeping more stock, but of increasing the yield of corn. Already some of the drawbacks of broad-cast sowing had been pointed out, and the advantages of setting in regular rows suggested. Already the foreign use of oil-cake for cattle had been observed and recommended to English farmers. But, as Mortimer [1] notices, Lincolnshire farmers, after pressing out the " oyl " from their coleseed, preferred to " burn the cakes to heat their Ovens." Already also the field-cultivation of potatoes had been suggested, and it is a coincidence that the suggestion was made only a few years after the drainage of those fens, on the clover-sick soil of which, two centuries and a half later, the adoption of the crop worked a revolution. Already the use of silos and of ensilage, the storage of water in tanks for dry districts, the value of coverings to rick-stands, even the utility of the incubator for rearing poultry—a box heated by a candle or a lamp—had been urged on Stewart agriculturists. In a tentative fashion the " Rustick Authors " were feeling after improved agricultural machinery. Googe's reaping car, the double-furrow plough of the " ingenious yeoman of Kent," Plattes' corn-setter, the corn drill depicted by Worlidge, which made the furrow, sowed the seed, and deposited the manure, were the ancestors of many useful inventions. Still more vaguely Stewart writers were looking for the aid of science. Its future benefits could not, of course, be foreseen. But the demand for an Agricultural College, the recognition of the work of the Royal Society, the study of such books as Willis' *De fermentatione* or Glauber's *Miraculum Mundi*, in which an attempt was made to analyse the elements that contribute to vegetation, show that expectations had been aroused. Already a Land Registry, by which land could be made to pass as freely as money, had been suggested by Andrew Yarranton. Already also the abolition of " slavish customs," and of " *Ill Tenures* as *Copyhold, Knight-Service* etc.," which " much discourage *Improvements* and are (as I suppose) *Badges* of our Norman Slavery " was demanded. The Hares and Rabbits legislation had been foreshadowed in the out-

[1] *The whole Art of Husbandry; or the Way of Managing and Improving of Land* (1707).

cry against the destruction of growing crops by "coneys," and hares which in 1696, according to Gregory King's minute calculation, numbered 24,000. The necessity for General Highway and Enclosure Acts had been urged on the country. The prelude to the long struggle for compensation for unexhausted improvements had been sounded. Even the twentieth century agitation for pure bread had been anticipated in the protest that "the corruption of the best aliments, as bread, and which are in most use with us, causeth the worst Epidemicall Diseases."

Here and there some changes in farming practices had been made for the better. But such progress was purely local, and rarely survived the individual by whom it was effected. Traditional methods were jealously guarded as agricultural heirlooms. Even ocular proof of the superior advantages derived from improvements failed to drive the John Trot geniuses of farming from the beaten track in which their ancestors had plodded. Circumstances combined to render the force of custom tyrannical. The agrarian partnerships on village farms opposed a natural obstacle to change. On open-fields, where the rotations of crops were fixed by immemorial usage, based on the common rights of the whole body of associated farmers, no individual could move hand or foot to effect improvements. Unless a large number of joint occupiers, often ignorant, suspicious, and prejudiced, agreed to forgo common rights and adopt turnips and clover, it was impossible to introduce their cultivation. The enterprise of twenty farmers might be checked by the apathy or caution of one. It was for this reason mainly that Worlidge addresses his treatise to the "gentry and yeomanry," and that he thinks the moment opportune for improvement, because so many farmers had been obliged to give up their holdings owing to "the great Plenty and Smallness of Value of the *Ordinary Productions* of the *Earth*," which left no profit to those who "exercised onely the *Vulgar Methods of Agriculture*." Even if the new materials for agricultural wealth were successfully introduced by some energetic landlord or tenant on an enclosed farm, the result of the experiment was rarely known beyond the immediate neighbourhood. Each village was at once isolated and self-sufficing. Communication was difficult; frequented roads were often impassable except for a well-mounted horseman or a coach drawn by eight horses. Education had not spread to the class to which farmers generally belonged. Letters were rarely interchanged. Visits were seldom paid. The only form in which information could be disseminated was in books or pamphlets, and in remote villages buyers were few or none. Newspapers had hardly begun to exist. The first attempt to found a scientific agricultural paper was made by John Houghton, whose *Collection of Letters for the Improvement of Husbandry and Trade* appeared in a weekly series from 1681 to 1683, and again from 1692 to 1703. It is improbable that the circulation could have been extensive even among the wealthiest of the country gentry. Rumours of the progress of the outside world scarcely penetrated to distant villages. Farmers of one district knew little more of the practices of the next than they did of those of Kamchatka. Beyond the limited range of their horizon, their neighbours were only

> "Anthropophagi, and men whose heads
> Do grow beneath their shoulders."

In this extreme isolation must be sought a fruitful cause for the slow diffusion of agricultural improvements. Another cause lay in the absence of any strong incentive to raise more produce from the soil than was requisite for the immediate wants of the producers. Markets were, in many parts of England, not only difficult of access but few in number. From vast and crowded haunts of labour and trade the cry of the artisan had not yet arisen for bread and meat. As soon as the farmer had satisfied the needs of himself, his family, and his rent, his work was done. Till a wider demand for agricultural produce had been created by the rapid growth of population which resulted from the development of manufacturing industries, and till the new markets had been brought to the farmer's door by improved means of communication, the supply was mainly regulated by the wants of the producer himself.

Another cause for the neglect of improvements has been already mentioned. A contempt for book-farmers, which was not wholly unjustifiable, partially explains the slow adoption of new methods and new crops. Of this class of agricultural writers, Thomas Tryon affords an interesting example. Like most men of his kind, he was a "Jack of all trades." He was a voluminous writer on a miscellaneous variety of subjects—against drinking brandy and "smoaking tobacco," upon brewing ale and beer, upon medical topics, upon dreams and visions, on the benefit of clean beds on the generation of bugs, on the pain in the teeth. He also composed a " short discourse " of a Pythagorean and a mystic. His agricultural book, *The Countryman's Companion* (1681), is chiefly noticeable for its account of that " Monsterous, Mortifying Distemper, the Rot," and for the strange remedies which he suggests for the preservation of sheep from that disorder. Thomas Tryon is an admirable representative of the class of writers who brought the book-farmer into disrepute. But already true science was coming to the aid of agriculture. The *Sylva* (1664) and *Terra* (1676) of John Evelyn are known to all well-read agriculturists, and John Ray's *Catalogus Plantarum Angliae* (1670) marks an epoch in the history of botanical science.

All these conditions combined to raise formidable obstacles to the diffusion of improvements in farming. Agricultural writers scarcely expected that the changes they suggested would be adopted. Donaldson, for instance, says that people will probably answer him with " Away with your fool Notions ; there are too many Bees in your Bonet-case. We will satisfie ourselves with such Measures as our Fathers have followed hitherto." Farmers, says Hartlib's *Legacie*, did not venture to attempt innovations lest they should be called " projectors." Bradley, Professor of Botany at Cambridge, complains in his *Complete Body of Husbandry* (1727), that if he were to advise farmers " about improvements, they will ask me whether I can hold a plough, for in that they think the whole mystery of husbandry consists." It was long before clover emerged " from the fields of gentlemen into common use " ; it did not penetrate into Suffolk villages till the eighteenth century. In Worcestershire and adjoining districts the personal efforts of Andrew Yarranton in 1653-77 had for the time established its use. But " farmers," says Jethro Tull, writing in the reign of George II., " if advised to sow clover would certainly reply, ' Gentlemen might sow it if they pleased, but they (the farmers) must take care to pay their rents.' " Even more obstinate was the resistance to turnips. It was of little use that Worlidge in his *Systema* (1669) urged upon farmers the cultivation of roots ; or that Reeve (1670) reprinted Weston's advice to use turnips as the best methods of improving " barren and heathy land " ; or that Houghton (1684) described the benefits which had resulted in Norfolk and Essex from growing them as winter food for sheep. Even their advocates had not yet appreciated the full value of roots. Worlidge [1] in 1683 had observed that " sheep fatten

---

[1] Houghton's *Collections on Husbandry and Trade* (ed. 1728), vol. iv. p. 142.

very well on turnips, which prove an excellent nourishment for them in hard winters, when fodder is scarce ; for they will not only eat the greens, but feed on the roots in the ground, and scoop them hollow even to the very skin." Houghton [1] in 1694 writes that " Some in Essex have their fallow after turneps, which feed their sheep in winter, by which means their turneps are scooped, and so made capable to hold dews and rain water which, by corrupting, imbibes the nitre of the air, and when the shell breaks, it runs about and fertilizes. By feeding the sheep, the land is dung'd as if it had been folded ; and these turneps, tho' few or none be carried off for human use, are a very excellent improvement ; nay, some reckon it so, tho' they only plough the turneps in, without feeding." They made but slow progress. Sir John Cullum, in his *History of the Manor of Hawsted*, preserves the name of Michael Houghton as the first man in that Suffolk parish, who about 1700 raised a crop of turnips on two acres of his land. " I introduced turnips into the field," says Tull, " in King William's reign ; but the practice did not travel beyond the hedges of my estate till after the Peace of Utrecht " (1713). Potatoes were even less successful. John Forster (1664) had, as has been already noticed, urged their adoption as a field crop. Houghton notices that they had been brought from Ireland " to *Lancashire*, where they are very numerous, and now they begin to spread all the Kingdom over. They are a pleasant food boiled or roasted, and eaten with butter and sugar." [2] But Mortimer (*Whole Art of Husbandry*, etc., 1707) despised them even in the garden as " very near the Nature of the *Jerusalem Artichoak*, which is not so good or wholesome. These are planted either of the Roots or Seeds, and may probably be propagated in great Quantities, and prove a good food for Swine." Neither clover nor turnips became general in England before the latter half of the eighteenth century, and potatoes were not extensively grown till fifty years later, when their value was urged on the country by the Board of Agriculture.

The widest differences existed between the farming of various districts. The general level was extremely low. But in individual cases a high standard was attained, and the best possible use made of such resources as agriculturists could command. In natural fertility the Vale of Taunton, which Norden calls the " Paradise of England," was pre-eminent. The best pastures, according to the

[1] *Ibid.* vol. i. p. 213.   [2] *Collections*, etc. vol. ii. p. 469.

same authority, were at Crediton and Welshpool. In arable farming, says Mascall, or his editor, Ruscam, the seasons for the operations of agriculture, as well as the choice of implements must depend on the character of the soil. Thus on the " stiffe clayes of *Huntingdonshire, Bedfordshire, Cambridgeshire,*" on " mixt soils that are good and fruitful, as *Northamptonshire, Hartfordshire,* most parts of *Kent, Essex, Barkshire,*" on " light and dry grounds which have also a certain natural fruitfulness in them as in *Norfolk, Suffolk,* most parts of *Lincolnshire, Hampshire* and *Surrey* "— farmers will adapt themselves to circumstances. On " the barren and unfruitful earths, as in *Devonshire, Cornwall,* many parts of *Wales, Darbyshire, Lancashire, Cheshire, Yorkshire,*" they must profit by experience. "The best corn land in Europe," in the opinion of Gabriel Plattes, was the Vale of Belvoir. The best cheeses were made at Banbury, in Cheshire, or in the Chedder district. But the latter, says Hartlib's *Legacie,* were " seldom seen but at Noblemans tables or rich Vintners Sellars." In some places the new crops recommended by the Stewart writers had been tried. Liquorice was grown with success at Pontefract in Yorkshire and at " Godliman " in Surrey ; saffron was established in Essex and Cambridgeshire ; canary seed and caraways were tried in Kent and Oxfordshire ; hops were not confined to Kent, but had spread into Suffolk, Essex, Surrey, and other counties ; sainfoin had been tested at Cobham in Kent ; weld, used for dyeing of " bright Yellows and Limon-colours," flourished near Canterbury ; madder and woad had been proved to be profitable crops ; the best flax and hemp [1] were grown near Maidstone, where a thread factory had been recently established, at Bow and Stratford in Essex, and in Nottinghamshire. At a later date the district round Beccles in Suffolk was famous for its hemp ; rape and cole-seed were established in Kent, Lincolnshire, and elsewhere. Kent, Worcestershire, Herefordshire, Gloucestershire, and the neighbourhood of London were famous for their apples, and as many as 200 varieties were collected in a single orchard. The cherries of Kent and the quinces of Essex were in chief repute. "There are now," writes William Hughes,[2] " in *Kent* and other places of this Nation,

---

[1] *England's Improvement, and Seasonable Advice,* etc. (London, 1691) is an anonymous treatise on the growth of hemp and flax.

[2] *The Compleat Vineyard,* by William Hughes, 1665. A second and enlarged edition appeared in 1670, and *The Flower Garden and Compleat Vineyard* in 1683.

such Vineyards and Wall-vines as produce great store of excellent good wine."

Increased attention was also being paid to live-stock, and the values of distinctive breeds of horses, cattle, sheep, and pigs were discussed. If Gervase Markham's *Cheape and good Husbandry* (edition of 1631) is compared with Mortimer's *Whole Art of Husbandry* (1707), some idea may be formed of the views of the seventeenth century on stock-breeding.

On horses, Markham, in spite of the criticism of Child already quoted, was reputed an authority. "Now for the choyse of the best Horse," he writes, "it is divers according to the use for which you will imploy him." Of "Horses for the Warre," he says, "the courser of *Naples* is accounted the best, the *Almaine*, the *Sardinian,* or the *French*." "For a Prince's Seat, any supreame Magistrate, or for any great Lady of state," he recommends a "milkewhite" or "faire dapple gray" steed of English breed : failing that, a "*Hungarian, Swethland, Poland*, or *Irish*" horse. The best hunter he finds in "the English horse, bastardized with any of the former Races first spoake of." The finest race-horses are "the *Arabian, Barbary*, or his bastard-Jennets, but the *Turkes* are better." "For travaile or burthen" the best is the English horse, and "the best for ease is the *Irish-hobby*." "For portage, that is for the Packe or Hampers," and "for the Cart or Plough," he makes no selection. For coach horses, he chooses the large English gelding, or the Flemish mare, or the Flemish or Frisian horse. There were doubtless already distinctive breeds in England, such as the Yorkshire saddle-horses of the Cleveland district, the heavy Black Horse of the Midlands, the Suffolk Punch, or the West-country packhorse ; but they are not mentioned by Markham. Nor does Mortimer refer to any English breeds. He tells us, however, that Leicestershire was in his day one of the great horse-breeding counties, and that Hertfordshire farmers bought the colts as two-year-olds, and sold them "at about six Years old to Gentlemen at London for their Coaches."

Among cattle, the best breeds "for meat" were the long-horned cattle of Yorkshire, Derbyshire, Lancashire, and Staffordshire. The tall long-legged Lincolns, generally "pide," with more white than any other colour, were reckoned the best for "labour and draught." "Those in Somersetshire and Gloucestershire are generally of a blood-red colour, in all shapes like unto those in Lincolne-shire, and fittest for their uses." So far Markham. Mortimer adds other breeds. "A good hardy Sort for fatting on barren or middling Sort of Land are your *Angleseys* and *Welch*. The hardiest are the *Scotch.*" The best breed for milking, in his opinion, was "the longlegged short-horn'd Cow of the *Dutch* breed," chiefly found in Lincolnshire and Kent.

Both Markham and Mortimer have much to say about sheep, which were reckoned as the most profitable of live-stock. Their manifold uses inspired Leonard Mascall [1] to rhyme in " praise of sheep " :

> " These cattle (sheep) among the rest,
> Is counted for man one of the best,
> No harmful beast, nor hurt at all ;
> His fleece of wool doth cloath us all,
> Which keeps us all from extream cold ;
> His flesh doth feed both young and old :
> His tallow makes the candles white,
> To burn and serve us day and night :
> His skin doth pleasure divers ways,
> To write, to wear, at all assaies ;
> His guts, thereof we make wheel-strings ;
> They use his bones for other things ;
> His horns some shepherds will not lose,
> Because therewith they patch their shooes ;
> His dung is chief, I understand,
> To help and dung the Plowman's land ;
> Therefore the Sheep among the rest,
> He is for man a worthy beast."

But Mascall makes no attempt to distinguish varieties of breed. Like many of the Stewart writers, he would probably have answered as the Cumberland shepherd replied to the question—where he got his rough-legged, ill-formed sheep—" Lor', sir, they are sik as God set upon the land ; we never change any." Markham, however, distinguishes the various breeds by the quality of their wool. The finest short wool came from the small black-faced Herefordshire sheep in the neighbourhood of Leominster, and in parts of Worcestershire and Shropshire. The Cotswold breed was heavier, but the wool was longer and straighter in the staple, and the fleece coarser. Parts of Warwickshire and Worcestershire, " all *Leicestershire, Buckinghamshire,* and part of *Northamptonshire,* and that part of *Nottinghamshire* which is exempt from *Sherwood Forest* " produced " a large-boned Sheep, of the best shape and deepest

---

[1] Mascall's book on the *Government of Cattell*, originally published in 1591, was still in circulation nearly a century later, under the title of *The Countreyman's Jewel*. The edition of 1680 is said to be " Gathered at first by *Leonard Mascal*, but much Inlarged by *Richard Ruscam*, Gent."

staple." These were pasture sheep, and their wool was coarse in quality. The Yorkshire breed was " of reasonable bigge bone, but of a staple rough and hairie." Welsh sheep were to be " praised only in the dish, for they are the sweetest mutton." The Lincolnshire salt marshes bore the largest animals ; but " their legges and bellies are long and naked, and their staple is coarser than any other." Mortimer practically repeats Markham's list. But he adds one significant remark. Speaking of Lincolns and the coarseness of their wool, he says : " they are lately much amended in their Breed." Some local pioneer of Bakewell and his Leicesters was already attempting the improvement of Lincolns. Both Markham and Mortimer condemn horned sheep, and advise buyers to choose animals with plenty of bone. Both also repeat the warning of Fitzherbert and Tusser that on open-field farms lambs must be timed to fall in January.

Pigs naturally take a prominent place in the books of " Rustick Authors." They are, says Markham, " troublesome, noysome, unruly, and great ravenours," yet they are " the *Husbandmans Best Scavenger*, and the *Huswifes* most wholesome sinke," and, " in the dish, so lovely and so wholesome, that all other faults may be borne with." Mascall quotes as a proverb the common saying : " The hog is never good but when he is in the dish." The natural cleanliness of the animal is strongly urged by all the seventeenth century writers. As to breed, no English county could be said to have a better sort than any other. But Markham thinks the best pigs are raised in Leicestershire, some parts of Northamptonshire, and the clay countries bordering on Leicestershire. As to colour, he recommends white or " sanded," or black. But these last are said to be rare. Pied pigs he considers to be more subject to measles. Both he and Mortimer attribute the superiority of Leicestershire and the surrounding districts to the great quantities of beans and pulse which were raised in those counties, and Mortimer adds that the pigs from those parts of the country were mostly sold in London for use at sea.

At the Restoration, the greatest need of English farming was the leadership of practical men, possessed of the leisure, the education, and the capital, to test by experiments the value of a mass of theoretical advice, to adopt new crops, introduce new methods, improve the live-stock of the country. Such pioneers were found, at a later date, among the large landowners. In 1660 they were not forthcoming from that or from any other class, and this want of leadership to a great extent explains the reluctance of farmers to put in practice many of the improvements which not only book-farmers but practical agriculturists were recommending. The state of society was still too unsettled, the title to land too insecure, to tempt expenditure. The number of men who could afford the necessary outlay was relatively few. Landed property in 1660 was distributed in smaller quantities among more numerous owners than it was a century later. The events of the Commonwealth period had further increased this wide distribution of ownership. Large quantities of land, confiscated by the Parliament, had been thrown on the market. Many estates had also been forfeited to the Government and sold, often in small parcels, because the royalist owners either refused or neglected to compound for their " delinquencies." Portions of other properties had been sold by their owners to pay the composition or the Decimation Tax. In all these cases, numbers of the purchasers were small men. At the Restoration, the estates of the Crown and of the Church, and the confiscated lands of eminent royalists were restored to their original owners, without compensation to purchasers who had bought under the authority of the Commonwealth Government. But no attempt was made to cancel the purchase of lands which had been sold under forfeitures to the Parliament, or under the pressure of the taxation imposed by the victorious Puritans on the vanquished royalists. All claims of this nature were barred by an Act, which disappointed Cavaliers condemned as an act of indemnity to the King's enemies and of oblivion to his friends. But whether the Republicans were deprived of their purchases, or confirmed in their possession, the example was not lost on their contemporaries. The nature of the compromise effected at the Restoration necessarily impaired the sense of security. When titles were precarious, outlay of capital seemed too speculative a risk. Moreover, many of the royalists who were fortunate enough to retain or regain possession of their estates, found themselves too impoverished to spend money on their improvement, or too formed in their habits to endure the tediousness of directing them. The generations which knew the Civil War, the Commonwealth, the Restoration, the rebellion of Monmouth, and the Revolution had passed away, before landowners, in widely different circumstances, assumed the lead in agricultural progress.

Changes were already at work which, within the next half century, not only restored the position of the landed gentry, but gave them an influence which they had never before possessed. Parliament gained control over the Government, and the House of Commons over Parliament. At the same time the jurisdiction of magistrates was greatly extended. Controlling the House of Commons through the county elections, administering local justice, allied with the Church as the bulwark of Protestantism, recruiting from its wealthiest members the order of the peerage, absorbing into its own ranks their younger sons, the landed gentry became the predominant class in the country. How great was the increase in their power may be illustrated by the difference in the attitude which Elizabethan and Hanoverian Parliaments assumed towards enclosures. Many of the seeds of this growth in the political and social ascendancy of the landed aristocracy were sown during the period under notice.

One of the first questions which came before the Restoration Parliament was that of finance. Some permanent provision had to be made for the ordinary charges of Government. A Committee was appointed which reported that the average yearly income of Charles I. for the period 1637-41 had been £900,000, but that of this sum £200,000 were derived from sources no longer available. Parliament decided to raise the annual income of the Crown to £1,200,000. In providing this sum the lines laid down by the Republican financiers were in the main followed. The cost of the Civil War and the subsequent expenses of the Commonwealth Government had been met by the old device of customs duties, and by the new expedients of monthly assessments on lands and goods, and of excise duties, borrowed from the Dutch financiers, on a large range of products which at one time included meat and salt. The old feudal dues, exacted by the Crown on all lands held by military tenure, had dwindled in importance and value, in spite of the attempts made by Henry VIII. and Charles I. to enforce them with greater rigour. To a large extent their place had been taken by parliamentary grants of subsidies on lands and goods. Those which remained in operation were comparatively unproductive ; they were besides uneconomical, uncertain, and inconvenient. They were also not granted by Parliament, and thus provided the Crown with funds which were not under national control. Their abolition had been recommended in the reign of James I. ; it had been carried by a resolution of both Houses of Parliament in 1645 ; it was one of the terms of the Treaty of Newport in 1648, when Charles I. agreed to surrender the dues for the payment of £100,000 a year ; it had been demanded by Puritan agriculturists like Hartlib and Blith ; finally, in 1656 the abolition had been passed into law with the consent of Cromwell. Technically speaking, the legislation of the Commonwealth was annulled by the Restoration ; practically, however, the question was not whether the abolished dues should be *continued*, but whether they should be *revived*. Against this revival it was argued in 1660 that much land had changed hands in the previous fifteen years without any provision for the possible revival of the liability. The income voted for Charles II. had to be provided, the problem of ways and means to be solved. The Restoration Parliament might have abandoned the excise duty, or revived the feudal dues, or substituted for them a land tax. They retained the excise introduced by Republican financiers, but reduced it by a half ; they confirmed Cromwell's abolition of the emoluments which the Crown had derived from lands held in chivalry ; [1] they declined by a majority of two votes to impose a land tax. At the same time the Crown surrendered its oppressive prerogatives of purveyance and pre-emption. No doubt the immediate result of these fiscal changes was that the landed aristocracy continued to be relieved from a burden, and that, from motives of self-interest, they refused to revive, either in its original or in a substituted form, a system of taxation which, before the Commonwealth, had once attached to land held in chivalry.

The abolition of military tenures reduced to some extent the necessary outgoings of many of the landed gentry. At the same time the commercial policy adopted by the Restoration Government maintained, if it did not swell, their incomes. The steady rise in the price of wool during the past century had begun to hamper the clothing trade. In order to lower prices for home manufacturers, an Act passed in 1647, and re-enacted in 1660, prohibited its exportation. Still further to stimulate the clothing industry, a series of Acts,[2] from 1666 onwards, ordered the burial of the dead in woollen fabrics. Partly for revenue, partly in compensation for these concessions to manufacturing industries, partly to meet the claims of impoverished adherents, partly to maintain the balance between pasture and tillage, partly, no doubt, to make England

[1] 12 Car. II. c. 24.   [2] 18 and 19 Car. II. c. 4.

self-supporting in its food supplies, important changes were made in the laws which regulated the trade in corn.[1] In the reign of Philip and Mary, home-grown corn could not be exported if home-prices for wheat rose above 6s. 8d. per quarter, and for cheaper grains in proportion. This limit was raised by subsequent legislation. Thus the home price for wheat, at which exportation was prohibited, was raised in 1593 to 20s., in 1604 to 26s. 8d., in 1623 to 32s., in 1660 to 40s.,[2] in 1663 to 48s.[3] In 1660 duties were also imposed on the importation of foreign wheat. These duties were at first nominal. Thus they started at 2s. per quarter on imported wheat, when home-prices exceeded 44s. In 1663 they were raised to 5s. 4d. per quarter, when home-grown wheat rose above 48s. In 1670[4] the corn laws became more frankly protective. No limit of price was fixed above which the exportation of home-grown corn was prohibited, and a heavy duty of 16s. a quarter was imposed on foreign wheat when home prices did not exceed 53s. 4d. per quarter. Similar duties were imposed on the importation of other foreign grain at proportionate prices. A further change was made in 1688.[5] The Act of that year offered a bounty on the export of home-grown corn of 5s. per quarter of wheat, whenever the home-price fell below 48s. per quarter, and on other grain in proportion. On these two principles, namely a duty on the importation of foreign corn and a bounty on the exportation of home-grown corn, combined with frequent prohibitions of exports, the corn trade was regulated throughout the eighteenth century. Similar measures were adopted to encourage the raising of cattle, and importations from Ireland were prohibited. Legislation did not, however, raise prices ; it only succeeded in maintaining them. Increased production at home counteracted the effect which the restriction of imports might otherwise have produced. England, says Sir William Petty,[6] " doeth so abound in Victuals as that it maketh Laws against the Importation of Cattle, Flesh and Fish from abroad ; and that the draining of Fens, improving of Forests, inclosing of Commons, Sowing of St. Foyne and Clover-grass be grumbled against by Landlords, as the Way to depress the price of Victuals." Elsewhere he adds :

---

[1] See Appendix III. The Corn Laws.   [2] 12 Car. II. c. 4.
[3] 15 Car. II. c. 7.   [4] 22 Car. II. c. 13.
[5] 1 William and Mary, c. 12.
[6] *Several Essays in Political Arithmetic,* ed. 1755, pp. 150-169.

" it is manifest that the land in its present Condition is able to bear more Provision and Commodities, than it was forty years ago."

Throughout the period from the Restoration to the Revolution, except for one disastrous year of plague, fire, and war, the country prospered. The receipts from customs steadily advanced. Trade was expanding. As Amsterdam decayed, and Portuguese and Spanish Jews fled to England to escape the Inquisition, money flowed into the country. Other religious refugees brought with them useful arts and manufactures. The development of banking stimulated commercial undertakings. Between 1661 and 1687 the receipts from the customs duties more than doubled. Fortunes, made in the city were often invested in land, which now was beginning to confer on its possessors a new political and social influence. The landed gentry shared in the growing prosperity, either through its general effects on the country, or by wealthy marriages, or by sending their sons—as Rashleigh Osbaldistone was sent by Sir Hildebrand—into business. Between 1675 and 1700, said Sir William Temple " the first noble families married into the City." [1] Latimer had preached against landlords becoming " graziers," and aldermen turning " colliers," and disquietude at this commercial tendency had influenced the legislation of Edward VI. But times had changed. Though Heralds still distinguished between " foreign Merchants " and retail shopkeepers, on the ground apparently that " Navigation was the only laudable part of all buying and selling," yet they [2] had solemnly decided that " if a Gentleman be bound an Apprentice to a Merchant, or other Trade, he hath not thereby lost his Degree of Gentility."

Closely united with the nobility, the Church, and the merchant princes, sharing in the general prosperity, and, in virtue of their property, exercising new political and social powers, the landed gentry were beginning to acquire that predominant influence which was so marked a feature in the eighteenth century. The change necessarily added an artificial value to the ownership of land : it not only arrested the tendency towards its wider distribution, but encouraged its accumulation in fewer hands. Once acquired, estates were held together by the introduction of family settle-

---

[1] Quoted by Toynbee, *Industrial Revolution*, ed. 1887, p. 63.

[2] Logan's *Treatise of Honor* at the end of Gwillim's *Display of Heraldry* (ed. 1679), p. 155.

ments. On the eve of this change, it may be of interest to note a contemporary estimate of the agricultural population and wealth of the country at the close of the eighteenth century.

Gregory King, whose training and experience specially qualified him for the task, drew up a statistical account of the " State and Condition " of England and Wales in 1696. His estimates of the actual numbers of the population are the result of an investigation by a competent and careful observer, who made the fullest use of the information supplied by such figures as those contained in the Hearth-office, the assessments on Births, Marriages, and Burials, the Parish Registers, and Public Accounts. The substantial accuracy of this part of his work has stood the test of subsequent criticism, in spite of his prophecy that in 1900 the population would have risen to 7,350,000. For the rest of his estimates he mainly depended on guess-work. Confidence is scarcely created by his laborious calculation of the numbers of hares, rabbits, and wild fowl in the country. King's figures were largely used by Davenant,[1] but his actual manuscript remained unpublished till 1801.[2]

King estimated the total acreage of England and Wales at 39 million [3] acres ; of which 11 million acres were arable, averaging a yearly rent per acre of 5s. 10d. ; and 10 million were meadow or pasture, averaging 9s. an acre. Of the 11 million arable acres, ten million were under the plough for corn, pease, beans, and vetches ; one million acres were allotted to flax, hemp, saffron, woad and other dyeing weeds, etc. He goes on to calculate the live-stock of the country thus : " horses (and asses)," 600,000 ; cattle, 4½ million ; sheep, 11 million ; pigs, 2 million. The total population in 1696 is estimated at 5,500,000 persons, distributed into 1,400,000 urban, and 4,100,000 rural, inhabitants. The total yearly income of the nation in 1688 is calculated at £43,500,000. Of this total, con-

---

[1] *An Essay upon the Probable Methods of making a People Gainers in the Ballance of Trade*, by Charles Davenant, 1698 (Section I. " Of the People of England," and Section II. " Of the Land of England and its Product ").

[2] Published in *An Estimate of the Comparative Strength of Great Britain*, by George Chalmers (1802), under the title of " Natural and Political Observations and Conclusions upon the State and Condition of England, 1696 ; by Gregory King, Esq., Lancaster Herald."

[3] The actual figure is 37,319,221 acres.

siderably more than half (£24,480,000) belonged to the following families :

	Average Yearly Income.		
40,000 Freeholders [1] of the better sort -	£84	0	0
140,000 Freeholders of the lesser sort - -	50	0	0
150,000 Farmers - - - - - -	44	0	0
364,000 Labouring People and Out-servants -	15	0	0
400,000 Cottagers and Paupers - - -	6	10	0 [2]

King's estimates bring into strong relief the vast revolution which the eighteenth and nineteenth centuries produced in the distribution of population and of wealth.   The same point is illustrated from a different point of view by a comparison of the wealth of the different counties in 1696 and at the present day.   Material for such a comparison is found in the frequent assessments which were made of the counties during the seventeenth century for various fiscal purposes.   The central counties are the richest ; then follow in order of wealth the south, the east, the west.   Poorest of all is the north.   Throughout the whole period, Middlesex is the richest and Cumberland the poorest county.   The most conspicuous change was that of Surrey, which rose from the eighteenth place in 1636 to the second in 1693.   Excluding Middlesex, and excepting Surrey, the wealthiest district throughout the whole period was formed by a block of six agricultural counties north of the Thames—namely Hertfordshire, Bedfordshire, Buckinghamshire, Berkshire, Oxfordshire, and Northamptonshire.   Their position illustrates the importance of London as a market for agricultural produce.   Already its rapid growth was exciting alarm, lest " the Head " should become " too big for the Body."   According to Gregory King, its population was 530,000 souls out of an urban population of 1,400,000, and a total population, urban and rural, of 5½ millions.   Throughout the whole period, again, the seven poorest counties, though their order in the list varies, were Cheshire, Derbyshire, Yorkshire, Lancashire, Northumberland, Durham, and Cumberland.   The assessment of the whole district north of the Humber, comprising one-fifth of the total area of England, was not greater than that of

[1] It should be noted that freeholders included not only owners and occupying owners, but tenants for life and lives, as well as copyholders.

[2] For tables of estimates drawn up by King and Davenant, see Appendix IV.

Wiltshire. In the latter half of the following century not only wealth but population migrated northwards, and the inhabitants of rural districts began to flow into the centres of trade and manufacture which crowded round the coal and iron fields and water-power of the northern counties.

CHAPTER VII

JETHRO TULL AND LORD TOWNSHEND. 1700-1760

Agricultural progress in the eighteenth century; enclosures necessary to advance; advocates and opponents of the enclosing movement; area of uncultivated land and of land cultivated in open-fields; defects of the open-field system as a method of farming; pasture commons as adjuncts to open-field holdings; the necessary lead in agricultural progress given by large landowners and large farmers; procedure in enclosures by Act of Parliament: varying dates at which districts have been enclosed: influence of soil and climate in breaking up or maintaining the open-field system: the East Midland and North Eastern group of counties: improved methods and increased resources of farming; Jethro Tull the "greatest individual improver"; Lord Townshend's influence on Norfolk husbandry.

THE gigantic advance of agriculture in the nineteenth century dwarfs into insignificance any previous rate of progress. Yet the change between 1700 and 1800 was astonishing. England not only produced food for a population that had doubled itself, as well as grain for treble the number of horses, but during the first part of the period became, as M. de Lavergne has said, the granary of Europe. Population before 1760 grew so slowly that the soil, without any great increase in farming skill or in cultivated area, produced a surplus. Under the spur of the bounty, land which had been converted to pasture was again ploughed for corn, and proved by its yield that it had profited by the prolonged rest. The price of wheat, between the years 1713 and 1764, in spite of large exports, averaged 34s. 11d. per quarter; poor-rates fell below the level of the preceding century; real wages were higher than they had been since the reign of Henry VI. In England, at least, there was little civil war or tumult, no glut of the labour market, no sudden growth of an artisan class. The standard of living improved. Instead of the salted carcases of half-starved and aged oxen, fresh meat began to be eaten by the peasantry. Wheaten bread ceased to be a luxury of the wealthy, and, at the accession of George III. had become the bread-stuff of half the population. Politically and morally, the period was corrupt and coarse ; materially, it was one of the Golden Ages of the peasant. The only drawbacks to the general prosperity of agriculture during the first half of the century were the visitations of the rot, and of the cattle plague. Ellis [1] speaks of the rot in 1735 as " the most general one that has happened in the memory of man . . . the dead bodies of rotten sheep were so numerous in roads, lanes, and fields, that their carrion stench and smell proved extremely offensive to the neighbouring parts and the passant travellers." A newer and more mysterious scourge was the cattle plague. Starting in Bohemia, it travelled westward, devastated the north of France, and three times visited England. The only remedy was to slaughter infected animals ; in a single year the Government, paying one-third of the value, expended £135,000 in compensation.

The great changes which English agriculture witnessed as the eighteenth century advanced, and particularly after the accession of George III. (1760), are, broadly speaking, identified with Jethro Tull, Lord Townshend, Bakewell of Dishley, Arthur Young, and Coke of Norfolk. With their names are associated the chief characteristics in the farming progress of the period, which may be summed up in the adoption of improved methods of cultivation, the introduction of new crops, the reduction of stock-breeding to a science, the provision of increased facilities of communication and of transport, and the enterprise and outlay of capitalist landlords and tenant-farmers. The improvements which these pioneers initiated, taught, or exemplified, enabled England to meet the strain of the Napoleonic wars, to bear the burden of additional taxation, and to feed the vast centres of commercial industry which sprang up, as if by magic, at a time when food supplies could not have been provided from another country. Without the substitution of separate occupation for the ancient system of common cultivation, this agricultural progress was impossible. But in carrying out the necessary changes, rural society was convulsed, and its general conditions revolutionised. The divorce of the peasantry from the soil, and the extinction of commoners, open-field farmers, and eventually of small freeholders, were the heavy price which the nation ultimately paid for the supply of bread and meat to its manufacturing population.

[1] *Shepherd's Sure Guide*, 1749.

Neither the reclamation of wastes, nor the break-up of open-field farms, nor the appropriation of commons, were novelties. For the last three centuries the three processes, which are generally spoken of as enclosures, had all been proceeding at varying rates of progress. But in the period from 1760 to 1815 each received an immense impetus, partly from the rise in the price of corn, partly from the consequent increase in rental values, partly from the pressure of a growing population, partly from the improved standard of agriculture. The literary struggle in advocacy or condemnation of enclosures still continued. But the advocates were gaining the upper hand. In the first half of the eighteenth century, there are at least two notable contributions to the literature of the subject by champions of enclosures, and only one of any importance by an opponent.

By the new writers, the unprofitable nature of the use of land under common tillage or common pasture is insisted upon. Thus Timothy Nourse, *Gent.*, in his *Campania Foelix ; or Discourse of the Benefits and Improvements of Husbandry* (1700), vigorously attacks commons as " Seminaries of a lazy Thieving sort of People." In his opinion their live-stock were as unprofitable to the community as the commoners themselves. Their sheep are described as " poor, tatter'd, and poyson'd with the Rot," their cattle " as starv'd, Tod-bellied Runts, neither fit for the Dairy nor the Yoke." So, also, an anonymous author in a short and pithy tract, *An Old Almanack* (*with some considerations for improving commons*) *printed in* 1710. *With a Postscript* (1734-5), suggests that, if the landowner and two-thirds, in number and value, of those interested in an open-field farm and common agreed to an enclosure, their consent should override the opposition of the minority. " *Will the Commoners complain,*" he asks, " *for want of their Commonage ?* This they can't do, for few of them have any Cattle, and whether they have or not, there is Recompence out of the Inclosures will more than treble their Loss ? *Will the Incumbents complain ?* What ! for converting the dry Commons into Corn, and the Fenns into Hemp and Flax. *Will the Ingrossers of Commons complain, who eat up their own Share and others too ?* This they dare not. *But won't those honest Men complain who now live upon the Thefts of Common ?* And not with the least Reason, but then there will be Work for them." But the two important advocates of enclosures were the brothers John and Edward Laurence. In *A New System of Agriculture* (1726) a note is struck which sounded more loudly as towns grew, as, with their growth, the demand increased for meat, milk, and butter, as agriculture improved, as communication was facilitated. The author, the Rev. John Laurence, Rector of Bishops Wearmouth, treats open-field farms as obstacles to agricultural progress. He insists on enclosures and separate occupation as the best means of increasing produce and of raising rents. He dwells on the rapid progress which enclosures were then making, points out the great rise in rental value consequent on increased produce, and argues that so far from injuring the poor, enclosures will rather create a new demand for labour by the introduction of improved tillage and pasture-farming, will give employment in fencing and ditching, and remove the attractions of wastes and open spaces, which " draw to them the poor and necessitous only for the advantage of pilfering and stealing." In *The Duty of a Steward to his Lord* (1727) Edward Laurence, himself a land-surveyor, and apparently agent to the Duke of Buckingham, argues the case from the point of view of better and more economical management. A new skilled profession was growing up. It is prophetic of future changes that Laurence points out the evils of employing " country-Attorneys (not skilled in Husbandry) " in the management of landed property, and argues that the gentry should allow handsome salaries to their stewards, who could, if inadequately paid, adopt other means of enriching themselves. A champion of " engrossing," he insists on the advantages of consolidating small holdings in larger farms. He urges stewards to prevent piecemeal enclosures by individuals, to substitute leaseholds for copyholds, to buy up any freeholds on the estate which lie in intermixed strips, as necessary preliminaries to any successful and general scheme for the enclosure of open-fields and commons. The other side to the picture is vigorously painted by John Cowper in his *Essay proving that Inclosing Commons and Common-Field-Lands is Contrary to the Interest of the Nation* (1732). He answers the arguments of the two Laurences, arguing that enclosures necessarily injure the small freeholder and the poor, and pleading that, so far from encouraging labour, they depopulate the villages in which they have been carried out. Speaking of the small freeholder, he says that " none are more industrious, none toil and labour so hard. . . . I myself have seen within these 30 years, above 20 Lordships or parishes enclosed, and everyone of them has thereby been in a manner depopulated. If any one can shew me where an Inclosure has been made, and not at least half the inhabitants gone, I will throw up the argument."

In the passages quoted from these five books are outlined some of the principal points in the dispute which was fought out in the next eighty years. On the one side are pleaded the pernicious effects of commons on the inhabitants of the neighbourhood and their live-stock; the absence of any legal title to many of the rights claimed over pasture commons, and their frequent abuse by commoners; the obstacles to farming improvement which were presented by open arable fields; the unprofitable use of land occupied in common; the commercial and productive advantages of enlarged, separate holdings. On the other side is urged the injury which the break-up of open-field farms and the partition of commons inflicted on small owners and occupiers of land. Much was to be said from both points of view. Many sweeping assertions were made, both by advocates and opponents, which were true of one district but untrue of another. Both socially and economically, the reclamation of wastes, the extinction of open-field farms, the appropriation of commons, might be justified by the urgent necessity of developing the productiveness of the soil, and of increasing to the fullest extent the food resources of the country. In favour of the first two changes, most agricultural writers are agreed; in dealing with the commons, it is at least doubtful whether the best possible course was always adopted.

From the productive point of view, the amount of waste land was a standing reproach to agriculture. The disappearance of the wild boar and the wolf in the reign of Charles II. suggests some diminution of the area in which those animals had harboured. But in 1696 Gregory King had estimated the heaths, moors, mountains, and barren lands of England and Wales at ten million acres, or more than a quarter of the total area. In all probability, the estimate is wholly inadequate. But, assuming the calculation to be approximately correct, it affords some measure of comparison with conditions at the close of the eighteenth century. In 1795 the Board of Agriculture [1] stated that over 22 million acres in Great Britain were uncultivated, of which 7,888,977 acres were in England and Wales. Here too there is probably a gross under-estimate.

[1] *Report of the Committee of the Board of Agriculture* (1795). The total acreages are over-estimated.

Arthur Young,[1] twenty years before (1773), had called attention to the extent of land lying waste in Great Britain. " There are," he says, " at least 600,000 acres waste in the single county of Northumberland. In those of Cumberland and Westmoreland, there are as many more. In the north and part of the West Riding of Yorkshire, and the contiguous ones of Lancashire, and in the west part of Durham, are yet greater tracts ; you may draw a line from the north point of Derbyshire to the extremity of Northumberland, of 150 miles as the crow flies, which shall be entirely across waste lands : the exception of small cultivated spots very trifling." It was across this district that Jeanie Deans travelled in the days of George II., when great districts of Northumberland were covered with forests of broom, thick and tall enough to hide a Scottish army. Lancashire in 1794 still had 108,500 acres of waste, and Rossendale remained a chace. As late as 1794, three-quarters of Westmoreland, according to Bishop Watson, lay uncultivated. In 1734 the forest of Knaresborough had surrounded Harrogate so thickly that " he was thought a cunning fellow that could readily find out those Spaws." Even in the last decade of the eighteenth century, 265,000 acres of Yorkshire were lying waste, yet largely capable of cultivation. Up to the accession of George III., that part of the East Riding which was called the Carrs, from Bridlington Quay to Spurn Point, and inland as far as Driffield, was an extensive swamp producing little but the ague ; willow trees marked out the road from Hull to Beverley, and the bells rang at dusk from the tower of Barton-upon-Humber to guide belated travellers. Great tracts of Derbyshire were " black regions of ling." From Sleaford to Brigg, " all that the devil o'erlooks from Lincoln Town," was a desolate waste, over which wayfarers were directed by the land lighthouse of Dunstan pillar. No fences were to be seen for miles— only the furze-capped sand-banks which enclosed the warrens. The high ground from Spilsby to Caistor was similarly a bleak unproductive heath. Robin Hood and Little John might still have sheltered in Sherwood Forest, which occupied a great part of Nottinghamshire. The fen districts of the counties of Cambridge, Huntingdon, Lincoln, and Northampton continued to defy the assaults of drainage. Even in the neighbourhood of London similar

[1] *Observations on the Present State of Waste Lands of Great Britain*, 1773. Young's calculations are also based on an exaggerated estimate of the acreage of England and Wales.

conditions prevailed. Nathaniel Kent, writing in 1775 (*Hints to Gentlemen of Landed Property*), says " that within thirty miles of the *capital*, there is not less than 200,000 acres of waste land." As late as 1793, Hounslow Heath and Finchley Common were described as wastes, fitted only for " Cherokees and savages." In 1791, the Weald of Surrey still bore evidence of its desolation in the posts which stood across it as "guides to letter-carriers." In Essex, Epping and Hainault Forests were in 1794 " known to be a resort of the most idle and profligate of men ; here the undergraduates in iniquity commence their career with deer stealing, and here the more finished and hardened robber retires from justice." Counties more remote from London had a still larger area of wastes. When Young made his Farmer's Tours in the first decade of the reign of George III., Sedgmoor was still one vast fen, the Mendip Hills were uncultivated, and eighteen thousand acres on the Quantock Hills lay desolate. Over Devonshire, Cornwall, and the whole of Wales, stretched, in 1773, " immense " tracts of wastes. To bring some of these wastes into cultivation was part of the work which agriculturists undertook in the eighteenth century, and if the estimates of Gregory King (1696) and of the Board of Agriculture (1793) are approximately correct, upwards of two million acres were added to the cultivated area before the close of the period.

It is possible that in 1700 at least half the arable land of the country was still cultivated on the open-field system—that is, in village farms by associations of agricultural partners who occupied intermixed strips, and cultivated the whole area under common rules of cropping. Out of 8,500 parishes, which in round numbers existed at the Reformation, 4,500 seem to have been still laid out, in whole or in part, on this ancient method. John Laurence in 1726 had calculated that a third of the cultivated area " is what we call Common Fields." The agricultural defects of the open-field system were obvious and numerous. So long as farming had been unprogressive, and population had remained stationary, the economic loss was comparatively unimportant. When improved methods and increased resources were commanded by farmers, and when the demand for food threatened to outstrip the supply, the need for change became imperative. Under the primitive system, the area under the plough was excessive, and much land, which might have been more profitably employed as pasture, was tilled for corn. A quantity of the arable land was wasted in innumerable balks and footpaths. All the occupiers were bound by rigid customary rules, compelled to treat all kinds of soil alike, obliged to keep exact time with one another in sowing and reaping their crops. Freeholders on open-field farms were only half-owners. No winter crops could be grown so long as the arable fields were subjected to common rights of pasture from August to February. It meant financial ruin, if any member of the community grew turnips, clover, or artificial grasses for the benefit of his neighbours. The strips of land occupied by each partner were too narrow to admit of cross-ploughing or cross-harrowing, and on heavy land this was a serious drawback. Drainage was practically impossible, for, if one man drained or water-furrowed his land, or scoured his courses, his neighbour might block his outfalls. It was to carry off the water that the arable land was heaped up into high ridges between two furrows. But the remedy was almost as bad as the disease. The richness of the soil was washed off the summit of the ridge into the trenches, which often, as Kent [1] records, contained water three yards wide, dammed back at either end by the high-ridged headlands. The cultivated fields were generally foul, if not from the fault of the occupier, from the slovenliness of his neighbours ; the turf-balks harboured twitch ; the triennial fallows left their heritage of crops of docks and thistles. The unsheltered, hedgeless open-fields were often hurtful to live-stock, though the absence of hedges was not without its advantages to the corn. The farm-buildings were gathered together in the village, often a mile or more from the land. As each man's strips lay scattered over each of the open-fields, he wasted his day in visiting the different parcels of his holding, and his expenses of manuring, reaping, carting, and horse-keeping were enormously increased by the remoteness of the different parts of his occupation. Vexatious rights interfered with proper cultivation. One man might have the right to turn his plough on another's strip, and the victim must either wait his neighbour's pleasure or risk the damage to his sown crops. " Travellers," as Joseph Lee [2] remarked in 1656, " know no highwaies in the common fields " ; each avoided his predecessor's ruts, and cattle trespassed as they passed. For twenty yards on either side of the track the growing corn was often spoiled. The sheep were driven to the commons by day, and in the summer folded at night on the fallows.

---

[1] *Hints to Gentlemen of Landed Property*, by Nathaniel Kent, 1775.
[2] *Vindication of a Regulated Enclosure*, p. 24.

Otherwise the manure of the live-stock was wasted over the wide area, which the animals traversed to find their scanty food. Unable to provide winter keep, and fettered by the common rights of pasture which each of the partners enjoyed over the whole of the arable land, farmers reared lambs and calves under every disadvantage. During the summer months, when the horses and cattle were tethered on the unsheltered balks, they lost flesh and pined in the heat. Ill-fed all the year round, and half-starved in the winter, the live-stock dwindled in size. The promiscuous herding of sheep and cattle generated every sort of disorder. The common pasture was pimpled with mole-heaps and ant-hills, and, from want of drainage, pitted with wet patches where nothing grew but rushes. The scab was rarely absent from the crowded common-fold, or the rot from the ill-drained plough-land and pasture. No individual owner could attempt to improve his flock or his herd, when all the cattle and sheep of the village grazed together on the same commons.

The open-field system was proverbially the source of quarrels. Litigation was incessant. It was easy for men to plough up a portion of the common balks or headlands, to shift their neighbour's landmarks, or poach their land, by a turn of the plough, or filch their crops when reaping. Robert Mannyng in his *Handlyng Synne* (1303) had condemned the " fals husbandys " that " ere aweye falsly mennys landys," and William Langland in *Piers Plowman* (1369) had denounced the ploughman who " pynched on " the adjoining half-acre, and the reapers who reaped their neighbour's ground. Tusser repeats the complaint of the mediaeval moralists against the ' champion ' or open-field farmer :—

> " The Champion robbeth by night,
> And prowleth and filcheth by day :
> Himself and his beasts out of sight,
> Both spoileth and maketh away
> Not only thy grass but thy corn,
> Both after and e'er it be shorn."

Gascoigne in *The Steel Glasse* (1576) condemns the open-field farmer who

> " . . . set debate between their lords
> By earing up the balks that part their bounds."

Joseph Lee repeats the charge. " It is," he says, " a practice too common in the common fields, where men make nothing to pull up their neighbour's landmark, to plow up their land and mow their grasse that lyeth next to them." For open-field farmers the curse in the Commination Service had a real meaning. Edward Laurence [1] (1727) dwells on the temptations to dishonesty which the unfenced lands and precarious boundaries of open-fields offered to the needy, and the same point is repeatedly insisted upon by the Reporters to the Board of Agriculture at the end of the eighteenth century. Hence it was that open-field farmers agreed among themselves as little as " wasp doth with bee." Hence also came the numerous law-suits. " How many brawling contentions," says Lee, " are brought before the Judges every Assizes by the inhabitants of the common fields."

Speaking generally, enclosure meant the simultaneous processes of consolidating the intermixed strips of open-field farms and of dividing the commons attached to them as adjuncts of the arable holdings. But this was not universally the case. Sometimes the arable farm had been enclosed, and only the pasture common remained to be divided. Sometimes the reverse was the case; the common had gone, and only the arable land remained to be enclosed. Sometimes land, previously enclosed by agreement or piecemeal by individuals, was re-enclosed under a general scheme, probably for purposes of redistribution. Sometimes the acreage mentioned in Inclosure Acts, as tested by the awards, is exaggerated, more rarely under-estimated. All these differences make accurate calculations of the actual area affected by the appropriation of pasture-commons and the extinction of open-field farms extremely difficult, if not impossible. Now that the commons as adjuncts of arable farming have greatly contracted in area, their comparative disappearance is deplored on both economic and social grounds, in accordance with ideas which are of recent growth. It might have been possible to regulate their use to greater profit, or to preserve them as open spaces for recreation and as the lungs of large towns, or to divide them on methods which recognised more fully the minor rights claimed by small commoners, and would thus have benefited a larger section of the community. But so long as the herbage of the commons, both in legal theory and historical origin, formed an essential part of the arable farm, and was subject to rights claimed against all the world by the privileged occupiers of the tillage land, there were practical difficulties in the way of each of these possible courses. Agriculturists scarcely looked

[1] *Duty of a Steward to his Lord.*

beyond the undrained and impoverished condition of the pastures ; lawyers held that rights of common, claimed apart from the tenure of arable land or ancient cottages, were in the nature of encroachments or trespass ; economists condemned their occupation in common as a wasteful and unprofitable use of the land ; social reformers pointed to the attractions which commons possessed for idlers, and deplored their influence on morals and industry. All these classes may have been, consciously or unconsciously, self-interested. There were few, certainly, who realised the full consequences of enclosures, or appreciated the strength of the impulse which the enclosing movement would give to capitalist farming, and the immediate success of the agricultural change removed the hesitation even of the most far-seeing.

Custom in the course of centuries had dealt hardly with the commons. Many of them were unstinted, and were consequently overcharged with stock, which often belonged to jobbers and not to the commoners. Even in good seasons, there was barely enough grass to keep the cattle and sheep alive. In bad seasons, when the weather was cold or wet, and the grass late and scanty, many died from want of food. In other cases, while the main body of commoners were restricted in the number of their stock, one or more commoners, not always lords of adjacent manors, were restrained by no limit, and not only turned out as many of their own sheep and cattle as they could, but also took in those of strangers. The poorer the commoner, the less was the benefit he derived. If the commons were stinted, every commoner, who occupied other pasture land in severalty, saved his own grass till the last moment by keeping his sheep and cattle on the common, and the small man, who had no other refuge for his live-stock, was the sufferer. Where the commons, again, were stinted, the richer men frequently turned out more than the custom allowed, and the smaller commoners had lost the protection of the old Courts Baron, where the offenders, before the decay of those tribunals, would have been " presented." Monied men turned stock-jobbers or dealers, hired land at double rents on the edge of the commons, and so obtained grazing rights which they exercised by overstocking the land with their own sheep and cattle or by agisting the live-stock of strangers. It was thus that, in 1793, " an immense number of greyhound-like sheep, pitiful half-starved-looking animals, subject to rot," crowded Hounslow Heath, and that in 1804 the common of Cheshunt was grazed not by the poor but by a parcel of jobbers. The poverty of the pasture was often proved by the condition of the stock. " It is painful to observe the very wretched appearance of the animals," writes an anonymous author in *The Farmer's Magazine* for May, 1802, " who have no other dependence but upon the pasture of these commons, and who, in most instances bear a greater resemblance to living skeletons than anything else." " The stock," he continues, " turned out yearly into these commons consists of a motley mixture of all the different breeds of sheep and cattle at present known in the island ; many of which are *diseased, deformed, small,* and in every respect unworthy of being bred from." In theory, the commons enabled the cottagers, who occupied at higher rents the ancient cottages which legally conferred the rights, to supplement their wages by keeping a cow or two. But the theory did not always agree with the practice. Often, if the cottager had money enough to buy a cow, the cow could barely find a living on land already overrun with sheep. The cottager's profits from the commons mainly consisted in the use or sale of turf, gorse, and brushwood which he cut for fuel, the run for a few geese and a " ragged shabby horse " or pony. In theory, again, the value of the commons to a small farmer, whose holding, whether freehold, copyhold, or leasehold, was mainly arable, was inestimable —provided that he was near enough to make good use of the grassland. But, in fact, the value was often minimised by distance, by the wretched condition of the undrained and over-stocked pasture, and by the risk of infection to the live-stock. There can be no question that, from an agricultural point of view, five acres of pasture, added in individual occupation to the arable holding of a small occupier, and placed near the rest of his land, would have been a greater boon than pasture rights over 250 acres of common.

Some of the practical evils of open-fields and their attendant pasture-commons might have been, with time, skill, and patience, mitigated. In some districts the village farms were better managed than in others. But even if the pressure of increasing population and the difficulties of a great war had not necessitated immediate action, the inherent defects of the system could not be cured. The general description which has been given of open-field farming applies to every part of the country. Scotland formed no exception to the rule. Scottish farmers, who are now reckoned among the most skilful, were, in 1700, inferior in their management of land

to those of England, and their methods of raising crops had remained
unchanged since the Battle of Bannockburn. Advocates of enclosure in England might legitimately argue that the rapid progress
of Scottish farming dates from the General Enclosure Act for
Scotland which was passed in 1695. The south-eastern counties
were the first to be improved. Forty years before (1661), John
Ray [1] had painted an unfavourable picture of the condition of the
inhabitants. "The men seem to be very lazy, and may be
frequently observed to plow in their cloaks. . . . They have
neither good bread, cheese, or drink. They cannot make them, nor
will they learn. Their butter is very indifferent, and one would
wonder how they could contrive to make it so bad. They use
much pottage made of coal-wort, which they call *keal*, sometimes
broth of decorticated barley. The ordinary country houses are
pitiful cots, built of stone, and covered with turves, having in them
but one room, many of them no chimneys, the windows very small
holes, and not glazed." Alexander Garden of Troup describes the
farming system which was followed in 1686. The land was divided
into in-field and out-field. The in-field was kept "constantly
under corne and bear, the husbandmen dunging it every thrie years,
and, for his pains, if he reap the fourth corne, he is satisfied." The
out-field was allowed to grow green with weeds and thistles, and,
after four or five years of this repose, was twice ploughed and sown
with corn. Three crops were taken in succession ; then, when the
soil was too exhausted to repay seed and labour, it reverted to its
weeds and thistles. Sir Archibald Grant,[2] of Monymusk in Aber-
deenshire, says that in 1716 turnips grown in fields by the Earl of
Rothes and a few others were objects of wonder to the neighbourhood, that, except in East Lothian, no wheat was grown, that on
his own estate there were no enclosures, no metalled roads, and no
wheel-carriages. On the family property, when his father allowed
him to undertake the management—"there was not one acre
inclosed, nor any timber upon it, but a few elm, cycamore, and ash
about a small kitchen garden adjoining to the house, and some
stragling trees at some of the farm-yards, with a small cops-
wood, not inclosed, and dwarfish, and broused by sheep and
cattle. All the farmes ill disposed, and mixed ; different persons
having alternate ridges ; not one wheel-carriage on the esteat, nor

[1] *Select Remains of John Ray*, London, 1760.
[2] *Miscellany of the Spalding Club*, Aberdeen, 1841-2, vol. ii. p. 96 etc.

indeed any one road that would alow it. . . . The whole land raised and uneven, and full of stones, many of them very large, of a hard iron quality, and all the ridges crooked in shape of an S, and very high and full of noxious weeds, and poor, being worn out by culture, without proper manure or tillage. . . . The people poor, ignorant, and slothfull, and ingrained enimies to planting, enclosing, or any improvements or cleanness."

Neither in Scotland nor in England were open-field farmers, or tenants-at-will, or even leaseholders for lives, likely to initiate changes in the cultivation of the soil. It was almost equally idle to expect that small freeholders would attempt experiments on the agricultural methods of their forefathers, which, in a single season, might bring them to the verge of ruin. In both countries, it was the large landlords who took the lead in the agricultural revolution of the eighteenth century, and the larger farmers who were the first to adopt improvements. Both classes found that land was the most profitable investment for their capital. Their personal motives were probably, in the main, self-interested, and a rise in rental value or in the profits of their business was their reward. But though philanthropy and farming make a fractious mixture, the movement was of national value. When the sudden development of manufacturing industries created new markets for food-supplies, necessity demanded the conversion of the primitive self-sufficing village-farms into factories of bread and meat. For more than half a century the natural conservatism or caution of agriculturists resisted any extensive change. Down to 1760 the pressure of a growing population was scarcely felt. Nor were the commercial advantages of scientific husbandry so clearly established, even in 1790, as to convince the bulk of English landlords of the wisdom of adopting improved methods.

The comparatively slow progress of the movement is illustrated by the variations in the number of Enclosure Acts passed before and after 1760. But it must always be remembered that an Act of Parliament was not the only method of enclosure, and that counties had been enclosed, either entirely or mainly, without their intervention. In Tudor times open-field arable lands and common pastures had been sometimes enclosed not only by agreement or purchase, but by force or fraud. Sometimes they had been extinguished, in whole or in part, by one individual freeholder, who had bought up the strips of his partners. Sometimes, where

there was no other freeholder, they had been consolidated by the
landlord, who allowed the leases to expire, and re-let the land in
several occupation. Sometimes they had been enclosed piecemeal by a number of separate owners ; sometimes all the partners
had united in appointing commissioners, or arbitrators, who distributed the open-field in individual ownership. By these private
arrangements large tracts of land had been enclosed without the
intervention of the law, and some of these processes continued in
active operation throughout the eighteenth century. But it was
difficult to make a voluntary agreement universally binding.
Modifications of the open-field system, which were introduced
without Parliamentary sanction, were liable to be set aside by
subsequent action. Instances of breaches of voluntary agreements
are quoted by the Reporters to the Board of Agriculture. Thus,
in one Buckinghamshire parish, the inhabitants, who had obtained
an Act of Parliament for the interchange and consolidation of
intermixed holdings, but not for their enclosure, ploughed up the
dividing balks, and grew clover. But, several years later, one
of the farmers asserted his legal right to the herbage of the balks
by turning his sheep into the clover crops which had taken their
place. In another parish in the same county, the inhabitants
agreed to exchange the dual system of one crop and a fallow for a
three-year course of two crops and a fallow. But, after a few years,
the agreement was broken by one of the farmers exercising his
common rights over the fallows by feeding his sheep on the growing
crops. Such breaches of voluntary arrangements could only be
prevented by obtaining the sanction of Parliament, and so binding,
not only dissentients, but those who were minors, possessed limited
interests, or were under some other legal disability to give valid
assent.

In the seventeenth century, it had to some extent become the
practice to obtain confirmation of enclosing agreements from the
Court of Chancery, or, where the Crown was concerned, the Royal
sanction. There is some evidence that the threat of a Chancery
suit was used as a means of obtaining consents, and that an attempt
was made to represent the decision as a legal bar to claims of common
by those who were not parties to the suit. After the Restoration
a change of practice was made, which marks, perhaps, the growing
desire to curb the power of the Crown. The jurisdiction of the
Court of Chancery was at first supplemented, then ousted, by the

private Act of Parliament. If four-fifths or sometimes a smaller proportion, in number and value, of the parties interested, together with the landowner and the tithe-owner, were agreed, the Enclosure Bill received Parliamentary sanction. Commissioners were appointed who proceeded to make an award, consolidating the intermixed lands of the open farm and dividing up the commons. Of these private Enclosing Acts the earliest instance occurs in the reign of James I. (4 Jac. I. c. 11). But it was not till the reign of Anne that they became the recognised method of proceeding. Even then the Acts were sometimes only confirmatory of arrangements already made between the parties. In the reigns of George I., George II., and George III., the number increased, at first slowly, then rapidly. Acts for enclosing only wastes, in which pasture commons were often included, must be distinguished from those Acts which dealt, not only with pasture-commons, but also with open arable fields and meadows, mown and grazed by the partners in common. Of the first class, there were, in the first sixty years of the eighteenth century, not more than 70 Acts, while from 1760 to 1815 there were upwards of 1000. Before 1760 the number of Acts dealing more specifically with the open-field system did not exceed 130. Between 1760 and 1815 the number rose to upwards of 1800. Of the area of waste, open-field and common, actually enclosed for the first time, it is impossible to speak with any certainty. The quantity of land is often not mentioned in the Enclosure Act, or can only be calculated from uncertain data. No record is available for the area enclosed by private arrangement or individual enterprise. It may, however, be safely estimated that not less than 4 million acres were enclosed in England and Wales within the period. Probably this figure was in reality considerably exceeded ; possibly it might be, without exaggeration, increased by two-thirds.

Before 1790, in many parts of England, the process of enclosing open-field farms and commons had been practically completed by private arrangement without the expensive intervention of Parliament. At different dates, and with little or no legislative help, the ancient system of cultivation, if it ever existed, had been almost extinguished in the south-eastern counties of Suffolk and Essex ; in the southern counties of Kent and Sussex ; in the south-western counties of Somerset, Devon, and Cornwall ; in the western counties of Hereford, Monmouth, Shropshire, and Stafford ; in the northern counties of Cheshire, Lancashire, Westmoreland, Cumberland, Northumberland, and Durham. No generalisation will explain why these districts should have been enclosed sooner or more easily than elsewhere.[1] The facts remain that no Parliamentary enclosures took place in Kent, Devonshire, Cornwall, or Lancashire ; that as early as the middle of the sixteenth century Kent, Essex, and Devonshire were stated by a Tudor writer to be the most enclosed and wealthiest counties ;[2] that in 1602 Carew, the historian of Cornwall, recorded that his countrymen " fal everywhere from Commons to Inclosure, and partake not of some Eastern Tenants' envious dispositions, who will sooner prejudice their owne present thrift, by continuing this mingle-mangle, than advance the Lords expectant benefit, after their terme expired " ;[3] that in 1656 Joseph Lee[4] mentions Essex, Hereford, Devonshire, Shropshire, Worcester as " wholly enclosed " ; that in 1727 the Rev. John Laurence says that " as to the Bishoprick of *Durham*, which is by much the richest Part of the North, Nine Parts in Ten are already inclosed." [5]

Since the last half of the fifteenth century the enclosing movement had been continuously in operation. Why, in the eighteenth and nineteenth centuries, was more land enclosed by Act of Parliament in some districts than in others ? The answer depends on

[1] The question may be stated in figures, which are collected from Dr. Slater's *The English Peasantry and the Enclosure of Common Fields* (1907), Appendix B.

During the eighteenth and nineteenth centuries, enclosures by Act of Parliament were made of the following areas of open-fields (arable and meadow) and commons, in the South-East and South-West, in the West, the North-West, and North : Suffolk, 22,206 acres ; Essex, 17,393 acres ; Kent, none ; Sussex, 15,185 acres ; Somerset, 30,848 acres ; Devon, none ; Cornwall, none ; Hereford, 8,168 acres; Monmouth, 1,293 acres; Shropshire, 2,310 acres ; Stafford, 16,925 acres ; Cheshire, 3,326 acres ; Lancashire, none ; Westmoreland, 3,237 acres ; Cumberland, 8,700 acres ; Northumberland, 22,348 acres ; Durham, 4,637 acres.

During the same period the following areas of open-fields and commons were enclosed by Act of Parliament in the Midlands, the East, and the North-East : Bedfordshire, 91,589 acres ; Buckinghamshire, 111,427 acres ; Oxfordshire, 142,238 acres ; Northamptonshire, 308,722 acres ; Warwickshire, 131,104 acres ; Rutland, 43,901 acres ; Leicestershire, 185,176 acres ; East Riding of Yorkshire, 274,479 acres ; West Riding, 172,944 acres ; Lincolnshire, 445,777 acres ; Norfolk, 106,043 acres ; Cambridgeshire, 87,413 acres ; Huntingdonshire, 93,366 acres.

[2] *Compendious Examination*, etc., by W. S. (1549).

[3] *Cornwall* (1602).

[4] *Vindication of a Regulated Enclosure* (1656).

[5] *A New System of Agriculture* (1727).

local circumstances or agricultural conditions. Disturbances on the northern and western borders were unfavourable to settled agriculture, and village farms and commons never throve extensively in the counties adjoining the borders of Scotland and Wales. In districts which abounded in fens, marshes, moorlands or hills, the space occupied by open-fields was necessarily limited, although the inhabitants of the neighbourhood may have exercised over these waste tracts rights of goose-pasture, of cutting fuel, turf, or reeds, or, where possible, of grazing. But the land, when enclosed, was taken in from the wild, and was, from the first, cultivated in separate holdings. Other districts, which naturally were clothed with extensive woodlands or forest, were enclosed piecemeal by individual enterprise for individual occupation. After the end of the fourteenth century, it is unlikely that any cleared land would have been cultivated in common. Other districts, lastly, which were industrially developed by the neighbourhood of large towns, or by the existence of some manufacturing industry, were early enclosed, either because of the demand for animal food and dairy produce, or because of the scarcity of purely agricultural labour.

On these general principles, before the era of Parliamentary enclosure, may be partially explained the comparative absence or disappearance of open-fields and pasture commons in the border counties, in the Wealds of Surrey, Kent, Sussex, in the forest districts of Hampshire, Essex, Warwickshire, or Nottinghamshire, in the neighbourhood of London or Bristol, or in the clothing districts of Devon and Somerset, of Essex, and Suffolk, or of parts of Norfolk. No doubt enclosure of cultivated land by agreement was at this period chiefly made for grazing and dairying purposes. But at the same time a large addition was being continuously made to the arable area of the country, partly by the reconversion of grass-land to tillage after fertility had been restored by rest, partly by the reclamation and enclosure of new land well adapted for grain. "Consider," writes Blith, "the Wood-lands who before Enclosure were wont to be releeved by the Fieldon with Corn of all sorts, And now are grown as gallant Corn Countries as be in England." [1] This addition to tillage necessarily affected the whole of the old corn-growing districts, where a large acreage, more fitted for pasture than for tillage, was kept under the plough by the open-field system. The effect was more and more felt when in-

[1] *The English Improver*, chap. xiii.

creasing facilities of communication enabled farmers to put their land to the best use by relieving them from the old uniform necessity of growing corn for the locality.

Elsewhere, the early or late enclosure of land was in the main determined by such agricultural reasons as climate or soil. Enclosure took place first, where it paid best agriculturally. In the moister climate of the South-west and West the rigid separation of arable from pasture was unnecessary. In some parts of the country the suitability of the land for hops or fruit necessitated early enclosure. Blith's reference to the plantation of the hedgerows with fruit-trees in " *Worcestershire, Hereford,* and *Glostershire* and great part of the county of *Kent* " points to separate occupation in the first half of the seventeenth century.[1] In other parts, if corn-land was more adapted to pasture, it was, under the new conditions, enclosed and laid down to grass. It was thus that the grazing districts on the water-bearing pasture belt of the Midlands, or the dairying districts of Gloucestershire or Wiltshire came into separate occupation. So also, where the soil was of a quality to respond quickly to turnips, clover, and artificial grasses, it was enclosed in order that it might profit by the new discoveries. This was the case on the light soils of Norfolk, where, as Houghton noted, turnip husbandry had been introduced with success before the close of the seventeenth century. This early use of roots is confirmed by Defoe,[2] who says of Norfolk ; " This part of *England* is remarkable for being the first where the Feeding and Fattening of Cattle, both Sheep as well as black Cattle, with Turnips, was first practis'd in *England.*"

Where land did not appear to be so immediately susceptible to the influence of these improvements, which were still imperfectly understood, the question of enclosure, and of the use to which the land was put, became mainly one of expense. Only the best and strongest land was able to endure the open-field system without exhaustion. To separate occupiers, eighteenth century improvements offered new means of restoring the fertility of exhausted soil. At the same time the revolution in stock-breeding held out new temptations to graziers. Much worn-out arable land of indifferent or medium quality was enclosed because its produce was declining.

---

[1] *English Improver,* chap. xix.

[2] *A Tour thro' the whole Island of Great Britain* (2nd edition, 1738), vol. i. pp. 60-61. Defoe began his tour in 1722.

If the price of corn was low, it was cheaper, and more profitable for the time, to lay it down to grass. If prices were high, the increased margin of profits from arable farming under separate management might cover the heavy cost of legislation and adaptation. Throughout the eighteenth century the number of Enclosure Acts fluctuated considerably with the advance or decline in the price of wheat. Thus the serious scarcity of corn from 1765 to 1774 produced a great crop of legislation. During the next fifteen years, the number of Acts was kept in check by the comparative abundance of the harvests. Once more, during the famine years of the Napoleonic war, the Acts rapidly multiplied under the pressure of necessity and with the progress of agricultural skill. The need was too urgent to admit of those private arrangements for the break-up of open-field farms which could often only be carried out after years of preparation. Private Acts of Parliament were more speedy in their operation. Still the quality of the soil to a great extent controlled the course of legislation. Open-fields continued longest in the districts where the soil was chalk, or where the village farm occupied rich corn-growing land, or where the soil was so unsuited for grass that the prospects of increased profits from arable farming, even in separate occupation, were doubtful. A geological map of the country would, it is believed, supply the key to many difficulties in the history of enclosure.

The parts of England which were most affected by the Enclosure Acts of the Hanoverian era were the corn-growing districts of the East, North-east, and East Midlands. Within this area are fourteen out of the fifteen counties which, in proportion to their size, contained the largest acreage enclosed by Act of Parliament. The ease with which in other districts individual occupation was substituted for common cultivation renders it difficult to answer the question, why in these particular groups of counties the cheaper process of private arrangement was not adopted ? No completely satisfactory answer can be given. It was from these districts that the greatest opposition to the enclosing movement of the Tudor and Stewart periods had come. It was also in these districts that, in the closing years of Elizabeth, enclosures were proceeding so rapidly as to be restricted by a special Act of Parliament.[1] The effect of popular outcry and consequent legislation may have been to confine the enclosing movement to Northamptonshire, Leicester-

[1] 39 Eliz. c. 2 (1597).

shire, and Warwickshire, where it continued to run its course in
the seventeenth century. Elsewhere in the Midlands, the counties
that formed the area from which London drew its chief supplies
of corn, could not have been converted into pasture without raising
a storm of opposition. Yet throughout the seventeenth century
and during the first three quarters of the eighteenth, enclosures
had been most profitable where arable land had been converted to
grass, and large tracts of Midland pasture were the result of this
movement before Parliamentary intervention had begun. Leicestershire is a conspicuous example of this conversion. It was, notes
Marshall in 1786, "not long ago an open arable county ; now it is
a continuous sheet of greensward." The vale of Belvoir, which, in
the days of Plattes, was considered to be the richest corn-district
in the country, had been laid down to grass before the time of Defoe
(1722-38).[1] He describes the whole county as given over to grazing.
" Even most of the Gentlemen are Grasiers, and in some Places the
Grasiers are so rich that they grow Gentlemen." Yet in the first
half of the seventeenth century it had been a county of open-
fields, famous for the pigs that were fattened on its beans and
pease.[2] Apart from difficulties arising from local peculiarities ot
tenure, or of the shape of open-field farms, or from want of roads,
from public opinion, or special legislation, the Midland corn counties
perhaps owed some of their immunity to the interested opposition
of tithe-owners, whose assent was necessary to Parliamentary
enclosure. For the sake of the great tithes, they would always
strenuously resist any attempt by private Act to turn open-fields
into pasture farms. It was not till after 1765 that their views
underwent a change. The improvements in arable farming, which
were now possible on separate holdings, together with the high
price of corn, made it probable that, even when open-fields and
commons were enclosed, the area of tillage would not be diminished.
These considerations were strengthened during the French wars of
1793-1815, which by the stoppage of foreign corn supplies added
new reasons for seeking legislative aid in enclosure.

Up to the accession of George III. (1760) prices of corn ruled low.
More than once in the preceding period (1700-60) loud complaints
were heard of agricultural depression, of farmers unable to pay

---

[1] *Tour*, vol. ii. pp. 332, 335.

[2] The same remark is made by Professor Bradley in his *Gentleman and
Farmer's Guide for the Increase and Improvement of Cattle* (1729), p. 75.

their rents, of the small gentry forced to sell their estates, of landlords compelled by loss of income to curtail their establishments. As yet there was no scarcity caused by population outstripping production, no increased demand for food supplies from great industrial centres. But without these spurs to farming progress, preparations for advance were being made, and far-reaching improvements in the cultivation of arable land had been already tested or initiated by men like Jethro Tull and Lord Townshend.

In the progress of scientific farming Tull is one of the most remarkable of pioneers. His method of drilling wheat and roots in rows was not generally adopted till many years after his death. But the main principles which he laid down in his *Horse-Hoeing Husbandry* (1733) proved to be the principles on which was based an agricultural revolution in tillage. The "greatest individual improver" that British agriculture had ever known, he sought to discover scientific reasons for observed results of particular practices. He was thus led to strike out for himself new and independent lines of investigation. The chemistry of plant-life was in its infancy, the science of vegetable physiology an almost untrodden field of knowledge. Into these comparatively unexplored regions Tull advanced alone, and, by minute observation of nature and stubborn tenacity of purpose, he advanced far. Considering his difficulties and disadvantages, it is a remarkable proof of his real genius that he should have discovered so much. He lived in a solitary farmhouse, remote from such scientific aid as the age afforded, or from friends in whom he could confide. His microscope was "very ordinary"; his appliances were self-made; his experiments thought out for himself. He made his observations and notes, tortured by the "stone, and other diseases as incurable and almost as cruel." His labourers, by whom he was, metaphorically, "insulted, assaulted, kicked, cuffed and bridewelled," tried his patience beyond endurance. His son turned out an extravagant spendthrift who ended his days in the Fleet Prison. Ill-health and misfortune made him irritable. His sensitive nature was galled alike by the venomous criticism of the book,[1] in which he published the results of his thirty years' experience as a farmer, and by its shameless plagiarism. Yet he never lost his confidence that his

[1] *The new Horse-Houghing Husbandry,* 1731. (Five chapters of the subsequent book which were pirated and re-printed in Ireland.) *New Horse-Hoing Husbandry,* 1733. Supplement, 1740. William Cobbett edited and published the *Horse Hoeing Husbandry* in 1822 : 2nd edition, 1829.

" practice would one day become the general husbandry of England."

The son of a Berkshire landowner, Jethro Tull was born at Basildon in 1674. From Oxford, which he left without taking a degree, he entered Gray's Inn as a law student, made the grand tour of Europe, and was called to the Bar in 1699. Scholar, musician, traveller, lawyer, he became a farmer not by choice but from necessity. In 1699 he settled down with his newly married wife at " Howberry " Farm [1] in the parish of Crowmarsh, near Wallingford. There he lived ten years. In 1709 he moved to Mount Prosperous, a hill-farm in the parish of Shalbourn, on the borders of Berkshire and Wiltshire. Two years later, the failure of his health drove him abroad to save his life. Returning in 1714 to Mount Prosperous, he remained there till his death in 1740, living in a house, covered with home-made glazed tiles, which Arthur Young, who visited the place fifty years later, described as a " wretched hovel."

At Crowmarsh Tull invented his drill. As a gentleman-farmer he found himself at the mercy of his farm-servants. From his own experience he verified the truth of the saying :

" He who by the plough would thrive
Must either hold himself or drive."

He determined to plant his whole farm with sainfoin. But " seed was scarce, dear and bad, and enough could scarce be got to sow, as was usual, seven bushels to the acre." He set himself to conquer the difficulty. By constant observation and experiment he learned the difference between good and bad seed, as well as the advantages of care in selection, of cleaning, steeping, and change ; he also proved that a thin sowing produced the thickest crop, and discovered the exact depth at which the seed throve best. " So," he says, " I caused channels to be made, and sowed a very small proportion of seed, covered exactly. This was a great success." But it was also an innovation, and his labourers struck in a body. Tull refused to be beaten. He set his inventive faculty to work " to contrive an engine to plant sainfoin more faithfully than hands would do." His knowledge of the mechanism of an organ stood him in good stead. The groove, tongue, and spring of the sounding board suggested the idea of an implement which delivered the seed through notched barrels. Behind was attached a bush harrow

[1] It is remarkable that this farm now (1912) contains one of the most highly cultivated pieces of land in the world.

which covered the seed. The machine answered its purpose, and he afterwards introduced several improvements of his original plan. The originality of his invention cannot justly be disputed, though his enemies, and he had many, asserted that he brought the machine from abroad or had been preceded by Plat, Plattes or Worlidge. All four inventors saw the advantage of sowing not broadcast but in rows. Both Plat and Plattes were setters, rather than drillers, of corn, and they took for their model the dibbing of beans or peas. Plat seems to have invented a board, to which were fixed iron dibbers. Something of this sort is depicted on the title-page of Edward Maxey's *New Instruction* [1] (1601). Gabriel Plattes designed a machine to punch holes in the land as it went along. But, as is pointed out in Hartlib's *Legacie*, the author of which suggested hoeing the furrows by hand, the machine would have been practically useless in wet and heavy land. Neither Plat nor Plattes contemplated a mechanical sowing ; both intended the seed to be deposited by hand. In this respect Worlidge's drill was an advance on his predecessors. He placed coulters in front of the seed-boxes, from which the seed was deposited through barrels into furrows. But he never made or tried his implement. When Professor Bradley in 1727 constructed a machine from Worlidge's drawing, he found that it would not work. To Tull, therefore, belongs the credit of the first drill which served any practical purpose.

Tull's many mechanical inventions were less valuable than the reasons which he gave for their employment. His implements were speedily superseded ; his principles of agriculture remain. During his foreign travels he was impressed with the cultivation of vineyards in the south of France, where frequent ploughings between parallel rows of vines not only cleaned the land, but worked and stirred the food-beds of the plants until the vintage approached maturity. Tull determined to extend the principles of vine-culture to the crops of the English farm. He argued that tillage was equally necessary before and after sowing. When crops were sown, nature at once began to undo the effect of previous ploughings and sowings. The earth united, coalesced, consolidated, and so shut out the air and water from the roots, and decreased the food supply at the moment when the growing plants most needed increased nourishment. To some extent the use of farm-

---

[1] *A New Instruction of Plowing and Setting of Corne, handled in manner of a Dialogue betweene a Ploughman and a Scholler.*

yard manure kept the land friable ; but it also stimulated the growth of weeds. The better course, therefore, was to keep the land pulverised by tillage and so to prevent the contraction of the food area of the growing crops. So long as wheat and turnips were sown broadcast, this method could not be satisfactorily employed. But if they were drilled in rows, divided by sufficiently wide intervals, the principles of vine culture could be profitably applied. In two ways the crops benefited by constant tillage. In the first place, the land was kept clean from weeds, and so saved from exhaustion. In the second place, the repeated pulverisation of the soil admitted air, rain-water, and dews to the roots of the plants, and extended the range from which their lateral growths drew their food supplies. In some respects Tull's system failed. His rows of thinly sown wheat, for instance, were drilled so far apart that the plants were slow to mature, and therefore, if sown late in the year, were more susceptible to blight. But for turnips his method was admirable. Incidentally also he found that his " drill husbandry " was a substitute not only for fallows, but for farm· yard dung, which he dreaded as a weed-carrier. Without fallows or manure, he grew on the same land, by constant tillage, for thirteen years in succession heavier wheat crops, from one-third of the quantity of seed, than his neighbours could produce by following the accepted routine. By this discovery he anticipated one of the most startling results of the Rothamsted experiments.

The chief legacies which Jethro Tull left to his successors were clean farming, economy in seedings, drilling, and the maxim that the more the irons are among the roots the better for the crop. It was along these lines that agriculture advanced. On open-field farmers who sowed their seed broadcast, thickly, and at varying depths, Tull's experiments were lost. Equally fruitless, so far as his immediate neighbours were concerned, was his demonstration of the value of sainfoin and turnips, or the drilling of wheat and roots. Even his system of drilling roots was neglected in England, till it had been tested and adopted in Scotland.

It was not till Tull's principles were put in practice by large landlords in various parts of the country that their full advantages became apparent. In England this was the work of men like Lord Townshend at Raynham in Norfolk, Lord Ducie at Woodchester in Gloucestershire, or Lord Halifax at Abbs Court near Walton-on-Thames. In Scotland the " Tullian system " was enthusiasti-

FARMING A FASHION 173

cally preached by the Society of Improvers in the Knowledge of Agriculture in Scotland (founded 1723, dissolved 1745), by Lord Cathcart, and by Mr. Hope of Rankeillor. In the *Heart of Midlothian*, Scott is true to the spirit, if not to the details, of history when he credits the Duke of Argyll with a keen interest in all branches of farming and the introduction into Inverness-shire of a herd of Devonshire cattle. Agriculture had for the moment become a fashion in society, a part, perhaps, of the artificial movement which in gardening created the Landscape School. Tull's system was discussed at Court. It was explained to George II., and therefore interested Lady Suffolk. The practical Queen Caroline subscribed to the publication of the *Horse-Hoeing Husbandry*. Pope loved to " play the philosopher among cabbages and turnips." Sir Robert Walpole, it is said, opened the letters of his farm steward before he broke the seals of correspondence on State affairs ; Bolingbroke caused Dawley Farm to be painted with trophies of ricks, spades, and prongs, and, propped between two haycocks, read Swift's letters, uplifting his eyes to heaven, not in admiration of the author but in fear of rain. " Dawley," said his political opponents, " has long been famous for a Great Cry and little Wool. Tup *Harry* become Mutton master." [1]

Other landowners threw themselves energetically into the practical work of agricultural improvement. Charles, second Viscount Townshend, may be taken as a type of the reforming landlords who took the lead in farming their estates. Born in 1674, he died in 1738, having succeeded to the title and estates of his father when a child of thirteen years old. In his early life, he had played a prominent part in the political history of the country at a critical period. Lord Privy Seal under William III., he served as a Commissioner to treat for the Union of England and Scotland, and, as a joint plenipotentiary with Marlborough, signed the Peace of Gertruydenberg in 1709. In the same year, as Ambassador at the Hague, he negotiated the famous Barrier Treaty. Under George I. and George II., he acted as Secretary of State, was appointed Lord Lieutenant of Ireland, and, as joint Secretary of State with Walpole, directed the foreign policy of Great Britain.

In 1730 Lord Townshend retired from political life to Raynham in Norfolk. There he devoted himself to the care of his estates, experimenting in the farming practices which he had observed

[1] *The Hyp Doctor*, No. 32, July 20, 1731.

L—E F P P

abroad, and devoting himself, above all, to improvements in the rotation of crops, and to the field cultivation of turnips and clover, which, in the preceding half century, had been successfully introduced in the county. His land mainly consisted of rush-grown marshes, or sandy wastes where a few sheep starved and "two rabbits struggled for every blade of grass." The brief but exhaustive list of its productions is "nettles and warrens." Townshend revived the ancient but almost obsolete practice of marling the light lands of Norfolk. Farmers believed that marl was "good for the father, bad for the son," till he proved its value on the sandy soil of the county. The tide of fashion set once more in its favour, and farmers found another proverbial saying for their purpose :

> " He who marls sand
> May buy the land ;
> He that marls moss
> Suffers no loss ;
> He that marls clay
> Throws all away."

By the use of marl alone Young calculates that "four hundred thousand acres have been turned into gardens." Following the lines of Jethro Tull, Townshend drilled and horse-hoed his turnips instead of sowing them broadcast. He was also the initiator of the so-called Norfolk, or four-course, system of cropping, in which cereals, roots, and artificial grasses were alternated.[1] The introduction of roots and grasses encouraged the farmer to observe the useful rule of never taking two corn crops in succession, saved him from the necessity of leaving a portion of land every year in unproductive fallow, enabled him to carry more stock and maintain it without falling off during the winter months. For the light sands of Norfolk turnips possessed a special value. Roots, fed on the ground by sheep, fertilised and consolidated the poorest soil. Another portion of the crop, drawn off and stored for winter keep, helped the farmer to keep more stock, to obtain more manure, to enrich the land, to increase its yield, to verify the truth of the proverb " A full bullock-yard and a full fold make a full granary." Farming in a circle, unlike arguing, proved a productive process.

So zealous was Townshend's advocacy of turnips as the pivot of agricultural improvement, that he gained the nickname of " Turnip " Townshend, and supplied Pope with an example for his Horatian Illustrations, (Bk. ii. Epist. ii. ll. 270-9) :

---

[1] See footnote on next page.

" Why, of two brothers, rich and restless one
Ploughs, burns, manures, and toils from sun to sun ;
The other slights, for women, sports, and wines,
All Townshend's turnips and all Grosvenor's mines,

  *   *   *   *   *   *   *   *   *   *

Is known alone to that Directing Power
Who forms the genius in the natal hour."

Townshend's efforts to improve his estates were richly rewarded.
On the sandy soil of his own county, his methods were peculiarly
successful. Furze-capped warrens were in a few years converted
into tracts of well-cultivated productive land. Those who followed
his example realised fortunes. In thirty years one farm rose in
rental value from £180 to £800 ; another, rented by a warrener at
£18 a year, was let to a farmer at an annual rent of £240 ; a farmer
named Mallett is said to have made enough off a holding of 1500
acres to buy an estate of the annual value of £1800. Some farmers
were reported to be worth ten thousand pounds. But the example
only spread into other counties by slow degrees. Outside Norfolk,
both landlords and farmers still classed turnips with rats as
Hanoverian innovations, and refused their assistance with Jacobite
indignation. Even in Townshend's own county, it was not till the
close of the century that the practice was at all universally adopted ;
still later was it before the improved methods were accepted which
converted Lincolnshire from a rabbit-warren or a swamp into cornfields and pasture.

¹ Though Townshend made the four course rotation widely known, it had
probably been worked out before his time by the farmers of the Eastern
Counties. Ellis (*Chiltern and Vale Farming*, 1733) describes it as regular
Hertfordshire practice.

CHAPTER VIII

THE STOCK-BREEDER'S ART AND ROBERT BAKEWELL
(1725-95)

Necessity for improving the live-stock of the country ; sheep valued for
their wool, cattle for power of draught or yield of milk ; beef and mutton
the growing need : Robert Bakewell the agricultural opportunist ; his
experiments with the Black Horse, the Leicester Longhorns, and the New
Leicesters ; rapid progress of stock-breeding : sacrifice of wool to mutton.

WITHOUT the aid of turnips the mere support of live-stock had
been in winter and spring a difficult problem ; to fatten sheep and
cattle for the market was in many districts a practical impossibility.
The introduction, therefore, of the field cultivation of roots, clover,
and artificial grasses proved the pivot of agricultural progress. It
enabled farmers to carry more numerous, bigger, and heavier stock ;
more stock gave more manure ; more manure raised larger crops ;
larger crops supported still larger flocks and herds. Thus to the
hopeful enthusiasts of the close of the eighteenth century the agricultural circle seemed capable of almost indefinite and always profitable expansion.

But recent improvements in arable farming could not yield their
full profits till the live-stock of the country was also improved.
The necessary revolution in the breeding and rearing of stock was
mainly the work of Robert Bakewell (1725-95), a Leicestershire
farmer, living at Dishley, near Loughborough. Its results were
even more remarkable than those which followed from the new
methods of Tull and Townshend. Bakewell's improvements were
also more immediately accepted by agriculturists. The slow
adoption of improved practices in tillage was mainly due to caution ;
in some degree, also, it was due to the fact that the innovators
were, if not amateurs, gentlemen-farmers.[1] On the other hand,

[1] In 1756 or 1757 Mr. Pringle, a retired army surgeon, introduced the
drilling of turnips on his estate near Coldstream in Berwickshire. His crops

the improved principles of stock-breeding were more readily accepted, not only because their superiority was at once manifest to the eye, but because they emanated from the practical brain of a professional farmer. Yet for open-field farmers they were of little value. As sheep and cattle increased in size and weight, and were bred for more speedy conversion into mutton and beef, they needed better and more abundant food than village farms could supply. Thus the improvements of Bakewell, like those of Tull and Townshend, added a new impulse to the progress of enclosures.

Up to the middle of the eighteenth century sheep had been valued, agriculturally for their manure in the fold, commercially for their skins and, above all, for their wool. Wool was in fact the chief source of trading profit to English farmers. Other forms of agricultural produce were raised as much for home consumption as for sale. But the trade in raw or manufactured wool, both at home and abroad, had been for centuries the most important of English industries. To the golden fleece the carcase was sacrificed ; the mutton as food was comparatively neglected. As wool-producing animals sheep were classified into short wools and long wools. Of these two classes, short-wooled sheep were by far the most numerous, and were scattered all over England. Small in frame, active, hardy, able to pick up a living on the scantiest food, patient of hunger, they were the sheep of open-field farmers ; they were the breeds formed by centuries of far travelling, close feeding on scanty pasturage, and a starvation allowance of hay in winter. Such were the " heath-croppers " of Berkshire—small ill-shaped sheep which, however, produced " very sweet mutton." In some counties, as, for instance, Buckinghamshire, open-field farmers hired sheep, with or without a shepherd, for folding on their arable land. The flocks, hired from Bagshot Heath, were fed, partly on the commons, partly on the arable fallows, where they were folded every night from April to October. No money passed. The flockmaster was paid by the feed ; the farmer by the folding. The one made his profit by the wool, the other by the manure. Sometimes

were superior to those of the neighbouring farmers. But none followed his example. In 1762 a farmer named William Dawson adopted the practice on his farm at Frogden in Roxburghshire. "No sooner did Mr. Dawson (an actual farmer) adopt the same system, than it was immediately followed, not only by several farmers in his vicinity, but by those very farmers adjoining Mr. Pringle, whose crops they had seen for ten or twelve years so much superior to their own " (*General view of the Agriculture of the County of Northumberland*, by J. Bailey and G. Culley, 3rd edition (1805), p. 102).

small men who had rights of common, or had acquired them from commoners, drove their flocks from open-field to open-field, folding them on the fallow lands of the village farm and receiving from the occupiers of the fallows 1s. a week per score or leave to graze on the commons during some part of the winter. Nearly every breeding county in England had its local favourites, adapted to their environment of soil, climate, and geographical configuration. For fineness of wool, the Ryeland or Herefordshire sheep now held the first place in the manufacture of superfine broad-cloth, though in the fourteenth century the fleece of the Morfe Common sheep of Shropshire had commanded the highest prices. Sussex South Downs, inferior in size and shape to their present type, were also famed for the excellence of their soft, fine, curly wools. Dorsets, already prized for their early lambs, supplied Ilminster with the material for its second, or livery, cloths. West-country clothiers drew their supplies, partly from Wales, partly from the large, horned, and black-faced Wiltshires, from the Exmoors, Dartmoors, or Devonshire Notts, the Mendips of Somerset, the Dean Foresters of Gloucester, or the Ryelands of Herefordshire. The eastern counties had their native short-wooled Norfolk and Suffolk breeds. The North had its Cheviots, its Northumberland Muggs, its Lancashire Silverdales, its Cumberland Herdwicks, its Cheshire Delameres. Here and there, some local breed was especially famous for the quality of its mutton, like that of Banstead or of Bagshot in Surrey, of Portland in Dorsetshire of Clun Forest in Shropshire, or of the mountain sheep of Wales. But, speaking generally, it was by their fleeces only that sheep were distinguished.[1] The local varieties of short-wools differed widely from one another. In appearance the long-wools were more uniform in type ; all were polled, white-faced, and white-legged ; all were large-framed, and, from more abundant food, heavier in carcase and in fleece ; in all the wool was long, straight, and strong. Less widely distributed than the other class, they were also by far the least numerous. In the eighteenth century they probably did not exceed more than one-fourth of the total number of sheep in the country. But the superior weight of their fleeces made their produce more than one-third of the total clip. Among the long-wools the Cotswolds were, at this time, pre-eminent. Other varieties, better adapted to the special conditions of their respective counties, were the Lincolns, Leicesters, Devonshire Bamptons, and the Romney Marsh sheep of Kent.

---

[1] The brown-faced, short-woolled " heath " sheep of southern England, from which the Down breeds were selected, have some claim to be regarded as the autochthonous sheep of England. The large white-faced long-wools may

With these different breeds, both short and long wools, there was abundant scope for experiment and improvement. Some effort had been made at the close of the seventeenth century, as has been already noticed, to improve the lustrous fleeces of Lincolns, and to remedy the bareness of their legs and bellies. But, from the grazier's point of view, no breeder had yet attempted to obtain a more profitable shape. If any care was shown in the selection of rams and ewes, the choice was guided by fanciful points which possessed no practical value. Thus Wiltshire breeders demanded a horn which fell back so as to form a semicircle, beyond which the ear projected ; Norfolk flockmasters valued the length and spiral form of the horn and the blackness of the face and legs ; Dorsetshire shepherds staked everything on the horn projecting in front of the ear ; champions of the South Downs condemned all alike, and made their grand objects a speckled face and leg and no horn at all.

In cattle, again, no true standard of shape was recognised. Size was the only criterion of merit. "Nothing would please," wrote George Culley in 1786, "but Elephants or Giants." [1] The qualities for which animals were valued were not propensity to fatten or early maturity, but their milking capacity or their power of draught. The pail and the plough set the standard ; the butcher was ignored. Each breeding county, however, had its native varieties, classified into Middle-horns, prevailing in the South and West of England, in Wales and in Scotland ; Long-horns, in the North-west of England and the Midlands ; and Short-horns, in the North-east, Yorkshire, and Durham.

The Middle-horns in the South and West of England were red cattle of a uniform type ; the North Devons, nimble and free of movement, were unrivalled in the yoke ; the Herefords, not yet bred with white faces, were heavier animals which fattened to a greater weight ; the Sussex breed came midway in size between the two. None of the three were remarkable for the quantity of their milk. Other middle-horned breeds were the black Pembrokes, like their Cornish relatives, excellent for the small farmer, and the Red Glamorgans, which in the eighteenth century were highly esteemed as an all-round breed. Every year thousands of the black Angleseys were swum across the Menai Straits to the main-

have had a Roman origin : the Cotswolds may go back to the Roman occupation, the Leicesters, Lincolns, and Romney Marsh breeds presumably represent later introductions from Flanders.

[1] Quoted by Arthur Young in his Lecture on the *Husbandry of Three Famous Farmers* (1811), pp. 10-11.

land. Scotland had its West Highlanders ; its Ayrshires, second to none as milkers ; its Galloways and its Anguses, originally middle-horned but now becoming polled, which were driven southwards to the October and November fairs of Norfolk and Suffolk to be fattened for the London markets. On these imported Galloways were founded the Norfolk breed of polled cattle, and the Suffolk Duns once famous all over England for their milking qualities. The North-west of England and the Midlands were occupied by the Long-horns. Of these the most celebrated were the Lancashires, or Cravens, so called from their home in the corner of the West Riding of Yorkshire which borders on Lancashire and Westmoreland. To this breed some attention had, as is noticed in the *Legacie*, been paid in the seventeenth century. To the same stock belonged the brindled or grizzled Staffordshires, valued, like the Cravens, for the dairy and for meat. The North of Lincolnshire, the East Riding of Yorkshire, and Durham were famous for the enormous size of their short-horned cattle, which were extraordinary milkers. The Holderness breed, as it was called before its establishment on the banks of the Tees, were "more like an ill-made black horse than an ox or cow." [1] The cattle were badly shaped, long-bodied, bulky in the coarser points, small in the prime parts. But they satisfied the taste of the eighteenth century grazier, because their gigantic frames offered plenty of bone on which to lay flesh. They were undoubtedly a breed of foreign origin. Tradition relates that, towards the end of the seventeenth century, a bull and some cows were introduced into the Holderness district from the Low Countries. But the introduction must have been of an earlier date. Lawson in his *New Orchard* (1618) says : " The goodnesse of the soile in Howle, or Hollow-, derness in Yorkshire is well knowne to all that know the River Humber and the huge bulkes of their Cattell there." It is probably to this introduction of foreign blood that Child alludes in his Letter in Hartlib's *Legacie* (1651), when he says that little attention was paid to breeding except in the north-western and north-eastern counties. To the same stock belonged the " long-legged short-horn'd Cow of the Dutch breed," which Mortimer (1707) selected as the best breed for milking. Probably, also, the famous " Lincolnshire Ox " was one of these Holderness Dutch-crossed animals. This beast was exhibited, as the Advertisement sets out, " with great satisfac-

[1] Culley's *Observations on Live Stock* (1786), p. 30.

tion, at the University of Cambridge," in the reign of Queen Anne. He was "Nineteen Hands High, and Four Yards Long from his Face to his Rump. The like Beast for Bigness was never seen in the World before. *Vivat Regina!*"

Stock-breeding, as applied to both cattle and sheep, was the haphazard union of nobody's son with everybody's daughter. On open-field farms parish bulls were only selected for the quality in which Mr. Shandy's pet, so strenuously denounced by Obadiah, was alleged to be wanting. When prizes were offered for. the longest legs, it is not surprising that all over the country were scattered tall, raw-boned, wall-sided cattle, and lean, leggy, unthrifty sheep. Our ancestors, however, were not unwise in their generation. Length of leg was necessary, when animals had to traverse miry lanes and "foundrous" highways, and roam for miles in search of food. Size of bone served the ox in good stead when he had to draw a heavy plough through stiff soil. But a time was rapidly approaching when beef and mutton were to be more necessary than power of draught or fineness of wool. Bakewell was the agricultural opportunist who saw the impending change, and knew how it should be met. By providing meat for the million, he contributed as much to the wealth of the country as Arkwright or Watt. There is some foundation for the statement that many monuments have been reared in Westminster Abbey to the memory of men who less deserved the honour than Robert Bakewell.

Cart-horses also shared Bakewell's attention. Before his day principles of breeding had been little studied except in the interests of sport. In the reign of Richard II. the principal breeding counties had been Lincolnshire, Cambridgeshire, and the East and West Ridings of Yorkshire. Men in armour needed big weight-carrying horses. But in the fifteenth century, horses, like the rest of English live-stock, seem to have dwindled in size. The legislature was alarmed ; Henry VIII. attempted to improve their height by the importation of the best foreign breeds, and by sumptuary laws which prescribed the number and height of the horses that were to be kept by various classes of his subjects. Elizabeth's introduction of coaches created a new need ; if the invention of gunpowder and the disuse of armour displaced the "great horse" in war, he found a new place between the shafts. Shakespeare's plays illustrate some of the changes which approximated the Stewart standard of horse-flesh to modern ideals. The courser,. which in time of war had endured " the shock of wrathful iron arms," [1] and in peace was the " foot-cloth " horse,[2] and three times stumbled under Lord Hastings,[3] gives place to the " prince of palfreys " who " trots the air " and makes the earth sing as he touches it with his elastic tread.[4] As highways improved, travellers journeyed more easily and more often. The ambling roadster, whose artificial gait was comparatively easy, was supplanted by the hack ; the coach-horse and the waggon-horse began to dispute the monopoly of the lumbering " great horse " and the pack-horse. Sport was also adapting itself to the changing conditions of society. Racing and hunting became fashionable. Though Shakespeare had heard

> " of riding wagers,
> When horses have been nimbler than the sands
> That run i' the clock's behalf." [5]

and was aware that " switch and spur " [6] were plied in a " wild-goose chase " on the Cotswold Hills, he knew nothing of the modern race-course. Races, then, were trials rather of endurance than of speed. Nor was pace much needed in Tudor hunting ; a " good continuer," [7] or, as we might say, a good stayer, was more necessary. In coursing the hare, only the greyhounds must be fleeter than " poor Wat." The red deer was followed by hounds " slow of pursuit " [8] and by men armed with leaping-poles, except on those rare occasions when the great hart was hunted " at force." At hawkings, unless the long-winged peregrine flew down wind, horsemen were not pushed to the gallop ; the short-winged goshawk exacted from his pursuers no turn of speed. But as agriculture advanced, the red deer's covert was destroyed, and his extermination demanded as an inveterate foe to the crops. So, too, the sport of falconry was doomed, when hedgerows and enclosures displaced the broad expanse of open-fields, and the partridge no longer cowered in the stubble by the edge of the turf-balk under the tinkling bells of the " towering " falcon.[9] Another beast of the chase and other means of capture were needed. Shakespeare stood on " no quillets how to slay " [10] a fox with snares and gins. But

---

[1] *Ric. II.* Act i. Sc. 3, l. 136. [2] *2 Hen. VI.* Act iv. Sc. 7, l. 52.
[3] *Ric. III.* Act iii. Sc. 4, l. 83.
[4] *Hen. V.* Act iii. Sc. 7, l. 17. Comp. also *Ven. and Ad.* st. 50, where Blundevill is closely copied.
[5] *Cymb.* Act iii. Sc. 2, ll. 72-4. [6] *Rom. and Jul.* Act ii. Sc. 4, l. 75.
[7] *Much Ado*, Act i. Sc. 1, l. 149. [8] *Mid. N. D.* Act iv. Sc. 1, l. 129.
[9] *Macb.* Act ii. Sc. 4, l. 12. [10] *2 Hen. VI.* Act iii. Sc. 1, l. 261.

the fox was no foe to crops ; hedgerows only added zest to his pursuit ; the new sport satisfied the new conditions, and demanded the production of the modern hunter.

The seventeenth century saw some of the conditions created which have developed the various types in horses of to-day. James I. reduced racing to rules ; Charles I. established races at Newmarket ; Oliver Cromwell kept his stud ; Charles II. introduced the "Royal Mares." Changes in the art of war demanded a lighter and more active cavalry. Fox-hunting had become a passion with the country gentry. Coaches travelled more rapidly. Oxen were less used on the farm. During the same century, foreign breeds were extensively imported. Arabs were favourites of James I. But the authority of the Duke of Newcastle, who disliked the breed, was paramount in matters of horse-flesh.[1] Barbs, or Turks were preferred till the Godolphin and Darley Arabians proved worthy rivals to the Byerly Turk. Other breeds were largely imported from Naples, Sardinia, Spain, Poland, Germany, Hungary, Flanders, and Libya. So great was the admixture of blood, that Bradley, writing in 1727, thinks the true-bred English horse hardly exists, "unless we may account the Horses to be such that are bred wild in some of our Forests and among the Mountains."[2] Horses intended for "the Course, the Chase, War or Travel" were already carefully studied. But horses for farm use were as yet despised. De Grey[3] speaks with contempt of horses for the cart, the plough, the pack-saddle, and Bradley ignores them altogether.

It was with the heavy Black-horse of the Midland counties that Bakewell conducted his experiments. The breed had long been known, and had doubtless helped to supply mounts to mediaeval knights. Early in the eighteenth century the breed had been improved by the importation of six Zealand mares. But the long back and long thick hairy legs were still characteristic. Defoe speaks of the Leicestershire horse as the "largest in England, being generally the great black Coach-Horses and Dray-Horses, of which so great a Number are continually brought up to London." Bakewell's object was to correct the type to that which was best suited

---

[1] *Methode et Invention Nouvelle de dresser les Chevaux* (1658). Newcastle's experiments were made with Barbs. The Duke also published in 1667 *A New Method and Extraordinary Invention to Dress Horses,* etc.

[2] *Gentleman and Farmers Guide*, p. 249.

[3] *The Compleat Horseman and Expert Ferrier,* by Thomas de Grey (5th ed. 1684), p. 8.

for draft. Strength and activity rather than height and weight were his aim. In his hands the Black Horse developed a thick short carcase on clean short legs. Marshall, who visited Dishley in 1784, grows enthusiastic over " the grandeur and symmetry of form " displayed in the stallion named K. " He was, in *reality, the fancied* war-horse of the German painters ; who, in the luxuriance of imagination, never perhaps excelled the grandeur of this horse." The Midland horses were generally sold as two-year-olds to the farmers of Buckinghamshire, Hertfordshire, Berkshire, and Wiltshire, who broke them into harness, worked them lightly on the land, and sold them at five or six to London dealers. The practice may account for some of the extravagant plough-teams, which agricultural writers of the eighteenth century often notice and condemn.

Born in 1725, Bakewell was barely twenty when he began his experiments in stock-breeding. He succeeded to the sole management of his father's farm in 1760. Ten years later, when Arthur Young, armed with an introduction from the Marquis of Rockingham, visited Dishley, Bakewell must have somewhat resembled the typical English yeoman who figures on jugs of Staffordshire pottery : " a tall, broad-shouldered, stout man of brown-red complexion, clad in a loose brown coat, scarlet waistcoat, leather breeches, and top-boots." Visitors from all parts of the world assembled to see his farm—his water-canals, his plough-team of cows, his irrigated meadows on which mowers were busy from May to Christmas, and, above all, his live-stock—his famous black stallion, his bull " Two-penny," and his ram " Two-pounder." All who came were astonished at the results which they saw, at the docility of the animals, at the kindness with which they were treated. But, if they hoped to learn from Bakewell's lips the principles which are now the axioms of stock-breeding, they went away disappointed. He was a keen man of business. The secrets of his success were jealously guarded, except from the old shepherd to whom they were confided. So careful was he to keep the lead in his own hands that he adopted the practice of only letting his stallions, bulls, and rams by the season, and, when his best bred sheep were past service and fatted and sold to the butcher, he is said to have infected them with the rot in order to prevent their use for breeding purposes. So reports Arthur Young.[1] Round

---

[1] *Farmers Tour through the East of England* (1771), vol. i. p. 118.

the hall of his house were arranged skeletons of his most celebrated animals ; from the walls hung joints, preserved in pickle, which illustrated such points as smallness of bone or thickness of fat. As there was no inn in the village, he seems to have kept open house for his visitors. He was never married. In his kitchen he entertained Russian princes, French and German royal dukes, British peers, and sightseers of every degree. Yet he never altered the routine of his daily life. " Breakfast at eight ; dinner at one ; supper at nine ; bed at eleven o'clock ; at half-past ten, let who would be there, he knocked out his last pipe." Very large sums of money passed through his hands. Yet, if the entry in the *Gentleman's Magazine* [1] refers to him, he was bankrupt in 1776, and so lavish was his hospitality that he is said to have died in poverty.[2]

In the treatment of live-stock for the butcher Bakewell's object was to breed animals which weighed heaviest in the best joints and most quickly repaid the cost of the food they consumed. He sought to discover the animal which was the best machine for turning food into money. " Small in size and great in value," or the Holkham toast of " Symmetry well covered," was the motto of his experiments. In his view the essentials were the valuable joints, and he swept away as non-essentials all the points on which fashion or prejudice had hitherto concentrated, such as head, neck, horn, leg, or colour. The points which he wished to develop and perpetuate were beauty combined with utility of form, quality of flesh, and propensity to fatness. To attain these objects he struck out a new line for himself. Crossing was then understood to mean the mixture of two alien breeds, one of which was relatively inferior. Bakewell adopted a different principle, because he regarded this form of crossing as an adulteration rather than as an improvement. He bred in-and-in, using not merely animals of the same native breed and line of descent, but of the same family. He thus secured the union of the finest specimens of the breed which he had chosen as the best, selected for the possession of the points which he wished to reproduce or strengthen.

---

[1] In the *Gent. Mag.* for Nov. 1776, appears the following entry in the list of bankrupts : " R. Bakewell, Dishley, Leicestersh. dealer." (p. 531).

[2] Other contemporary references to Bakewell, besides those quoted, will be found in the *Gent. Mag.* vols. lxiii. pt. ii. p. 792, and lxv. pt. ii. pp. 969-70 ; Marshall's *Midland Counties*, vol. i. pp. 292-493 etc. ; *Annals of Agriculture*, vol. vi. (1786), pp. 466-98 ; Arthur Young's *Husbandry of Three Famous Farmers* (1811) ; George Culley's *Observations on Live Stock* (1786).

It was with sheep that Bakewell achieved his greatest success. When he began his stock-breeding experiments, he selected his sheep from the best animals in the neighbourhood, and a guinea, or even half a guinea, secured him his choice from the fold. The breed from which they were chosen were the Leicestershire or Warwickshire long wools. The " true old Warwickshire ram " is thus described by Marshall in 1789 : " His frame large, and remarkably loose. His bone, throughout, heavy. His legs long and thick, terminating in large splaw feet. His chine, as well as his rump, as sharp as a hatchet. His skin might be said to rattle upon his ribs . . . like a skeleton wrapped in parchment." Even this animal was handsomer than a ram of the " true old Leicestershire sort," which Marshall saw in 1784. " A naturalist," he says, " would have found some difficulty in classing him ; and, seeing him on a mountain, might have deemed him a nondescript ; a something between a sheep and a goat." Out of these unpromising materials Bakewell succeeded in creating a new variety. His " new Leicesters " became the most profitable sheep for arable farmers. As by degrees the compactness of form, smallness of bone, fattening propensities, and early maturity were perpetuated, the breed was established, and for a time swept all competitors before them. While other breeds required three or four years to fit them for market, the New Leicesters were prepared in two. Those who tried the Dishley sheep found that they throve where others pined, that while alive they were the hardiest, and when dead the heaviest. In 1750 Bakewell let rams for the season at 16s. or 17s. 6d. apiece. In 1789 he let none under 20 guineas, and received 3000 guineas for the total of that season's letting. The New Leicesters were the first breed of sheep which were scientifically treated in England, and though they were less adapted for the southern, eastern, and northern counties, their supremacy on enclosed land in their own Midland districts was undisputed.

Bakewell raised the New Leicesters to the highest perfection. But this was not all. His breed in weight of fleece could not compete with Lincolns, and was less suited to hills or mountains than for enclosed arable land. He had, however, shown the way in which other breeds might be improved ; imitation was easy. In a less immediate sense he was the creator, not only of the New Leicesters, but of the improved Lincolns, South Downs and Cheviots. Before these breeds, fitted for the most fertile grasslands and plains as well as suited to hills and mountains, native races died away, like Red Indians before the civilised intruders. But gradually supporters rallied round other varieties. Bakewell's weapons were turned against himself. Native sheep of other districts, improved on his principles, began to hold their own, and, though on historical grounds precedence will always be given to the New Leicesters and the South Downs (improved by John Ellman of Glynde, 1753-1832), it may be questioned whether they have not been rivalled and surpassed by other breeds in the qualities for which they were once pre-eminent.[1]

In cattle-breeding Bakewell was less successful. It was his material not his system which failed. He endeavoured to found his typical race on the Lancashires or Craven Longhorns, which were the favourite cattle in Leicestershire, and, in his opinion, the best breed in England. He based his improvements on the labours of two of his predecessors. Sir Thomas Gresley of Drakelow, near Burton-on-Trent, began about 1720 the formation of a herd of Longhorns. On this Drakelow blood Webster of Canley, near Coventry, worked, and to his breed all the improved Longhorns traced their descent. Bakewell founded his experiments on a Westmoreland bull and two heifers from the Canley herd. To them he applied the same principles which he followed in sheep-breeding, and with great success. As graziers' stock, the breed was greatly improved. But as milkers, the new Longhorns were deteriorated by their increased propensity to fatness. In a county like Leicestershire, which depended not only on feeding stock but on dairy produce,[2] this poverty of milking quality was a fatal objection. Even in his Longhorns Bakewell did not long retain the lead. It soon passed away from him to Fowler of Rollright, in Oxfordshire. But the breed itself was beaten by one which possessed superior natural qualities. Almost throughout England the Durham Shorthorns, founded on the Holderness and Teeswater cattle, jumped into the first place, as the best rent-payers, both as milkers and meat-producers. The Ketton herd of Charles

---

[1] It is probable that towards the close of the eighteenth and the beginning of the nineteenth century there was some introduction of Merino blood which had its effect upon the short-wooled breeds. George the Third kept a flock of pure-bred merinos.

[2] Mrs. Paulet of Wymondham, in the Melton district of Leicestershire, is said to have been the first maker of Stilton cheeses. She supplied them to Cooper Thornhill, who kept the Bell Inn at Stilton (Hunts) on the great north road from London to Edinburgh, and they became famous among his customers, and throughout England. The manufacture of Stilton cheeses became an industry of the district. Mrs. Paulet was still living in 1780.

Colling became to cattle-breeders what Bakewell's Dishley flock of New Leicesters were to sheep-masters. It was as necessary for a superior Shorthorn to claim descent from Colling's bull "Hubback" as for a race-horse to boast the blood of the Godolphin Arabian. From "Hubback" was descended the famous Durham ox, which travelled through England in a specially constructed carriage from 1801 to 1810, exhibiting to the eyes of thousands of farmers a truer standard of shape than any their ancestors had conceived, and convincing them by personal interviews of the excellence of the improved breed. The example was followed in many parts of the country. Other breeds, notably the Herefords and North Devons, were similarly improved. The formation of herds became a favourite pursuit of wealthy landlords. Flora MacIvor herself might have lived to see the day, when country gentlemen could become breeders of cattle, without being "boorish two-legged steers like Killancureit."

Bakewell's success and the rapidly increasing demand for butcher's meat raised up a host of imitators. Breeders everywhere followed his example ; his standard of excellence was gradually recognised. The foundation of the Smithfield Club in 1798 did much to promote the improvement of live-stock. Some idea of the effect produced may be gathered from the average weights of sheep and cattle sold at Smithfield Market in 1710 and in 1795. In 1710 the average weights for beeves was 370 lbs., for calves 50 lbs., for sheep 28 lbs., for lambs 18 lbs. In 1795 beeves had risen in average weight to 800 lbs., calves to 148 lbs., sheep to 80 lbs., lambs to 50 lbs.[1] This enormous addition to the meat supply of the country, was due partly to the efforts of agriculturists like Tull, Townshend, Bakewell, and others, partly to the enclosure of open-fields and commons which their improvements encouraged. On open-fields and commons, owing mainly to the scarcity of winter keep, the live-stock was dwarfed in size and weight. Even if the number of animals which might be grazed on the commons was regulated by custom, the stint was often so large that the pasture could only carry the smallest animals. Where the grazing rights were unlimited, as seems to have been not unusually the case in the eighteenth century, the herbage was necessarily still more impoverished, and the size of the live-stock more stunted. On

---

[1] Sir John Sinclair's note for the use of the Select Parliamentary Committee appointed in 1795 to consider "the Waste, Uninclosed and unproductive Lands of the Kingdom." Appendix B, section 1, pp. 17, *note*. Sir John is not, however, always a reliable witness.

enclosed land, on the other hand, the introduction of turnip and clover husbandry doubled the number and weight of the stock which the land would carry, and the early maturity of the improved breeds enabled farmers to fatten them more expeditiously. But one of the consequences of this change in sheep-farming was not at first foreseen. The wool was sacrificed to the mutton. A large sheep paid better than a small. But as the size of the animal increased, its fleece grew heavier, and the staple longer. The supply of fine fleeces from the light, poorly-fed, short-wooled sheep of the commons diminished so rapidly that, before the end of the century, a new classification of sheep was introduced. Instead of being divided into long wools and short wools, they were now classified as long wools and middle wools. Improvements in machinery and the introduction of new fabrics utilised the produce of the heavier breeds of sheep ; but, for the better kinds of cloth, home manufacturers became increasingly dependent on foreign supplies of short wool, brought from Spain, Saxony, and New South Wales. A change of fashion intensified the need of wool for a finer quality of cloth than could be obtained in this country. The coarser fabrics of manufacture from English material, which had contented our ancestors, could not retain their hold on the home or foreign markets. During the Napoleonic wars, the full effect of this change in the raw material of woollen manufactures was concealed by the suspension of continental rivalry. When peace was finally proclaimed, it was at once felt. A pitched battle began between the manufacturer and the agriculturist ; the one demanded the free import of foreign short wool, the other the free export of English long wools, which made better prices abroad. Each resisted the demand of the other. Home manufacturers opposed the free export of British long-wools, because they feared the competition of foreign cloth. British farmers opposed the free import of foreign short wool, because they dreaded lest its introduction would force down the price of their home produce. Finally, in 1826, Lord Liverpool's government took off the duties both on the import and the export of the raw material. To advocates of enclosures, the last agricultural defence of the open-field farmer and commoner seemed to be destroyed, when the removal of the import duty deprived the fleeces of their half-starved sheep of all artificial advantages over the finer and cheaper wools of foreign countries.

CHAPTER IX

ARTHUR YOUNG AND THE DIFFUSION OF KNOWLEDGE. 1760-1800

The counties distinguished for the best farming : Hertfordshire, Essex, Suffolk, Norfolk, Leicestershire : the low general standard ; Arthur Young ; his crusade against bad farming, and the hindrances to progress ; waste land ; the "Goths and Vandals " of open-field farmers : want of capital and education ; insecurity of tenure ; prejudices and traditional practices ; impassable roads ; rapid development of manufacture demands a change of agricultural front : Young's advocacy of capitalist landlords and large tenant-farmers.

DURING the first three quarters of the eighteenth century many advances had been made in the theory, and some in the practice, of agriculture. Alternations of crops and the management of live-stock were better understood. But progress was still confined to localities, if not to individuals. Only in such counties as Hertford-shire, Essex, Suffolk, Norfolk, and Leicestershire was a fair standard of farming generally established. The superior enterprise of these favoured districts was due to various causes, and was displayed in different directions.

Without any special fertility of soil, Hertfordshire had for the last hundred years enjoyed the reputation of being the best corn county in England. To some extent it owed its superiority to the neighbourhood of London. But Middlesex, which shared the same advantage, was relatively backward. In Hertfordshire roads were above the average. In Middlesex turnpike roads, in spite of a large revenue from tolls, are described as " very bad." On the main road from Tyburn to Uxbridge, in the winter of 1797-8, there was but " one passable track, and that was less than six feet wide, and was eight inches deep in fluid mud. All the rest of the road was from a foot to eighteen inches deep in adhesive mud." Hert-fordshire, which had been to a great extent covered with forest, contained, at the close of the eighteenth century, few open-field farms and an inconsiderable area of commons, which were practically confined to the chalk districts in the north of the county. In Middlesex, on the other hand, 17,000 acres, or one-tenth of the county, were commons, and, out of 23,000 arable acres, 20,000 were cultivated in open-field farms. The neighbourhood of London probably accounts for the predominance of pasture.[1] Hertfordshire had been, for many years, an enclosed county, divided into small estates, and small farms conveniently varied in size. Unlike Middlesex, it was almost entirely arable. Its farmers had at once appreciated the value of turnips and clover, for which the soil was well adapted. Both crops must have been adopted within a few years after their first introduction into the country, if there is any truth in the tradition that Oliver Cromwell paid £100 a year to a Hertfordshire farmer named Howe for their successful cultivation.[2] Other useful practices were established at an early date. William Ellis of Gaddesden [3] (died 1758), a Hertfordshire farmer whose writings enjoyed a short-lived popularity, attributed the reputation of " this our celebrated county " to four principal means of improvement : " good ploughings, mixing earths, dunging and dressing, resting the ground with sown grasses." The Hertfordshire men were clean farmers. Their ploughmen were so celebrated that the county was " accounted a Nursery for skill in that Profession." Chalk was largely used on heavy clays, and red clay on sandy or gravelly soils. Nor were the advantages gained by neighbourhood to a great city neglected. London refuse was liberally bought and freely employed. Large quantities " of soot, coney-clippings, Horn-shavings, Rags, Hoofs-hair, ashes " were purchased from " Mr. *Atkins* in *Turnmill-Street* near *Clerkenwel*." To these were added, when Walker [4] wrote his report on the county, bones—boiled or burned—sheep-trotters, and malt-dust. Great numbers of sheep were also folded, mostly bought at Tring Fair from West-country drovers. But the peculiar practice of Hertfordshire farmers, in which Ellis took the greatest pride, was the sowing of tares on the turnip fallows as green fodder for horses in May.

---

[1] The pasture and bad roads of Middlesex may be explained by the fact that the greater part of its surface is covered by the unadulterated " London Clay " formation, which gives place in Hertfordshire to superficial gravels and further north the chalk.

[2] *General View of the Agriculture of Hertfordshire*, by Arthur Young (1804), p. 55.

[3] E.g. *Chiltern and Vale Farming explained* (1733); *The Modern Husbandman*, 8 vols. (1750).

[4] *General View of the Agriculture of the County of Hertford*, by D. Walker (1795).

Young (1770) states that, while in other counties the land lay idle, these crops fed five horses to the acre for a month, at 2s. 6d. each a week. It was on these crops that Hertfordshire farmers reared the horses which they bought as two-year-olds in Leicestershire. Yet at the beginning of the nineteenth century the example had been rarely followed in other counties.

Suffolk and Essex also afforded good examples of the best English farming as it was practised at the close of the eighteenth century. Both counties had, as a whole, been enclosed for many years. Only on the poor and chalky soil of the north-western district had open-fields held their own. As early as 1618,[1] East Suffolk and Mid Suffolk were enclosed, and only " the westerne parts ether wholly champion or neer." In both counties yeomanry abounded, and in Essex the class was in 1807 still increasing. " For twenty or thirty years past scarcely an estate is sold, if divided into lots of forty or fifty to two or three hundred a year but is purchased by farmers."[2] Both counties were centres of manufacturing industries, and in addition enjoyed the advantage of access to a great market. Suffolk supplied London with butter, Essex with calves, for which it had been famous in the seventeenth century. In both counties large quantities of manure were now used on the land. Farmers were not always so energetic. Under a lease of 1753 a tenant of the Suffolk manor of Hawsted was allowed two shillings for every load of manure which he brought from Bury and laid on the land. In a tenancy of twenty-one years only one load was charged to the landlord. Sixty years later, agriculturists had become more energetic. On the light sands of East Suffolk, marl and a calcareous shelly mixture of phosphates called " crag " were freely employed as fertilisers. Chalk from the Kentish quarries for use on the clays, as well as London refuse, were purchased by Essex farmers, conveyed by sea up the estuaries, and thence distributed in the county. Probably this traffic partly explains the condition of the Essex roads, which were as bad as the Suffolk highways were good. In both counties hollow drainage was practised earlier than elsewhere. The drains were wedge-shaped, filled with branches, twisted straw, or stone, and covered in with earth. Bradley[3] speaks of the " Essex practice " of making drains

---

[1] *Breviary of Suffolk*, by Robert Reyce, 1618, edited by Lord F. Hervey, 1902.
[2] *North-East Essex*, by Arthur Young (1807), vol. i. p. 40.
[3] *Complete Body of Husbandry* (1727), p. 133-4.

two feet deep, at close and regular intervals throughout a whole field, filled with rubble or bushes, and he derives the term " thorough-drainage " from an Essex word " thorow," meaning a trench to carry off the water. Ploughing was in both counties economically conducted. The Suffolk swing-plough, drawn by two horses, was the common implement. Oxen were seldom used : " no groaning ox is doomed to labour there " is the evidence of Bloomfield. Turnips and clover were firmly established as arable crops. Suffolk had been for two centuries famous for its field cultivation of carrots. Cabbages were a later introduction, but extensively grown. Hemp was cultivated in the neighbourhood of Beccles, and hops flourished round Saxmundham. In Essex a peculiar crop, grown, generally together, on the same land for three years in succession, consisted of caraway, coriander, and teazels. The teazels were bought by woollen manufacturers, and fixed in a revolving cylinder to catch the surface of bays, says, etc., and so raise the nap of cloth to the required length. Suffolk was also famous for its live-stock. The Suffolk Punch was a short compact horse of about fifteen hands high, properly of a sorrel colour, unrivalled in its power of draught, though, as Cullum wrote in 1790, " not made to indulge the rapid impatience of this posting generation." In the dairy the " milch kine " of Suffolk are said by Reyce (1618) to be as good as in any other county, and he notes the beauty of their horns. In later times the Suffolk Dun was renowned for the quantity of her milk. Suffolk cheese, however, had an evil reputation. It was " so hard that pigs grunt at it, dogs bark at it, but none dare bite it." The mystery of its interior inspired Bloomfield to sing of the substance, which

> " Mocks the weak effort of the bending blade,
> Or in the hog-trough rests in perfect spite,
> Too big to swallow and too hard to bite."

As the eighteenth century drew to a close, it was to Norfolk and to Leicestershire that men had begun to look for the best examples of arable and pasture farming. In both counties progress had been largely due to the character of the farmers, and in Norfolk to the alertness and industry of the labourers. In Norfolk, Marshall (1787) says that farmers were " strongly marked by a liberality of thinking," that they were men who had " mixed with what is called the World, of which their leases render them independent . . occupying the same position in society as the clergy and smaller squires." Many of them had prospered enough to buy their holdings, and to add to them "numerous small estates of the yeomanry." Nor is this surprising in view of the productiveness of their land under the Norfolk system of husbandry. At the end of the eighteenth century the average annual number of live-stock sent from the county to Smithfield was 20,000 cattle and 30,000 sheep. It was also stated in 1795. that as much corn was exported from the four Norfolk ports of Yarmouth, Lynn, Wells, and Blakeney, as was sent abroad from the whole of the rest of England. In Leicestershire, again, "yeomanry of the higher class" abounded. "Men cultivating their own estates of two, three, four or five hundreds a year are thickly scattered over almost every part of the country"; they had "travelled much and mixed constantly with one another." In both Leicestershire and Norfolk the special branches of farming which were generally followed brought agriculturists into contact with their rivals, compelled them to be wide-awake, and sharpened their intelligence. Both were occupied in fattening stock for town markets, the Leicestershire men on pasture breeding their own stock, the Norfolk farmers on arable land buying their cattle from Scottish drovers. In one important respect there was a wide difference in their development. In Norfolk, great landowners, like Lord Townshend and, later, Coke of Norfolk, took the lead in improvement, tested for the benefit of their tenants the value of the new arable methods, encouraged them by long leases to follow their example, and by high rents made imitation compulsory. In Leicestershire, on the other hand, large landlords were few and had given no lead ; the example was set by large tenant-farmers or substantial yeomen.

Other counties had adopted other useful practices which had scarcely spread beyond their borders. Thus Lancashire excelled in the cultivation of potatoes ; Middlesex was celebrated for the art and practice of haymaking ; Wiltshire for the irrigation and treatment of water-meadows ; Cheshire for its management of dairy produce ; Yorkshire farmers round Sheffield had tested the value of bone-dust, many years before the value of the manure was known in other districts. But there is some evidence that other counties had rather fallen back than advanced. This is especially true of Cambridgeshire, which enjoyed the reputation of being the worst cultivated county in England. It will probably be true to say that the country as a whole had made no general advance

on the agriculture of the thirteenth century. The stagnation was
mainly due to the prevalence of wastes, the system of open-field
farming, the risk of loss of capital in improvements made under
tenancies-at-will, the poverty and ignorance of hand-to-mouth
farmers, the obstinacy of traditionary practices, the want of markets, and difficulties of communication. Till these obstacles were
to some extent overcome, agricultural progress could not become
general. It is with the removal of these hindrances that the name
of Arthur Young is inseparably connected.

Born in London in 1741, Arthur Young was the younger son of
the Rev. Arthur Young, who owned a small estate of 200 acres at
Bradfield in Suffolk. From his father he inherited his literary
tastes, a habit of negligence in money matters, and ultimately a
landed property. Out of Lavenham School he passed, at the age
of seventeen, into a wine merchant's office at Lynn. A youthful
fop and gallant, he there began his literary career in order to pay
for books and clothes. Before he was nineteen, he had published
four novels and two political pamphlets. On his father's death in
1759, he abandoned trade for literature, and Lynn for London,
where he launched a monthly magazine called *The Universal
Museum*, which only ran for six months. The venture was unprofitable. Without profession or employment, he drifted back, in
1763, to his mother's home at Bradfield, married, and settled down
to farming as a business. As a practical farmer he failed, and the
impression left by his writings is that he always would have done
so. On three farms, which he took in rapid succession, he lost
money. Meanwhile he was succeeding better as a writer. Books
and pamphlets flowed from his pen with prodigious rapidity, and
his income was considerable. In 1767 he began those farming
tours, in the course of which he drew his graphic sketches of rural
England, Ireland, and France.[1] His careless ease of style, his racy
forcible English, his gift of happy phrases, his quick observation,
his wealth of miscellaneous detail, make him the first of English
agricultural writers. Apart from the value of the facts which they

---

[1] *A Six Weeks' Tour through the Southern Counties of England and Wales*
(1768) ; *A Six Months' Tour through the North of England* (1770), 4 vols. ;
The Farmer's Tour through the East of England (1771), 4 vols. ; *Tour in
Ireland*, 1776-7-8 (1780), 2 vols. ; *Travels during the Years* 1787, '88, '89 *and
1790, undertaken more Particularly with a view of ascertaining the Cultivation,
Wealth, Resources, and National Prosperity of the Kingdom of France* (1792-4),
2 vols.

contain, his tours, with their fresh word-pictures, their gossip, their personal incidents, and even their irrelevancies, have the charm of private diaries. His *Ireland* was described by Maria Edgeworth as " the first faithful portrait of the inhabitants," and his *France* was recognised by Tocqueville as a first-hand authority on the rural conditions of the country on the eve of the Revolution. In 1784 he began his *Annals of Agriculture*, a monthly publication to which George III., under the name of his shepherd at Windsor, " Ralph Robinson," occasionally contributed. The magazine was continued till 1809, when, owing to failing eyesight, Young discontinued its publication. He had written more than a quarter of the forty-six volumes himself.

Young had now succeeded, on the death of his mother in 1785, to the Bradfield estate, his elder brother having broken his neck in the hunting-field. His *Travels* in France show that he sympathised with the peasants in their early efforts to free themselves from the *ancien régime*. But the subsequent course of the Revolution filled him with horror. In 1793, he wrote an effective pamphlet on *The Example of France a Warning to Great Britain*, urged the formation of a " militia of property," and himself joined the Suffolk yeomanry. In the same year Pitt established the Board of Agriculture, with Sir John Sinclair as President. Arthur Young was appointed Secretary with a salary of £400 a year and, later, an official residence in Sackville Street, London. One of the first objects of the Board was to collect information respecting the agricultural conditions of each county. For this purpose Commissioners were appointed. They were not always wisely selected ; but for this choice, against which Young protested, the President was responsible. Their Reports were severely criticised by William Marshall [1] (1745-1818), an embittered, disappointed man, who had

---

[1] Marshall's *General Survey* . . . *of the Rural Economy of England* has been frequently quoted. His valuable records fill twelve volumes published between 1787 and 1798, two volumes being allotted to each of the six departments into which he divides the country : (1) the Eastern : *Norfolk*, 2 vols. (1787) ; (2) the Northern : *Yorkshire*, 2 vols. (1788) ; (3) the West Central : *Gloucestershire, North Wilts*, and *Herefordshire*, 2 vols. (1789) ; (4) the Midland : *Leicestershire*, etc., 2 vols. (1790) ; (5) the Western : *Devonshire and parts of Somersetshire, Dorsetshire, and Cornwall*, 2 vols. (1796) ; (6) the Southern : *Kent, Surrey, Sussex and Hampshire*, 2 vols. (1798). Of the first ten volumes a second edition was published in 1796. A second edition of the Southern volumes was published in 1799, with the prefix of a sketch of the *Vale of London*.
Marshall has none of the charm of Young. He is a heavy, didactic writer.

himself originally suggested the establishment of the Board and the compilation of the surveys. But, with all their faults, the reporters collected a mass of valuable information on the state of farming from 1793 to 1813. Six of the surveys were by Young himself,[1] and his *Report on Oxfordshire* was almost his last literary work.

Young was a man of strong prejudices. He was also wanting in power of generalisation. But he worked untiringly for what he believed to be the progress of good farming. On this object were concentrated the chief labours of his life—his enquiries, experiments, researches, his collections of statistics, his notes of useful practices, his observations on new methods. His eager face, with its keen eyes and aquiline features, expressed the vivacity of his temperament, just as his tall slender figure indicated the restless activity of his body. A gay and charming companion, his enthusiasms were infectious. He was the soul and inspiration of the progressive movement. To him, more than to any other individual, were due the dissemination of new ideas on farming, the diffusion of the latest results of observation and experiment, the creation of new agencies for the interchange of experiences, the establishment of farmers' clubs, ploughing matches, and agricultural societies and shows. His married life was not happy ; but his wife was not entirely to blame. An affectionate father, his whole heart was given to his youngest daughter (Martha Ann, born 1783, died 1797) nicknamed " Bobbin." Versailles did not afford him so much pleasure as giving to the child a French doll. Her death broke down his health and spirits. Grief deepened into religious melancholy. His gloom was intensified by failing eyesight. In 1811 he became totally blind. Nine years later (1820), he died in London.

When Young began to write on agriculture, vast districts, which might have been profitably cultivated, still lay waste. Of the area already under tillage, a large proportion lay in open-fields.

But his system is better ; his generalisations are more conclusive, and less contradictory ; his facts are better arranged ; he was, also, a better farmer. A zealous collector of " provincialisms " of speech, he gives lists of the local words which he found in use in the Northern, Midland, and West Central departments, and appends them, with a glossary, to the volumes to which they relate. Besides the *Rural Economy*, he published numerous other works, chiefly on agriculture.

[1] Young wrote the *General View of the Agriculture of the County of Suffolk* (1797), *County of Lincoln* (1799), *of Hertfordshire* (1804), *of Norfolk* (1804), *of Essex*, 2 vols. (1807) ; *of Oxfordshire* (1809).

Under this system, whatever might be the differences or capacities of the soil, the whole of the land, with rare exceptions, was placed under the same unvarying rotation. It was this inability to put land to its best use which especially roused Young's indignation. When he made his Eastern Tour in 1770, he found nearly all the Vale of Aylesbury cultivated in arable open-fields, lying in broad, high, crooked ridges. The course of cropping was (1) Fallow, (2) Wheat or Barley, (3) Beans. The land was ploughed from two to four inches deep, and five horses were used to each plough. Beans were sown broadcast, and never hoed. Drainage was badly needed, for the ridge system had failed. But the lands were so intermixed that any other system was difficult, if not impossible. Even in June, only the tops of the ridges were dry, and, in the winter, most of the land, crops and all, were soaked with water. As a result, the products were as bad as the land was good. The Vale of Aylesbury farmers, whom Ellis (1733) describes as " one of the most obstinate bigotted sort," " reap bushels where they should reap quarters." Both in Buckinghamshire and in Northamptonshire, the cow-dung was collected from the fields, mixed with short straw, kneaded into lumps, daubed on the walls of buildings, and, when dry, used as fuel. " There cannot," says Young, " be such an application of manure anywhere but among the Hottentots." [1] Naseby Field in 1770 consisted of 6000 acres, all cultivated on the open-field system, on the same course of cropping which Young found established on village farms from the Vale of Aylesbury to the north of Derbyshire. Round the mud-built village lay a few pasture enclosures. The three arable fields were crossed and re-crossed by paths to the different holdings, filled with a cavernous depth of mire ; the pastures were in a state of nature, overrun with nettles, furze, and rushes. The farm-houses and buildings, all collected in the village, were two miles distant from a great part of the fields. When Young visited the village again in 1785, he found that the land in tillage for spring corn was " perfectly matted with couch." Marshall, a less prejudiced observer than Young, visited the Vale of Gloucester in 1789. There he found half the arable land unenclosed. Near Gloucester, and in other parts of the district, there were extensive tracts of land, called " Every

---

[1] It was no uncommon practice. Edward Laurence suggests (1727) that "Cow-dung not to be burnt for fuel " should be inserted as a restrictive covenant in all leases. He mentions Yorkshire and Lincolnshire as counties where dung was frequently used as fuel.

Year's Land," which were cropped year after year without any fallows. Only the cleanest farming could have made such a system productive. But here Marshall found beans hidden among mustard growing wild as a weed; peas choked by poppies and corn marigolds; every stem of barley fettered with convolvulus; wheat pining in thickets of couch and thistle. It is not surprising that the yield of wheat was anything from 18 bushels an acre down to 12 or 8 bushels.

Other instances might be quoted to show the general condition of open-field farms. But the system had its champions, even among practical agriculturists, especially if they were flock-masters. It cannot, therefore, always have been characterised by the worst farming. No doubt lower depths might be reached. If severalty made a good farmer better, it also made a bad farmer worse. Nor was the system altogether incapable of improvement. Here and there Young or Marshall alludes to some useful practice adopted on village farms. For instance, Young speaks of the drainage of common pastures by very large ploughs belonging to the parish, cutting 16 inches in depth and the same in width, drawn by 12 horses; of the introduction of clover by common consent into the rotation of crops, or of the adoption of a fourth course instead of the old two- or three-shift system. So also Marshall notes the open-field practice of dibbing and hoeing beans in Gloucestershire, where beans commanded a ready market among the Guinea traders of Bristol as food for negro slaves on the voyage from the African coast to the West Indies. But, speaking generally, any rotation of crops in which roots formed an element was with difficulty introduced on arable land which was pastured in common during the autumn and winter months; drainage was impracticable on the intermixed lands of village farms; among the underfed, undersized, and underbred flocks and herds of the commons the principles of Bakewell could not be followed. That open-field farmers were impervious to new methods is certain. "You might," says Young, " as well recommend to them an Orrery as a hand-hoe." That they had not the capital to carry out costly improvements is also obvious. They could not bring into cultivation the sands of Norfolk, the wolds of Lincolnshire, or the ling-covered Peak of Derbyshire. From a purely agricultural point of view Young's intemperate crusade against village farms was justified, and he had reason on his side when he said that " the Goths and Vandals of open-field

farmers must die out before any complete change takes place."
To some extent the same arguments applied to small·farmers
occupying their holdings in severalty. "Poverty and ignorance,"
says Marshall, speaking of the Vale of Pickering in 1787, "are the
ordinary inhabitants of small farms ; even the smaller estates of
the yeomanry are notorious for bad management." It was on the
larger farms that he found the spirit of improvement and the best
practice. In Gloucestershire (1789) he looked to the "few men
of superior intelligence" to raise the standard of the profession.
Nor did enclosures necessarily mean an improvement of methods.
In Derbyshire, at the time of Young's tour in 1770, many farmers
on new enclosures pursued the same course of cropping to which
they had been restricted by the "field constraint" of village farms.
Sometimes·the landlord, and not the tenant, was the Vandal or the
Goth. Thus in Cambridgeshire farmers on freshly enclosed land
were bound by their leases to continue the old course of fallow,
corn, and beans.

Even when a tenant-farmer possessed both enterprise and capital,
the method of land-tenure discouraged improvement. Without
some security for his outlay, no tenant could venture to spend
money on his land. At the same time he was often expected to
make improvements which now are considered the duty of a landlord and parts of the necessary equipment of a farm. Yet the
commonest forms of tenure were lettings from year to year, voidable
on either side, as they then were, at six months' notice. In the
eastern counties leases for terms of years, with covenants for
management, were in the last half of the century becoming a usual
form of letting. But elsewhere long leases were regarded with
justifiable suspicion by both parties. Tenants objected to them,
because they bound them to take land for a long period before they
knew what the land would do, and to make fixed annual payments
based on current prices which might not be maintained. Landlords also objected to them, because they deprived owners of the
advantages of a rise in prices, and "told the farmer when he might
begin systematically to exhaust the land." Where a good understanding existed between landlords and tenants, leases were not
indispensable. Land was often farmed on verbal agreements.
Ordinary tenancies-at-will secured Berkshire and Nottinghamshire
farmers in their holdings from géneration to generation. Under the
same tenancy, on the Duke of Devonshire's estates in Derbyshire,

tenants even carried out costly and permanent improvements. Often, however, the uncertainty of this form of tenure checked enterprise ; because of it, also, tenants fell into the routine of the district and plodded along in the beaten track trodden by their ancestors. Sometimes the uncertainty was a real insecurity. Thus, in Yorkshire, in 1787, Marshall notices that confidence between landlord and tenant had been destroyed by successive rises in rents. " Good farming ceased, for fear the fields should look green and the rent be raised." Local rhymes expressed the popular belief that he " that havocs may sit," while the improving tenant must either pay increased rent or " flit." Leases for lives were common, especially in the south-western counties. They gave a fixity of tenure ; but they were necessarily, both for tenant and landlord, somewhat of a gambling speculation. Fourteen years' purchase of the rental value was the usual price for a lease of three lives. The initial outlay crippled the first tenant, and, only if the lives proved good, was the purchase remunerative. On the other hand, the landlord was often obliged, as the third life drew towards its close, to put himself in as sub-tenant to save his land from exhaustion and his buildings from ruin. Leases for very short terms were not infrequent. On open-field farms in Bedfordshire and Huntingdon the term was three years, in Durham six years, corresponding to the completion of one or two courses of the ordinary three-shift routine. But in the last twenty years of the eighteenth century, leases for 7, 14, and 21 years became more common. Even longer terms were often granted, as the enthusiasm for improvement extended. Tenants under long leases throve on rents fixed before the high prices during the Napoleonic war ; but after 1813 the position was disastrously reversed. Prudent men had taken their money out. The sufferers were new men, who had enjoyed none of the advantages of the system ; they were its victims, never its beneficiaries. Two of the difficulties by which the tenure is embarrassed were already becoming important, if not burning, questions—the compensation for unexhausted improvements, and the covenants imposed by landlords. Some of the restrictions imposed by leases were a bar to progress. Leicestershire graziers, for example, were crippled by the absolute prohibition of arable farming ; they were forced either to sell off their stock at Michaelmas when it was cheapest, or to buy winter-keep from Hertfordshire. On the other hand, covenants of a reasonable nature proved invaluable in lifting the standard of a stationary agriculture, and raising farming to a higher level.

Other formidable obstacles to progress lay in the mass of local prejudices and the obstinate adherence to antiquated methods. All over the country there were men like the "round-frocked" farmers of Surrey, who prided themselves on preserving the practices and dress of their forefathers, men of "inflexible honesty," enemies equally to "improvements in agriculture" and to the commercial morality of a new generation. Reforming agriculturists no doubt were too ready to ignore the solid basis of sound sense and experience which often underlay practices that in theory were objectionable. In their excuse it may be urged that their patience was sorely tried. Traditional methods were treasured with jealous care as agricultural heirlooms ; even ocular proof of the superiority of other systems failed to wean farmers from the routine of their ancestors. In 1768 turnips and clover were still unknown in many parts of the country ; and their full use only appreciated in the eastern counties. In some districts, as in Essex (1808), clover had been adopted with such zeal that the land was already turning sick ; in others it was scarcely tried. In Westmoreland, for instance, in 1794, "the prejudice that exists almost universally against clover and rye-grass" was said to be "a great obstacle to the improvement of the husbandry of the county." In Cumberland, where clover had been introduced in 1752, it was still rare in 1797. Turnips remained, at the close of the eighteenth century, an "alien crop" in many counties, such as Wiltshire, Dorsetshire, Hampshire, Staffordshire, Herefordshire, Shropshire, Glamorganshire, and Worcestershire. Even where they were grown, they were generally sown broadcast, and seldom hoed. In 1780 a Norfolk farmer settled in Devonshire, where he drilled and hoed his roots. His crops were far superior to those of other farmers in the district ; yet, at the close of the century, no neighbour had followed his example. In 1794 many Northumberland sheep-masters still milked their ewes, though the more intelligent had discontinued the practice. Another illustration of the tyranny of custom may be taken from ploughing. In many districts the Norfolk, Rotherham, or Small's ploughs had been introduced at a great economy of cost. But elsewhere farmers still clung to some ancestral implement. In Kent, at the time of Cromwell, it was not unusual to see six, eight, or twelve oxen attached to a single plough. On the dry land of East Kent, on stony land, on rough hill-sides, the implement undoubtedly had, and has, its uses. But on all soils alike, a century and a half later, the same huge machine, looking at a distance more like a cart than a plough, with a beam the size of a gate-post, remained the idol of the men of Kent. In Middlesex, in 1796, it was no uncommon sight to see ploughs drawn by six horses, with three men in attendance. In Berkshire (1794), four horses and two men ploughed one acre a day. In Northampton-shire Donaldson (1794) found in general use a clumsy implement, with a long massive beam, drawn by four to six horses at length, with a boy to lead and a man to hold. By immemorial custom in Gloucestershire two men, a boy, and a team of six horses were usually employed in ploughing. Coke of Norfolk sent into the county a Norfolk plough, and ploughman, who, with a pair of horses, did the same work in the same time. But though the annual cost of the operation was thus diminished by a half, it was twenty years before the neighbours profited by the lesson.

The backwardness of many agricultural counties was to some extent due to difficulties of communication. By the creation of Turnpike Trusts (1663 and onwards) portions of the great high-ways were placed in repair.[1] Yet in the eighteen miles of turnpike road between Preston and Wigan, Young in 1770 measured ruts "four feet deep and floating with mud only from a wet summer," and passed three broken-down carts. "I know not in the whole range of language," he says, "terms sufficiently expressive to describe this infernal road. Let me most seriously caution all travellers who may accidentally propose to travel this terrible country to avoid it as they would the devil, for a thousand to one they break their necks or their limbs, by overthrows or breakings down." The turnpike road to Newcastle from the south seems to have been equally dangerous. "A more dreadful road," he says, "cannot be imagined. I was obliged to hire two men at one place to support my chaise from overturning. Let me persuade all travellers to avoid this terrible country, which must either dislocate their bones with broken pavements, or bury them in muddy sand." The turnpike road from Chepstow to Newport was a rocky lane, "full of hugeous stones, as big as one's horse, and abominable holes." Marshall says that the Leicestershire roads, till about 1770, had been "in a state of almost total neglect since the days

[1] For further details as to roads, see chap. xiii.

of the Mercians." The principal road from Tamworth to Ashby lay, in 1789, " in a state almost impassable several months in the year." Waggons were taken off their wheels and dragged on their bellies. Essex, in the time of Fitzherbert, was famous for the badness of its roads. In the eighteenth century it worthily maintained its reputation. " A mouse could barely pass a carriage in its narrow lanes," which were filled with bottomless ruts, and often choked by a string of chalk waggons, buried so deeply in the mire that they could only be extricated by thirty or forty horses. " Of all the cursed roads that ever disgraced this kingdom in the very age of barbarism none ever equalled that from Billericay to the ' King's Head ' at Tilbury " was the suffering cry of Young in 1769. The roads of Herefordshire, says Marshall, twenty years later, were " such as you might expect to find in the marshes of Holland or the mountains of Switzerland." In Devonshire, which Marshall considered to be agriculturally the most benighted district of England, there was not in 1750 one single wheeled carriage ; everything was carried in sledges or on pack-horses. The latter were still in universal use in 1796. Crops were piled between willow " crooks," to which the load was bound ; manure was carried in strong panniers, or " potts," the bottom of which was a sort of falling door ; sand was slung in bags across the wooden pack-saddle. Even where efforts were made to improve the highways, the attempt was often rendered useless by ignorance of the science of road-making. Some roads were convex and barrel-shaped. But the fall from the centre of the road to the sides was so rapid that carts could only t avel in the centre with safety. Many roads were concave, constructed in the form of a trough, filled in with sand. In wet weather this deposit became porridge. On a road of this formation between Woodstock and Oxford, Marshall, in 1789, encountered labourers employed in " scooping out the batter." Yet in spite of the difficulty of communication, distant counties carried on a considerable trade in agricultural produce. Thus calves, bred in Northamptonshire, were sent to Essex to be reared. The animals travelled in carts with their legs tied together, were eight days on the road, and during the journey were fed with " gin-balls," *i.e.* flour and gin mixed together. Off the main lines of communication, highways were unmetalled tracks, which spread in width as vehicles deviated to avoid the ruts of their predecessors. By-roads were often zigzag lanes, engineered on the principle that one good or bad turn deserved another.   In narrow ways the bells on the teams were not merely ornaments ; they were warnings that the passage was barred by the entry of another vehicle.   When rural districts were thus cut off from one another, their isolation was not only a formidable obstacle to agricultural progress, but made a uniform system of growing corn on every kind of land a practical necessity.   Yet the days when Gloucester seemed " in the Orcades," and York a " Pindarick flight " from London had their advantages.   In 1800 it required fifty-four hours, and favourable circumstances, for " a philosopher, six shirts, his genius, and his hat upon it," to reach London from Dublin.

Shut off from neighbours by impassable roads, impeded in their access to markets, not ambitious of raising from the soil anything beyond their own needs and the satisfaction of the local demand for bread, farmers felt no spur to improvement.   Hitherto the slow increase of a rural population was the only effective incentive to increased production.   But as the eighteenth century drew to its close, Watt, Hargreaves, Crompton, Arkwright, and other mechanical geniuses were beginning to change the face of society with the swiftness of a revolution.   Population was shifting from the South to the North, and advancing by leaps and bounds in crowded manufacturing towns.   Huge markets were springing up for agricultural produce.   Hitherto there had been few divisions of employment because only the simplest implements of production were used ; spinners, weavers, and cloth-workers, iron-workers, handicraftsmen, had combined much of their special industries with the tillage of the soil.   But the rapid development of manufacture caused its complete separation from agriculture, and the application of machinery to manual industries completed the revolution in social arrangements.   A division of labour became an economic necessity. Farmers and manufacturers grew mutually dependent.   Self-sufficing farming was thrown out of date.   Like manufacture, agriculture was ceasing to be a domestic industry.   Both had to be organised on a commercial footing.   The problem was, how could the inevitable changes be met best and most promptly ?   How could a country at war with Europe raise the most home-grown food for a rapidly growing population, concentrated in the coal and iron fields ?   How could agriculture supply the demand for artisan labour, and yet increase its own productiveness ? ' Arthur Young was, at this period of his career, ready with an unhesitating answer

—large farms, large capital, long leases, and the most improved methods of cultivation and stock-breeding. His object was to develop to the utmost the resources of the soil. To this end all social considerations must be subordinated. Every obstacle to good farming must be swept away—wastes reclaimed, commons divided, open-fields converted into individual occupations, antiquated methods abandoned, obsolete implements scrapped, improved practices uniformly adopted. " "Where," he asks, with perfect truth, " is the little farmer to be found who will cover his whole farm with marl at the rate of 100 or 150 tons per acre ? who will drain all his land at the expense of £2 or £3 an acre ? 'who will pay a heavy price for the manure of towns, and convey it thirty miles by land carriage ? who will float his meadows at the expense of £5 an acre ? who, to improve the breed of his sheep, will give 1000 guineas for the use of a single ram for a single season ? who will send across the Kingdom to distant provinces for new implements, and for men to use them ? who will employ and pay men for residing in provinces where practices are found which they want to introduce into their farms ? "

Young's spirited crusade against bad or poor farming would probably have fallen on deaf ears, if it had not been supported by the prospect of financial gain and by the impulse of industrial necessities. As he put the case, more produce from the land meant higher rents for the landlord, larger incomes for farmers, better wages for labourers, more home-grown food for the nation. Under the pressure of war-prices and of the gigantic growth of a manufacturing population, the system which he advocated made rapid progress. Years after his death, it was established with such completeness that men forgot not only the existence of any different conditions, but even the very name of the most active pioneer of the change. In the agricultural literature of the early and middle Victorian era, he is almost ignored. The article on English agriculture in the *Encyclopaedia Britannica*, for example, devotes only a few lines to his career. Recently his memory has been revived in England by the renewal under different circumstances of the struggle between large and small farmers. In France, on the other hand, where the contest between capitalist farmers and peasant proprietors was never decisively terminated, the discussion has always centred round his name. In the words of Lesage, his latest editor and translator, France has made an adopted child of Arthur Young.

CHAPTER X

LARGE FARMS AND CAPITALIST FARMERS

1780-1813

Agricultural enthusiasm at the close of the eighteenth century; high prices
of agricultural produce; the causes of the advance; increased demand
and cessation of foreign supplies; the state of the currency; rapid advance
of agriculture on the new lines of capitalist farming; impulse given to
enclosing movement and the introduction of improved practices; Davy's
Lectures on Agricultural Chemistry; the work of large landlords: Coke
of Norfolk.

THE enthusiasm for farming progress, which Arthur Young zeal-
ously promoted, spread with rapidity. A fashion was created
which was more lasting, because less artificial and more practical,
than it had been in the days of Pope. Great landlords took the
lead in agricultural improvements. Their farming zeal did not
escape criticism. Dr. Edwards [1] in 1783 expressed a feeling which
was prevalent two centuries before: "Gentlemen have no right
to be farmers; and their entering upon agriculture to follow it as
a business is perhaps a breach of their moral duty." But it was
now that young men, heirs to landed estates as well as younger
sons, began to go as pupils to farmers. George III. rejoiced in
the title of "Farmer George," considered himself more indebted to
Arthur Young than to any man in his dominions, carried the last
volume of the *Annals* with him in his travelling carriage, kept his
model farm at Windsor,[2] formed his flock of merino sheep, and
experimented in stock-breeding. The Duke of Bedford at Woburn,
Lord Rockingham at Wentworth, Lord Egremont at Petworth,
Coke at Holkham, and numerous other landlords, headed the

---

[1] *Plan of an Undertaking for the Improvement of Husbandry* etc., by Dr.
Edwards of Barnard Castle (1783).

[2] The King's Windsor Farm is described by Nathaniel Kent in Hunter's
Georgical Essays (1803), vol. iv. Essay vii.

reforming movement. Fox, even in the Louvre, was lost in consideration whether the weather was favourable to his turnips at St. Anne's Hill. Burke experimented in carrots as a field crop on his farm at Beaconsfield, though he pointed his sarcasms against the Duke of Bedford for his devotion to agriculture. Lord Althorp, in the nineteenth century, maintained the traditions of his official predecessors. During a serious crisis of affairs, when he was Chancellor of the Exchequer, John Grey of Dilston called upon him in Downing Street on political business. Lord Althorp's first question, eagerly asked, was " Have you been at Wiseton on your way up ? Have you seen the cows ? " The enthusiasm for farming began to be scientific as well as practical. No new book escaped the vigilance of agriculturists. Miss Edgeworth's *Essay on Irish Bulls* (1802) had scarcely been published a week before it was ordered by the secretary of an agricultural society. Nor were the clergy less zealous. An archdeacon, finding a churchyard cultivated for turnips, rebuked the rector with the remark, " This must not occur again." The reply, " Oh no, Mr. Archdeacon, it will be barley next year," shows that, whatever were the shortcomings of the Church, the eighteenth century clergy were at least devoted to the rotation of crops.

Every department of agriculture was permeated by a new spirit of energy and enterprise. Rents rose, but profits outstripped the rise. New crops were cultivated ; swedes, mangel-wurzel, kohl rabi, prickly comfrey were readily adopted by a new race of agriculturists. Breeders spent capital freely in improving live-stock. New implements were introduced. The economy and handiness of ploughs like the Norfolk, or the Rotherham ploughs as improved by James Small of Blackadder Mount, were gradually recognised, and the cumbrous mediaeval instruments with their extravagant teams superseded. Meikle's threshing machine (1784) began to drive out the flail by its economy of human labour. Numerous patents were taken out between 1788 and 1816 for drills, reaping, mowing, haymaking, and winnowing machines, as well as for horse-rakes, scarifiers, chaff-cutters, turnip-slicers, and other mechanical aids to agriculture. In the northern counties iron gates and fences began to be used. The uniformity of weights and measures [1] was eagerly

---

[1] Under the Act of Union with Scotland (clause 17) it had been provided that the same weights and measures which were established in England should be used throughout the United Kingdom. But the clause remained

discussed and recommended. Cattle-shows, wool-fairs, ploughing-matches were held in various parts of the country. Counties, like Durham, Northumberland, Cheshire, and Leicestershire, started experimental farms. The short-lived Society of "Improvers in the Knowledge of Agriculture" had been formed in 1723. The Society for the "Encouragement of Arts, Manufactures and Commerce" was instituted in London in 1754. Other associations, more exclusively agricultural, speedily followed: The Bath and West of England Society was founded in 1777, the Highland Society in 1784, the Smithfield Club in 1798. The creation of the Board of Agriculture in 1793 has been already mentioned. The Farmers' Club was established in 1793. The first number of the *Farmer's Magazine* which appeared in January, 1800, rapidly passed through five editions. Provincial societies multiplied. At Lewes, in 1772, Lord Sheffield had established a Society for the "Encouragement of Agriculture, Manufacture and Industry"; but it does not seem to have survived the war with France and the United States. Few counties were without their organisations for the promotion of agricultural improvement. One of the first was established at Odiham in Hampshire. Kent had its agricultural society at Canterbury (1793) and the Kentish Society at Maidstone. In Cornwall (1793), Berkshire (1794), Shropshire (1790), at Shifnal and at Drayton in Leicestershire (1794), in Herefordshire (1797), provincial societies were founded. The West Riding of Yorkshire had its society at Sheffield, Lancashire at Manchester, Worcestershire at Evesham (1792), Huntingdonshire at Kimbolton. In Northamptonshire similar associations were formed at Peterborough, Wellingborough and Lamport. The list might be enlarged. But, though many of these societies were short-lived, their foundation illustrates the new spirit which animated farming at the close of the eighteenth century.

a dead letter. The establishment of uniformity was difficult. In 1758 a Parliamentary Committee reported that there were in use in England four different legal measures of capacity, the respective quantities being in the case of the bushel 2124, 2150, 2168, and 2240 cubic inches. The widest differences existed between the weights and measures of the same county. Thus in Cornwall, for instance, wheat was sold either by the double Winchester of 16 gallons or the treble Winchester of 24 gallons; oats were sold in the eastern district by the hogshead of 9 Winchesters, in the west by a double Winchester of 17 gallons; a bushel of seed-wheat bought from a western farmer ran short of the eastern measure by between one and two gallons. Butter was sold at 18 oz. to the pound. The customary perch was 18 feet in length instead of the statutory length of 16½ feet.

The period from 1780 to 1813 was one of exceptional activity in agricultural progress. Apart from the flowing tide of enthusiasm, landlords and farmers were spurred to fresh exertions and a great outlay of capital and labour by the large returns on their expenditure. All over the country new facilities of transport and communication began to bring markets to the gates of farmers ; new tracts of land were reclaimed ; open arable farms and pasture commons were broken up, enclosed, and brought into more profitable cultivation ; vast sums of money were spent on buildings and improvement. In spite of increased production, prices rose higher and higher, and carried rents with them. " Corn," says Ricardo, " is not high because a rent is paid ; but a rent is paid because corn is high." In certain circumstances—if the State is landlord, or if landowners could combine for the purpose—rents might raise prices. But the general truth of Ricardo's view was illustrated during the French War. From 1790 to 1813, rents rose with the rise in prices, until over a great part of Great Britain they were probably doubled. Even the larger yield from the land under improved methods of cultivation did not cheapen produce, reduce prices, and so cause lower rents. On the contrary, prices were not only maintained, but continued to rise.

This continuously upward tendency in prices was unprecedented. It cannot be attributed to the operation of the Corn Laws.[1] Down to 1815 that legislation had scarcely affected prices at all, and therefore could not influence rents. The rise was rather due to a variety of causes, some of which were exceptional and temporary. A series of unprosperous seasons prevailed over the whole available corn-area of Northern Europe. In England deficient harvests, though the shortage was to some extent mitigated by the increased breadth under corn, reduced the home supply at a time when the growth of an artisan population increased the demand. The country throughout these years either stood, or thought that it stood, on the verge of famine. Prices were raised by panic-stricken competition. As the area of the war extended, foreign supplies became less and less available. The enormous increase in the war-charges for freight and insurance made Great Britain more and more dependent on her own produce. Necessity compelled the full development of her existing resources, as well as the resort to inferior land. Larger supplies of home-grown corn could only be

[1] See chapter xii.

obtained either by improved methods of cultivation or by bringing untilled land under the plough. The one method powerfully stimulated the progress of agriculture, which may be summed up in increasing the yield and lowering the cost of production ; the other was the valid justification of the rapid enclosure of wastes, open-fields, and commons. Much of the land that now was sown with corn could only be tilled at a profit when prices were high, because the outlay on its tillage was greater, and the return from its cultivation was less, than on ordinary land. Yet, as prices then stood, even this inferior soil was able to bear a rent, and by each step towards the margin of cultivation, the rental value of land of better quality was enhanced. Thus Napoleon proved to be the Triptolemus or patron saint not only of farmers but of land-lords.

Another cause of the high prices of the time was the state of the currency. When gold is cheap, commodities are dear. Any great increase in the production of gold for a time raises prices ; the sovereign becomes of less relative value ; it buys less than before, and more gold has to be paid for the same quantity. But this direct effect of gold discoveries was not then in operation ; it had spent its force, and at the close of the eighteenth century did not materially affect prices. Similar results were, however, produced by the immense extension of that system of deferred payment which is called credit. Paper money was issued in excessive quantities, not only by the Bank of England but by the private banks all over the country. A new medium of exchange was created. This addition to the circulating medium raised prices in the same kind of way as an actual addition to the quantity of coin. But there was this important difference. Paper money is only a promise to pay ; it is only representative money, and, unless it is convertible into gold, the credit which it creates is fictitious and may be excessive. The immense development of manufacturing industries and of the canal system, in the years 1785-92, required increased facilities for carrying on commercial transactions. But bankers, in their eagerness to create business, made advances on insufficient or inconvertible securities, discounted bills without regard to the actual value of the commodities on which the trans-actions were based, and issued notes far beyond the amount which their actual funds justified. In 1793 came the first crash. The Bank of England, warned by the fall of the exchanges and the outflow of gold, restricted their issue of notes. A panic followed. Out of 350 country banks in England and Wales, more than 100 stopped payment; their promises to pay were repudiated; and their paper was destroyed at the expense of the holder. The ruin and the loss of confidence were widespread; those who escaped the crash hoarded their money instead of making investments in mercantile undertakings. But the destruction of so much paper temporarily restored the proportion between the gold in the country and the paper by which it was represented.

In 1797 a second crisis occurred. Alarmed at a prospect of invasion, country depositors crowded to withdraw deposits and realise their property. There were runs on the country banks, and such heavy demands for their support were made on the Bank of England that, on Saturday, February 25, 1797, the stock of coin and bullion had fallen to under £1,300,000, with every prospect of a renewal and an increase of the run on the following Monday. On Sunday, February 26, an Order of Council suspended payments in cash until Parliament could consider the situation. The merchants of London came to the rescue of the bank. They guaranteed the payment of its notes in gold; the national credit was saved, and the worst of the threatened crisis was averted. But the failures of country banks were again numerous. Once more the same process was repeated. Paper money in large quantities was destroyed at the cost of its holders, and the balance between the promise and the ability to pay was again readjusted. The experience was not lost on agriculturists, who found that their land was not only the most remunerative but the safest investment.

Under the Bank Restriction Act of 1797, the Bank of England suspended payment in coin. In other words a paper currency was created which was not convertible into gold. The Act was originally a temporary expedient. But it was not till 1821 that the bank completely resumed payment in specie. No doubt the effect of the Act was to aggravate the tendency of prices to rise. Yet the measure was probably justified by the exceptional circumstances of the war and of trade. It supplied the Government with gold for the expenses of our own expeditionary forces, as well as for the payment of subsidies to our allies. It also enabled the country to carry on the one-sided system of trade to which we were gradually reduced by the Continental blockade. Our exports of manufactured goods were excluded from European ports. Consequently the materials which we imported were paid for in cash instead of in goods, and the vessels which conveyed them to our ports returned in ballast. There was thus a constant drain of gold from the country. So long as the power to issue inconvertible notes was sparingly used, the paper currency maintained its nominal value. But from 1808 onwards such large quantities of paper were issued, not only by the Bank of England but by country banks, that it rapidly depreciated as compared with gold. It is probable that from 1811 to 1813 one-fifth of the enormous prices of agricultural produce were due to the disordered state of the currency. In 1814, owing partly to the abundant harvest of the previous year, partly to the collapse of the Continental blockade, prices rapidly fell. A financial crash followed which caused even more widespread ruin in country districts than the paroxysm of 1793. Of the country banks, 240 stopped payment, and 89 became bankrupt. The result was a wholesale destruction of bank-paper, the reduction of thousands of families from wealth to destitution, and the gradual restoration of the equilibrium of the currency.

The seasons, the war, the growth of population, the disorders of the currency, combined to raise and maintain at a high level the prices of agricultural produce in Great Britain. At the same time the prohibitive cost of transport prevented such foreign supplies as were then available from reducing the prices of home-grown corn. Circumstances thus gave British agriculturists a monopoly, which, after 1815, they endeavoured to preserve by legislation. Land was not only a most profitable investment, but the fate of speculators had again and again convinced both landlords and tenants that land was the safest bank. Thus business caution, as well as business enterprise, prompted the outlay of capital on agricultural improvement. Economic ideas pointed in the same direction. The doctrine of John Locke,[1] that high rents were a symptom of prosperity still prevailed among politicians. It was also maintained that high rents were a necessary spur to agricultural progress. So long as land remained cheap, farmers rested satisfied with antiquated practices ; the dearer the land, the more energetic and enterprising they necessarily became. Young went so far as to say that the spendthrift, who frequented London club-houses and

---

[1] " An infallible sign of your decay of wealth is the falling of rents, and the raising of them would be worth the nation's care " (*Works*, ed. 1823, vol. v. p. 69).

raised rents to pay his debts of honour, was a greater benefactor to agriculture than the stay-at-home squire who lived frugally in order to keep within his ancestral income. No economist of the day had conceived any other method of satisfying the wants of a growing population except by improving the existing practices of farmers or bringing fresh tracts of land under the plough. Advanced Free Traders like Porter [1] never imagined that a progressive country could become dependent on foreign nations for its daily food. It was to the continuous improvement in agricultural methods that he looked for the means of supplying a population, which, he calculated, would, at the end of the nineteenth century, exceed 40 millions. Nor did he entertain any doubt that, by the progress of skill and enterprise, the quantity raised in 1840 could be increased by the requisite 150 per cent.

Encouraged by high profits, approved by economists, justified by necessity, agriculture advanced rapidly on the new lines of large farms and large capital. The change was one side of a wider movement. In the infancy of agriculture and of trade, self-supporting associations had been formed for mutual defence and protection. Manorial organisations like trade guilds had begun to break up, when the central power was firmly established. Now, once more, agriculture and manufacture were simultaneously reorganised. Division of labour had become a necessity. Domestic handicrafts were gathered into populous manufacturing centres, which were dependent for food on the labour of agriculturists. Farms ceased to be self-sufficing industries, and became factories of beef and mutton. The pressure of these conditions demanded the utmost development of the resources of the soil. The cultivation of additional land by the most improved methods grew more and more necessary. Enclosures went on apace. Yet, even in favourable seasons, it was a struggle to keep pace with growing needs ; scarcity, if not famine, resulted from deficiency. During part of the period, foreign supplies might be relied on to avert the worst. But throughout the Napoleonic wars this resource grew

---

[1] " To supply the United Kingdom with the single article of wheat would call for the employment of more than twice the amount of shipping which now annually enters our ports, if indeed it would be possible to procure the grain from other countries in sufficient quantity ; and to bring to our shores every article of agricultural produce in the abundance which we now enjoy, would probably give constant occupation to the mercantile navy of the whole world " (*Progress of the Nation*, ed. 1847, p. 136).

yearly more uncertain and more costly. The pace of enclosure was immensely accelerated. In the first 33 years of the reign of George III., there were 1355 Acts passed ; in the 23 years of the wars with France (1793-1815) there were 1934. It is easy to attribute the great increase of enclosures during this last period solely to the greed of landlords, eager to profit by the high prices of agricultural produce. That the land would not have been brought into cultivation unless it paid to do so, may be admitted. But it must in justice be remembered that an addition to the cultivated area was, in existing circumstances, one of the two methods, which at that time were alone available, of increasing the supply of food, averting famine, and reducing prices. Economically, enclosures can be justified. But the processes by which they were sometimes carried out were often indefensible, and socially their effects were disastrous. On these points more will be said subsequently. Here it will be enough to reiterate the statement that enclosure meant not merely reclamation of waste ground, but partition of the commons and extinction of the open-field system. It has been suggested, on the authority of passages in his tract on *Wastes*, that Arthur Young learned to deplore his previous crusade against village farms, when he saw the effect of enclosures on rural life. What Young deplored was the loss of a golden opportunity of attaching land to the home of the cottager. But he never faltered in his conviction of the necessity of breaking up the open-fields and dividing the commons. In the tract on *Wastes* he emphatically asserts his wish to see all commons enclosed, and he was too great a master of his subject not to know that without pasture the arable village farms must inevitably perish.

The other method of increasing the food supplies of the country consisted of agricultural improvements. Here also the preparation of the ground involved changes which bore hardly on small occupiers of land. The new system of farming required large holdings, to which a new class of tenant of superior education and intelligence was attracted. It was on these holdings that capital could be expended to the greatest advantage, that meat and corn could be grown in the largest quantities, that most use could be made of those mechanical aids which cheapened production. Costly improvements could not be carried out by small hand-to-mouth occupiers, even if their obstinate adherence to antiquated methods would have allowed them to contemplate the possibility of change.

But this consolidation of holdings threw into the hands of one tenant land which had previously been occupied by several. If the land was laid down to grass, and in the case of heavy land, down to 1790, this was the most profitable form of enclosure,—there was also a diminution in the demand for labour, and a consequent decrease in the population of the village. If, on the other hand, the land was cultivated as an arable farm, there was probably a greater demand for labour and possibly an increase in the numbers of the rural population. Arthur Young in 1801 [1] shows that, out of 37 enclosed parishes in an arable county like Norfolk, population had risen in 24, fallen in 8, and remained stationary in 5. It cannot therefore be said that either enclosures, or the consolidation of holdings, necessarily depopulated country villages. Whether this result followed, or did not follow, depended on the use to which the land was put, though even on arable farms the gradual introduction of machinery, at present limited to the threshing machine, tended to diminish the demand for labour.

If the country was to be fed, more scientific methods of farming were necessary. The need was pressing, and both enclosures and the consolidation of large farms prepared the way for a new stage of agricultural progress. Hitherto bucolic life had been the pastime of a fashionable world, the relaxation of statesmen, the artificial inspiration of poets. But farmers had neither asked nor allowed scientific aid. The dawn of a new era, in which practical experience was to be combined with scientific knowledge, was marked by the lectures of Humphry Davy in 1803. In 1757 Francis Home [2] had insisted on the dependence of agriculture on "Chymistry." Without a knowledge of that science, he said, agriculture could not be reduced to principles. In 1802 the first steps were taken towards this end. The Board of Agriculture arranged a series of lectures on "The Connection of Chemistry with Vegetable Physiology," to be delivered by Davy, then a young man of twenty-three, and recently (July, 1801) appointed Assistant Professor of Chemistry at the Royal Institution of Great Britain. He had already made his mark as the most brilliant lecturer of the day, attracting round him by his scientific use of the imagination such men as Dr. Parr and

---

[1] *Inquiry into the Propriety of applying Wastes to the Better Maintenance and Support of the Poor.*

[2] *The Principles of Agriculture and Vegetation,* by Francis Home, M.D, 1757,

S. T. Coleridge, and the talent, rank, and fashion of London, women as well as men. His six lectures on agricultural chemistry, commencing May 10, 1803, were delivered before the Board of Agriculture. So great was their success that he was appointed Professor of Chemistry to the Board, and in that capacity gave courses of lectures during the ten following years. In 1813 the results of his researches were published in his *Elements of Agricultural Chemistry*. The volume is now out-of-date, though the lecture on " Soils and their Analyses," in spite of the progress of geological science and the adoption of new classifications, remains of permanent interest. Many passages that were then listened to as novelties are now commonplaces ; others, especially those on manures, have been completely superseded by the advance of knowledge. But if the book has ceased to be a practical guide, it remains a historical landmark, and something more. It is the foundation-stone on which the science of agricultural chemistry has been reared, and its author was the direct ancestor of Liebig, Lawes, and Gilbert, to whose labours, in the field which Davy first explored, modern agriculture is at every turn so deeply indebted. It was Davy's work which inspired the choice by the Royal Agricultural Society (founded in 1838) of its motto " Practice with Science."

In Thomas Coke of Norfolk [1] the new system of large farms and large capital found their most celebrated champion. In 1776, at the age of twenty-two, he came into his estate with " the King of Denmark " as " his nearest neighbour." Wealthy, devoted to field sports, and already Member of Parliament for Norfolk, it seemed improbable that he would find time for farming. But as an ardent Whig and a prominent supporter of Fox in the House of Commons, he was excluded by his politics from court life or political office. In 1778 the refusal of two tenants to accept leases at an increased rent threw a quantity of land on his hands. He determined to farm the land himself. From that time till his death in 1842, he stood at the head of the new agricultural movement. On his own estates his energy was richly rewarded. Dr. Rigby,[2] writing in 1816, states that the annual rental of Holkham rose from £2,200 in 1776 to £20,000 in 1816.

When Coke took his land in hand, not an acre of wheat was

---

[1] *Coke of Norfolk and his Friends*, by A. M. W. Stirling, 2 vols. 1910.

[2] *The Pamphleteer*, vol. xiii. pp. 469-70 ; *Holkham and its Agriculture*, 3rd edition, 1818, pp. 25, 28.

to be seen from Holkham to Lynn. The thin sandy soil produced but a scanty yield of rye. Naturally wanting in richness, it was still further impoverished by a barbarous system of cropping. No manure was purchased ; a few Norfolk sheep with backs like rabbits, and, here and there, a few half-starved milch cows were the only live-stock ; the little muck that was produced was miserably poor. Coke determined to grow wheat. He marled and clayed the land, purchased large quantities of manure, drilled his wheat and turnips, grew sainfoin and clover, trebled his live-stock. On the light drifty land in his neighbourhood the Flemish maxim held good : " Point de fourrage, point de bestiaux ; sans bestiaux, aucun engrais ; sans engrais, nulle recolte." " No keep, no live-stock ; without stock, no manure ; without manure, no crops." It is, in fact, the Norfolk proverb, " Muck is the mother of money." In the last quarter of the eighteenth century the value of bones as fertilisers was realised.[1] The discovery has been attributed to a Yorkshire fox-hunter who was cleaning out his kennels ; others assign it to farmers in the neighbourhood of Sheffield, where refuse heaps were formed of the bones which were not available for the handles of cutlery. By the use of the new discovery Coke profited largely. He also introduced into the county the use of artificial foods like oil-cake, which, with roots, enabled Norfolk farms to carry increased stock. Under his example and advice stall-feeding was extensively practised. On Bullock's Hill near Norwich, during the great fair of St. Faith's, drovers assembled from all parts of the country, especially from Scotland, with herds of half-fed beasts which were bought up by Norfolk farmers to be fattened for London markets. The grass lands, on which the beef and mutton of our ancestors were raised, were deserted for the sands of the eastern counties, from which under the new farming practice, the metropolis drew its meat supplies. Numbers of animals fattened on nutritious food gave farmers the command of the richest manure, fertilised their land, and enabled them not only to grow wheat but to verify the maxim " never to sow a crop unless there is condition to grow it luxuriantly."

In nine years Coke had succeeded in growing good crops of wheat on the land which he farmed himself. He next set himself to improve the live-stock. After patient trial of other breeds, and

---

[1] Blithe and Evelyn both mention the value of bones, and in a York edition of Evelyn's *Terra* of 1778 may be read, " At Sheffield it has now become a trade to grind bones for the use of the farmer."

especially of Shorthorns among cattle and of the New Leicesters and Merinos among sheep, he adopted Devons and Southdowns. His efforts were not confined to the home-farm. Early and late he worked in his smock-frock, assisting tenants to improve their flocks and herds. Grass lands, till he gave them his attention, were wholly neglected in the district. If meadow or pasture wanted renewal, or arable land was to be laid down in grass, farmers either allowed it to tumble down, or threw indiscriminately on the ground a quantity of seed drawn at haphazard from their own or their neighbour's ricks, containing as much rank weed as nutritious herbage. It was a mere chance whether the sour or the sweet grasses were aided in their struggle for existence. Stillingfleet, in 1760, had distinguished the good and bad herbage by excellent illustrations of the kinds best calculated to produce the richest hay and sweetest pasture. The Society of Arts, Manufacture and Commerce had offered premiums for the best collections of the best kinds, and in Edinburgh the Lawsons were experimenting on grasses. But Coke was the first landlord who appreciated the value of the distinctions by applying them to his own land. In May and June, when the grasses were in bloom, he gave his simple botanical lessons to the children of his tenantry, who scoured the country to procure his stocks of seed.

Impressed with the community of interest among owners, occupiers, and labourers, Coke stimulated the enterprise of his tenants, encouraged them to put more money and more labour into the land, and assisted them to take advantage of every new invention and discovery. Experiments with drill husbandry on 3,000 acres of corn land convinced him of its value in economy of time, in saving of seed, in securing an equal depth of sowing, and in facilitating the cleaning of the land. He calculated that he saved in seed a bushel and a half per acre, and increased the yield per acre by twelve bushels. As with the drill, so with other innovations. He tested every novelty himself, and offered to his neighbours only the results of his own successful experience. It was thus that the practice of drilling turnips and wheat, and the value of sainfoin, swedes, mangelwurzel, and potatoes were forced on the notice of Norfolk farmers. His farm-buildings, dwelling-houses, and cottages were models to other landlords. On them he spared no reasonable expense. They cost him, during his tenure of the property, more than half a million of money. By offering long leases of twenty-one years, he guaranteed to improving farmers a return for their energy and outlay. Two years before the expiration of a lease, the tenant was informed of the new rent proposed, and offered a renewal. " My best bank," said one of his farmers, " is my land." At the same time he guarded against the mischief of a long unrestricted tenancy by covenants regulating the course of high-class cultivation. Though management clauses were then comparatively unknown in English leases, his farms commanded competition among the pick of English farmers. " Live and let live " was not only a toast at the Holkham sheep-shearings, but a rule in the control of the Holkham estate. Cobbett was not prejudiced in favour of landlords. Yet even he was compelled to admit the benefits which Coke's tenants derived from his paternal rule. " Every one," he writes in 1821, " made use of the expressions towards him which affectionate children use towards their parents."

One great obstacle to the improvement of Norfolk farming remained. Farmers of the eighteenth century lived, thought, and farmed like farmers of the thirteenth century. Wheat instead of rye might be grown with success ; turnips, if drilled, were more easily hoed and yielded a heavier crop than those which were sown broadcast ; marl and clay might help to consolidate drifting soil. But the neighbouring farmers were suspicious of new methods, and distrusted a young man who disobeyed the saws and maxims of their forefathers. Politics ran so high that Coke's Southdowns were denounced as " Whiggish sheep." It was nine years before he found anyone to imitate him in growing wheat. " It might be good for Mr. Coke ; but it was not good enough for them." As to potatoes, the best they would say was, that " perhaps they wouldn't poison the pigs." Even those who had given up broad-cast sowing still preferred the dibber to the drill. Sixteen years passed before the implement was adopted. Coke himself calculated that his improvements travelled at the rate of a mile a year. The Holkham sheep-shearings did much by ocular demonstration to break down traditions and prejudices. These meetings originated in 1778, in Coke's own ignorance of farming matters ; small parties of farmers were annually invited to discuss agricultural topics at his house and aid him with their practical advice. Before many years had passed, the gatherings had grown larger, and Coke had become a teacher as well as a learner. The Holkham sheep-shearing in June, 1806 is described in the *Farmer's Magazine* [1] in the stilted language of the day, as " the happy resort of the most distinguished patrons and amateurs of Georgic employments." In 1818 open house was kept at Holkham for a week ; hundreds of persons assembled from all parts of Great Britain, the Continent, and America. The mornings were spent in inspecting the land and the stock ; at three o'clock, six hundred persons sate down to dinner ; the rest of each day was spent in discussion, toasts, and speeches. The Emperor of Russia sent a special representative, and among the learners was Erskine, who abandoned the study of Coke at Westminster Hall to gather the wisdom of his namesake at Holkham. At the sheep-shearings, year after year, were collected practical and theoretical agriculturists, farmers from every district, breeders of every kind of stock, who compared notes and exchanged experiences. In many other parts of England similar meetings were held by great landlords, like the Duke of Bedford at Woburn,[2] or Lord Egremont at Petworth, who in their own localities were carrying on the same work as Coke.

At Holkham and Woburn sheep-shearings, both landlords and farmers were learners ; both required to be educated in the new principles of their altered business. It was by no means uncommon to find landlords who prevented progress by refusing to let land except at will, or bound their tenants by restrictive covenants to follow obsolete practices. There was, moreover, a tendency among the land-owning class to expect from rent-paying tenants a greater outlay on the land than a farmer's capital could bear or an occupier was justified in making. The question of improvements had not yet assumed the complicated forms which have developed under modern agricultural methods. But it had already been raised in the simpler shape. The liability for improvements of a permanent character required to be defined ; no distinction was yet drawn between changes which added some lasting benefit to the holding and those whose effects were exhausted within the limits of a brief occupation. Expenditure which might legitimately be borne by landlords was often demanded from tenants at will or even from year to year. Thousands of acres still lay unproductive because owners looked to occupiers for the reclamation of waste, the drainage of

---

[1] *Farmer's Magazine*, August, 1806.

[2] For a description of a Woburn sheep-shearing, or "this truly rational Agricultural Fete," see *Farmer's Magazine* for July, 1800.

swamps, or an embankment against floods. It was one of the lessons which were taught by the agricultural depression after the peace of 1815 that landowners must find the money for lasting improvements effected on their property. That farmers should have realised the possibility of improving traditional practices was a great step in advance. The new race of men, who were beginning to occupy land, were better educated, commanded more capital, were more open to new ideas and more enterprising than their predecessors. Their holdings were larger, and offered greater scope for energy and experiment. The Reporters to the Board of Agriculture on Northumberland (1805) lay stress on the size of the farms, and on the spirit of enterprise and in dependence which now animated the tenants. " Scarcely a year passes without some of them making extensive tours for the sole purpose of examining modes of culture, of purchasing or hiring the most improved breeds of stock, and seeing the operations of new-invented and most useful implements." The Reporter on Middlesex (1798) emphasises the stagnation of farming among small occupiers. " It is rather the larger farmers and yeomen, or men who occupy their own land, that mostly introduce improvements in the practice of agriculture, and that uniformly grow much greater crops of corn, and produce more beef and mutton per acre than others of a smaller capital." The Oxfordshire Reporter (1809) says : " If you go into Banbury market next Thursday, you may distinguish the farmers from enclosures from those from open fields ; quite a different sort of men ; the farmers as much changed as their husbandry—quite new men, in point of knowledge and ideas." Elsewhere in the same Report,—it is Arthur Young who writes,—occurs the following passage : The Oxfordshire farmers " are now in the period of a great change in their ideas, knowledge, practice, and other circumstances. Enclosing to a greater proportional amount than in almost any other county in the kingdom, has changed the men as much as it has improved the country ; they are now in the ebullition of this change ; a vast amelioration has been wrought, and is working ; and a great deal of ignorance and barbarity remains. The Goths and Vandals of open-fields touch the civilisation of enclosures. Men have been taught to think, and till that moment arrives, nothing can be done effectively. When I passed from the conversation of the farmers I was recommended to call on, to that of men whom chance threw in my way,

I seemed to have lost a century in time, or to have moved a thousand miles in a day. Liberal communication, the result of enlarged ideas, was contrasted with a dark ignorance under the covert of wise suspicions ; a sullen reserve lest landlords should be rendered too knowing, and false information given under the hope that it might deceive, were in such opposition, that it was easy to see the *change*, however it might work, had not done its business. The old open-field school must die off before new ideas can become generally rooted." In Lincolnshire, in the early years of George III., Arthur Young had found few points in the management of arable land which did not merit condemnation. The progress, which he noted as Reporter to the Board of Agriculture in 1799, was largely due to the changed character of the farmers. " I have not," he says, " seen a set more liberal in any part of the kingdom. Industrious, active, enlightened, free from all foolish and expensive show, . . . they live comfortably and hospitably, as good farmers ought to live ; and in my opinion are remarkably void of those rooted prejudices which sometimes are reasonably objected to this race of men. I met with many who had mounted their nags, and quitted their homes purposely to examine other parts of the kingdom ; had done it with enlarged views, and to the benefit of their own cultivation."

CHAPTER XI

OPEN-FIELD FARMS AND PASTURE COMMONS

(1793 -1815)

Condition of open-field arable land and pasture commons as described by the Reporters to the Board of Agriculture, 1793-1815 ; (1) The North and North-Western District ; (2) West Midland and South-Western District : (3) South-Eastern and Midland District ; (4) Eastern and North-Eastern District ; (5) the Fens ; the cumulative effect of the evidence ; procedure under private Enclosure Acts ; its defects and cost ; the General enclosure Act of 1801 ; the Inclosure Commissioners ; the new Board of Agriculture.

IT might perhaps be supposed that in 1793 the agricultural defects of the ancient system of open arable fields and common pasture had been remedied by experience ; that open-field farmers had shared in the general progress of farming ; that time alone was needed to raise them to the higher level of an improved standard : that, therefore, enclosures had ceased to be an economic necessity. In 1773, an important Act of Parliament had been passed,[1] which attempted to help open-field farmers in adapting their inconvenient system of occupation to the improved practices of recent agriculture. Three-fourths of the partners in village-farms were empowered, with the consent of the landowner and the titheowner, to appoint field-reeves, and through them to regulate and improve the cultivation of the open arable fields. But any arrangement made under these powers was only to last six years, and, partly for this reason, the Act seems to have been from the first almost a dead letter. At Hunmanby, on the wolds of the East Riding of Yorkshire,[2] the provisions of the Act were certainly put in force, and it is

[1] 13 Geo. III. c. 81.

[2] Isaac Leatham's *General View of the Agriculture of the East Riding of York-shire* (1794), p. 45. Thomas Stone, in his *Suggestions for Rendering the Inclosure of Common Fields and Waste Lands a source of Population and Riches* (1787), says that he knew of no instance in which the Act had been put in force.

possible that it was also applied at Wilburton in Cambridgeshire. With these exceptions, little, if any, use seems to have been made of a well-intentioned piece of legislation.

Small progress had in fact been made among the cultivators of open-fields. Here and there, the new spirit of agricultural enterprise had influenced the occupiers of village farms. In rare instances improved practices were introduced. But the demand for increased food supplies had become, as our ancestors were experiencing, too pressing for delay. Any continuous series of adverse seasons created a real scarcity of bread, and more than once during the Napoleonic wars, famine was at the door. Unless food could be produced at home, it could not be obtained elsewhere. An extension of the cultivated area was the quickest means of adding to production. Agriculturists at the close of the eighteenth century were convinced that no adequate increase in the produce of the soil could be obtained, unless open-field farms were broken up, and the commons brought into more profitable cultivation. If they were right in that belief, the great agricultural change was justified, which established the uniform system with which we are familiar to-day. The point is one of the greatest importance. The uncritical praises lavished by sixteenth and seventeenth century travellers on open-field farming are of little value because they had no higher standard with which to compare its results. Such a standard had now been to some extent created. It may therefore be useful to illustrate, from the contemporary records supplied by the Reports to the Board of Agriculture,[1] the condition of open arable land and of pasture-commons in the years 1793-1815. The material is arranged according to the four districts into which, for statistical purposes, the English counties are usually divided. The cumulative force of the evidence is great. But some of it relates to wastes which were not attached to village farms, although common of pasture and fuel was often claimed over the area by the inhabitants of the neighbourhood. As to the reliability of the whole evidence, it

[1] The Reports to the Board are extant in two forms. The quarto editions were drafts, intended for private circulation and for correction by practical agriculturists belonging to the district under survey. They all belong to the years 1793-94-95. The octavo editions are the Reports in their final form. They were published at various dates, ranging from 1795 in the case of Holt's *Lancashire*, to 1815 when Quayle's *Channel Islands* was issued. In some cases the Reports are practically the same in their draft and final forms. Sometimes, on the other hand, they were re-written by other Reporters with scarcely any reference to the original Survey.

would be only fair to add that the Reporters were not likely to be prejudiced in favour of open-field farms or unappropriated commons.

1. In the *North and North-Western District*, enclosure had gone on apace since 1770. In Northumberland, for instance (1805), very little common land was left which could be made profitable under the plough. 120,000 acres were said to have been enclosed " in the last thirty years." [1] In Durham, it is stated that " the lands, or common fields of townships, were for the most part inclosed soon after the Restoration." The Reporter laments " that in some of the rich parts of the county, particularly in the neighbourhood of the capital of it, large quantities of land should still lie totally deprived of the benefit of cultivation, in commons ; and that ancient inclosures, by being subject to the perverse custom of intercommon, be prevented from that degree of fertilization, to which the easy opportunity of procuring manure, in most cases, would certainly soon carry the improvement of them ; in their present state, little or no benefit is derived to any person·whatsoever, entitled either to common, or intercommon, from the use of them." [2] The waste lands of the West Riding of Yorkshire [3] are calculated at 265,000 acres capable of cultivation. The Reporter proposes to " add to these the common fields which are also extensive, and susceptible of as much improvement as the wastes." The man on inclosed land " has not the *vis inertiae* of his stupid neighbour to contend with him, before he can commence any alteration in his management . . . he is completely master of his land, which, in its open state, is only *half* his own. This is strongly evident in the cultivation of turnips, or other vegetables for the winter consumption of cattle ; they are constantly cultivated in inclosures, when they are never thought of in the open fields in some parts." In the North Riding " few open or common fields now remain, nearly the whole having long been inclosed." [4] But on the commons the practice of surcharging is said to have increased to " an alarming degree." It had become a frequent custom for persons, often dwelling in distant townships, to take single fields which were entitled to common rights, and stock the commons with an excessive quantity of cattle. In Cumberland (1794), [5] there were still 150,000

---

[1] *Northumberland*, by J. Bailey and G. Culley (3rd edition, 1805), p. 126.
[2] Granger's *Durham* (1794), p. 44.
[3] Brown's *West Riding of Yorkshire* (1799), pp. 131, 133.
[4] Tuke's *North Riding* (1800), pp. 90, 199.
[5] *Cumberland*, by J. Bailey and G. Culley (1794), pp. 202, 215. 236.

acres of improvable common, which were " generally overstocked."
" No improvement of breed was possible, while a man's ewes mixed
promiscuously with his neighbour's flocks." There were "few
commons but have parts which are liable to rot, nor can the sheep
be prevented from depasturing it." " If any part of the flock
had the scab or other infectious disease, there was no means of
preventing it from spreading." A large part of these commons
was good corn-land ; if enclosed, and part ploughed for grain crops,
not only would there be an increased supply of corn, but, instead
of " the ill-formed, poor, starved, meagre animals that depasture
it at present," there might be " an abundant supply of fat mutton
sent to our big towns." In Cheshire (1794),[1] there were said to
be of " common fields, probably not so much as 1000 acres."
Staffordshire [2] in 1808 contained little more than 1000 acres of open-
fields, which " are generally imperfectly cultivated, and exhausted
by hard tillage." Since the reduction of their area, the general
produce of the county is stated to be greater, the stock better, and
the rent higher by 5s. an acre. The county was " emerging out of
barbarism." But, thirty years before, on some of the " best land
of the county," the rotation had been " (1) fallow ; (2) wheat ;
(3) barley ; (4) oats ; and often oats repeated, and then left to
Nature ; the worst lands left to pasture and spontaneous rubbish ;
turnips and artificial grasses scarcely at all known in farming."
In Derbyshire [3] (1811), a list of the thirteen open arable fields which
remained is given. " Many of them," says the Reporter, " must
remain in their present open, unproductive, and disgraceful state,
(though principally in the best stratum in the County) " owing to
the expense of enclosure. There were, however, still thirty-six
open commons, such as Elmton, with its " deep cart-ruts, and
every other species of injury and neglect that can, perhaps be
shown on useful land ; part of it has been ploughed at no distant
period, as completely exhausted as could be, and then resigned to
Weeds and Paltry " ; or Hollington, which, " though overgrown
with Rushes through neglect, is on a rich Red Marl soil " ; or
Roston, " miserably carted on, cut up, and in want of Draining ;
in wet seasons it generally rots the sheep depastured on it ; . . . probably injurious, rather than beneficial, in its present state, both to
the Parishioners and the Public."

---

[1] Wedge's *Cheshire* (1794), p. 8.   [2] Pitt's *Staffordshire* (1808), pp. 13, 51, 313.
[3] Farey's *Derbyshire* (1813), vol. ii. p. 77.

2. In the *West Midland and South-Western District*, Shropshire (1794) [1] " does not contain much common field lands, most of these having been formerly enclosed, and before acts of parliament for that purpose were in use ; but the inconvenience of the property being detached and intermixed in small parcels, is severely felt, as is also the inconvenience of having the farm buildings in villages." There still remained large commons of which the largest were Clun Forest and Morfe Common, near Bridgnorth. The Reporter strongly advocates their enclosure. " The idea of leaving them in their unimproved state, to bear chiefly gorse bushes, and fern, is now completely scouted, except by a very few, who have falsely conceived that the inclosing of them is an injury to the poor ; but if those persons had seen as much of the. contrary effects in that respect as I have, I am fully persuaded their opposition would at once cease. Let those who doubt, go round the commons now open, and view the miserable huts, and poor, ill-cultivated, impoverished spots erected, or rather *thrown together*, and inclosed by themselves, for which they pay 6d. or 1s. per year, which, by loss of time both to the man and his family, affords them a very trifle towards their maintenance, yet operates upon their minds as a sort of independence ; this idea leads the man to lose many days work by which he gets a habit of indolence ; a daughter kept at home to milk a poor half-starved cow, who being open to temptations, soon turns harlot, and becomes a distressed, ignorant mother, instead of making a good useful servant."

Herefordshire [2] (1794) contained a great number of open field farms, occupying some of " the best land of the county," and pursuing the " invariable rotation of (1) fallow, (2) wheat, (3) pease or oats, and then fallow again." Speaking of the waste lands at the foot of the Black Mountains above the Golden Valley, the Reporter says : " I do appeal to such gentlemen as have often served on Grand Juries in this county, whether they have not had more felons brought before them from that than from any other quarter of the county." He attributes this lawlessness to the right, which the cottager possessed in virtue of his arable holding, of turning out stock on the hills, and to the encouragement which this right afforded him of living by any means other than his labour.

[1] Bishton's *Shropshire* (1794), pp. 8, 24.

[2] Clark's *Herefordshire* (1794), pp. 69, 28.

Worcestershire [1] (1794) contained from 10,000 to 20,000 acres of wastes, " in general depastured by a miserable breed of sheep belonging to the adjoining cottagers and occupiers, placed there for the sake of their fleeces, the meat of which seldom reaches the market, a third fleece being mostly the last return they live to make." Yet, adds the Reporter, " most of the common or waste land is capable of being converted into tillage of the first quality." Considerable tracts still lay in open-fields, especially in the neighbourhood of Bredon, Ripple, and to the east of Worcester. " The advantages from inclosing common fields . . . have been very considerable ; . . . the rent has always risen, and mostly in a very great proportion ; the increase of produce is very great, the value of stock has advanced almost beyond conception ; . . . indeed it is in inclosures alone, that any improvement in the line of breeding in general can be made." Speaking of the district towards the Gloucestershire border, it is stated that " the lands being in common fields, and property much intermixed, there can be of course but little experimental husbandry ; being, by custom, tied down to three crops and a fallow. . . . The mixture of property in our fields prevents our land being drained, and one negligent farmer, from not opening his drains, will frequently flood the lands of ten that lie above, to the very great loss of his neighbours and community at large. Add to this, that although our lands are naturally well adapted to the breed of sheep, yet the draining etc. is so little attended to in general, that, out of at least 1000 sheep, annually pastured in our open fields, not more than forty, on an average, are annually drawn out for slaughter, or other uses ; infectious disorders, rot, scab, etc. sweep them off, which would not be the case if property were separated." Of the pasture commons, it is said that they are " overstocked," " produce a beggarly breed of sheep," and " are of little or no value." Again, it is stated that, where enclosures " have been completed fifteen or twenty years, property is trebled ; the lands drained ; and if the land has not been converted into pasture, the produce of grain very much increased ; where converted into pasture, the stock of sheep and cattle wonderfully improved. Where there are large commons, advantages are innumerable, to population as well as cultivation, and instead of a horde of pilferers, you obtain a skilful race, as well of mechanics as other labourers."

[1] Pomeroy's *Worcestershire* (1794), pp. 17, 16, App. pp. 2, 3, 5.

In Gloucestershire [1] (1794) common fields and common meadows still prevailed over extensive districts. Of the Cotswold district the Reporter says : " probably no part of the kingdom has been more improved within the last forty years than the Cotswold Hills. The first inclosures are about that standing ; but the greater part are of a later date. Three parishes are now inclosing ; and out of about thirteen, which still remain in the common field state, two, I understand, are taking the requisite measures for an inclosure : the advantages are great, rent more than doubled, the produce of every kind proportionably increased." Of the Vale of Gloucester he says : " I know one acre which is divided into eight lands, and spread over a large common field, so that a man must travel two or three miles to visit it all. But though this is a remarkable instance of minute division, yet, it takes place to such a degree, as very much to impede all the processes of husbandry. But this is not the worst ; the lands shooting different ways, some serve as headlands to turn on in ploughing others ; and frequently when the good manager has sown his corn, and it is come up, his slovenly neighbour turns upon it, and cuts up more for him, than his own is worth. It likewise makes one occupier subservient to another in cropping his land ; and in water furrowing, one sloven may keep the water on, and poison the lands of two or three industrious neighbours." Lot meadows were numerous in the county, on which the herbage was common after hay-making. Several tracts such as Corse Lawn, Huntley and Gorsley Commons were practically wastes, " not only of very little real utility, but productive of one very great nuisance, that of the erection of cottages by idle and dissolute people, sometimes from the neighbourhood, and sometimes strangers. The chief building materials are store-poles, stolen from the neighbouring woods. These cottages are seldom or never the abode of honest industry, but serve for harbour to poachers and thieves of all descriptions." In the Vale of Tewkesbury the common fields were " very subject to rot. . . . Though it is reckoned they (farmers) lose their flocks once in three years on average, there is a considerable quantity kept, the farmers being persuaded they could not raise corn without them. The arable fields after harvest are stocked without stint. When spring seedtime commences, they are confined to the fallow quarter of the field, and stinted in proportion to the properties ; they are folded every night, and kept

[1] Turner's *Gloucestershire* (1794), pp 10, 39, 49.

so hard, that scarce a blade of grass or even a thistle escapes them ; and this management is thought essentially necessary, especially on the stiff soils, to keep them in good order, such soils being too hard to plough in very dry weather, and, of course, not eligible in wet. The grass and weeds, without this expedient, would often get so much ahead as not to be afterwards conquered."

Another agricultural Report on Gloucestershire [1] was presented in 1807. The Reporter mentions that, in the reign of George III., " more than seventy Acts have passed the Parliament for inclosing or laying into severalty." " By these proceedings, the landlord and occupier are benefited ; the former in an advance of rent, the latter in the increase of crops. On the Cotswolds, many thousand acres are brought into cultivation, which before were productive of little more than furze and a few scanty blades of grass. In the Vale, by the inclosure of common fields, lands have been laid together, and rescued from the immemorial custom, or routine of crops—wheat, beans, and fallow ; and the farmers have found, to their great advantage, that clover, vetches and turnips may be raised in the fallow year, which was before attended only with labour and expense." The Reporter enumerates five advantages resulting from enclosure of common field farms :—(1) an increase of crops and rent ; (2) the commutation of tithes ; (3) the drainage of the land ; (4) the removal of the injury and cause of disputes occasioned by turning on the head- and fore-lands of neighbours ; (5) the encouragement of population. Of the advantages of enclos- ing common pastures or wastes he is equally convinced ; " the common or waste lands in the Vale are seldom stinted to a definite quantity of stock in proportion to the number of acres occupied ; but the cottager claims by custom to stock equally with the largest landholder. It is justly questioned whether any profit accrues to either from the depasturing of sheep, since the waste commons, being under no agricultural management, are usually poisoned by stagnated water, which corrupts or renders unwholesome the herbage, producing rot, and other diseases in the miserable animals that are turned adrift to seek their food there." Since 1794 Corse Common had been enclosed. From the results the Reporter of 1807 illustrates some of the benefits of enclosure. " The supposed advantages derived by cottagers, in having food for a few sheep and geese on a neighbouring common, have usually been brought for-

[1] Rudge's *Gloucestershire* (1807), pp. 89, 250.

ward as objections to the enclosing system. This question was much agitated with regard to the inclosure of Corse Chace in this county ; but if the present state and appearance of it, since the inclosure in 1796, be contrasted to what it was before, or its present produce of corn to the sheep that used to run over it, little doubt can remain of the advantageous result in favour of the community ; 1350 acres of wet and rushy waste were inclosed, and, in the first year of cultivation, the produce was calculated at 20,250 bushels of wheat, or of some other crop in equal proportion. If it could even be proved that some cottagers were deprived of a few trifling advantages, yet the small losses of individuals ought not to stand in the way of certain improvements on a large scale." The Reporter also quotes two Cotswold parishes, formerly open-fields, but now enclosed, as examples of increased produce. In Aldsworth, the annual produce of corn rose from 720 quarters to 2300 quarters ; in Eastington, it increased from 690 quarters to 2100 quarters. He adds that enclosures encouraged labour. " Labourers, who formerly were under the necessity of seeking employment in London and other places, now find it in sufficient quantity at home in their respective parishes."

In Somersetshire [1] (1797) the two largest districts of waste land were the Brent Marsh and King's Sedgmoor. The Reporter describes the Brent Marsh as a country which had " been heretofore much neglected, probably on account of the stagnant waters, and unwholesome air. But of late many efforts have been made to improve the soil, by draining and enclosing, under a variety of Acts of Parliament. The benefit resulting therefrom has been astonishing." The total area was over 20,000 acres, of which many thousands, " heretofore overflown . . . and of little or no value, are become fine grazing and dairy lands." Besides the general improvement to the health of the district, " scarcely a farmer can now be found who does not possess a considerable landed property ; and many whose fathers lived in idleness and sloth, on the precarious support of a few half-starved cows, or a few limping geese, are now in affluence." On the South Marsh, chiefly formed by the river Parret, " near thirty thousand acres of fine land are frequently overflown for a considerable time together, rendering the herbage unwholesome for the cattle, and the air unhealthy to the inhabitants." An Act of Parliament had been recently (1791)

[1] Billingsley's *Somersetshire* (1797), pp 167-73, 188.

obtained for draining a portion of this fen called King's Sedgmoor, containing " about 20,000 acres."

The Dorsetshire [1] commons in 1794 were "generally overrun with furze and ant-hills," worth 8s. an acre unenclosed, but "highly proper to cultivate, and, if converted, would be worth from 18s. to 20s. an acre." A second Report on Dorsetshire was issued in 1812.[2] The Reporter calls attention to the "half year meads." One person has the hay, and another person the "after-shear." These meadows were not near commonable fields, and the origin of the claim is not clear. Obviously, neither of the persons who shared the produce was likely to attempt to improve the herbage.

In Wiltshire [3] (1794) the Reporter fixes on four disadvantages of open-field husbandry : (1) the obligation to plough and crop all soils alike ; (2) the impossibility of improving sheep ; (3) the difficulty of raising food for their winter keep ; (4) the expense, trouble, and excessive number of horses required to cultivate detached dispersed lands. On the south-east side of the county lay a considerable tract of open-fields, and in the north-west, in the centre of the richest land of the district, were scattered numerous commons. The open arable fields are said to be in "a very bad state of husbandry," and the common pastures in a "very neglected unimproved " condition. "There are," says the Reporter, " numerous instances in which the common-field arable land lets for less than half the price of the inclosed arable adjoining ; and the commons are very seldom reckoned worth anything, in valuing any estate that has a right on them." For the last half-century very little land had been enclosed, "although the improvement on the lands, heretofore inclosed, has been so very great." "The reason seems to have been the very great difficulty and expence of making new roads in a country naturally wet and deep, and where the old public roads were, till within the last few years, almost impassable." Good turnpike roads had now been introduced ; villages were energetic in repairing the approaches to them ; and "it is to be hoped that so great an improvement as that of inclosing and cultivating the commonable lands will no longer be neglected." The crying need was the want of drainage. The common pastures

---

[1] Claridge's *Dorsetshire* (1793), p. 43.

[2] Stevenson's *Dorsetshire* (1812), p. 307.

[3] Davis' *Wiltshire* (1794), p. 136. This is, perhaps, the best of all the agricultural Reports.

from Westbury to Cricklade were in a " wet rotten state," depastured by an " unprofitable kind of stock," but " wanting only inclosing and draining to make them as good pasture land as many of the surrounding inclosures." Some of the cold arable fields would have been much more valuable if turned to pasture, and, in their undrained state, even the driest were " not safe for sheep in a wet autumn."

3. From the *South-Eastern and Midland District* the evidence is the fullest, because the district was still in a great measure farmed on the open-field system.

In Berkshire [1] (1794) there were 220,000 acres of open-fields, and downs, to 170,000 acres of inclosed land. Half of the county " is still lying in common fields ; and though it is not divided into such very small parcels as in some other counties, the farmer labours under all the inconvenience of commonable land ; and by that, is withheld from improving or treating his land, so as to return the produce which it ought to do, if entire, and under a good course of husbandry." " We generally see on all the commons and waste lands, a number of miserable cattle, sheep, and horses, which are a disgrace to their respective breeds, and the cause of many distempers."

In Buckinghamshire [2] (1794) 91,906 acres remained in open fields. The Reporters point out that " the slovenly operations of one man are often of serious consequence to his neighbours, with whose property his lands may lie, and generally do lie, very much intermixed. Every one is aware of the noxious quality of weeds, whose downy and winged seeds are wafted by every wind, and are deposited upon those lands which are contiguous to them ; and which before were perhaps as clean as the nature of them would admit, to the manifest injury of the careful and attentive farmer. Inclosures would, in a certain degree, lessen so great an evil ; they would also prevent the inroads of other people's cattle, as particularized in the parish of Wendover, and in which one man held eighteen acres in thirty-one different allotments."

Oxfordshire [3] in 1794 contained " upwards of an hundred uninclosed parishes or hamlets." The Reporter enumerates several advantages of enclosure. " The first of these is getting rid of the

[1] Pearce's *Berkshire* (1794), pp. 13, 49, 59.
[2] James' and Malcolm's *Buckinghamshire* (1794), pp. 32, 58.
[3] Davis's *Oxfordshire* (1794), pp. 22, 30.

restrictions of the former course of husbandry, and appropriating each of the various sorts of land to that use to which it is best adapted. 2. The prevention of the loss of time, both as to labourers and cattle, in travelling . . . from one end of a parish to another ; and also in fetching the horses from distant commons before they go to work. 3. There is a much better chance of escaping the distempers to which cattle of all kinds are liable from being mixed with those infected, particularly the scab in sheep. This circumstance, in common fields, must operate as a discouragement to the improvement of stock. . . . 5. The great benefit which arises from draining lands, which cannot so well, if at all, be done on single acres and half acres, and would effectually prevent the rot amongst sheep, so very common in open field land. 6. Lastly the preventing of constant quarrels, which happen as well from the trespasses of cattle, as by ploughing away from each others' land." Otmoor, near Islip, containing " about four thousand acres," is mentioned as the largest and most valuable tract of waste in the county. " This whole tract of land lies so extremely flat, that the water, in wet seasons, stands on it a long time together, and of course renders it very unwholesome to the cattle, as well as the neighbourhood. The sheep are thereby subject to the rot, and the larger cattle to a disease called the moor evil. The abuses here (as is the case of most commons where many parishes are concerned) are very great, there being no regular stint, but each neighbouring householder turns out upon the moor what number he pleases. There are flocks of geese likewise kept on this common, by which several people gain a livelihood."

In 1809, Arthur Young reported on Oxfordshire,[1] where he found that, in proportion to its extent, more land had been enclosed since 1770 in the county than in any other part of England. Otmoor and Wychwood Forest were still uninclosed wastes. Apart from the question of productiveness, he urged that the enclosure of the latter district was necessary on moral grounds. " The vicinity is filled with poachers, deer-stealers, thieves, and pilferers of every kind ; offences of almost every description abound so much, that the offenders are a terror to all quiet and well-disposed persons ; and Oxford gaol would be uninhabited, were it not for this fertile source of crimes." Nearly one hundred parishes still remained in open-fields. " It is," says Young, speaking of open-field practices,

[1] Young's *Oxfordshire* (1809), pp. 87, 236, 239, 102.

" a well-known fact that men have ploughed their land in the night for the express purpose of stealing a furrow from their neighbour ; and at all times it is a constant practice in some to plough from each other." "I have known," says one of his informants, " years wherein not a single sheep totally kept in the open field has escaped the rot." Yet on this same land, enclosed and drained, not one sheep died from the rot in nineteen years.

In 1770, the South and East of Warwickshire had mainly consisted of open-fields. Now (1794)[1] there still remained 50,000 acres. But in 1813[2] it is reported that a very small area continued in an unenclosed state.

Northamptonshire,[3] in 1794, contained 89 parishes still in open-fields. There was, therefore, " above one third of the whole (county) by no means in the best state of cultivation of which it is susceptible." The commons did not " yield pasturage," " at the highest computation," which was worth more than " 5s. an acre. Indeed, if the calculation was fairly made, the occupiers are not benefited to the extent of half that sum, as the stock which they send to depasture upon these commons is liable to so many diseases and accidents, as, one year with another, nearly counterbalances any advantages which can be derived from possessing this right. . . . By every information that could be procured, it appears that the stock is not kept with a view to any profit that can possibly arise from the sales, but merely as the means of cultivating and manuring the soil. Indeed, long experience has evinced, that no species of stock kept in these open fields can be carried to market on terms nearly so advantageous as the same articles raised by those farmers who occupy inclosed lands ; nor is it to be supposed, considering the manner in which the stock is treated, that the owners will pay much attention to the improvement of the different breeds." As to the arable land, " the several occupiers must conform to the ancient mode of cultivation of each division or field in which their lands are respectively situated ; from which it will appear that one obstinate tenant (and fortunate must that parish be accounted, where only one tenant of that description may be found) has it in his power to prevent the introduction of any improvement. . . . The tillage lands are divided into small lots of two or

---

[1] Wedge's *Warwickshire* (1794), p. 20.

[2] Murray's *Warwickshire* (1813), pp. 62, 144.

[3] Donaldson's *Northants* (1794), pp 24, 29, 58.

three old-fashioned, broad, crooked ridges (gathered very high towards the middle, or crown, being the only means of drainage that the manner in which the lands are occupied will admit of), and consequently the farmer possessing 100 acres must traverse the whole extent of the parish, however large, in order to cultivate this small portion."

In Leicestershire [1] (1800) very little open-field land was left "not more than 10,000 acres." In Nottinghamshire [2] (1798) enclosure was proceeding rapidly. "Good land, with extensive commons," is said to be most capable of improvement; "clay land with small commons," to have been the least capable. Midway between the two came "clay land with large commons." But "even the worst" may be increased in value by a fourth, after deducting all improvements.

In Middlesex [3] (1794) many thousands of acres of wastes lay unenclosed—"an absolute nuisance to the public." The commons of Enfield, Edmonton, and Tottenham were frequently flooded; but no effort was made to keep the ditches scoured. In 1798 there were still 17,000 aqres of "common meadows, all capable of improvement, not producing to the community in their present state more than 4s. an acre." To the Reporter's eyes the commons were "a real injury to the public," partly because they tempted the poor man to settle on their borders, build a cottage out of the material they afforded, and trust to his pigs and poultry for a living; partly because they became "the constant rendezvous of gypseys, strollers and other loose persons . . . the resort of footpads and highwaymen." The arable land of the county is estimated at 23,000 acres, of which, in 1798, 20,000 were in open-fields.

In Hampshire [4] (1813) the Reporter found the commons so overstocked as to produce little or no substantial benefit to those who enjoyed the grazing rights, and the surface "shamefully deteriorated" by the exercise of rights of turbary or paring turf for fuel. He hopes to see "every species of intercommonable rights extinguished," and, with them, "that nest and conservatory of sloth, idleness, and misery, which is uniformly to be witnessed in

---

[1] Pitt's *Leicestershire* (1800), p. 68.
[2] Lowe's *Nottinghamshire* (1798), pp. 19, 165.
[3] Foot's *Middlesex* (1794), pp. 30, 32, and Middleton's *Middlesex* (1798), pp. 98, 103, 138.
[4] Vancouver's *Hampshire* (1813), pp. 318, 496.

the vicinity of all commons, waste-lands, and forests throughout the kingdom."

4. In the *Eastern and North-Eastern* counties, neither Essex nor Hertfordshire possessed many commons or open-field farms. A description of the inhabitants of the neighbourhood of Epping and Hainault Forests in Essex (1795) has been already quoted.[1] In Hertfordshire [2] (1795) the Reporter notes that the few remaining open-fields had been freed from the old restraints, and were cultivated as if they were held in separate occupation. Speaking of pasture commons, he says : " Where wastes and commons are most extensive, there I have perceived that cottagers are the most wretched and worthless ; accustomed to relie on a precarious and a vagabond subsistence, from land in a state of Nature, when that fails they recur to pilfering. . . . For cottagers of this description the game is preserved and by them destroyed." Of Cheshunt Common [3] (1813) it is stated that " the common was not fed by the poor, but by a parcel of jobbers, who hired cottages, that they might eat up the whole."

Two-thirds of the county of Huntingdon [4] in 1793 lay in open-fields. Proprietors rarely had more than two or three acres contiguous. " The residue lies in acres and half acres quite disjointed, and tenants under the same land-owner cross each other continually in performing their necessary daily labour. . . . The sheep of the common fields and commons are of a very inferior sort, except in some few instances, and little if any care is taken either in the breeding, feeding or preserving them ; and from the neglected state of the land on which they are depastured, and the scanty provision for their support in winter, and the consequent diseases to which they are liable, their wool is also of a very inferior quality."

On the uplands of Lincolnshire [5] (1794) there were but few open-field farms. " The sheep of the common fields," says the Reporter, " I do not bring into this account from the circumstances of hardship, attending the scantiness of their food, the wetness of their layer, the neglect of a proper choice in their breed, their being overheated in being (where folded) dogged to their confinement,

---

[1] See p. 154.
[2] Walker's *Hertfordshire* (1795), pp. 48, 53.
[3] Young's *Hertfordshire* (1804), p. 45.
[4] Stone's *Huntingdonshire* (1793), pp. 8, 17, 18.
[5] Stone's *Lincolnshire* (1794), p. 62.

where they are often too much crowded ; the scab, the rot, and
every circumstance attend them, which can delay their being
profitable ; so that it may be reasonably concluded, that they are
of less value than those bred in inclosures, from 10s. to 15s. per
head, and their fleeces are equally unproductive." Five years
later Arthur Young reported on this part of the county.[1] He
describes the true Lincolnshire cattle which he found on open-
field farms as a " wretched " breed ; " they all run together on a
pasture, without the least thought of selection." At three years
old, they were worth little more than half what they fetched on
enclosed land. Open-field farmers " breed four or five calves
from a wretched cow before they sell it, so that a great quantity
of food is sadly misapplied." It was from this " post-legged,
square-buttocked breed of demi-elephants," to use Marshall's
description, that the Navy beef of England was chiefly provided.
The open-field sheep had not improved. " I never," says Young,
apparently with surprise, " saw a fold in the county, except in a
few open fields about Stamford ; . . . but the sheep are miserably
bad ; in wool 8 or 9 to the tod." In the East Riding [2] of Yorkshire
(1794) the pasture commons varied " in extent from two hundred
to two thousand five hundred acres, and all of them may be converted into useful land by drains. sub-divisions, plantations, and
other improvements. . . . When commons are not stinted in
proportion to the stock they are capable of keeping, very little
benefit is derived from them. . . . It is not a little extraordinary
to see a starving stock upon a common of five hundred acres soaked
with water, when the expense of a few shillings for each right,
prudently laid out in drains and bridges, would double its value.
Such is the obstinacy of men, and so difficult is it to induce them
to form the same opinion ; though an union of sentiment would
much more materially promote their interest."

Norfolk [3] in 1796 contained 80,000 acres of unimproved commons,
and about one-fourth of the arable area of the county was tilled
on the common or open-field system. " There is," says the Reporter,
who was the well-known Nathaniel Kent, " still a considerable deal
of common-field land in Norfolk, though a much less proportion
than in many other counties ; for notwithstanding common rights

---

[1] Young's *Lincolnshire* (1799), pp. 303, 374.
[2] Leatham's *East Riding* (1794), p. 39.
[3] Kent's *Norfolk* (1796), pp. 6, 32, 72, 73, 81, 158.

for great cattle exist in all of them, and even sheep-walk privileges in many, yet the natural industry of the people is such, that, whenever a person can get four or five acres together, he plants a whitethorn hedge round it, and sets an oak at every rod distance, which is consented to by a kind of general courtesy from one neighbour to another." "Land," he elsewhere remarks, "when very much divided, occasions considerable loss of time to the occupier, in going over a great deal of useless space, in keeping a communication with the different pieces. As it lies generally in long narrow slips, it is but seldom it can receive any benefit from cross-ploughing and harrowing, therefore it cannot be kept so clean ; but what is still worse, there can be but little variety observed in the system of cropping ; because the right which every parishioner has of commonage over the field, a great part of the year, prevents the sowing of turnips, clover, or other grass seeds, and consequently cramps a farmer in the stock which he would otherwise keep." Commons of pasture lay "in all parts of the county, and are very different in their quality. Those in the neighbourhood of Wymondham and Attleborough are equal to the finest land in the county, worth, at least, twenty shillings an acre ; being capable of making either good pasture, or producing corn, hemp or flax. There are other parts which partake of a wet nature and some of a furze and heathy quality ; but they are most of them worth improving, and all of them capable of producing something ; and it is a lamentable thing, that those large tracts of land should be suffered to remain in their present unprofitable state." Under the head of Poor Rates, the Reporter observes "that the larger the common, the greater the number and the more miserable are the poor." In the parishes of Horsford, Hevingham, and Marsham, which "link into each other, from four to nine miles from Norwich, there are not less than 3,000 acres of waste land, and yet the average of the rates are, at least, ten shillings in the pound. This shows the absolute necessity of doing something with these lands, or these, uncultivated, will utterly ruin the cultivated parts,—for these mistaken people place a fallacious dependence upon these precarious commons, and do not trust to the returns of regular labour, which would be, by far, a better support to them." Of Wymondham Common, Arthur Young [1] wrote in 1801. The area was 2,000 acres ; but "the benefit to the poor is little or nothing further

[1] *Inquiry into the propriety of applying Wastes, etc.,* 1801.

than the keeping a few geese ; as to cows there are very few. The common is so overstocked with sheep that cows would be starved on it ; and these sheep are mostly in the hands of jobbers, who hire small spots contiguous [to the common] for no other purpose. These men monopolise almost the whole."

Bedfordshire in 1794 [1] was famous for its backward farming. It still disputed with Cambridgeshire the reputation of being the Boeotia of agriculture. It contained 217,000 acres of open or common fields, common meadows, common pastures, and waste lands, to 68,000 acres of enclosure and 22,000 acres of woodlands. As a rule, the enclosed land was as badly farmed as the open-fields. Hence the practice of enclosing had fallen into disrepute. The Reporter seems to suggest another reason for the reluctance of landlords to enclose. " It has," he says, " frequently occurred to me in practice, that some of the occupiers of a common field are pursuing the best possible mode of management the situations are capable of, whilst others are reducing land intermixed therewith to the lowest state of poverty, beggary and rubbish. . . . Upon the inclosure of common fields it frequently occurs that commissioners are obliged to consider such worn-out land of considerably less value than such parts as have been well-farmed ; of course, the proprietors, whose misfortune it has been to have their land badly occupied, have had a smaller share, upon the general division of the property, than they otherwise would have had, in case their land had been better farmed." In one respect enclosed land had the advantage. Sheep in Bedfordshire were practically only used as manure-carriers. They were " generally of a very unprofitable quality, but more especially those bred in the common fields, where the provision intended for their maintenance is generally unwholesome and scanty. . . . From the undrained state of the commons and common fields, the stock of sheep depastured upon them is but too frequently swept away by the rot ; and, it being absolutely necessary, according to the present system of farming, that their places should be constantly supplied with others for the folding of the land, under such circumstances of casualty and necessity, the healthiness of the animal when purchased is the first and almost the only object of consideration with the farmers." Sheep, from any county, of any breed, and of any description, were therefore bought indiscriminately. Nine-

[1] Stone's *Bedfordshire* (1794) pp. 11, 61, 31.

tenths of the sheep of the common fields of the country are " coarse in their heads and necks, proportionately large in their bones, high on the leg, narrow in their bosoms, shoulders, chines and quarters, and light in their thighs, and their wool is generally of a very indifferent quality, weighing from three to four pounds per fleece. . . . The sheep bred upon the inclosures are generally of a much superior quality . . . very useful and profitable." Thirteen years later (1807),[1] 43 parishes, or about a third of the county, were farmed on the open-field system. To the rapid spread of enclosures and to the influence and example of great landlords, the Reporter attributed the material improvement in the sheep stock of the county.

Out of 147,000 acres of arable land in Cambridgeshire [2] (1794) 132,000 lay in open-fields. The rental of the enclosed land averaged 18s. per acre, and that of the open-fields 10s. On the uplands of the county, as distinguished from the fen districts, there were 2,000 acres of half-yearly meadow lands which were grazed by the village partners from hay-harvest till Easter ; 7,500 acres of highland common ; 8,000 acres of fen or moor common, which, though easily drained, " contribute little to the support of the stock, though greatly to the disease of the rot in the sheep and cows." The Reporter considered that no general improvement of the farming of the county was possible until the intermixed lands of " the common open fields " were laid together and occupied in severalty. He made it part of his business to enquire into the feeling of " the yeomanry in their sedate and sober moments . . . as to this important innovation upon the establishment of ages. A few have given an unqualified dissent, but they were flock-masters ; others have concurred under certain limitations, but the mass of the farmers are decidedly for the measure in question." He estimates that the general average produce per acre of enclosed land exceeded that of the open-fields in the following proportions : wheat, 3 bushels 1 peck ; rye, 3 pecks ; barley, 15 bushels 1 peck ; oats, 1 bushel 1 peck ; peas, 2 bushels 1 peck. " But, if a single instance be adverted to, and a comparison made between the parishes of Childersley, which is enclosed, and Hardwicke, which remains in open common field, and which parishes appear by the journal to consist of a perfectly similar soil," the result is much more favourable to enclosures. Childersley produced 24 bushels of wheat to

---

[1] Batchelor's *Bedfordshire* (1808), pp. 217, 537
[2] Vancouver's *Cambridgeshire* ' 1794), pp. 193, 203, 195, 112, 111.

Hardwicke's 16 bushels ; 36 bushels of barley to 18 bushels ; 36 bushels of oats to 18 bushels, or 20 bushels of oats to 8 bushels. To this increase of produce must be added another advantage. Childersley and Knapwell, both enclosed, were entirely exempt from the rot among their sheep, while the neighbouring parishes were desolated by the disease. The ravages of the rot which are chronicled may probably have been exceptional. On the open-fields of Gamlingay a fourth of the flock, or 340 sheep, perished in 1793. The mortality is attributed to the want of drainage in the arable land. At Croxton in 1793 1,000 sheep were rotted on the unenclosed lands, and, in the same year, 700 on the open-fields of Eltsley. In 1813 another Report on Cambridgeshire [1] was issued. In the interval of twelve years, the area of open-field and common had been greatly lessened. In consequence, says the Reporter, Cambridgeshire farmers " have an opportunity of redeeming the county from the imputation it has so long lain under, of being the worst cultivated in England, and of proving (the fact) that the same industry, spirit and skill which have been manifested in other parts of the Kingdom, exist also in this, the open-field state and system precluding the possibility of exercising them."

To the Eastern and North Midland districts mainly belonged the fen-lands. This vast tract of waterlogged land still included Peterborough Fen in Northamptonshire, embraced small portions of both Norfolk and Suffolk, and extended over a considerable part of Huntingdonshire, Cambridgeshire, and Lincolnshire. At a moderate computation, the total area, which at the best was imperfectly drained, and lay to a great extent unenclosed, comprised 600,000 acres. The drainage works of the seventeenth century had only partially succeeded. Where the system had been carefully watched and maintained, the land had been greatly improved. But the neglected outfalls were once more choked with silt ; the porous banks admitted the water almost as fast as it was removed by the draining-mills ; in some instances they had been broken down by floods and not repaired ; in some they had been wilfully damaged or destroyed by the commoners. Yet much of this drowned area, either actually or potentially, consisted of some of the richest land in Great Britain. Some portions of the drier ground were cultivated on the open-field system, and the commons were numerous and extensive.

[1] Gooch's *Cambridgeshire* (1813), pp. 2, 56.

Peterborough Fen [1] (1793) consisted of from 6,000 to 7,000 acres of " fine level land, of a soil equal to any perhaps in the kingdom of Great Britain, and susceptible of the highest cultivation." In its present wet state it was dangerous to stock. Farmers living in the neighbourhood never turned their cattle on it except in very dry seasons. It was, however, depastured by the horses, cattle, and sheep of 32 parishes in the Soke of Peterborough. " Considering the present mode of management," says the Reporter, " it is impossible that any advantage can arise to the persons having right therein." But, in his opinion, the land, if properly drained, enclosed, and tilled, might yield a greatly increased produce and employ from 1300 to 1400 hands.

The Huntingdonshire fens [2] contained (1793) 44,000 acres. Marshall speaks of " the disgraceful state in which some of these lands were suffered to remain (a blank in English territory)." The Reporter says that the fen is " generally unproductive, being constantly either covered with water, or at least in too wet a state for cultivation." Of so little value was it that those who exercised rights over it frequently preferred relinquishing their claims to paying the drainage taxes. Very considerable portions of the fen districts were occupied by meres—shallow lakes filled with water which was often brackish. Their only value lay in the reeds, which were used for thatching or in malting, and in the fishing. But many of the meres were so silted up with mud that the fish had diminished in numbers. Their drainage. says the Reporter in 1811, [3] would be of inestimable service to the health of the inhabitants. " They are awful reservoirs of stagnated water, which poisons the air for many miles round about, and sickens and frequently destroys many of the inhabitants, especially such as are not natives."

In Cambridgeshire [4] (1794) there were " 50,000 acres of improved fen, and 200,000 acres of wastes and unimproved fen." Vancouver, who was the Reporter to the Board, walked over every parish in the district in order to obtain reliable information. Except on foot, he could not penetrate into the recesses of the district. Neighbouring parishes were ignorant of each other's condition. The roads were often impassable, and at their best were only repaired

[1] Donaldson's *Northamptonshire* (1793), p. 30.
[2] Stone's *Huntingdonshire* (1793), pp. 8, 13.
[3] Parkinson's *Huntingdonshire* (1813), p. 21.
[4] Vancouver's *Cambridgeshire* (1794), pp. 25, 36, 151, 154, 184, 186, 187, 149.

with a silt which resembled "pulverised sand." Almost everywhere he speaks of the "deplorable condition of the drainage," and consequently of the "miserable state of cultivation" which prevailed on the open-field lands. The fen-lands of Chatteris, Elm, Leverington, Parson Drove, Wisbech St. Mary's and Thorney, amounting to about 50,000 acres yield "a produce far beyond the richest high lands in the county, averaging a rent of more than fifteen shillings per acre. Whereas the waste, the drowned, and partially improved fens, amounting on a moderate computation to 150,000 acres, cannot be fairly averaged at more than four shillings per acre." Very rarely were the open-fields and commons even in a fair state of cultivation. Wilburton was a favourable example. There field-reeves had been appointed by the parish, with power to open up neglected drains at the expense of those to whom they belonged. But almost universally the common pasture was deteriorated by turf-cutting ; the marsh lands, if tilled, were exhausted by barbarous cropping ; and effective drainage was prevented by the intermixed condition in which the land was occupied. At Snailwell, an open upland parish, there was a flock of 1,200 Norfolk sheep, which were only "kept healthy by being prevented from feeding upon the wet moory fen common." The general attitude of the ague-stricken, opium-eating fen-men towards the drainage of the district may be illustrated by the example of Burwell, a chalkland parish on the Suffolk border. "Any attempt in contemplation of the better drainage" of Burwell fen, already "greatly injured by the digging of turf," and "constantly inundated," is "considered as hostile to the true interests of these deluded people."

In 1794 the principal Lincolnshire [1] commons were the East and West (29,000 acres), the Wildmore Fen (10,500 acres), the East and West Deeping Fens (15,000 acres). The East and West and Wildmore Fens were "under better regulations than any others in the fen country." "Yet," says the Reporter, "they are extremely wet and unprofitable in their present state, standing much in need of drainage, are generally overstocked, and dug up for turf and fuel. The cattle and sheep depastured upon them are often very unhealthy, and of an inferior sort, occasioned by the scantiness, as well as the bad quality of their food, and the wetness of their lair. Geese, with which these commons are generally stocked . . . are often subject to be destroyed. It is not a constant prac-

[1] Stone's *Lincolnshire* (1794), pp. 18, 22.

tice with the commoners to take all their cattle off the fens upon the approach of winter ; but some of the worst of the neat cattle, with the horses,—and particularly those upon Wildmore Fen,—are left to abide the event of the winter season ; and it seldom happens that of the neat cattle many escape the effects of a severe winter. The horses are driven to such distress for food that they eat up every remaining dead thistle, and are said to devour the hair off the manes and tails of each other and also the dung of geese." A second Reporter [1] (1799), Arthur Young, speaks of " whole acres " in Wildmore Fen as " covered with thistles and nettles four feet high and more. There are men that have vast numbers of geese, even to 1000 and more. . . . In 1793 it was estimated that 40,000 sheep, or one per acre, rotted on the three fens (*i.e.* on East and West and Wildmore Fens). So wild a country nurses up a race of people as wild as the fen ; and thus the morals and eternal welfare of numbers are hazarded and ruined for want of an inclosure. . . . In discourse at Louth upon the characters of the poor, observations were made upon the consequences of great commons in nursing up a mischievous race of people ; and instanced that, on the very day we were talking, a gang of villains were brought to Louth gaol from Coningsby, who had committed numberless outrages upon cattle and corn ; laming, killing, cutting off tails, and wounding a variety of cattle, hogs, and sheep ; and that many of them were commoners on the immense fens of East, West, and Wildmore."

These descriptions apply to commons under the best regulations. Deeping Fens may be taken as examples of the ordinary management of Lincolnshire commons in the fen districts. " They stand," thinks the Reporter of 1794,[2] " very much in need of inclosing and draining, as the cattle and sheep depastured thereon are very unhealthy. The occupiers frequently, in one season, lose four fifths of their stock. These commons are without stint, and almost every cottage within the manors has a common right belonging to it. Every kind of depredation is made upon this land in cutting up the best of the turf for fuel ; and the farmers in the neighbourhood, having common rights, availing themselves of a fine season, turn on 7 or 800 sheep each, to ease their inclosed land, whilst the mere cottager cannot get a bite for a cow ; but yet the cottager, in his turn, in a colourable way, takes the stock of a foreigner as his own, who occasionally turns on immense quantities of stock in good

[1] Young's *Lincolnshire* (1799), p. 223.   [2] Stone's *Lincolnshire*, p. 22.

seasons. The cattle and sheep, which are constantly depastured on this common, are of a very unthrifty ill-shapen kind, from being frequently starved, and no attention paid to their breed. Geese are the only animals which are at any time thrifty ; and these frequently, when young, die of the cramp, or, when plucked, in consequence of the excessive bleakness and wetness of the commons. A goose pays annually from 1s. to 16d. by being 4 times plucked. These commons are the frequent resort of thieves, who convey the cattle into distant Counties for sale."

The North Fens round the Isle of Axholm formed in 1794 another large area (12,000 acres) of commons and wastes. If " divided and inclosed," says the Reporter,[1] they " would for the most part make very valuable land . . . in their present state, they are chiefly covered with water, and in summer throw forth the coarsest of productions ; the best parts, which are those nearest the enclosed high lands, are constantly pared and burnt to produce vegetable ashes. . . . The more remote parts of the common are dug up for fuel. On account of the general wetness of those commons, and their being constantly overstocked by the large occupiers of contiguous estates, or in such seasons as the depasturage is desirable in summer, to ease the inclosed land, the cattle and sheep necessarily depastured thereon at all seasons being those of the cottagers, who are for the most part destitute of provision for them in winter, are always unthrifty, and subject to various diseases, which render them very unprofitable to the occupiers." The farming of the open arable fields had, in the Reporter's opinion, deteriorated rather than progressed. " If," he says,[2] " those gentlemen, whether proprietors or agents, who have any concern in the management of common fields, will examine into the present mode of occupancy of the different classes of them . . . they will in most cases find them in a weak impoverished state ; and that the original systematic farming of them is either lost or laid aside, and that the agriculture of the common fields of this county has rather declined than improved." The Cambridgeshire Reporter,[3] it may be added, formed the same opinion of the open-fields in that county, and he produces some evidence to prove that the rental of open farms had fallen since the seventeenth century.

The general impression left by this mass of evidence is that the agricultural defects of the intermixture of land under the

[1] Stone, p. 29.   [2] Stone, p. 56.   [3] Vancouver, p. 97.

open-field system were overwhelming and ineradicable ; that as an instrument of land cultivation it had probably deteriorated since the thirteenth century ; that no increased production or general adoption of improved practices could be expected under the ancient system. But the Reporters note exceptions, from which other conclusions may possibly be drawn. In some districts the customary rotations had been abandoned for independent cultivation, or modified so as to admit some variation of cropping. Thus, by agreement, in Berkshire a portion of the fields was " hitched," or, according to the Wiltshire equivalent, " hooked." In other words, common rights of pasture on the arable land were suspended so as to allow the cultivation of turnips, clover, or potatoes. Elsewhere, again, portions of the arable land were withdrawn from tillage to serve as cow-commons. Nor must it be supposed that enclosed land was always better cultivated than open-field farms. The Bedfordshire and Lincolnshire Reporters, for example, state that in certain cases enclosure had produced no improvement, and in Wiltshire the Reporter hints that open-field regulations at least prevented some abuses to which land held in severalty was liable. In some districts landlords imposed upon tenants of separate holdings the same restrictions and course of cropping by which they had been fettered as occupiers of land in open-fields. Without a large expenditure on equipment the agricultural conditions of enclosed land were often worsened, rather than bettered. Thus the Somersetshire Reporter quotes an example from the Mendip Hills, where, when land had been enclosed, the landlord refused to erect the necessary buildings. Similar cases might have been collected from many other parts of the country. In these respects, as well as in others, landlords had yet to be taught the business of owning and letting land. There were " Goths and Vandals," not only among tenants, but also among owners.

Before any accurate estimate can be formed of the agricultural advantages or defects of arable farming on intermixed strips of land subject to common grazing rights, and of stock breeding and rearing on pasture commons, it is necessary to allow for some possibilities of improvement by the cultivators of open-fields and for some neglected opportunities by landlords and tenants of enclosed land. But, when every reasonable allowance has been made, it is clear that the balance was overwhelmingly in favour of separate occupation. As an instrument of production the ancient system was inferior. Every advance in science made by agriculture, and every new resource which is adopted, only served to accentuate the relative disadvantages of open-field farming. Change was, in the circumstances, necessary. It was generally effected by obtaining Parliamentary sanction for an enclosure.

The ordinary procedure,[1] by which open-fields or commons were enclosed under Parliamentary authority, opened with a Petition presented to Parliament by persons locally interested. The Petition was signed by the owner of the land or lord of the manor, by the owner of the tithes, and by a majority of the persons interested. No fixed rule seems to have been followed, as to the proportion of consents and dissents. But Parliamentary Committees looked to the values as well as to the numbers which were represented. On this Petition, by leave of the House, a Bill was introduced, read a first and second time, and then referred to a Committee, which might consist of the whole House or of selected members. The Committee, after receiving counter-petitions and hearing evidence, reported to the House, that the standing orders had, or had not, been complied with ; that the allegations were, or were not, true ; that they were, or were not, satisfied that the parties concerned had consented to the Bill. On the Committee's Report, the Bill either was rejected, or was read a third time, passed, sent to the Lords, and received the Royal Assent. If the Bill passed, the Commissioners, or Commissioner, named in the Act, arrived at the village. There they heard the claims of the persons interested, and made their award, distributing the property in separate ownership among those who had succeeded in establishing their claims, with due regard to the " quality, quantity, and contiguity " of the land.

The procedure was open to abuses. Even if it is assumed that a Parliamentary Committee, largely composed of landed proprietors, was always disinterested on questions affecting land, little trouble seems to have been taken to elicit the opinions of small claimants. Schemes of enclosure rarely began with a public meeting of the parish. The principal owners generally met in secret, arranged the points in which their own interests conflicted, selected the solicitor and surveyor, nominated the Commissioners,

---

[1] *An Essay upon the nature and method of ascertaining the specific shares of proprietors upon the inclosure of common fields*, by the Rev. Henry Sacheverell Homer (1766).

settled the terms of the petition. Even the next step—that of obtaining signatures—might be taken privately. Sometimes it happened that the first intimation which the bulk of the inhabitants received of the scheme was that the petition had been presented, and that leave to bring in an enclosure Bill had been granted. To prevent so flagrant an abuse, clauses as to notice had been generally inserted in Bills from 1727 onwards. But, in order to secure the necessary publicity of proceedings, the House of Commons in 1774 made it a standing order that notice of the scheme must be affixed to the door of the church of the parish affected, for three Sundays in the months of August or September. Other standing orders corrected other abuses in the procedure. They regulated the payments of the Commissioners, required them to account for all monies assessed or expended by them, restricted the choice of men who could fill the office, limited their powers of dealing with the titles of claimants, and laid down the principle that the allotments to titheowners and lords of manors should be stated in the Bill.

At all stages of the proceedings heavy costs were incurred. The fees paid to Parliamentary officials were considerable. If a tract of common land was to be enclosed, over which several parishes claimed rights, fees were charged for each parish. On this ground, partly, the Lincolnshire Reporter explains the delay in enclosing the East and West and Wildmore Fens. Forty-seven parishes were there affected, and the general Act would be charged as forty-seven Acts, with fees in proportion. Witnesses had to attend the Committee of the House of Commons and subsequently of the House of Lords. There might be postponements, delays, and protracted intervals ; but the witnesses, often professional men, had either to be maintained in London or to make two or more costly journeys to town. Such an expenditure was generally prohibitive for the opponents of the Bill. Unable to fee lawyers, produce witnesses, or urge their claims in person, they were obliged to content themselves with a counter-petition, which, possibly, might not be referred to the Committee. Nor did the cost cease when the Bill was passed. There were still the expenses of the Commissioners and their clerk ; the fees for the surveyor and his survey, and the valuer and his valuation ; the charges of the lawyers in proving or contesting claims, preparing the award, and other miscellaneous business ; the outlay on roads, gates, bridges drainage, and other expenditure necessitated by the enclosure of the land. Where the area was large, a portion of the land was usually sold to pay the necessary expenses. But the cost of fencing the portions allotted to individuals was thrown upon the owners, and the smaller the allotment, the greater the relative burden. Small men might well hesitate, apart from the uncertainty of proving their title, to support an enclosure scheme, since the value of their allotment might be almost swallowed up in the expense of surrounding it with a hedge.

Many small tracts of common land were left unenclosed, because the extravagant cost threatened to absorb the possible profits of the undertaking. A general Enclosure Act would, it was urged, reduce the cost of enclosing small areas, promote uniformity of legislative action by embodying the best methods of procedure and the most requisite safeguards which experience suggested, and provide means for overcoming opposition by modifying the existing powers of resistance. On all these grounds, a Bill was framed by the Board of Agriculture. It was strongly opposed in Parliament.[1] Many persons were interested in the continuance of the existing procedure. " What," asks one of the Board's Reporters, " would become of the *poor* but *honest* attorney, officers of Parliament, and a long train of etc, etc, who obtain a *decent* livelihood from the *trifling* fees of every individual inclosure Bill—all these of infinite use to the community, and must be encouraged whether the wastes be enclosed or not ? . . . The waste lands, in the dribbling difficult way they are at present inclosed, will cost the country upwards of 20 millions to these gentry etc. which on a *general* Inclosure Bill would be done for less than *one*." [2] The first Bill proposed by the Board was rejected mainly through the influence of these private interests. A further attempt was made in May, 1797, when two Bills were introduced. The first was wrecked by the opposition of titheowners. One of the chief advantages of enclosures was that tithes were usually extinguished by an allotment of land in lieu. This commutation of tithe was favoured by the Board, which in consequence incurred the suspicion of being hostile to the Established Church. The House of Lords seems to have been particularly influenced by this view. Though the first of the two Bills passed the Commons, it was rejected in the Upper House. The second Bill did not advance beyond the Committee stage in

[1] Arthur Young's *Lecture before the Board of Agriculture*, May, 1809.
[2] Brown's *West Riding*, App. I., p. 14.

the House of Commons. Finally, in 1801, the first General Enclosure Act (41 Geo. III. c. 109) was passed " for consolidating in one Act certain provisions usually inserted in Acts of Inclosure, and for facilitating the mode of proving the several facts usually required on the passing of such Acts." No alteration in the machinery of enclosure was made. Private Acts of Parliament were still required. But they were simplified, and to some extent the expense was reduced. The effect was at once seen in an increase in the number of private Acts and a diminution in the size of the areas which each enclosed.

The Act of 1801 was mainly applied to commons. Open-fields were specifically dealt with by subsequent legislation. In 1836, an Act (6 and 7 Wm. IV. c. 115) was passed " for facilitating the inclosure of open and arable fields." It empowered two-thirds of the possessors of open-field rights, in number and value, to nominate commissioners and carry out enclosure ; or seven-eighths, in number and value, to enclose without the intervention of commissioners. The debate in Parliament is chiefly noticeable for the stress which, for the first time since the days of Elizabeth, was laid on the desirability of preserving commons as breathing-places and play-grounds. In the Bill itself the point was not really raised. But, as the nineteenth century advanced, this aspect of the question of enclosing commons and wastes became increasingly important. It was prominent in the General Inclosure Act of 1845 (8 and 9 Vic. c. 118). The principal change made in this Act was the substitution of Inclosure Commissioners for the Parliamentary Committee as a local tribunal of enquiry, before which the necessary examination could be conducted on the spot. But Parliamentary control was not abandoned. All the schemes framed by the Commissioners in each given year were embodied in a general Act, and submitted to Parliament for sanction. The administration of the Inclosure Acts is now entrusted to the Board of Agriculture. As a State department, the Board can deal with open-fields and commons on broader lines than the strict interpretation of the statute, which constituted their authority, allowed to the Inclosure Commissioners.

CHAPTER XII

THE ENGLISH CORN LAWS.[1]

Difficulty in deciding on the good or bad influence of the Corn Laws ; restric
tions on home as well as on foreign trade in corn ; gradual abandonment of
the attempt to secure just prices by legislation ; means adopted to steady
prices ; prohibition both of exports and of imports : the bounty on home-
grown corn ; the system established in 1670 and 1689 lasts till 1815 ; its
general effect ; influence of seasons from 1689 to 1764, and from 1765 to
1815 ; difficulty of obtaining foreign supplies during the Napoleonic wars ;
practical monopoly in the home market : small margin of home supply
owing to growth of population ; exaggerated effect on prices of good or
bad harvests ; protection after 1815 ; demand by agriculturists for fair
profits ; changed conditions of supply ; repeal of the Corn Laws, 1846.

MEN are apt to pass a hasty judgment on the Corn Laws in accordance with their political prejudices.   One party condemns them as
mischievous ;   another party approves them as salutary.   Neither
troubles to consider their practical effect.   Yet, from 1689 to 1815,
it is probable that the marked deficiency or abundance of the
harvest in any single year produced a greater effect on prices than
was produced by the Corn Laws in the 125 years of their existence as
a complete system.

It is almost impossible to decide whether the total effect of the
Corn Laws has been to promote or to retard agricultural progress.
Probably the balance of their influence in either direction would
be found to be inconsiderable.   The utmost nicety of calculation
would be required in order to measure with any degree of accuracy
the extent to which, before 1815, they affected prices of corn.
Before the balance can be correctly struck, the advance in price,
which was due to the increased demand consequent on the growth
of population and to the gradual depreciation of gold and silver,
must be discounted ;   the fall in price, which resulted from economy
in the cost and increase in the yield of production, must be

[1] See Appendix III.

eliminated ; an explanation must be offered of the facts that in England, during the seventeenth century, wheat averaged only a halfpenny the bushel cheaper than during the eighteenth century,[1] and that the general prices of Europe, under different fiscal systems, did not, during the period, materially differ from those of England. Still more difficult would it be to determine whether, taken as a whole and over the entire period of their existence, they have benefited or injured consumers, so far as these can be distinguished from producers. If they aggravated evils in some directions, they compensated them in others. Whatever else the legislation effected, it did, except during the last few years of its operation, steady prices, and to consumers steadiness was perhaps as great a boon as a spasmodic cheapness which alternated with excessive dearness. At a time when England was practically dependent on home-grown supplies, prices of corn were extravagantly sensitive to fluctuations in the yield of harvests. The reason is obvious. Average harvests provided bread enough for the population ; but there was often little margin to spare. A partial failure, therefore, meant the prospect of dearth, if not of famine. In prolonged periods of scarcity, like that of the Napoleonic wars, our ancestors might pass self-denying ordinances to reduce their domestic consumption by one-third, dispense with flour for their own wigs or the hair of their lackeys, substitute clay imitations for the pastry of their pies, forbid the sale of bread till it was twenty-four hours old, prohibit the use of corn in the making of starch or in distilleries. Yet, in the case of a necessary like corn, it was impossible to exercise such economies as would make good any considerable shortage. Hence corn, when a deficient harvest was anticipated, was specially liable to panic-stricken competition. Any falling off in the annual yield caused a far greater advance in price than was justified by the actual shortage. Somewhat similar, though less exaggerated, was the effect of an anticipated abundance. The fall in price was wholly disproportionate to the real surplus. These violent alternations between dearness and cheapness, if they had not been steadied and regulated by the legislature, would have been disastrous to both consumers and producers.

Beginning in the early Middle Ages, and ending in 1869, the English Corn Laws lasted for upwards of six centuries. Attention

[1] Seventeenth century, 38s. 2d. the quarter ; eighteenth century, 38s. 7d. the quarter (Arthur Young's *Progressive Value of Money*, p. 76).

has been so exclusively concentrated on one side only of their pro-visions, that the regulation of the inland trade in corn and the restrictions on its exportation have been long forgotten. Yet, except during the period 1815-46, the duties on foreign grain, which are now regarded as the principal feature of the old Corn Laws, were of minor importance. The successive Governments which framed and revised the legislation on corn were not more enlightened than their contemporaries, for whose direction the regulations were passed ; the ultimate effect of their measures was sometimes miscalculated ; their policy varied from time to time ; different objects were prominent at different periods. But it is impossible to pass any summary sentence of condemnation on the Corn Laws as a system selfishly designed to enrich, at the expense of consumers, a ruling class of landowning aristocrats. On the contrary, if the legislation is treated as a whole, and the restrictions on both exports and imports are examined together, it will be found that, up to 1815, the interests alike of consumers, producers, and the nation were collectively and continuously considered. The general aim of legislators was to maintain an abundant supply of food at fair and steady prices ; to assist the agricultural industry in which, up to the middle of the eighteenth century, the great mass of the people were engaged as producers ; to prevent the depopulation of rural districts, build up the commercial and maritime power of the nation, make it independent of foreign food supplies, and foster the growth of the infant colonies.

Mediaeval Corn Laws were based on principles of morality, if not of religion. They were akin to the laws against usury. It was considered immoral to prey on human needs, or to take advantage of scarcity by exacting more than a moderate profit on the production of necessaries of life. The object of legislation was, therefore, to establish " just " prices, and in the interest of con-sumers to restrict the liberty of sellers. The idea that British corn might be cheapened by bringing the granaries of Europe into competition with home supplies had either not suggested itself, or been rejected as impracticable. In order to establish just prices, the methods of early legislators were various. They endeavoured to attain their end, and, incidentally, to secure better profits to producers, by keeping home-grown corn in the country, by regulating the inland trade, by penalising the intervention of 'middlemen between farmers and their customers, by protecting buyers against the craft of bakers, by preventing monopolies and speculations in grain which, in days when difficulties of transport restricted competition to narrow areas fed by local supplies, were a real danger. To this class of laws belong prohibitions against selling corn out of the country, or transporting it from one district to another; statutes [1] against corn-dealers who " forestalled," " engrossed," or " regrated " grain ; and the Assizes of Bread,[2] which, down to the reign of George II., regulated the actual size of the loaf by the price of corn, instead of proportioning its cost to that of its material. Eventually this class of legislation defeated its own object. It hampered the natural trade in corn, locked up the capital of farmers, and so tended to reduce the area under the plough. But the national dread of corn speculation, of which many laws were the expression, was only paralleled by the national horror of witchcraft, and lasted longer among educated classes. As facilities for internal transport increased, opportunities for local monopolies diminished. Successive steps were taken towards freedom of inland trade. Thus in 1571 corn was permitted to be transported from one district to another on payment of a licence duty of 1s. a quarter ; in 1663 liberty to buy corn in order to sell it again was conceded, when it was below a certain limit, provided that it was not resold for three months in the same market ; in 1772 the statutory penalties against corn-dealers were repealed as tending to " discourage the growth and enhance the price " of corn ; in 1822 the practice of setting out Assizes of Bread was by Act of Parliament discontinued in London ; in 1836 an Act, similar in terms to that of London, abolished Assizes in provincial towns and country districts. Instead of attempting to secure just prices by multiplying laws in restraint of speculation, or by regulating the cost of corn and bread, the modern tendency has been to enforce honest dealing by increasing the protection of consumers against false weights and adulteration.

Other means were adopted to maintain steady prices in the interest of consumers and, indirectly, of producers. Thus the erection of public granaries, in which farmers might store the surplus of one year against the shortage of the next, was borrowed from Holland, and urged on the country by royal proclamation.

---

[1] *E.g.* 5 and 6 Edward VI. c. 14 (1552); 15 Car. II. c. 7 (1663); 12 Geo. III. c. 71 (1772).

[2] See Appendix III. C.

In 1620 the King's Council " wrote letters into every shire and some say to every market-town, to provide a granary or storehouse, with stock to buy corn, and keep it for a dear year." A similar object inspired the subsequent institution of bonded warehouses under the King's lock (1663), in which foreign grain might be stored, free of duty, until withdrawn for consumption. Restrictions on the exportation of home-grown corn were governed by the same desire to prevent excesses in surplus or deficiency, and to save the country from violent oscillations between cheapness and dearness. The much debated bounty on exports of grain was designed to produce the same result. Even the regulation of imports of foreign corn was partly governed by the same desire to secure a steady level of price.

At an early date prohibitions against exporting corn were influenced by political motives of retaliation on the king's enemies, just as the corresponding permission was affected by considerations of the needs of the public treasury. Revenue, though never the first aim of the Corn Laws, was, in mediaeval times and again under the Stewarts, a secondary object. To this extent the special interests, not only of consumers and producers, but also of the nation, were thus early brought into play. Originally, corn was only exported by those who had obtained, and in most cases bought, the king's licence. But the exercise of the royal prerogative in the grant of licences provoked ·a constitutional struggle, which for three centuries was fought with varying fortunes. The principle at stake was the control of Parliament over all taxation. In 1393 freedom of export was allowed by statute ; but the statutory liberty might be overridden by the king in Council. Seventy years later (1463) the royal power to prohibit or permit exportation was taken away, and, instead of the sovereign's discretion, a scale of prices was fixed below which trade in corn was allowed. More despotic than their immediate predecessors, the Tudor sovereigns reasserted the royal right to grant licences.[1] Special circumstances may have justified the claim and its exercise. Agriculturally, the general aim of the Tudors was to encourage tillage in order to counteract the depopulating tendencies of sheep-farming. Commercially, they desired to build up a foreign trade as the chief support of sea power, and English corn was one of the commodities which they hoped

[1] *E.g.* 25 Henry VIII. c. 2 (1533); 1 and 2 P. and M. c. 5 (1554). See Appendix III., B.

to exchange for foreign produce.[1] On both grounds they fostered
a trade in exported grain. But uncontrolled liberty of sending
corn out of the country might have raised home prices by depleting
home supplies. It may therefore have seemed essential to Tudor
statesmen that the royal power of prohibiting exports should be
revived in the interest of consumers. The emptiness of the royal
treasury drove the Stewarts to seek in this control of the corn-
trade an independent source of income, and it was one of the
complaints against Charles I. that he had exercised the royal prerogative in order to swell his revenue. Ultimately the constitutional principle triumphed. From the Restoration down to 1815
freedom to export home-grown corn was controlled and regulated
by the legislature in accordance with scales of prices current in the
home market. At the same time the power of the king in Council
to suspend the laws regulating both exports and imports of grain
was retained in use and sanctioned by Parliament.

At the Restoration the fiscal policy of the country towards
corn assumed a more definite shape. Statutes were passed in 1660,
1663, and 1670, which regulated both exports and imports of corn.
The two sets of regulations cannot henceforward be considered
separately. The one was the complement of the other. The Act
of 1660 [2] allowed home-grown corn to be exported when prices at
the port of shipment did not exceed, for wheat, 40s. per quarter ;
for rye, peas, and beans, 24s. ; for barley and malt, 20s. ; for oats,
16s. The same Act levied a duty of 2s. a quarter on imports of
foreign wheat, when home prices were at or under 44s. a quarter :
above that price, the duty was reduced to 4d. Proportionate
duties were imposed on other foreign grains according to their
prices in the home market. These scales of duties and prices were
revised in the Act of 1663.[3] In the Acts both of 1660 and 1663 the
object of the Government seems to have been revenue, for the
scales of duties on foreign imports are remarkably low. In 1670,
however, this policy was changed. In this Act " for the Improvement of Tillage " [4] corn might be exported, though the home prices
rose above the limit fixed in 1663. At the same time prohibitive
duties were levied on imports of foreign corn. When wheat, for

---

[1] An Act for the *Maintenance of the Navy* passed in 1562 (5 Eliz. c. 5), permitted the export of corn when the price of wheat was at or under 10s. per
quarter ; of rye, beans, and peas, 8s. ; of barley, 6s. 8d. See also Appendix
III., B.

[2] 12 Car. II. c. 4.   [3] 15 Car. II. c. 7.   [4] 22 Car. II. c. 13.

instance, stood at under 53s. 4d. a quarter, a duty of 16s, a quarter was imposed on foreign corn ; when the home price was between 53s. 4d. and 80s., the duty was reduced to 8s. ; when prices rose above 80s., the ordinary poundage of 4d. a quarter only was chargeable. On other foreign grains, at proportionate prices, similar duties were levied.

In the reign of William and Mary [1] an addition was made to the system. When the home price of wheat was at, or under, 48s. a quarter, a bounty of 5s. a quarter was allowed on every quarter of home-grown wheat exported. Similar bounties were allowed on the export of other grains at proportionate prices. In the Parliamentary debates on this measure the interests both of consumers and producers were avowedly considered. On the one side, the Act was unquestionably framed for the benefit of producers, to relieve them of accumulated stock, and so to enable them to bear increased public burdens. On the other side, it was expected that the stimulus of the bounty would promote production, bring a larger area of land under the plough, increase the quantity of home-grown grain, and so provide a more constant supply of corn at steady prices and a lower average. For the first sixty-five years of the eighteenth century results seemed to justify the argument. But it is difficult to determine how far the low range of prices which prevailed from 1715 to 1765 was due to prosperous seasons, or how far it was the effect of the stimulus to employ improved methods on an increased area of land. In years of scarcity, the direct effect of the bounty was inconsiderable, because not only was that encouragement withdrawn, but the liberty to export any home-grown corn was also suspended. In years of abundance, the bounty, by stimulating exportation, may have checked the natural fall of prices. But it was urged that this advantage to producers was a reasonable compensation for the loss they sustained in years of scarcity from the frequent prohibitions of exports ; that prices were steadied ; that no violent fall drove parts of the corn-area out of cultivation ; that the home-supply, on which alone the country could depend, was therefore more abundant than it otherwise would have been ; that, as the bounty was paid without regard to the quality of the exported grain, English consumers benefited by the retention of the superior qualities for home consumption. Possibly consumers may have found that these advantages counter-

[1] 1 W. and M. c. 12.

balanced the loss sustained in years of abundance by the interference with natural cheapness, and were a set off to the loss
of the six million pounds,[1] which, between 1697 and 1765,
were raised by taxation, and in the shape of bounties paid over to
producers.

The fiscal policy on which the Government embarked in 1689
practically governed the corn trade down to 1815. Scales of regulating prices were often revised ; but the principles remained the
same. On one side, the import of foreign corn was in ordinary
years practically prohibited by heavy duties. On the other side,
home production was artificially stimulated in order that a larger
area might be maintained under corn cultivation than was required
in average seasons for the maintenance of the population. In the
125 years during which this system prevailed, two periods may be
distinguished ; the first lasting from 1689 to 1765, the second
extending from 1765 to 1815.

In considering the results of the fiscal policy of the Government
during the first of these two periods, it must be remembered that
both sets of laws were in operation at the same time. When prices
were below a certain level, foreign imports were practically prohibited, exports of home-grown corn permitted, and the quantity of
production stimulated by bounties. When home prices rose above
a certain level, the bounties ceased, exports were prohibited, and
imports of foreign grain admitted duty free or at reduced rates.
It is, therefore, not easy to decide, whether consumers gained most
by the laws which kept corn in the country, or lost most by those
which kept it out. In the twentieth century, when there is a large
additional or alternative supply of grain, produced under different
climatic conditions to our own, there could be no question that the
loss inflicted by the prohibition of imports would be incomparably
the greatest. But the conditions of the corn-markets of the world
in the seventeenth and eighteenth centuries were so widely different,
that the policy of the Government may not then have been unreasonable. Additional supplies were only obtainable from Northern
Europe. But the north of France, the Netherlands, Denmark,
North-west Germany, and, to a less extent, North-east Germany and
Poland, were affected by similar climatic conditions to those of
England. Thus in unfavourable seasons the whole corn-area then

---

[1] See Appendix III., E. for the bounties paid in the years 1697-1765 on exports
of grain under the Act of 1 William and Mary, c. 12.

available suffered simultaneously from deficient harvests. Through-out the period 1689-1765 the average price of wheat in England is stated to have been less by 4d. a quarter than the average price in Continental markets. Foreign corn, therefore, after bearing the cost of transport and insurance, seldom less than 12s. a quarter and often increased by war-risks, could not have reduced English prices, even if no import duties had been levied. Consumers were not shut out from an alternative and cheaper supply, because no other supply was available except at higher prices than were being paid for home-grown grain. On the other hand, they profited considerably by the results of the fiscal policy pursued in England. In average seasons England grew not only corn for her own people, but a surplus for exportation. It was only in adverse seasons that any deficiency was probable. When this was anticipated, the Government had two strings to its bow. The ports were closed against exports, and, if the supply continued inadequate, were opened to imports. It seemed probable, therefore, that consumers suffered no injury from the heavy duty on imports, or that, if they were injured at all, their loss was infinitesimal.

During the period 1689-1765, neither the bounties, nor the liberty of exportation, nor the restriction on imports, were continu-ously operative. In nine years [1] the bounty was suspended, or the exportation of home-grown corn altogether prohibited. Generally this expedient succeeded ; the unusual quantity of corn retained in the country met the deficiency. But in three years [2] out of the nine the further step was taken. In 1741, and both in 1757 and 1758 foreign corn was admitted duty free. The total amount of wheat imported into the country in those three years was 169,455 quarters. In these exceptional years, war and war-taxes, the restoration of the currency, or the gradual growth of the population may have specially affected English prices, and the bounty may, as its opponents asserted, have assisted their upward tendency. But all these causes in combination were comparatively unimportant. Throughout European markets the dearth or the abundance of grain, together with high or low prices, mainly depended on the weather, which generally affected the whole corn-area in the same way. The last seven years of the seventeenth century, for instance, were long remembered in Scotland as the " seven ill years," and in England

[1] 1698, 1699, 1700, 1709, 1710, 1741, 1757, 1758, 1759.
[2] In Scotland only.

they were almost equally disastrous. The winters of 1708-9 and 1739-40 were two of the three winters [1] which were famous in the eighteenth century for their prolonged severity. Both were followed by deficient harvests. The wet spring, summer, and autumn of 1756 produced a scarcity of corn, and the great heat of 1757 caused the crops to be too light to make good the previous shortage. These unfavourable seasons were not peculiar to England. They prevailed throughout Northern Europe, and the advance of prices was general. But in France, where the Government discouraged exports of grain and encouraged imports, the distress was acuter and more lasting than in England, where the opposite fiscal policy was adopted. England, in other words, profited in these years of scarcity by the large reserve which the bounty helped to maintain.

With the exception of the years in which these deficient harvests occurred, the period was generally prosperous for the labouring classes in England. The level of prices was low and steady. As compared with the average price of wheat in the seventeenth century, the first sixty-five years of the eighteenth century show a fall of 16 per cent., and this relative cheapness was accompanied by a rise of the same percentage in the wages of agricultural labour. It seems probable that the reign of George II. was the nearest approach to the Golden Age of the labouring classes. Necessaries of life were cheap and abundant ; population showed no rapid increase, but the standard of living improved. Complaints of the low prices [2] were loud. It was said that farmers could not pay their rents and landowners could " scarce support their families." The low range of prices quoted by Eden [3] for the years 1742-1756 is remarkable for a country which was entirely dependent upon home supplies, was a considerable exporter of grain, and in nine out of the fifteen years was engaged in war at home or abroad.

A succession of prices so low as those shown in the Table on page 263 would naturally have driven a considerable area out of cultivation for corn, and an advance of price would have been caused by a diminution of the supply. The practical effect of the bounty seems to have been that this natural result was to some degree counteracted, though throughout the

---

[1] The third winter was 1794-5.

[2] See *The Landlord's Companion*, by W. Allen (1736) ; *Considerations on the Present State of Affairs*, by Lord Lyttelton (1739).

[3] *History of the Labouring Classes*, Appendix, p. lxxx.

century a large area of corn-land was being converted to pasture.
Thus a surplus was provided which, in years of European scarcity,
mitigated the dearth at home. During the whole period from 1715
to 1765 the total imports of foreign corn did not exceed 300,000
quarters, while home-grown corn was sent out of the country to the
amount of 11¼ millions. The largest amount of wheat exported in
any single year was reached in 1750, when the quantity was 950,483
quarters.[1]

January Prices of Grain at Mark Lane and Bear Quay.

YEARS.	WHEAT.		BARLEY.		OATS.	
	s.	s.	s.	s.	s.	s.
1742	26 to 29		15 to 20		12 to 15	
1743	20 ,, 23		15 ,, 20		13 ,, 16	
1744	19 ,, 21		11 ,, 13		9 ,, 12	
1745	18 ,, 20		12 ,, 15		12 ,, 16	
1746	16 ,, 24		10 ,, 12		12 ,, 14	
1747	27 ,, 30		8 ,, 12		6 ,, 9	
1748	26 ,, 28		13 ,, 14		9 ,, 12	
1749	27 ,, 32		17 ,, 18		14 ,, 16	
1750	24 ,, 29		14 ,, 17		12 ,, 14	
1751	24 ,, 27		14 ,, 17		13 ,, 14	
1752	33 ,, 34		17 ,, 19		12/6 ,, 16	
1753	29 ,, 33		17 ,, 18		10/6 ,, 12	
1754	27 ,, 33		17 ,, 19		12/6 ,, 13	
1755	24 ,, 26		12 ,, 14		10 ,, 13	
1756	22 ,, 26		14 ,, 15		12 ,, 13/6	

During the second period (1765-1815) the Government maintained
the same fiscal policy of regulating both exports and imports, and
of encouraging exportation by means of bounties within a certain
range of prices. But in all other respects the two periods are
sharply contrasted. The first period was remarkable for low
prices, a large export trade in home-grown corn, and the prosperity
of the labouring classes ; the second period is equally remarkable
for high prices, a growing importation of foreign corn, and widespread misery among the wage-earning population. When the

[1] See Appendix III., D.

fiscal system was practically unaltered, to what causes must these differences be attributed ?

The average price of wheat during the half century which ended in 1764, was in the next fifty years practically trebled. The tendency is shown in the following decennial averages of the prices of wheat per quarter :

1765–74	51s.
1775–84	43s.
1785–94	47s.
1795–1804	75s.
1805–14	93s.
1815–24	68s.

It was now that England ceased to be a corn-exporting country and became a buyer of foreign grain. The year 1765 marks the first stage in this revolution in the English corn-trade. For some few years the balance hovered from side to side, inclining to excess now of exports, now of imports. After 1792 it definitely turned in favour of imports, which from that date increasingly preponderated. During the whole period which witnessed this change, the fiscal policy, though often revised, and notably in 1773 and 1791, remained in principle the same. But from 1765 to 1774, and again from 1792 to 1814, the liberty to export corn, as well as the bounty which encouraged exportation, was almost continuously suspended. Imports of foreign corn were also repeatedly admitted at reduced rates or duty free. This was the case in 1765, 1766-8, 1772-3, in 1783, in 1790, and practically from the commencement of the French war (1793) till its final close. Besides the frequent revisions and suspensions of the regulating prices, great efforts were made to increase home and foreign supplies. Thus in 1772 the inland trade was relieved from many restrictions by the repeal of the statutory penalties against " badgers, forestallers, engrossers, and regrators." To increase the area under corn, numerous enclosure Acts [1] were passed. To eke out the home produce, economies were enforced by Parliament. Thus the hair-powder tax was imposed in 1795, and the use of wheat and other grain in the making of starch or in distilleries was repeatedly prohibited.[2] Still more exceptional efforts were made to secure a supply of foreign corn. Government agents were employed to buy corn in the Baltic, as it was feared

---

[1] 1,593 Acts were passed between 1795 and 1812 inclusive.
[2] *E.g.* in 1795-6, 1800, 1801, 1809-12.

that private merchants would hesitate to pay the high prices which were demanded abroad. Corn in neutral ships, destined for foreign ports, was seized and carried to England. Bounties on imports of grain which had been offered in 1773 at the rate of 4s. a quarter by the City of London, were offered by the Government at the rate of from 16s. to 20s. a quarter in 1795-6 and again in 1800 and the years that followed. Substitutes for ordinary corn, such as rice and maize, were eagerly bought : the cultivation of the potato was greatly increased. But in spite of all these efforts to provide food, the scarcity continued until there seemed to be a real prospect of a failure in the supply of provisions. In 1812 the country stood on the very verge of famine. Shut out from Continental ports, at war not only with Napoleon but with America, England was reduced to acute and extreme distress. Conditions were at their worst. In August of that year the average price of wheat at Mark Lane was 155s. per quarter ; prices of other grains, as well as of meat, rose in proportion ; at the end of October the potato crop was found to have failed by one-fourth. The year was one of the most severe suffering. But 1813 brought relief. An abundant harvest lowered prices with extraordinary rapidity. In December wheat had fallen to 73s. 6d. In 1814 [1] the fiscal system which had lasted, though with many interruptions, since 1689, was finally abolished. After June of that year corn, grain, meal, and flour were allowed to be exported without payment of duty and without receiving any bounty. Henceforward the Corn Laws only survived in the one-sided form of restrictions on imports.

The high prices which prevailed in the second period (1765-1815) have been explained in various ways. They have been attributed to the improper practices of corn-dealers, the growth of population, the consolidation of holdings and diminution of open-field farming, the depreciation of the currency, unfavourable seasons, the war, or the fiscal system. Each of these causes may have contributed to the upward tendency of prices. But the most effective reasons for the dearness of corn were the weather and the war. These two causes alone would sufficiently explain the continued scarcity. Even under a system of absolute free trade, they would produce the same results to-day, if England still drew her supplementary supplies of corn from the same limited area at home and abroad.

The growth of the population is undoubtedly an important factor

[1] 54 Geo. III. c. 69.

in the problem. Between 1689 and 1815 the increase was considerable, though, like most of the political arithmetic which relates to
the eighteenth century, the actual numbers are largely a matter of
guess-work. In 1696 Gregory King estimated the population of
England and Wales at 5,500,000. At the accession of George III.
(1760), the numbers were supposed to have risen to between six and
seven millions.[1] As the reign advanced, the rate of increase was
accelerated. The first official census was taken in 1801. In that
year the population of England and Wales is stated to be 8,872,980.
In 1811 it had grown to 10,150,615. On these figures the population
had doubled itself in 125 years, and, even if no allowance is made for
an improved standard of living, it is probable that England during
the same period had doubled her production of food. The increased
supply required to feed double the numbers was certainly not obtained from abroad, for food imports, even at their highest, continued
to be infinitesimal in amount.[2] It was therefore produced at home.

In the case of wheat it would be difficult to prove the same rate
of progress. In abundant seasons the home supply would probably
have continued to feed the country, without risk of inadequacy or
panic-stricken competition, and therefore cheaply. But in ordinary
seasons the margin was at best a small one, and in unfavourable
weather a deficit was certain. It has been disputed whether six
bushels or eight bushels of wheat should be allowed as the average
quantity yearly consumed by each person. At the higher rate of
consumption, and assuming that wheat was the food of the whole
population, seven million quarters of wheat would be required in
1760, and ten million quarters in 1811. Arthur Young, in 1771,
calculated that 2,795,808 acres were then under cultivation for
wheat in England and Wales, and that the average produce per acre
was three quarters, giving a total yield of 8,387,424 quarters. In
1808, Comber[3] estimated the wheat area of England and Wales at

[1] Smith (*Tracts on the Corn Trade*) estimates the population of England and
Wales in 1766 at six millions, of whom 3,750,000 consumed wheat, the remaining 2,250,000 consuming rye, barley, or oats. Finlaison, of the National Debt
Office (M'Culloch's *Statistical Account of the British Empire*, vol. i. 399),
calculated the numbers in 1760 at 6,479,730. Porter (*Progress of the Nation*,
p. 146) gives the population in 1760-69 as 6,850,000. Nicholls (*Hist. of the
English Poor Law*, ed. 1904, vol. ii. p. 54) estimates it in 1760 as 7,000,000.

[2] From 1801 to 1810 the average amount of wheat annually imported was
600,946 quarters, or about 2 pecks per head ; from 1811 to 1820 it was only
458,578 quarters.

[3] *An Inquiry into the State of National Subsistence*, Appendix **xxv**.

3,160,000 acres, and the produce, adopting Young's average rate, would be 9,480,000 quarters. In other words, while the population had increased by three millions, the wheat production had increased by only one million quarters. This calculation, however, allows nothing for the increased productiveness of the soil under improved management, does not take into account the surplus wheat obtainable from Scotland and Ireland, and is at first sight contradicted by the large acreage which enclosures had added to the cultivated area. Evidence indeed exists to prove that the first effect of enclosures of open-field farms was often to diminish the corn area. Against this decrease must be set the quantity of land which, under the spur of the high prices of the Napoleonic war, were brought under the plough and tilled for corn. Comber's calculation of the wheat area appears to be extremely low ; but it is impossible to prove the suspected under-estimate. It is probably safe to say that, while in an average season enough wheat was grown in England and Wales to feed ten million people, the surplus was so small as to expose the country to panic prices whenever a deficiency in the normal yield was anticipated.

This conclusion is confirmed by a closer examination of the yield of corn harvests during the period. The seasons from 1765 to 1815 were far less favourable than those from 1715 to 1764 ; the former were as uniformly prosperous as the latter were uniformly adverse. Both in this country, and throughout Europe, the harvests of 1765-67, 1770-74, fell much below the average. Prices rose high. Exports dwindled, and imports increased in volume.[1] In the decennial period 1765-1774, for the first time in the history of English farming, imports of foreign wheat exceeded the home-grown exports. Since that period they have never lost their preponderance. For the next eighteen years (1775-1792) the seasons were irregular. Thus the harvest of 1779 was long famous for its productiveness. On the other hand, the years 1782-3-4 were most unfavourable, the winters unusually severe, and the spring and summer cold and ungenial. There was a general scarcity of food. In 1782 the imports of wheat (584,183 quarters) were the largest yet known, and the figure was only once (1796 : 879,200 quarters)

---

[1] 1765-74, Exports (in round numbers) 510,000 quarters ; imports, 1,341,000, 1775-84, exports, 1,366,100 ; imports, 1,972,000. 1785-94, exports, 1,305,385 ; imports, 2,015,000. 1795-1804, exports, 536,000 ; imports, 6,686,000. 1805-14, exports, 593,000 (nine years only, the records of 1813 having been destroyed) ; imports, 5,782,000.

exceeded in the eighteenth century. Writing in August, 1786, Arthur Young says : " Last winter, hay, straw, and fodder of all kinds were scarcer and dearer than ever known in this Kingdom. Severe frosts destroyed the turnips, and cattle of all kinds, and sheep suffered dreadfully ; many died, and the rest were in ill plight to fatten early in this summer." The crops of 1789 again were deficient. Exports were prohibited, and free imports permitted. But in France the scarcity almost amounted to famine. The Government spent large sums in the purchase of wheat, and Continental prices ruled considerably above those of England. Against the deficient harvests of 1790 and 1792 may be set the season of 1791, which was so favourable that, for the last time in the history of the corn-trade,[1] the exports of the following year exceeded the imports.

It will be seen that the yield of fourteen of the harvests during the twenty-eight years 1765-92 fell so far below the average as to create a scarcity ; that several others were defective ; and that only two (1779 and 1791) were really abundant. Yet, during the whole period, the total excess of imports of foreign wheat over the exports of home-grown produce only amounted to 1,661,000 quarters, or an average of little more than 59,000 quarters a year. It may, therefore, be reasonably assumed that, if England had enjoyed seasons as uniformly favourable as those of 1715-64, she would have been able to feed her growing population at low prices and yet to remain a grain-exporting country. The fact is a striking proof of her agricultural progress. It is more than doubtful whether such an expansion of her powers of production would have been possible if the open-field system of farming had been maintained.

In February, 1793, war was proclaimed with France. It continued with two brief intervals till 1815. As the struggle progressed the area of conflict was widened until it embraced America as well as Europe, and not only became a naval and military war in which all the Powers were engaged, but developed into a commercial blockade directed against this country. During the whole period the Corn Laws were practically inoperative. The progress of the war created conditions of supply which alone would suffice to explain an unprecedented rise of prices. But the situation was through-

---

[1] 1792, exports, 300,278 quarters ; imports, 22,417. The statement in the text is not literally true. In 1808 the exports exceeded the imports by 13,116 quarters (98,005 to 84,889). But the exportation was to the Peninsula for military purposes and for the supply of our own troops.

out aggravated by an unusual recurrence of unproductive seasons.

The wheat harvests in the twenty-two years 1793-1814 [1] may be thus analysed. Fourteen were deficient ; in seven out of the fourteen, the crops failed to a remarkable extent, namely, in 1795, 1799, 1800, 1809, 1810, 1811, 1812. Six produced an average yield. Only two, 1796 and 1813, were abundant ; but the latter was long regarded as the best within living memory. Towards the close of the period, the increased extent of the wheat area to some degree compensated for the comparative failure of the crops. But the repeated deficiencies created an almost continuous apprehension of real scarcity which was expressed in abnormal prices. To a generation which draws its supplies from sources so remote that climatic conditions vary almost infinitely, the panic may seem unintelligible. It was not so in the days of the Napoleonic wars. The quantity available from the United States was scanty, and over the corn areas of Europe a similar series of unproductive seasons seems to have prevailed. To this, however, there was one notable exception. The harvests of 1808 and 1809 were remarkably favourable in France and the Netherlands, and, at the very height of the struggle with Napoleon, it was from the French cornfields that England obtained her additional supplies.

The deficiency of the home harvests and the consequent fear of scarcity naturally raised prices of corn. The upward tendency was in various ways enormously increased by the progress of the war and the commercial blockade which it developed. No doubt the struggle in which the country was engaged quickened the activity and industry of the population, stimulated agricultural improvements, sharpened the inventive faculties to economise both in money and in labour. On the other hand, the war raised the rate of interest, added to the burden of taxation, increased the cost of corn-growing, and withdrew into unproductive channels a considerable portion of the capital and labour of the country. Besides these ordinary results, the peculiar character which the struggle gradually assumed threatened to deprive England of any alternative supply of foreign grain which could supplement the resources that she derived from her own soil, from Scotland, and from Ireland. Again and again the political situation was reflected in Mark Lane.

---

[1] Tooke's *History of Prices,* ɘd. 1857, Appendix vi. "Seasons 1792-1856" vol. vi. pp. 471-83, and vol. i. pp. 213-376, and vol. ii. pp. 1-3.

Thus, in 1800, it was not merely the prospect or subsequent certainty of an unproductive harvest which raised prices, for the actual deficiency had been greater in 1794-5. It was the further dread of being cut off from foreign supplies. It was the hostility of Russia and Denmark, the consequent fear that the Baltic would be closed against our grain-ships, and the almost simultaneous news that Prussia had imposed a heavy duty on all grain exports, which combined to send wheat to 130s. a quarter. At a later stage in the struggle, the deficiency in our home supply was less in 1811-12 than it had been in 1794-5 or in 1799-1800. But it was the threat of a complete stoppage of all foreign supplies by the Berlin and Milan decrees, which turned the dread of scarcity into a panic-stricken competition and carried the price of wheat in 1812 to 155s. a quarter. Even if the war never actually effected a commercial blockade, its risks, together with the restrictions on exports enforced by foreign Powers and the licences for navigation required by the British Government, forced up the rates of freight and insurance to a prodigious height. During the period 1810-12, this increase in the costs of conveyance culminated, and the charges for the transport of foreign corn rose to as much as 50s. a quarter. Thus, even if it was possible to obtain additional supplies from abroad, they could only be brought into the country at an unprecedented expense.

The history of the Corn Laws, thus briefly outlined, confirms the impression that, down to 1815, they exercised little or no influence on prices. If that is so, they were not the cause of the great rise of rents which the last quarter of a century had witnessed. Hitherto the only practical effect of the restrictions on imports had been to prevent corn from being brought into the country for the purpose of gaining the bounty on exportation. In ordinary years, no foreign corn could have been imported, even duty free, at prices which could reduce, or compete with, home-grown produce. In years of scarcity, the deficiency generally extended over Europe, and foreign supplies were either not obtainable, or obtainable only at prices at least as high as our own. During the frequent periods of war, these conditions were aggravated by the prodigious cost of transport. Great Britain had in the main fed her own population, and her prices had depended on the seasons. Consumers had not suffered from the Corn Laws, because no alternative cheaper supply was available from abroad.

After 1815 these conditions were to a great extent altered. The

bounty on exportation had been abolished. Freedom of export
was allowed, and was never suspended, because there was no margin
of produce which could be retained in the country by prohibiting
it from being sent abroad. Population was beginning to equal
production. So long as there had been a surplus of home-grown
grain, which could be kept in the country by suspending the licence
to export, the Corn Laws had steadied prices. Now, in times of
scarcity, they only increased the range of fluctuation in rise and
fall by excluding alternative supplies. Revenue was not their
object, because the duties were so high as to be prohibitory. They
were frankly protective, intended to shut out imports, and so
maintain the prices of home-grown produce above a permanent
level. Even so, the interests of consumers would not necessarily
have been sacrificed to those of producers, unless an additional
and cheaper supply had been obtainable. That condition was
now, in most years, fulfilled. The charges of transport had fallen
to their peace level ; throughout Northern Europe corn was once
more sown and reaped without fear of the ravages of war, and
Continental prices ruled below those of Great Britain : from the
New World came an increasing supply, which was not affected by
the same climatic conditions as those of the North of Europe.
Henceforth external sources existed, from which deficiencies in
the yield of home harvests might be supplied without raising prices
beyond the addition of the costs of conveyance. If to these costs
were added the payment of heavy duties, it might be said that the
price of bread was artificially raised to maintain the level of the
profits of landowners and farmers.

Another important change had taken place in the position of
the antagonists in the coming struggle over the prices of corn. The
issue was no longer centred on principles of abstract morality ;
it was transferred to the practical region of trade. Our ancestors
passed laws to establish just prices ; their successors legislated
to secure reasonable profits. The change may have been a change
rather of words than of ideas. But it was not without significance.
Down to the middle of the eighteenth century, the great preponderance of the nation had been interested in prices both as consumers
and producers of corn. Now the proportions were completely
altered, and the majority had permanently shifted. The new
manufacturing class was rapidly growing ; the mass of open-field
farmers had become agricultural labourers, whose real wages rose

with the cheapness or fell with the dearness of bread. On the other hand, the interests of producers of corn were now represented by a comparatively small and dwindling class of landowners and farmers, who in recent years had enormously raised their own standard of living. Numerically small, but politically powerful, this class was convinced that the war-prices yielded only reasonable profits. The great majority of the population was convinced to the contrary.

Yet it would be unfair to represent that the protective policy of the later Corn Laws was entirely maintained by a Parliamentary majority swayed by selfish motives. It was supported, up to a certain point, by many who stood outside the circle of the landed interests, and ranked as disciples of Adam Smith. It never entered into their calculations that Great Britain could ever become dependent for its food supply on foreign countries. On the contrary, the view was strongly held that every prosperous nation must in ordinary seasons rely for its means of subsistence on its own resources, and must meet the growth of numbers with a corresponding increase in the supply of food. This doctrine was almost universally accepted. Porter, the author of *The Progress of the Nation*,[1] was an advanced Free Trader. But he argued that " every country which makes great and rapid progress in population must make equal progress in the production of food." He quotes the example of Great Britain in support of his view. By comparing the growth of population with the increase in the quantity of imported wheat, he shows that improvements in agriculture had, to a remarkable extent, enabled the country to keep pace with its increasing needs. Thus in 1811, when the population of Great Britain was ascertained to be 11,769,725, only 600,946 were fed by foreign wheat. At the end of the next decade, 1811-20, the population had risen to 13,494,217, and the home supply was enough for all but 458,576. At the close of the third decade, 1821-30, the population had grown to 15,465,474 ; yet only 534,992 depended on the foreign supply. In 1841, the numbers had increased to 17,535,826 ; but home-grown wheat fed all but 907,638 persons. In other words, British wheat, in 1811, had fed a population of 11,168,779 ; in 1841, enough wheat was produced at home to feed a population of 16,628,188. Thus in thirty years British land had increased its pro-

---

[1] *The Progress of the Nation in its various social and economical relations from the beginning of the Nineteenth Century*, by G. R. Porter, ed. 1847, p. 136.

ductiveness by 5½ million quarters. Porter evidently expected that this proportionate progress would continue. He himself advocated the repeal of the Corn Laws ; but there were other Free Traders who hesitated to go this length, for fear that improvements should be discouraged, and that the country should mainly depend for its bread upon foreign wheat.

Agriculturists also argued, and no doubt conscientiously believed, that, if corn in any quantity were brought into the country from abroad, home-prices would cease to yield reasonable profits ; that agricultural land would be forced out of cultivation ; that rents and wages would fall ; that rural employment would diminish ; that the virility of the nation would be impaired by the influx into towns and the consequent depopulation of country districts. To these arguments Parliament lent a sympathetic ear. The limit of home-prices, at which the importation of grain was allowed at nominal duties, was raised in the case of wheat from 48s. in 1773 to 85s. in 1815. Below those limits, duties, so heavy as to be practically prohibitive, were levied on imported corn or on its removal from the bonded warehouses for consumption. In 1828 the evils of this restrictive legislation, though apparently modified, were really aggravated by the adoption of a sliding scale of duties, which varied with the prices of home-grown grain. The importation of corn became a gamble, and foreign importers combined to raise home-prices in order to pay the lower scale of duties. Yet in spite of this experience the graduated system was maintained in the legislation of 1842 and 1845.

Meanwhile the whole protective policy, of which the Corn Laws only formed a part, was gradually becoming discredited. In 1815 a minority of the Peers had entered a powerful protest against the exclusion of foreign corn. In 1820 the merchants presented their famous petition, which was drawn up by Thomas Tooke, the author of the *History of Prices*. A war of pamphlets raged con-tinuously. In the treatment of colonial produce especially, there were signs of the abandonment of a rigidly protective policy. The principle of colonial preference, already recognised in 1766, had been acted upon in 1791, 1804, and 1815. Corn from British possessions was allowed to be imported at a nominal duty at a lower limit of home-prices than that fixed for foreign produce. Ten years later, corn from the British possessions of North America was permitted to enter British ports at a constant duty of 5s.

without reference to home-prices. In 1843 this principle was carried yet further. A special concession was made to Canada. In return for a preference granted to British trade, Canadian corn, irrespective of home-prices, was admitted at a nominal duty of 1s. Encouraged by these concessions, the agitation against the Corn Laws gathered strength. It gradually extended from a demand for the relaxation of the stringent duties to a demand for their total abolition. For a brief period the pressure was reduced by the favourable seasons of 1831-36. In 1835, wheat fell to 39s. 4d., the lowest price at which it had been sold for 54 years. Hopes revived that the improvements in farming had again placed production on a level with the growth of population. The Corn Laws were for the moment forgotten. But the unfavourable cycle of 1837-41 again forced the question to the front. From 1839 onwards the Anti-Corn-Law League used its growing influence in favour of total repeal. The demand for cheap food grew more and more insistent from the labouring classes. Manufacturers echoed the cry, because cheap food meant a lower cost of production, and because food imports would be paid for by exported manufactures. Finally, the disastrous harvest of 1845 and the potato famine compelled the Government to yield. The " rain rained away " the Corn Laws. In 1846 the existing duties were modified according to a scale which was to continue in force till February 1, 1849. After that date all kinds of foreign corn were to be admitted at the nominal fixed duty of 1s. a quarter. That nominal duty was finally repealed in 1869.

CHAPTER XIII

HIGHWAYS

THE local progress of farming, at the close of the eighteenth century, had been great ; but its general advance was still hampered by numerous hindrances. In many parts of England the inveterate preference for old-fashioned practices was slowly yielding to experience of the results of more modern methods. Defects in the relations between owners and occupiers were mitigated by the grant of leases, which secured to improving tenants a return for their outlay of money and labour. Obstacles presented by soil and climate, so far as they were capable of remedy, were in process of removal. Experience had shown that sands might be fertilised, and the acidity of sour land corrected, by the use of the proper dressings, selected with judgment and applied with perseverance ; that considerable tracts of moor, heath, and moss might be brought into profitable cultivation ; that fens and swamps might be drained ; that even the disadvantages of climate might be ameliorated by plantations. But there remained a number of hindrances, which originated in the laws and customs of the country. To this class belonged difficulties of communication. The incidence of tithe on the produce of the land will be treated in a subsequent chapter.

A generation familiar with railways and good roads can hardly appreciate the obstacle to progress which was created by difficulties of transport and communication. Up to the middle of the eighteenth century, rivers had exercised the greatest influence on the development of inland trade centres. In few districts, and only in favourable seasons, could heavy goods be conveyed over the unmade roads. The command of water carriage was all-important. On straightening, deepening, or widening rivers so as to make them navigable, early legislators from the fifteenth century onwards, had mainly concentrated their efforts to improve internal communications. Not only inland towns, but seaports themselves, often owed their early prosperity to their situation at the mouths of rivers. Bristol, or Hull, or Boston, or Lynn, for instance, collected and distributed produce along the course of the Severn and the Wye, or the Trent and the Idle, or the Ouse, the Welland, and the Witham. Even London derived some of its pre-eminence from the produce which was carried over the Thames and its tributaries. To Liverpool the closing of the port of Chester by the sands which choked the Dee, and the opening up of the interior by making navigable the upper waters of the Mersey (1694), the Irwell and the Weaver (1720), proved the real starting-point of its trade. By means of these water-highways inland towns became seaports. They were the centres for collecting and distributing produce over the interior of the country. Fleets of trows, " billanders," floats, lighters, and barges were engaged in the trade. On the Severn, for instance, which was navigable as far as Welshpool, 376 vessels were employed in 1756. The famous Stourbridge Fair was supplied with heavy goods by the Ouse, which enabled boats, each carrying 40 tons of freight, to load and unload at Cambridge. York was accessible to vessels of from 60 to 80 tons, and claimed rights of wreckage as a seaport. Exeter and Taunton carried on a home and foreign trade by means of the Exe and the Parret. Coal reached Hereford by the Wye. Coventry communicated with the sea by means of the Warwickshire Avon. From Bawtry, on the Yorkshire Idle, were distributed the lead of Derbyshire, the edged tools of Sheffield, the iron goods of Hallamshire, as well as the foreign goods which entered the country at Hull. Cambridgeshire and Huntingdonshire shipped their barley and malt from Ware on the Lea. Gloucestershire cheesemakers sent their cheese to London down the Thames from Lechlade. Burslem wares were carried in pot-waggons or on pack-horses to Bridgnorth on the Severn.

From utilising the natural waterways of the country it seemed but a short step to supplementing them as arteries of trade by the construction of canals. Pioneers in the early stages of this movement were Sir Richard Weston, who in the reign of Charles I. canalised the Wey, and Sir William Sandys, of Ombersley in Worcestershire, who in 1661 obtained extensive powers to cut new channels, and build locks on the Wye and the Lugg. More extensive plans were floating in the minds of Francis Mathew [1] and Andrew Yarranton.[2] Mathew in 1655 had laid before Cromwell a scheme for connecting London with Bristol, by the construction of a canal to join the Thames and the Avon. No notice seems to have been taken of the plan. Nor was his project more successful fifteen years later. " Many Lords and Gentlemen," says Yarranton, " were ingaged in it. . . . But some foolish Discourse at Coffee-houses laid asleep that design as being a thing impossible and impracticable." Yarranton himself proposed to make Banbury a great distributing centre by connecting it with the Severn and the Thames. At an estimated cost of £10,000, he planned to make the Cherwell navigable from Oxford to Banbury, and to cut a new channel from the latter place to Shipton-on-the-Stour, whence goods might be carried by the Avon into the Severn below Tewkesbury. Both writers insist on the extreme isolation of inland districts, the need of supplying food to manufacturing centres, the prohibitive cost of conveying heavy goods by land, and the impassable nature of the roads for wheeled traffic.

In canal construction England lagged far behind foreign countries, though useful work continued to be done in making existing rivers navigable. Thus the clothiers of Leeds and Wakefield found new and cheaper markets when communication with Hull by the Aire and the Calder was opened up in 1699 ; Preston gained its opportunity for manufacturing development when the Douglas (1720) carried Wigan coal to the Ribble ; the connection of Sheffield with the Humber by means of the Don (1732) gave a fresh impulse to the cutlery trade. But rivers were unsatisfactory as carriers of goods. Subject to flood or drought, constantly liable to become choked, tortuous in their course, they were also limited in their range and left large districts untouched. If waterways were to be made efficient means of carriage, they must be permanently supplied with water, subject neither to deficiency nor excess, capable of being carried over or through natural obstacles in any direction required.

In 1755 the Sankey Brook Canal brought the St. Helens coalfields into direct communication with Liverpool by means of a

---

[1] *The Opening of Rivers for Navigation*, etc., by Francis Mathew, 1655. *A Mediterranean Passage by Water from London to Bristol*, etc., by Francis Mathew, 1670.

[2] *England's Improvement by Sea and Land*, by Andrew Yarranton *Gent.* 1677.

new channel, fed with a continuous supply of water, and provided with a system of locks which overcame the difficulties of the descent into the valley of the Mersey. This channel was the first true canal, as distinguished from straightening the courses of rivers. Before the work was completed, the Duke of Bridgwater obtained the sanction of the legislature (1759) for the famous canal which bears his name. Brindley's triumph was the real starting-point of the movement. He was the engineer of numerous similar works. The Mersey and Trent Canal, for example, joined Liverpool and Hull, and thus united the ports of the East and the West. Branches were thrown out, which gradually linked together Liverpool, London, Bristol, Birmingham, and Hull by water. The development of inland navigation which Brindley had begun was continued by Telford and others. The new means of transport powerfully influenced the progress of the industrial revolution. Between 1790 and 1794 alone, 81 Canal Acts were obtained, and a canal mania was started, which was only paralleled by the railway mania of the last century. By 1834 England had been covered with a network of more than 4000 miles of canals and navigable rivers.

To some extent the surface of the roads was saved by the substitution of water-carriage for the conveyance of heavy goods. But the development of canal traffic did not always improve internal communications. The increased carriage of heavy goods, such as coal, iron, timber, lime, stone, salt, and corn, to and from the wharves, destroyed the roads in the neighbourhood. To some extent this extraordinary traffic was carried on railways, laid down by the canal companies, as feeders to their trade.[1] But the range was limited. It was plain that, if full advantage was to be taken of the new means of inland navigation, roads must be scientifically constructed to bear the increased traffic. In McAdam and Telford were found the exponents of this necessary science. The progress of enclosures also favoured road-improvement. So long as land lay unenclosed, travellers were allowed to deviate from the track to avoid the ruts worn by their predecessors. Thomas Mace (1675)[2] describes how land was "spoiled and trampled down in all wide roads where coaches and carts take liberty to pick and

---

[1] See chapter xvii. pp. 350-3.

[2] *Profit, Conveniency, and Pleasure, to the Whole Nation, Being a Short Rational Discourse . . . concerning the Highways of England*, by Thomas Mace (1675).

chuse for their best advantages." A century later, a Reporter contrasts the state of a district near Norwich in the last decade of the eighteenth century with its condition before 1760 : " Thirty years ago," he says, " it was an extensive heath without either tree or shrub, only a sheep-walk to another farm. Such a number of carriages crossed it, that they would sometimes be a mile abreast of each other in search of the best track. Now there is an excellent turnpike road, enclosed on each side with a good quickset hedge, and the whole laid out in enclosures and cultivated in the Norfolk system in superior style." Instead of these common tracks, with their wide margins of deviation, enclosure Acts substituted defined and constructed roads. Not only was science needed for making new highways, but the existing machinery for maintaining those already in existence had broken down under the stress of modern needs.

Throughout the Middle Ages the great Roman roads were the main thoroughfares. Watling Street ran from Kent to Chester and York, branching northwards to Carlisle and Newcastle ; the Fosse Way crossed England from Bath to Lincoln ; Ermine Street led from London to Lincoln and thence to Doncaster and York ; Icknield Street, more difficult to trace, swept inland from Norwich, passed through Dunstable, and ultimately reached Southampton. For centuries they required and received little repair owing to the solidity of their construction. A firm foundation of beaten earth was secured. On this were laid, first, large stones, often embedded in mortar ; then a layer of small stones mixed with mortar ; above these two layers, lime mixed with chalk and pounded brick, or with gravel, sand, and clay ; and finally the paved surface.

Planned and built by the State, these Roman highways offered a striking contrast to the subsequent roads, which were laid out in haphazard fashion as need arose. The art of road-making was lost, or the cost beyond the reach of local effort. Unmetalled tracks crept along the edges of streams, which often afforded a better bottom than the ways themselves, or sought sound foothold for men and beasts across unenclosed land, or boldly kept on high ground to escape the bogs and quagmires. Gradually footways, horseways, and cartways [1] were levelled by traffic across the plains or hollowed

[1] The Romans recognised the same distinctions. The *iter, actus,* and *via* were the English footpath, bridle-way, and carriage road. Both in Roman and in English law the greater included the less, so that the *via* was open, not only to vehicles, but to foot-passengers and animals.

through the hills. Besides the highways between town and town, each manor had its by-roads, leading from the village to the open fields, the commons, the mill, or the church. The ordinary principle which governed the repair of thoroughfares was that they should be maintained by those who had the use of them. The duty of maintaining communication between market towns rested on the inhabitants of the parishes through which the roads passed ; within the limits of chartered towns it fell on the townsmen, on whom rates or tolls were sometimes levied. Local by-roads within the boundaries of manors were repaired by the manorial tenantry as one of the conditions of their tenure, and they were bound to provide the necessary implements and labour. These obligations were respectively enforced by county or municipal authorities or manorial courts. But road repair did not entirely depend on the performance of legal liabilities. It was also enjoined as a religious duty. Travellers were classed with the sick and poor as objects of Christian charity. Indulgences were granted to offenders who gave their money or their labour for the construction or repair of roads and bridges. For the same object pious bequests were encouraged. Gifts of this kind occur as late as the sixteenth century, and in the reign of Edward VI. one of the enquiries made at the Visitations of Bishops was whether these bequests were administered according to the intentions of the donors.

For a short period during the reign of Edward I., road improvement had received some attention from Government. When new ports, like those of Sandwich and Hull, were constructed, care was taken to provide good approaches by land. An attempt was also made to safeguard the lives and property of travellers on the king's highway. Adjoining landowners were compelled by statute to clear all roads between market towns from trees and underwood to a space of 200 feet on either side. The object was not the preservation of the roads by the admission of light and air, but the destruction of the lurking places of robbers. If any crime of violence was committed on a highway not properly cleared, the adjoining owner was held responsible. But the energies of Edward's successors were absorbed in other directions than the maintenance of rural roads. As the fourteenth century advanced, the general burden of taxation and the scarcity of labour increased the growing neglect of public highways. Agricultural changes told in the same direction. So long as lay and ecclesiastical nobles, in order to consume the produce of their estates, had travelled from manor to manor with their retinues, household furniture, and utensils, they had been interested in the means of transit. Their visits ceased when their lands were let on lease. At the same time the decay of the manorial organisation facilitated the evasion by the tenants of duties which had ceased to be personally valuable to their lords. Roads, made at will, were repaired, or not, at pleasure ; everybody's business was nobody's business ; the parochial liability, like the manorial obligation, was rarely and unsystematically enforced. Highways fell deeper into decay, and their neglect was increased by the cessation of voluntary efforts, when services which mediaeval piety recognised as religious duties came to be regarded only as civil burdens.

The condition of the roads across the Weald of Kent, at the opening of the Tudor period, was probably no worse than that of highways in other districts. Yet they are described as " right deep and noyous," only to be used at " great pains, peril and jeopardy." [1] The isolation of rural districts can hardly be pictured by the present generation. It restricted the agricultural use of the land, because the interchange of its products was difficult, and each district was compelled to grow its own corn. At the same time, it was recognised that, in the interests of expanding trade, the provision of better means of transit was necessary. The first general Act of Parliament applied to bridges and their approaches.[2] Passed in 1530, the statute placed the county on the same footing with regard to bridges as that in which the parish already stood to highways. It directed justices of the peace to enquire into the conditions of bridges in their districts, to ascertain what persons were liable for their maintenance, or to levy a rate on the inhabitants for their repair and that of their approaches for 300 feet on either side. In 1555 [3] another general Act was passed, dealing with the roads from market town to market town, which it describes as " verie noysome and tedious to travell in and dangerous to all Passengers and Carriages." It applied to the discharge of parochial liabilities the same methods by which manorial tenants had met their local obligations. Each parish was to elect two " honest persons " of the parish as " surveyors and orderers," for the repair of the roads within its boundaries

---

[1] 14 and 15 Hen. VIII. c. 6, Sections 1, 3.

[2] 22 Hen. VIII. c. 5.

[3] 2 and 3 Philip and Mary, c. 8.

by compulsory labour. Four days of eight hours each [1] were appointed for the work, the parishioners providing carts, teams, implements, and labour, according to their means. Like other Tudor legislation, the Act failed in its administration. Though neglect to discharge the liability was punishable by fines, little effect was produced.

After the Restoration further efforts were made to improve facilities of communication. Stage, or long, waggons had begun in 1564 to ply between the metropolis and the principal towns in the provinces ; private carriages were increasing ; about 1645 stage-coaches were established. Travelling on wheels was recommended for its " admirable commodiousness," and many of those who thus traversed the roads to London were " persons of quality " who could make their influence felt. Some means, in addition to statute labour, was required to maintain the roads in repair for the increasing traffic. During the first ten years of the reign of Charles II. it seemed probable that this supplement would be provided by the development of highway rates, which had been introduced in 1656. Eventually a new auxiliary to statute labour was devised, which arrested the growth of rates, and prolonged the life of the old system by a century and a half. In 1663 the first Turnpike Trust [2] was established on the Great North Road by the erection of toll-bars at Wadesmill, Caxton, and Stilton, and by the exaction of toll from those who used the highway. This portion of one of the principal roads in the country is described in the Act as " ruinous and almost impassable." The inhabitants of the adjoining parishes were too poor to put or keep the highway in repair, and though the Act did not relieve them from their liability, the tolls raised a fund towards the maintenance of the road. Other turnpike trusts were established on the same principle. Their creation was unpopular. Riots broke out, like those subsequently associated with the name of Rebecca in Wales ; toll-bars were frequently pulled down and burned ; and the opposition was only checked by an Act passed in the reign of George II. (1728) which made their destruction a felony. Turnpike Trusts multiplied rapidly, till in 1760 it was true to say that

> " no cit, nor clown,
> Can gratis see the country or the town."

---

[1] In 1562, by 5 Eliz. c. 13, the " statute labour," as it was called, was increased from 4 to 6 days.

[2] 15 Car. II. c. 1.

Travelling still continued to be a peril. The number of patents that were taken out to prevent coaches from overturning is some evidence of the risk. Nor were the inventions always effective. They did not prevent George II. and his queen from being upset in 1730 near Parsons Green on their way into London. In October, 1736, the queen was advised to leave Kensington Palace for St James's, because the road was so " infamously bad " as to separate her from her Ministers by " an impassable gulf of mud." If travelling was so difficult for royal personages over roads in the neighbourhood of London, the perils of penetrating rural districts may be imagined. In the winter months carriage traffic was suspended. Only horsemen could make their way. Judges and lawyers rode the circuits, chasing John Doe and Richard Roe from assize town to assize town on horseback. Few Quarter Sessions passed without some district being " presented " for non-repair of roads, and heavy were the fines inflicted by bruised and shaken judges, who, thinking that the majesty of the law was ill-supported by top-boots, endeavoured to reach their destination in carriages.

Even after Turnpike Trusts were generally established, travelling still continued to be neither swift, nor easy, nor safe. Guide-posts were almost unknown, and the way was frequently lost. In the reign of Charles II., the stage had taken two days to reach Oxford from London, and the journey to Exeter occupied four days. A century later, the one stage-coach, which plied once a month between Edinburgh and London, accomplished the journey in from twelve to fourteen days. Family coaches, lumbering and jolting over the uneven roads, for steel springs were not applied to carriages before the middle of the eighteenth century, made twenty miles a day. They set out provisioned and armed as if for a siege. When Sir Francis Headpiece travelled to London, he carried with him in his coach " the family basket-hilt-sword, the Turkish scimetar, the old blunder-buss, a good bag of bullets, and a great horn of powder."[1] Such precautions were not always effectual against a well-mounted highwayman, expert in the use of handier weapons ; and the slow pace at which vehicles travelled, unless they were defended with determination, made them easy victims.

Off the frequented lines of communication, and often even on these, the condition of the eighteenth century roads, as has been

[1] Vanbrugh's *Journey to London*, produced on the stage by Cibber, in 1728, under the title of the *Provoked Husband*.

shown in a previous chapter, rendered travelling in the winter months difficult, and sometimes, except for horsemen, impossible. But after 1760 a determined effort was made at improvement. Here and there some local genius, like " Blind Jack " Metcalf of Knaresborough, had already anticipated the methods of Telford and McAdam. In other parts of England, the turnpike trusts were placing portions of the highways in better repair. But the districts for which they were formed were often too small to be useful. Thus the main road from Shrewsbury to Bangor (85 miles) was in the care of six trusts, most of them in debt, all too poor to pay for skilled labour, and each too jealous of the others to co-operate. The multiplication of these turnpike trusts, though it often defeated its own object, affords strong evidence of the extent to which public attention had been called to the need for improved facilities of communication. Between 1760 and 1774 no less than 452 Turnpike Acts were passed, and in the sixteen years from 1785 to 1800 this number was increased by 643. Two General Highway Acts were passed in 1773 [1] which consolidated the previous legislation on the subject of parochial liability for road repair, transferred the appointment of surveyors to the justices of the peace out of lists of names submitted by each parish, allowed the compulsory statute labour for six days to be commuted by money payments, and authorised the levy of a rate, not exceeding 6d. in the pound, for the provision of road materials. In 1784 Palmer organised the service of mail-coaches. But letters were often still left at inns on main thoroughfares, where they remained in the bar till the ink had faded and the wrapper had turned the colour of saffron. The arrival of the pedlar was still eagerly expected in country villages, where he did not always appear as the philosophical enthusiast of the poet's licence. Rather he was the milliner of rural beauties, the arbiter of fashion to village bucks, the newsagent of the alehouse politician, the retailer of the most recent gossip, the vendor of smuggled tea, the purveyor of the latest amorous ditty. He was typical of the times when villages were isolated, self-sufficing, dependent on his summer and winter circuits for their knowledge of the world beyond the parish boundaries.

Both Young and Marshall note the improvement which was made during the last quarter of the eighteenth century in the roads of certain districts. Yet their writings, as well as the reports to the Board of Agriculture (1793-1815), afford abundant evidence that else-

[1] 13 Geo. III. cc. 78 and 84.

where much remained to be done.   For the slow progress made there
were many reasons.   Country gentlemen used the same arguments
against new roads which were afterwards employed against railways.
" Merry England " would be merry no longer if her highways ceased
to be miry.   They dreaded the disturbance of their game, feared the
intrusion of town manners, resented the sacrifice of their interests
to those of wealthy traders.   As magistrates they were reluctant
to enforce the law of road-repair against their own tenants.   Statute
labour was deservedly unpopular.   Surveyors, forced into office
against their will, only called upon their neighbours to fulfil their
liabilities as a last resource, and at seasons when agricultural work
was slack.   Urban and rural interests were opposed.   Market
towns might demand metalled roads for the transport of their
merchandise ; but self-sufficing villages were content with the drift-
ways which were sufficient to enable them to house their crops, and
to drag their flour from the mill through the same ruts which their
ancestors had worn.   Even when a parish was active in road-repair,
its energies were generally misdirected.   Roads were unguarded at
the sides.   Drainage was often provided by cutting open grips across
their surface.   If any convexity was attempted, it was so exaggerated
as to be dangerous ; the sides sloped like the roof of a house.   Hence
the whole traffic fell on the centre, which soon wore into ruts.   Many
roads were undrainable, because the continual scraping of mud from
the surface had sunk them below the level of the adjoining land.
Hence they were always wet, and, from the rapid decay of material,
expensive to maintain.   Where a parish was apathetic, the least
possible mending was done in the worst possible way.   A faggot, or a
bundle of broom or heather, powdered with gravel, served to stop
a bad hole ; if beyond repair by such means, mud, scraped from the
sides of the roads and ditches, was thrown on the centre of the road,
and into this bed was shot a cartload of large unbroken stones.   Not
infrequently the road material, raised and carted at the parish
expense, missed its destination, and made good, not the road, but
the gateways or the yard of some neighbouring farmer.
    The system of road maintenance was proving inadequate for
modern requirements.   Responsibility ceased at the parish boundaries, and no uniformity was possible.   The statute labour was
everywhere enforced with difficulty.   It was also exhausted at one
particular season, and nothing more was done till the period recurred.
It was a system of occasional outlay without continuous repair.

Surveyors were not appointed for their skill, but were compelled to serve against their will. The experience which they gained in their twelve months' service was wasted by their retirement at the end of the year when their successors were appointed. Already in France, Pierre de Trésaguet (1716-74) had set an example to European countries, laid down the principles of the construction of broken-stone roads, organised his corps of day-labourers, and substituted the principle of continuous upkeep for that of periodic repair. Already both Ireland and Scotland had gained a lead over England in the matter of road improvement. In Ireland statute labour was abolished in 1765,[1] and road-making entrusted to the County Grand Juries. Arthur Young says that before the Act was passed, Irish roads, " like those of England, remained impassable under the miserable police of the six days' labour ; . . . now the effect in all parts of the Kingdom is so great, that I found it perfectly practicable to travel upon wheels by a map. I will go here, I will go there ; I could trace a route upon paper as wild as fancy could dictate ; and everywhere I found beautiful roads, without break or hindrance, to enable me to realise my design."[2] In Scotland, in 1803, Commissioners were appointed for making roads in the Highlands. The expense was defrayed in equal portions by grants from Parliament and local contributions ; the assistance of Telford was secured, and more than 900 miles of good roads were constructed.

England, however, still lagged behind. Various alterations in the law were proposed and discussed. It was suggested that the labour service should be commuted for a money payment, and that, even if only a quarter of the equivalent were obtained in money, the roads would gain. On the other hand, it was said that commutation would be certainly unpopular with farmers, who would regard the pecuniary liability as a new tax. It was urged that large districts should be formed by uniting a number of parishes ; that surveyors should be appointed for their knowledge of road-making, and should be paid salaries ; or that, as Mace had suggested in 1675, " daymen " should be continuously employed upon the roads at weekly wages. It was not, however, till twenty years after the peace of 1815 that any substantial legislative changes were made. Before that time the science and practice, as well as the expense, of road-making and repair had made considerable advance. From 1811 onwards Parliamentary Committees sat almost continuously to hear evidence

---

[1] Irish Acts, 5 Geo. III. c. 14.　　　[2] *Tour in Ireland*, part ii. p. 40.

and to report. It was gradually realised that the construction of a good road required an unusual combination of practical and scientific knowledge, and that the task was not only above the abilities of inexperienced surveyors, but beyond the means of the inhabitants of an ordinary parish. Public money was voted for the improvement of national highways, and the services of the most celebrated engineer of the day were enlisted in the work. Telford in 1814 was employed to make good the road from Glasgow to Carlisle and in the following year to reconstruct the road from Shrewsbury to Holyhead. In his opinion and practice, it was necessary to make a regular bottoming of rough close-set pavement, on which a hard, smooth, inelastic surface could be laid, so as to minimise the labour of traction by offering the least resistance. The rival system was advocated by McAdam. To him the "Telford pavement" seemed unnecessary for the preparation of a suitable surface. In his view an elastic subsoil was even superior to a solid foundation; he preferred a bog to a rock, provided that the bog was sufficiently solid to bear a man's weight. As Surveyor-General of the Bristol roads (1815), he was already putting his theories into practice on an extensive scale. His practical success, his evidence before Parliamentary Committees, and his skill with the pen [1] persuaded the English public of the soundness of his theory. But the battle was hotly contested, and the very heat of the controversy served a useful purpose. It kept the improvement of English roads prominently before the public. Scientific opinion, here and abroad, was on the side of Telford; but McAdam was the popular favourite. In 1827 he was appointed Surveyor-General of roads in Great Britain. His influence was paramount, and men, in their gratitude for the unwonted luxury of safe and smooth travelling in fast coaches, were not disposed to criticise too closely the scientific principles of the road magician.

Turnpike tolls provided some of the cost of road maintenance, and served as auxiliaries to statute labour. For a time they satisfied the urgency of the need. But the heavy interest on the loans raised by the turnpike trustees, the excessive cost of management, the profits exacted by those who farmed the tolls, left, at the best, small margins for road expenditure. To increase the income, toll-bars were multiplied or scales of payment raised. The inequality of the burden was strongly felt. In one district, five tolls might be paid

[1] *A Practical Essay of the Scientific Repair and Preservation of Public Roads* (1819); *Remarks on the Present System of Road-making* (1820, 5th edition 1822).

in twelve miles ; in another, thirty might be travelled without a single payment. The financial chaos of the trusts, as well as the inadequacy of the statute labour, gave a fresh impulse to the ultimate triumph of the rival principle of a rate. Already Justices in Quarter Sessions had been empowered to levy a rate, assessed on the principle of the poor-rate, for general purposes of highway maintenance when other means proved insufficient, for the purchase of road material, and to buy land for the widening of highways. Already also the liability for statute labour might be compounded by the payment of a money equivalent. In 1835 these principles were extended by an Act[1] which abolished statute labour and substituted highway rates for the maintenance of all minor roads. The abolition of statute labour was a severe loss to the turnpike trusts, to whom the legislature still looked for the repair of important highways. In 1839, four years after the passing of the Highway Act, a Select Committee reported that in some instances the creditors of turnpike trusts had seized the tolls to secure payment of the interest on their mortgages, and that nothing was available from that source for road-repair. The development of railways struck the trusts another blow, for the decay of the coaching-traffic deprived them of one of the chief sources of their revenue. Their financial position went from bad to worse. Drastic action was needed. The powers of the Home Office to refuse the renewal of Turnpike Acts were in 1864 transferred to a Select Committee of the House of Commons. The new authority acted with vigour. Roads were dis-turnpiked at the average rate of 1,500 miles a year.

The extinction of turnpike trusts threw upon local ratepayers a heavy burden. Their existence had not relieved the parish from its old liability : their removal revived that liability in the form of increased rates. In rare instances, individuals were liable by tenure or prescription for the repair of portions of public roads. But, speaking generally, the parish was always responsible for the main-tenance of the highways within its area. For a time, turnpike roads had been partly maintained by the tolls which the trustees were authorised to raise. Yet whenever the trusts neglected their work, became bankrupt, or were extinguished, it was the inhabitants of the parish, not the trustees, who were subject to indictment for failure to maintain the roads. Tolls were subsidiary to local labour and local rates ; they were substitutes for neither. Now that they were

[1] 5 and 6 Wm. IV. c. 50.

withdrawn, the whole burden fell on the locality. Some relief was urgently needed. In order to distribute the burden more equitably, the parishes were grouped into Highway Districts. Within each area the cost was equalised. But parochial districts remained responsible for the maintenance of roads within their areas, legally liable for the extra burden if the expense was disproportionately heavy, legally entitled to the special benefit if the cost was disproportionately light. Further relief to local ratepayers was required. It came in the form of excepting main roads from the general law of district liability. Under the Highways and Locomotives Acts Amendment Act, 1878, the turnpike roads, whose trusts had been dissolved, were made main roads, and *half* the cost of their maintenance was transferred to the county authority, then Quarter Sessions. The remainder of the liability for the repair of main roads still rested on the parochial districts, a grant-in-aid being made by the Government. Under the Local Government Act, 1888, the County Council became the county authority, and parochial districts were relieved of the *remaining half* of their liability for the maintenance of main roads, wherever situated, and the cost of their upkeep was transferred to the county generally. But under the Public Health Act, 1875, the urban authorities were already responsible for the maintenance of highways within their areas. The effect of the two Acts of 1875 and 1888 was that urban authorities might elect either to maintain the main roads within their area themselves, or to call upon the County Council to do the work. If they elected to maintain the roads themselves, the measure of the County Council's liability was a contribution towards the cost properly incurred in the maintenance and reasonable improvement of the main roads within the area.

CHAPTER XIV

THE RURAL POPULATION. 1780-1813

Effect of enclosures on the rural population ; no necessary reduction in the number of small owners, but rather an increase ; consolidation of farms, either by purchase from small owners, or by throwing tenancies together ; the strict letter of the law ; small occupiers become landless labourers ; depopulation of villages when tillage was abandoned for pasture ; scarcity of employment in open-field villages ; the literary controversy ; the material injury inflicted upon the rural poor by the loss of the commons ; no possible equivalent in cash-value : the moral injury ; the simultaneous decay of domestic industries ; the rapid rise after 1790 in the price of provisions ; a substantial advance in agricultural wages.

DURING the thirty-three years from 1780 to 1813, the industrial revolution, which in agriculture was expressed by the new methods and spirit of farming, influenced rural life in two opposite directions. Far-reaching changes were made which were justified, and even demanded, by national exigencies. As, in trade, the capitalist manufacturer displaced the small master-workman and domestic craftsman, so, in agriculture, land was thrown together in large holdings at the expense of small occupiers. Both manufacture and agriculture became businesses which required the possession of capital. Without money, workers, whether in trade or on land, lost the prospect of themselves becoming masters or employers. But the same changes which brought unexampled prosperity to landowners and large tenant-farmers, combined with other causes to plunge the rest of the rural population into almost unparalleled misery. The rapid growth of manufacturing towns created a new demand for bread and meat ; it raised the rents of landowners ; it swelled the profits of farmers. For a long series of years the war, by practically excluding foreign corn, maintained a high level of agricultural prices in spite of increased production. But to labourers who neither owned nor occupied land, the rise of prices brought no compensating advantages. On the contrary, they paid more dearly for all necessaries of subsistence, and the increased cost of living was not adequately met by a corresponding rise in wages. At the same time, the steps which were required for the adoption of those agricultural improvements, by which the manufacturing industries as well as large owners and occupiers of land were profiting, multiplied the numbers and increased the sufferings of landless labourers. The extinction of open-field farms reduced numbers of small occupiers to the rank of hired wage-earners ; the appropriation of commons deprived many cottagers, not only of free fuel, but of the means of supplementing wages by the profits of their live-stock, their poultry, and their geese. In the eighteenth as in the sixteenth century it was still partially true that " enclosures make fat beasts and lean poor people."

The structure of rural society was affected to its very foundations by the agrarian revolution which was in progress. A great population, standing on the verge of famine, and beginning to gather in industrial centres, cried aloud for food. Technical improvements in farming had been tested, which promised to supply the new demand for bread and meat, if only free play were allowed to the modern methods of production. It was from this point of view that agricultural experts, almost to a man, were unanimous in requiring the removal of mediaeval obstacles to progress, and the addition of every possible acre to the cultivated area. As open-field arable farms were broken up, as pasture-commons were divided, as wastes were brought into cultivation, the face of the country altered. The enclosing movement was attacked on various grounds. To its effects were attributed the disappearance of the yeomanry, using the words in the strict sense of farmer-owners ; the monopoly of farms, or, in other words, the consolidation of a number of holdings into single occupations ; the depopulation of rural villages ; the material and moral loss which was alleged to be inflicted on the poor. Round these different points raged the contest of the latter half of the eighteenth century. Meanwhile the work of enclosure went on without interruption. At the present day the changes seem to have been surprisingly rapid ; but to men who were living under the stress of war and scarcity, they appeared almost criminally slow. They so appeared to William Marshall, perhaps the most experienced and the least bigoted of the agricultural observers of the day. Writing in 1801, before the full pressure of famine prices had been felt, he says : " Through the uncertainty and expense attending private acts, a great portion of the unstinted common lands remain nearly as nature left them; appearing in the present state of civilisation and science, as filthy blotches on the face of the country ; especially when seen under the threatening clouds of famine which have now repeatedly overspread it." [1]

It does not appear that the necessary result of the enclosing movement was to diminish the number of occupying owners. On the contrary, the first effect of an enclosure was to increase the freeholders, since rights of open arable field occupation and of pasture common were often replaced by allotments of land in separate ownership. After 1689, the decline in the number of owners of small estates begins to be noted by contemporary writers.[2] "At the Revolution," says a " Suffolk Gentleman,"[3] " there existed a race of Men in the Country besides the Gentlemen and Husbandmen, called Yeomanry, Men who cultivated their own property, consisting chiefly of farms from forty to fourscore pounds a year . . . the Pride of the Nation in War and Peace . . . hardy, brave, and of good morals." Their alleged disappearance can only have been remotely due to enclosure, if, as the " Suffolk Gentleman " says, " by the influx of riches and a change of manners, they were nearly annihilated in the year 1750." On the other hand, a considerable body of evidence exists to show that, after the accession of George III., a reaction had set in, and that small owners were not only numerous, but actually increasing in numbers. Thus Marshall, writing in 1790 of small freeholders both in Yorkshire (Vale of Pickering) and in Leicestershire, says : " Some years back, the same species of frenzy,—*Terramania*—showed itself here, as it did in other districts. Forty years purchase was, then, not unfrequently given."[4] The Reports to the Board of Agriculture (1793-1815) show that in many parts of the country small owners not only held their ground, but once more were buying land. Thus of the northeastern counties generally, Young [5] states that " farmers have been very considerable purchasers of land." Norfolk (1804) is said to

---

[1] *The Appropriation and Inclosure of Commonable and Intermixed Lands* (1801).

[2] Authorities are quoted in *The Disappearance of the Small Landowner*, by the Rev. A. H. Johnson (1909), pp. 136-8.

[3] *Letter to Sir T. C. Bunbury, Bart., on the Poor Rates and the High Price of Provisions* (1795).

[4] *Rural Economy of the Midland Counties*, vol. i. p. 16.

[5] Young's *Hertfordshire* (1804), p. 18.

contain " estates of all sizes, from nearly the largest scale to the little freehold ; one of £25,000 a year ; one of £14,000 ; one of £13,000 ; two of £10,000 ; many of about £5,000 ; and an increasing number of all smaller proportions."[1] In Suffolk (1797) " the rich yeomanry" are described as " very numerous . . . farmers occupying their own lands, of a value rising from £100 to £400 a year."[2] In Essex (1807), " there never was a greater proportion of small and moderate-sized farms, the property of mere farmers, who retain them in their own immediate occupation, than at present. Such has been the flourishing state of agriculture for twenty or thirty years past, that scarcely an estate is sold, if divided into lots of forty or fifty to two or three hundred a year, but is purchased by farmers. . . . Hence arises a fair prospect of landed property gradually returning to a situation of similar possession to what it was a hundred or a hundred and fifty years ago, when our inferior gentry resided upon their estates in the country."[3]

In the South-Eastern and East Midland counties, no marked decrease in the number of small estates is noticed. " One third " of Berkshire[4] is said to have been occupied in 1813 by the proprietors of the soil. Owners of landed property from £200 to £600 a year were " very numerous." Oxfordshire (1794) contained " many proprietors of a middling size, and many small proprietors, particularly in the open fields."[5] In Nottinghamshire (1798) " some considerable, as well as inferior yeomen occupy their own lands."[6] Of late years in Hampshire (1813) " a considerable subdivision of property has taken place." Speaking of the farmers on the chalk hills of the county, the Reporter says that " many of them are the possessors of small estates which their thrifty management keeps upon the increase."[7] In Kent, up to at least 1793, the number of owners of land seemed annually on the increase, " by the estates which are divided and sold to the occupiers. There is no description of persons who can afford to give so much money for the purchase of an estate as those who buy for their own occupation. Many in the eastern part of this county have been sold, within these few years, for forty, and some for fifty years purchase, and upwards."[8]

---

[1] Young's *Norfolk* (1804), p. 17.   [2] Young's *Suffolk* (1797), p. 8.
[3] Young's *Essex* (1807), vol. i. pp. 39, 40.
[4] Mavor's *Berkshire* (1808), p. 113.   [5] Davis' *Oxfordshire* (1794), p. 11.
[6] Lowe's *Nottinghamshire* (1798); p. 8.
[7] Vancouver's *Hampshire* (1813), pp. 51, 80.   [8] Boys' *Kent* (1796), p. 26.

In the West Midland and South-Western district, small owners were at least holding their own. In North Wilts (1794), where a considerable number of enclosures had been made, " a great deal of the property has been divided and sub-divided, and gone into the hands of the many." [1] Brent Marsh in Somersetshire (1797) was a district of 20,000 acres which the stagnant waters rendered unwholesome to man and beast. Within the last twenty years much of this land had been enclosed and drained under a variety of Acts of Parliament. " Scarcely a farmer," says the Reporter, " can now be found who does not possess a considerable landed property ; and many, whose fathers lived in idleness and sloth, on the precarious support of a few half-starved cows, or a few limping geese, are now in affluence, and blessed with every needful species of enjoyment." [2] Devonshire (1794) continued to be a county of small properties.[3] In Gloucestershire (1807), " the number of yeomen who possess freeholds, of various value, is great, as appears from the Sheriff's return of the poll at the election for a county member in 1776, when 5790 freeholders voted, and the number since that period is much increased." [4] Landed property in Shropshire (1803) is " considerably divided. . . . The number of gentlemen of small fortune living on their estates, has decreased ; their descendants have been clergymen or attornies, either in the country, or shopkeepers in the towns of their own county ; or more probably in this county emigrated to Birmingham, Liverpool, to Manchester, or to London ; but then the opulent farmer, who has purchased the farm he lives upon . . . is a character that has increased." [5]

The North and North-Western districts afford similar evidence, though in two counties a decrease is conspicuous. In Staffordshire (1813) the best and most improving farmers were " the proprietors of 200 or 300 acres of land, who farm it themselves." [6] Derbyshire (1794) possessed numerous small occupiers, who eked out the profits of the land by mining, spinning, and weaving ; but there were also occupiers of another description, " very properly styled yeomen ; men cultivating their own estates with a sufficient capital." [7] In Cheshire (1808) " the number of small land-owners is not apparently less than in other counties. The description of this latter class has,

---

[1] Davis' *Wiltshire* (1794), p. 8.   [2] Billingsley's *Somersetshire* (1797), pp. 166-73.
[3] Fraser's *Devonshire* (1794), p. 17.   [4] Rudge's *Gloucestershire* (1807), p. 34.
[5] Plymley's *Shropshire* (1803), p. 90.   [6] Pitt's *Staffordshire*, (1813), p. 20.
[7] Brown's *Derbyshire* (1794), p. 14.

however, been very much altered of late years. From the advantages which have been derived from trade, and from the effects of the increase of taxes, which have prevented a man living with the same degree of comfort on the same portion of land he could formerly, many of the old owners have been induced to sell their estates ; and new proprietors have spread themselves over the county, very different in their habits and prejudices." [1] In Lancashire (1795) " the yeomanry, formerly numerous and respectable, have greatly diminished of late, but are not yet extinct ; the great wealth, which has in many instances been so rapidly acquired by some of their neighbours, and probably heretofore dependants, has offered sufficient temptation to venture their property in trade, in order that they might keep pace with these fortunate adventurers. . . . Not only the yeomanry, but almost all the farmers, who have raised fortunes by agriculture, place their children in the manufacturing line." [2] "A large proportion of the county of Westmoreland," says the Reporter,[3] " is possessed by a yeomanry, who occupy small estates of their own, from £10 to £50 a year." These owners, as distinguished from tenant-farmers, were called ' statesmen. They live poorly and labour hard ; and some of them, particularly in the vicinity of Kendall, in the intervals of labour from agricultural avocations, busy themselves in weaving stuffs for the manufacturers of that town. . . . This class of men is daily decreasing. The turnpike roads have brought the manners of the capital to this extremity of the kingdom. The simplicity of ancient times is gone. Finer clothes, better dwellings, and more expensive viands, are now sought after by all. This change of manners, combined with other circumstances which have taken place within the last forty years, has compelled many a *statesman* to sell his property, and reduced him to the necessity of working as a labourer in those fields, which, perhaps, he and his ancestors had for many generations cultivated as their own." "A considerable part of the West Riding " (of Yorkshire), in 1799, " is possessed by small proprietors, and this respectable class of men, who generally farm their own lands, are as numerous in this district as in any other part of the Kingdom." [4] In the North Riding (1800), " the size of estates is very variable ; about one-third of it is possessed by yeomanry . . . much the largest proportion of

---

[1] Holland's *Cheshire* (1808), p. 79.　　[2] Holt's *Lancashire* (1795), p. 13.
[3] Pringle's *Westmoreland* (1794), pp. 18, 40.
[4] Brown's *West Riding of Yorkshire* (1799), p. 7.

the dales of the moorlands is in the possession of yeomanry, rarely amounting to £150 per annum." [1]   The Reporter asks " the common question, whether the number of the yeomanry increases or diminishes. . . . In a country like this, which is merely agricultural, I should suspect them to increase, in consequence of large properties having in late years been sold in parcels, and there being but few instances of gentlemen already possessed of considerable estates, making large purchases."

The Reports to the Board of Agriculture show that small owners were still numerous in many counties, and were increasing in Norfolk, Essex, in Hampshire, and Kent, in North Wilts, Somerset, Gloucestershire, Shropshire, and the North Riding of Yorkshire. They were dwindling in Lancashire, which was rapidly developing as a manufacturing centre, and in Westmoreland, where the hard penurious lives of the older race of statesmen were not congenial to their descendants.   In Hertfordshire farmers were not buying land, unlike their brethren in the eastern counties ; [2] but possibly the competition of city merchants gave land in the neighbourhood of London a residential value.   In Warwickshire (1794) it is definitely stated that consolidation of farms was driving occupiers off the land.   The Reporter is speaking of open fields " in the southern and eastern parts of this county," which had been enclosed, and mostly converted into pasture.   " These lands, being now grazed, want much fewer hands to manage them than they did in their former open state. Upon all enclosures of open-fields, the farms have generally been made much larger ; from these causes, the hardy yeomanry of country villages have been driven for employment into Birmingham, Coventry, and other manufacturing towns, whose flourishing trade has sometimes found them profitable employment." [3]   But though in this passage the word " yeomanry " [4] is used, it by no means

[1] Tuke's *North Riding* (1800), pp. 23, 28. [2] Young's *Hertfordshire* (1804), p. 18.
[3] Wedge's *Warwickshire* (1794), p. 20.

[4] The word " yeoman," which certainly included leaseholders for lives, and copyholders, was not confined to owners of land which they cultivated with their own hands, without being entitled to a crest.   Bacon (*Works*, vol. vi. p. 95) defines the English yeomanry as " the middle people between gentlemen and peasants," many of them living on " tenancies for years, lives and at will."   Latimer's " father was a yeoman, but had no land of his own."   He rented his occupation at £4 a year, and was a tenant-farmer.   Blackstone uses the word as equivalent to qualified rural voters (*Commentaries*, bk. i. ch. 12).   The definite restriction of the word to farmer-owners is a comparatively modern usage belonging to the nineteenth century.   See *Dictionary of Political Economy*, s.v. Yeoman.

follows that it is employed in the strict sense of occupiers who owned the land which they cultivated themselves. More probably it bears the looser meaning of " open-field farmers," with all their picturesque varieties of land tenure. Be this as it may, the evidence of the Reports is strong as to the general conditions of the country. In the period which they cover, and for a few years before, no great inroad had been made on the numbers of small owners. No necessary connection can, therefore, be established between the break-up of open-field farms and the alleged disappearance of farmer-owners.

The passage quoted from the Warwickshire Report indicates the lines on which the conflicting assertions of advocates and opponents of enclosures may be reconciled. The consequences, and often the objects, of the extinction of the system of intermixed arable strips on the open-fields, and of the partition of the pasture-commons were, generally speaking, the consolidation of larger holdings in the separate occupation of individuals. The village farm, as has been previously stated, consisted of two parts. There was the arable land, cultivated in intermixed strips ; there were the grazing rights exercised over the pasture commons. Both in legal theory and as a historical fact, only the partners in the cultivation of the tillage land were entitled to the pasture rights, which were limited to each individual by the size of his arable holding. Outside this close corporation any persons who turned in stock were trespassers ; they encroached, not only on the rights of the owner of the soil, but on the rights of those arable farmers to whom the herbage belonged. Strangers might be able to establish their rights ; but the burden of proof lay upon them. Similarly, it was only by long usage that occupiers who rented ancient cottages could exercise pasture rights, unless they also occupied arable land with their houses. The statute of Elizabeth (31 Eliz. c. 7, 1589) which ordered that four acres of land should be attached to each cottage let to agricultural labourers, evidently refers to four acres of tillage. If no arable land was attached to the cottage, the occupier might enjoy the right of providing himself with fuel, but he could not turn out stock. It was on these strict lines that enclosure proceeded, and one of its promised advantages was the power of dealing with compact blocks of land. In pursuance of this policy, Edward Laurence, in 1727, instructs his steward to purchase " all the Freeholders out as soon as possible " ; to " convert copyholds for lives into leaseholds for lives " ; to " get rid of Farms of £8 or £10 *per annum*, always supposing that some care be taken of the families " ; to " lay all the small Farms, let to poor indigent People, to the great ones," not forgetting that " it is much more reasonable and popular to stay till such Farms fall into Hand by Death." [1] This policy of substituting one large tenant for several small occupiers was generally pursued. Beyond possibility of dispute the Reports to the Board of Agriculture prove the tendency towards that " engrossment " of farms which Tudor writers denounced.[2]

The consolidation of holdings affected the old occupiers in very different ways. Where land was held by freeholders, copyholders of inheritance, or leaseholders for lives with outstanding terms, the process of collecting large areas in the hands of one owner could only be effected by purchase. On the enclosure of an open-field farm with a common attached, each proprietor had received a compact block, representing his intermixed arable strips, and an allotment corresponding in value to his pasture rights. Sometimes the area was so small as not to pay the cost of fencing ; it was sold at once, often before the award was published. In some cases, the rising standard of living, the loss of their domestic industries, the attractions of the rapid fortunes realised in trade, the temptation of the high prices which land commanded during the war, induced small owners to sell their estates. Others who for a time clung to their property found themselves, at a later stage, compelled to part with it by the increase in taxation, by the enormous rise in the poor rates, by the pressure of mortgages contracted for additional purchases, jointures, and portions, or by the fluctuations of agricultural prices, or by the failure of banks. The period at which farmer-owners diminished most rapidly in numbers was between the years 1813 and 1835.

Beyond the classes whose occupation of land or rights of common were of an independent or a permanent nature, no claim was, as a rule, recognised by enclosure commissioners. If any compensation was made, it was on voluntary and charitable lines. The strict letter of the law was generally followed. Occupiers of arable

---

[1] *Duty of a Steward to his Lord*, pp. 37, 60, 55, 35.

[2] Thomas Wright, in *The Monopoly of Small Farms a great cause of the present Scarcity* (1795), p. 9, urges the formation of societies to purchase large estates, divide them into small farms, and let or sell them to small farmers. " It is computed," writes the author of *A Plan for Relieving the Rates by Cottage Acres*, etc. (1817), " that since the year 1760 there have been upwards of forty thousand small farms monopolised and consolidated into large ones and as many cottages annihilated."

land, whose tenure depended on the will of the owner from whom their rights were derived, had no independent or permanent title to the strips which they cultivated, or to the common of pasture which they had enjoyed in virtue of their arable holdings. Many of them were offered no chance of renting land under the new system. If the holdings were thrown together, and let to a farmer with sufficient capital, the previous occupiers were at once reduced to landless labourers. If the open-field farmer was allowed to remain in the separate occupation of a compact holding, formed out of his arable strips and commuted common rights, he was often hampered by insufficient grass, by scanty capital, by the novelty of his new position, by ignorance of any but the traditional practices of farming. He went from bad to worse, and was in the end compelled to surrender his land and compete for employment for wages. Cottagers, who occupied at a yearly rent the ancient cottages to which common rights were attached, received no compensation for the loss of rights which they only exercised as tenants. Squatters, who had encroached on the wastes and commons, and had not made good their titles by prescriptive occupation, were evicted. Whether the village was depopulated by the change or not, mainly depended on the use to which the enclosed land was put.[1] If, as in the Warwickshire case, the tillage was converted into pasture, employment was reduced, and the rural population decreased. When, on the other hand, the breadth of tillage was either maintained or extended, and when the modern improvements in farming were introduced, there was an increase in employment and also in numbers.

It would be a mistake to suppose that village farms created a demand for agricultural labour, or offered facilities for acquiring land to increasing numbers. The contrary was the case. The open-field system was inelastic, adapted for a stationary population, dependent for the employment of surplus numbers on the large enclosed farms of the neighbourhood, or on the practice of domestic handicrafts, eked out by common-rights exercised under legal titles or by successful encroachments. The smaller occupiers, their wives and families, tilled their holdings for themselves ; the common herdsman, shepherd, and swineherd tended their live-stock.

---

[1] The question of depopulation is discussed in William Wales' *Inquiry into the Present State of Population in England and Wales* (1781), and in the Rev. John Howlett's *Enquiry into the Influence which Enclosures have had upon the Population of England* (1786).

On middle-sized occupations, servants in husbandry, annually hired at the fairs for fixed yearly wages, and boarded and lodged in the house, did the work of the farm. Except at harvest there was little demand for day-labour. Threshing, the most unwholesome of rural occupations, was practically the only winter employment. On the open-fields, there were no quickset hedges to plash, or trim, or weed ; no ditches to scour ; no drains to maintain. There were no drilled crops to keep clean ; turnips were seldom grown, and beans rarely hoed. This scarcity of constant, and especially of winter, employment, which will probably be reproduced under the rule of small holdings, partly explains the slow growth of rural population. It also emphasises the value to day-labourers of commons and domestic handicrafts. Without them it is difficult to understand how agricultural labourers, who were not partners in village farms, even existed. " In hay and harvest time," writes Forster,[1] " it is inconceivable what numbers of tradesmen and handicraftsmen flock into the country." " If," says Stone, " the farmers in the most unenclosed counties . . . where there are no manufactories, could get no further assistance during their harvest than from their own inhabitants, their grain would frequently be spoiled."[2] To the same effect wrote the Reporters to the Board of Agriculture. Open-field farmers were in harvest dependent on migratory labour. In unenclosed counties, says the Report to the Board for Huntingdonshire, very little employment of a constant kind was given to labourers, who stop with the farmers to help thresh out their grain in the winter, and " leave for more cultivated counties where labour is more required." The open-field farmer of the county depended for harvesting on " the wandering Irish, manufacturers from Leicestershire and other distant counties." The same was said to be true of Bedfordshire and Cambridgeshire. In Wiltshire, the crops were harvested by taskers " from the more populous parts of the county or from Somersetshire, or other neighbouring counties."[3] In the Isle of Wight, during the harvest of 1793, there were " from 600 to 700 " labourers employed " from Dorsetshire and West Somerset." It illustrates the times to add that, as there was a hot press out for the Navy, they came and went with a pass from the Government.[4] In Herefordshire the

---

[1] *Enquiry into the Present High Price of Provisions* (1767).

[2] *Suggestions*, etc., p. 31.   [3] Davis' *Wiltshire* (1794), p. 89.

[4] Driver's *Hampshire* (1794), p. 65.

crops were harvested by Welshmen from Cardiganshire, men owning one horse between four or five, riding bare-backed, turn and turn about, and "covering great distances with extraordinary speed." [1]

Enclosed counties where tillage was maintained, therefore, already afforded larger and more constant employment than unenclosed counties. Still greater was the demand for labour, where the improved practices had been adopted. If, however, the enclosed arable land was laid down to grass, the opposite effect was produced. Up to the last decade of the eighteenth century, it is probable that open arable farms, especially in the Midland counties, were mainly enclosed for conversion to pasture. In the later stages of the Napoleonic wars this tendency to grass-farming was not only checked, but violently reversed, and large tracts of pasture were ploughed for corn. Yet, during the first thirty years of the reign of George III., the occupiers of village farms had reason to fear, not only loss of their holdings, but scarcity of employment. Anonymous pamphlets are not the most reliable evidence ; but the "Country Gentleman" [2] is quoted by the Board of Agriculture with approval. His description of the dislike and alarm with which schemes of enclosure were regarded by the rural population may therefore be accepted as true :—"the great farmer dreads an increase of rent, and being constrained to a system of agriculture which neither his inclination or experience would tempt him to ; the small farmer, that his farm will be taken from him and consolidated with the larger ; the cottager not only expects to lose his commons, but the inheritable consequences of the diminution of labour, the being obliged to quit his native place in search of work." Their fears were often justified. Many an open-field farmer verified the truth of the "Country Gentleman's" conclusion that, after enclosure, "he must of necessity give over farming, and betake himself to labour for the support of his family." Hundreds of cottagers, deprived of the commons, experienced that lack of rural employment which drove them into the towns in search of work. To make the lot of these "reduced farmers" as easy as possible, he recommended that a "sufficient portion of land" should be attached to their cottages to enable them to keep a cow

---

[1] Clark's *Herefordshire* (1794), p. 29.

[2] *The Advantages and Disadvantages of enclosing Waste Lands and Common Fields*, by a Country Gentleman (1772), pp. 8, 32.

or two. With the gloomy forebodings of the "Country Gentleman" may be contrasted the triumphant hopefulness of Arthur Young. Both wrote at the same date; yet the gloom of the one and the hopes of the other were equally well founded in the districts to which they respectively refer. What, asks Young, will opponents "say to the inclosures in *Norfolk, Suffolk, Nottinghamshire, Derbyshire, Lincolnshire, Yorkshire*, and all the northern counties? What say they to the sands of *Norfolk, Suffolk*, and *Nottinghamshire*, which yield corn and mutton and beef from *the force of* INCLOSURE *alone*? What say they to the Wolds of *York* and *Lincoln*, which from barren heaths at 1s. per acre are *by* INCLOSURE *alone* rendered profitable farms? . . . What say they to the vast tracts in the peak of Derby which *by* INCLOSURE *alone* are changed from black regions of ling to fertile fields covered with cattle? What say they to the improvements of moors in the northern counties, where INCLOSURES alone have made those countries smile with culture which before were dreary as night?"[1]

In 1774, when both Arthur Young and the "Country Gentleman" were writing, improved methods of arable farming and the use of roots, clover, and artificial grasses had not extended beyond a few favoured districts; corn and cattle were still treated as distinct departments of farming, impossible on the same land; the tendency was still strong to convert arable land into pasture; the science of stock-breeding and stock-rearing was still in its infancy; improved means of communication had not relieved farmers in almost every district from the necessity of devoting the greater part of their holdings to corn-growing, or enabled them to put their land to the best use by facilitating the interchange of arable produce; above all, no urgent demand for meat and milk, as well as bread, was as yet made by a rapidly growing class of artisans. In another twenty years these conditions had been changed, or were altering fast. But it is to this early period, when arable land was being converted to pasture, and the superiority of the new agricultural methods was still disputed, that nearly all the writers belonged, who are most frequently quoted for or against enclosures. After 1790 no voice is raised against the movement on any other ground than the moral and social injury inflicted upon open-field farmers and commoners. The economic gain is

[1] *Political Arithmetic* (1774), p. 150.

admitted.[1] Individual occupation, as an instrument of scientific and practical farming and of increased production, had demonstrated its superiority over commonable fields. The supply of eggs and poultry may have dwindled ; but it was more than compensated by the larger supply of bread and meat. The arguments of the deserted village and of scarcity of employment were losing their force, when, under the strong pressure of necessity, the reaction had set in from pasture to extended tillage. In these directions the defence of the enclosing movement was immensely strengthened. But, during the same period, the social results of the agrarian revolution were rapidly revealing themselves, and were attracting increased attention. Those results, aggravated in their evil effects by

---

[1] The following works may be quoted in proof :

1. *Essay on the Nature and Method of ascertaining the Specifick Shares of Proprietors upon the Inclosure of Common Fields*, by H. S. Homer, 1766.
2. *An Enquiry into the Reasons For and Against Inclosing the Open Fields*, by a member of the Legislature, 1767.
3. *An Enquiry into the Causes of the Present High Price of Provisions*, by Nathaniel Forster, 1767.
4. *Reflections on Inclosing Large Commons and Common Fields*, by W. Pennington, 1769.
5. *Observations on Reversionary Payments, etc.*, by Richard Price, 1771.
6. *The Advantages and Disadvantages of inclosing Waste Lands and Open Fields*, by a Country Gentleman, 1772.
7. *An Inquiry into the Reasons For and Against inclosing Open Fields*, by Stephen Addington, 2nd edition, 1772.
8. *An Inquiry into the Connection between the present price of provisions and the Size of Farms*, by a Farmer [John Arbuthnot], 1773.
9. *Four Tracts, together with Two Sermons, on political and commercial subjects*, by Josiah Tucker, 1774.
10. *Hints to Gentlemen of Landed Property*, by Nathaniel Kent, 1777.
11. *An Enquiry into the Advantages and Disadvantages resulting from Bills of Inclosure, etc.*, Anon, 1780.
12. *Observations on a Pamphlet entitled An Enquiry into the Advantages, etc.*, Anon, 1781 (an answer to the foregoing).
13. *Cursory Remarks on the Importance of Agriculture*, by W. Lamport, 1784.
14. *A Political Enquiry into the Consequences of Enclosing Waste Lands, Being the sentiments of a Society of Farmers in ——shire*, 1785.
15. *An Enquiry into the Influence which Enclosures have had upon the Population of England*, by John Howlett, 1786.
16. *Cursory Remarks on Inclosures, etc.*, by a Country Farmer, 1786.
17. *Enclosures a Cause of Improved Agriculture, etc.*, by John Howlett, 1787.
18. *Suggestions for rendering the Inclosure of Common Fields and Waste Lands a source of Population and Riches*, by Thomas Stone, 1787.

An apparent exception is one of the most interesting works on the subject, namely :

19. *The Case of Labourers in Husbandry Stated and Considered*, by David Davies, 1795.

But the material was collected in 1787. The high prices of corn, 1765-74, seem to have given an impulse to enclosures and produced a crop of literature.

industrial changes and the operation of the Poor Law, were disas-trous to a large number of open-field farmers, cottagers, and commoners who had lost their hold upon the land. The strongest argument against enclosures was the material and moral damage inflicted upon the poor.

In comparatively rare instances commoners who exercised com-mon-rights were not put to strict proof of their legal title. Even where this lenient policy was adopted, or where the right was established at law, the claim was often supposed to be satisfied by the gift of a sum of money, or by an allotment of land. Money, to a man who had no power of investment, was a precarious pro-vision, which generally was soon spent. Land was a better sub-stitute ; but the allotment might be too small to repay the cost of fencing, or too distant to be of real benefit ; it was seldom enough for the summer and winter keep of a cow. The land and the cow were often sold together, as soon as, or sometimes before, the award was made. Sometimes, again, legal principles were set aside, and allotments of land, more or less inadequate, were made for cottage building, or for the benefit of the poor of the parish to supply pasture or fuel. But probably less than 5 per cent. of the enclosure Acts made any provision of this kind.

The injury inflicted on the poor by the loss of their common of pasture, whether legally exercised or not, was indisputably great. It was admitted by those who, on other grounds, were the strongest supporters of enclosures. Arthur Young himself, though he never swerved from his advocacy of large enclosed holdings, had been converted to the principle of an admixture of occupying ownerships for small farmers. His travels in France had shown him the " magic of property " at work. In England he had witnessed its effects in the Isle of Axholme. " In respect of property," he writes,[1] " I know nothing more singular respecting it, than its great division in the isle of Axholm. In most of the towns there, for it is not quite general, there is much resemblance of some rich parts of France and Flanders. The inhabitants are collected in villages and hamlets ; and almost every house you see, except very poor cottages on the borders of commons, is inhabited by a farmer, the proprietor of his farm, of from four or five, and even fewer, to twenty, forty, and more acres, scattered about the open-fields, and cultivated with all that minutiæ of care and anxiety, by

[1] *Lincolnshire* (1799), p. 17.

the hands of the family, which are found abroad, in the countries mentioned. They are very poor, respecting money, but very happy respecting their mode of existence. Contrivance, mutual assistance, by barter and hire, enable them to manage these little farms though they break all the rules of rural proportion."

On these lines, he urged in 1800 [1] that every scrap of waste and neglected land should be converted into possessions for the poor, and that all labourers should be assigned gardens and grass-land for the keep of a cow. In 1801 [2] he proposed that labourers should be allowed to absorb for themselves the small commons which were situated in the centre of enclosed districts, and that all Acts of Parliament for the reclamation of wastes should attach enough land to every cottage to provide summer and winter keep for a cow, the land to be inalienable and vested in the parish. He based these recommendations on his own personal observations of the effect of the enclosure of commons. " Many kept cows that have not since " is his frequent summary of results. Out of 37 parishes, he found only 12 in which the poor had not suffered.[3] " By nineteen Enclosure Acts out of twenty, the poor are injured, in some grossly injured. . . . The poor in these parishes may say, and with truth, *Parliament may be tender of property* ; *all I know is I had a cow, and an Act of Parliament has taken it from me.*" [4] The Board of Agriculture printed evidence to the same effect.[5] Out of 68 Enclosure Acts, 53 had injured the poor, who had lost their cows, and could no longer buy milk for their families. The same point is frequently noticed by the Reporters. Nathaniel Kent, for example, dwells upon it in his Report on Norfolk, and urges " all great farmers . . . to provide comfortable cottages for two or three of their most industrious labourers, and to lay two or three acres of grass land to each to enable such labourer to keep a cow and a pig." [6] Yet even when the opportunity to keep a cow occurred, it was not invariably used. " Cottagers," says Kent, "who live at the sides of the common generally neglect the advantage they have before them. There is not, perhaps, one out of six,

---

[1] *Question of Scarcity plainly stated* (1800).

[2] *Inquiry into the Propriety of applying Wastes to the better Maintenance and Support of the Poor* (1801).

[3] *Ibid.* p 19. [4] *Ibid.* p. 42.

[5] *General Report on Enclosures*, 1808, pp. 150-2.

[6] Kent's *Norfolk*, p. 172.

upon an average that keeps even a cow." [1] Nor was the disappearance of the cow invariably due to the loss of commons. Sometimes commercial motives operated. At Baldon, in Oxfordshire, "many cottagers had two, three or four acres, and they kept cows ; now, still having the land, they keep no cows ; their rent from 30s. to 42s. per acre and all applied as arable." [2] In this instance, at all events, the cheapness of butter and the high price of wheat had tempted these men to plough up their grass-land. Whatever exceptions there may have been, the loss of the cow generally followed the loss of the commons. Nor was this the only injury which the cottager suffered. He lost his free firing, and the run for his geese and poultry. It is, in fact, impossible to measure in terms of cash equivalents the benefits derived from the commons, or the loss inflicted by their withdrawal. The case is well summarised by Barnes, the Dorsetshire poet of rural life :

Thomas (*loq.*) :⠀⠀⠀⠀⠀⠀⠀⠀⠀⠀⠀⠀⠀Why, 'tis a handy thing
⠀⠀⠀⠀⠀⠀⠀To have a bit o' common, I do know,
⠀⠀⠀⠀⠀⠀⠀To put a little cow upon in spring,
⠀⠀⠀⠀⠀⠀⠀The while woone's bit ov orchard grass do grow.

John :⠀⠀⠀⠀⠀⠀Aye, that's the thing, you zee. Now I do mow
⠀⠀⠀⠀⠀⠀⠀My bit o' grass, an mĕake a little rick ;
⠀⠀⠀⠀⠀⠀⠀An' in the zummer, while do grow,
⠀⠀⠀⠀⠀⠀⠀My cow do run in common vor to pick
⠀⠀⠀⠀⠀⠀⠀A bleäde or two o' grass, if she can vind em,
⠀⠀⠀⠀⠀⠀⠀Vor tother cattle don't leäve much behind em.
⠀⠀⠀⠀⠀⠀⠀An' then, bezides the cow, why we do let
⠀⠀⠀⠀⠀⠀⠀Our geese run out among the emmet hills ;
⠀⠀⠀⠀⠀⠀⠀An' then, when we do pluck em, we do get
⠀⠀⠀⠀⠀⠀⠀Vor zeäle zome veäthers an' zome quills ;
⠀⠀⠀⠀⠀⠀⠀An' in the winter we do fat em well,
⠀⠀⠀⠀⠀⠀⠀An' car em to the market vor to zell
⠀⠀⠀⠀⠀⠀⠀To gentle-volks.⠀⠀⠀.⠀⠀⠀.⠀⠀⠀.
⠀⠀⠀⠀⠀⠀⠀An' then, when I ha' nothén else to do,
⠀⠀⠀⠀⠀⠀⠀Why, I can teäke my hook an' gloves, an' goo
⠀⠀⠀⠀⠀⠀⠀To cut a lot o' vuzz and briars
⠀⠀⠀⠀⠀⠀⠀Vor hetòn ovens or vor lightèn viers ;
⠀⠀⠀⠀⠀⠀⠀An' when the childern be too young to eärn
⠀⠀⠀⠀⠀⠀⠀A penny, they can g'out in zunny weather,
⠀⠀⠀⠀⠀⠀⠀An' run about, an' get together
⠀⠀⠀⠀⠀⠀⠀A bag o' cow-dung vor to burn.

The material loss inflicted on the poor was great : still more serious was the moral damage. It is probably true that the commons had attracted to their borders numbers of the idle and dissolute. But it is equally certain that they also afforded to hard-working and thrifty peasants the means of supplementing

---

[1] *Hints to Gentlemen* (1776), p. 112.⠀⠀⠀[2] Young's *Oxfordshire*, p. 23.

their weekly wages. They gave the man who enjoyed rights of common, and lived near enough to use them, an interest in the land and the hope of acquiring a larger interest. They encouraged his thrift and fostered his independence. Men who had grazing rights hoarded their money to buy a cow. They enabled wage-earners to keep live-stock, which was something of their own. They gave them fuel, instead of driving them to the baker for every sort of cooking. They formed the lowest rung in the social ladder, by which the successful commoner might hope to climb to the occupation of a holding suited to his capital. Now the commons were gone, and the farms which replaced them were too large to be attainable. Contemporary writers who comment on the increasing degradation of the labouring classes too often treat as its causes changes which were really its consequences. They note the increase of drunkenness, but forget that the occupation of the labourer's idle moments was gone ; they attack the mischievous practice of giving children tea, but forget that milk was no longer procurable ; they condemn the rising generation as incapable for farm labour, but forget that the parents no longer occupied land on which their children could learn to work ; they deplore the helplessness of the modern wives of cottagers who had become dependent on the village baker, but forget that they were now obliged to buy flour, and had lost their free fuel ; they denounce their improvident marriages, but forget that the motive of thrift was removed. The results were the hopelessness, the indifference, and the moral deterioration of the landless labourer. " Go," says Arthur Young, " to an ale-house kitchen of an old enclosed country, and there you will see the origin of poverty and the poor-rates. For whom are they to be sober ? For whom are they to save ? (such are their questions). For the parish ? If I am diligent, shall I have leave to build a cottage ? If I am sober, shall I have land for a cow ? If I am frugal, shall I have half an acre of potatoes ? You offer no motives ; you have nothing but a parish officer and a workhouse. Bring me another pot." [1] The same point is urged, with less vivacity and picturesqueness of statement, by the best writers of the day, especially by Howlett and Davies.

The displacement of numbers of cottagers, commoners, and open-field farmers came at a difficult crisis. Hitherto rural labourers in many parts of the country had regarded day work for wages

---

[1] *Annals of Agriculture*, vol. xxxvi. p. 508. *On Wastes* (1801), pp. 12, 13.

on the land of farmers as a by-employment, which eked out the profits of their other industries. Now the commons were gone, and at the same time their own domestic handicrafts were being superseded by manufactured goods. It was now that the industrial population was shifting from the South to the North ; that spinning and weaving deserted the home for the factory ; that old markets were exchanged for the new centres of trade which gathered round the water-power or the coal and iron fields of the North. In the closing years of the eighteenth century, widespread complaints are made of decaying industries, of the loss of employment in rural districts, of the mass of pauperism bequeathed to small towns and villages by the departure of trades.

Industries, which in 1800 were concentrating in the large towns of the North, had been previously scattered over a wide extent of country districts. Even where the trade maintained its ground, the introduction of machinery reduced the amount of employment, and transferred it from the cottage to the factory. At the same time many local manufactories were brought to the verge of ruin by the war, which limited the export trade. As the result of these changes in the conditions of rural life, poor-rates rose to an enormous height. Marshall, in his *Review of the Reports to the Board of Agriculture*, mentions the instance of Coggeshall in Essex, once a flourishing village, where the poor-rates, owing to the ruin of the baize trade, had risen to 16s. in the pound. This burden, increased as it was by the provision for the maintenance of the wives and families of militiamen, enlisted soldiers and sailors, crushed out of existence many small freeholders, who, because they employed no labour, derived no advantage from the operation of the Poor Law, but were assessed on the rental value of their land. As the local industries declined, or were concentrated in towns, or substituted machinery for manual work, the demand for labour was reduced in rural villages. Fewer opportunities for supplementing weekly wages by other employments were afforded. It was now that the South and South Midlands fell hopelessly behind the North.

It is difficult to give any adequate impression of the degree in which, under the dying system of self-contained communities, industrial employments other than those of agriculture had been distributed among rural villages. Counties which at first sight seem purely agricultural, possessed a number of local industries, which, in addition to dyeworks, malthouses, breweries, mills, and tanneries, gave considerable employment. Bedfordshire had its osier baskets' its reed matting, its straw plaiting ; its spinning of hemp had died out in 1803, but men as well as women still made pillow lace. Straw-plaiting extended along the borders of Buckinghamshire, Hertfordshire, and Cambridgeshire. The best, that is, the weakest straw, commanded high prices, and sold for from 2d. to 4d. the pound. In Cambridgeshire and Huntingdonshire, woollen and worsted yarn was also spun for Norwich and the northern markets. Lincolnshire wove fabrics for women's dresses ; Epworth made sack-cloth ; coarse linen or hempen cloth was woven in many parts of the county. Suffolk had its spinning and combing of wool ; and in the district round Beccles, where hemp was largely cultivated, quantities of hempen cloth were manufactured. Essex was famous for its baizes. But the trade was for the time ruined by the war. In the neighbourhood of Colchester, where during peace 20,000 persons had found employment, only 8,000 were now employed. At Halstead, Dedham, Bocking, and the surrounding villages, the industry had so decayed that numbers of hands were out of work, and the rates rose to over 20s. in the pound.

Hampshire was not a manufacturing county. But it had a variety of industries ranging from manufactories of cloth, shalloons and coarse woollens, to bed-ticking and earthenware pottery. Kent was the county of hops ; yet Canterbury and the villages round wove silk ; Dover and Maidstone made paper ; Crayford bleached linens and printed calicoes ; Whitstable had its copperas works, Sandwich its salt-works, Faversham and Deptford their powder mills. Along the banks of the Wandle in Surrey were paper, oil, snuff and flour mills, mills for grinding logwood, as well as leather, parchment, calico, and printing works. The Mole turned iron mills at Cobham and flatting mills at Ember Court. The Wey collected on its banks many paper mills. In the Weald there still lingered iron-workers and charcoal burners. Godalming and the neighbourhood had its patent fleecy hosiery, its works for wool-combing, for blankets, tilts, and collar-cloths. Sussex formerly sent every year quantities of iron by land-carriage to London ; but the trade was dying fast. It still remained one of the chief centres of the charcoal industry and of powder making. In Berkshire the woollen manufactures were dwindling. They were deserting Newbury, leaving behind a " numerous poor." But in the town and neighbourhood, kerseys, cottons, calicoes, linen and damask were still made, and the

P*—E F P P

introduction of the manufacture of blankets was attempted. At Oakingham there were established silk-spinning and silk-weaving manufactories, and a considerable trade was carried on in hat-bands, ribbons, watch-strings, shoe-strings, sarcenets, and figured gauzes for women's dresses. In Oxfordshire the shag-weavers of Bloxham and Banbury were out of employment and on the parish. The coarse velvet trade of Banbury was travelling north. The blankets of Witney still held their own ; but the introduction of machinery had thrown two-thirds of the workmen out of employment, and the rates had risen to 11s. in the pound. The glove trade of Woodstock flourished ; but the polished steel trade had migrated to Birmingham and Sheffield, and leathern breeches, no longer worn, had ceased to be made.

Northampton and the surrounding neighbourhood were already famous for boots ; Daventry manufactured whips and wove silk stockings ; in Wellingborough and the surrounding villages lace-making employed from 9,000 to 10,000 persons, the thread being imported from Flanders and distributed to the workers in their houses. Since the war, the worsted manufactories of Kettering had decayed ; instead of from 5,000 to 6,000 hands, only half were employed, and the remainder fell upon the rates. Warwickshire, Nottinghamshire, and Staffordshire were becoming manufacturing counties. Machinery was being introduced, and, as a consequence, their industries were being withdrawn from the villages and concentrated in towns. Outside Birmingham, there were the ribbon and tammy trades of Coventry, the horn combs of Kenilworth, the nails of Bromwich, the needles of Alcester, the worsted works of Warwick, the linen trade of Tamworth. For miles round Nottingham the villagers were stocking-makers ; in different parts of the county were scattered mills for combing and spinning wool, or silk spinning and weaving, for polishing marble, as well as works for the manufacture of pottery, starch, and sail-cloth. Few cottagers were without a web of home-spun cloth. Shropshire had a great variety of local industries, such as garden pots at Broseley ; fine china at Caughley ; china, ropes, and chains at Coalport ; glass works at Donnington ; dye-works at Lebotwood ; Shrewsbury and the neighbourhood maintained spinning and fulling mills, a trade in finishing Welsh flannels, manufactures of coarse linens and linen threads. In many cottages and farm-houses pieces of linen cloth were got up for sale. The glove trade of Worcester employed a large number of men and women in the city itself, and in the " county round to the extent of seven or eight miles." Kidderminster and the neighbourhood were carpet-makers. On the borders of Staffordshire and Warwickshire many were employed in making nails, needles, and fish-hooks.

Over the greater part of Gloucestershire, and especially in the Cotswold Hills, there was much spinning of wool. The trade in the fine broad-cloths of Stroud and the surrounding parishes was in 1794 at a stand-still ; but in the coarser quality required for army clothing it was brisk. Even here the introduction of machinery " has thrown many hands out of employment," and caused the poor-rates to rise " to six shillings in the pound and upward." At Cirencester in 1807 many labouring people were still employed in sorting wool from the fleece ; but the wool trade had much decreased in the last forty years, as also spinning woollen yarn and worsted since the introduction of machinery. Tewkesbury had its stocking-frame industry ; Dursley and Wotton-under-Edge made wire cards for the use of clothiers ; iron and brass wire, tin-plate, pins, rugs and blankets employed other districts of the county. But the decline of trade made itself felt in the great increase of rates. " In the clothing district," says Rudge, " the weight of parochial assessments falls uncommonly heavy on landed property. During the late scarcity, the average charge might be 4s. 6d. through the county ; while at the same time it amounted to at least three times that proportion in some of the parishes where the clothing manufacture is carried on." [1] In Somersetshire, the trade in woollen cloth and worsted stockings of Frome and Shepton Mallet had given employment, not only to the two towns, but to " a vast number of the lower order of people in the adjacent villages." But in 1797 the restriction of the export trade by the war, the introduction of machinery, and the competition of the North, had begun to injure the trade and lessen the demand for labour. Taunton had lost its woollen manufactures, though they still flourished at Wellington and Wiveliscombe.

In Cornwall, carding and spinning were in 1811 dying out, and " to the total decline of this business must, in some measure, be attributed the progressive increase of the rates of the county." [2] From Devonshire in 1808 came the same complaint of the failure of em-

---

[1] *Gloucestershire* (1807), p. 346.
[2] Worgan's *Cornwall* (1811), p. 33.

ployment, though in the eastern part of the county lace-making still flourished.[1] In Dorsetshire the principal manufactures were in the neighbourhood of Bridport and Beaminster, where in 1793, " all sorts of twine, string, packthread, netting, cordage, and ropes are made, from the finest thread used by saddlers, in lieu of silk, to the cable which holds the first-rate man of war."[2] In this neighbourhood also were made the sails for shipping, sacking for hammocks, and all kinds of bags and tarpaulin. Here too were braided nets for the Newfoundland fishery and for home use. At Loders sail-cloth was woven. At Shaftesbury and Blandford, and in the surrounding villages on all sides, to seven or eight miles distance, was carried on the manufacture of shirt-buttons.

Two other changes were in progress which in a minor degree added to the misfortunes of the labouring classes in country districts. In the first place, trade in agricultural produce was rapidly becoming wholesale instead of retail. Dairy-farms contracted for the supply of milk to towns, and milk was more easily obtained by the urban than the rural population. The produce of corn-farms was sold in bulk to corn-dealers or millers. Labourers could rarely purchase a bushel of wheat direct from the farmer. They could no longer carry their corn to the miller, pay for grinding, and take away the pure flour, and the offals for the pigs. Now they were obliged to buy from the miller or the baker, not only the flour, but the bran, with the profits of each trader added to the price of both. In the second place, a number of crops, some of which required much labour for their cultivation or special preparation, were dying out, because the industries which they served had migrated, or from some change of taste or fashion. As the linen trade became more concentrated in particular localities, flax was more rarely cultivated. The hempyards, which were once attached to many cottages and farm-houses, were similarly abandoned. The use of teasels by clothiers was displaced by machinery, and the crop was no longer cultivated. Woad, madder, and saffron found cheaper substitutes. Liquorice disappeared from Nottinghamshire, camomile from Derbyshire, canary seed from Kent, carraway seed from Essex.

The rapid increase in the price of provisions from 1793 onwards struck yet another, and a crushing, blow at the position of the landless labourer. The rise came with startling suddenness, and it

---

[1] Vancouver's *Devonshire* (1808), p. 464.
[2] Claridge's *Dorsetshire* (1793), p. 37.

found him defenceless. Without the commons he was entirely
dependent on purchased food ; without domestic industries he had
less money to buy the means of existence. The greater the distance
from London, the lower the wages and the higher the prices. This
was certainly true of the West and South-West of England. Thus,
the labourer had more to buy, less money to buy it with, and what
money he had did not go so far as formerly. In yet another way,
the great rise in prices affected the rural population for the worse.
It no longer paid the farmer to board servants in husbandry. In the
North, the system still survived, partly because of the high wages of
day-labour, partly because the diet which custom accepted was more
economical, and barley-broth and porridge were staple foods.
Elsewhere the number of servants who were boarded and lodged
in farm-houses dwindled ; they became day-labourers, living how
and where they could. Another opportunity for saving and another
restraint on improvident marriage were thus removed.

To a certain extent the rise in the prices was met by a substantial
advance in wages. It is always easy to raise wages ; it is extremely
difficult to lower them. The reluctance of farmers to increase wages,
when an advance in prices may be only temporary, is therefore
intelligible. How far wages rose is a difficult field of enquiry. The
remuneration of labour varies with the different seasons, with the
different occupations of the men, with different contracts of service,
with different districts of the same county. The one outstanding
point is that the real earnings of agricultural labourers are not now,
and, to a greater extent, were not then, represented only by the
weekly sums which are paid in cash. To these weekly payments
must be added earnings at piece-work, at hay and corn harvest,
perquisites, allowances in kind, cottages and gardens, either rent
free or rented below their economic value. On these points the
Reports to the Board of Agriculture, 1793-1815, supply no reliable
evidence. Most of them speak of a considerable rise in wages ;
they rarely mention the point from which the advance is
measured. They register the averages of the daily or weekly
payments ; they seldom give the method by which the rate is
calculated.

Failing the Reports to the Board of Agriculture, the enquirer is
thrown back on Young's generalisations. As the result of his
calculations in the *Farmers' Tours* of 1767-70, it may be estimated
that the average rate of wages was 1s. 2d. a day,—more in the

neighbourhood of London, less in more distant counties, least in the West and South. From 1770 to 1790 there does not seem to have been any appreciable and general rise. In the next twenty-five years a striking advance was made. Tooke states that agricultural wages were " doubled or nearly so." [1]   Young calculated that, taking the " mean rate " of wages in 1770 at 7s. 4½d., the " price of labour had in forty years about doubled." [2]   Both he and Tooke state that the wages of agricultural labourers had reached the level of those of artisans. It is difficult to accept these estimates. Few of the Reports to the Board of Agriculture really belong to the later part of the period 1793-1815, and the only county in which the Reporters to the Board state that wages had doubled between 1794 and 1812 is Warwickshire. In Essex, however, there is some indication of wages having doubled, if the 1s. 2d. of 1770 is taken as the starting-point. In the Report for 1794, the average of summer and winter wages is given as 9s. 1½d. a week ; in that for 1807, at 12s. 7d. The evidence of the subsequent rise comes from another source. On an Essex farm the rate of wages paid to an ordinary labourer, who had not the care of stock, rose from 10s. 6d. a week in 1800 to 12s. a week in 1802, and to 15s. a week in 1812.[3]   Whatever weight may be attached to the generalisations of Tooke and Young, it is certain that a very important advance in agricultural wages was made during the period of the Napoleonic wars. Unfortunately, it is equally certain that, even

[1] *History of Prices*, vol. i. p. 329.

[2] *Enquiry into the Rise of Prices in Europe during the last twenty-five years* (1815), p. 215.

[3] *Board of Trade Report on Agricultural Wages* (1900), Cd. 346, p. 238. In the *Communications to the Board of Agriculture* (vol. v. part i.), the average weekly wages of agricultural labourers in 1803 are stated at 11s. 11d. In Arthur Young's *Enquiry into the Rise of Prices in Europe*, the weekly wages in husbandry are stated to be 14s. 6d. in 1811. J. C. Curwen, M.P., moving in the House of Commons for a Committee to consider the Poor Laws (May 28, 1816), speaks of agricultural wages at that time as ranging from 10s. to 15s. (*Pamphleteer*, vol. viii. p. 9). A. H. Holdsworth, M.P. (*Letter on the Present Situation of the Country* (1816), *Pamphleteer*, vol. viii. p. 428), speaks of agricultural labourers receiving 2s. 6d. a day before the reductions of 1814. William Clarkson (*Inquiry into the Poor Rates* (1816), *Pamphleteer*, vol. viii. p. 392) gives the average rate of wages in 1812 at 15s. ; but thinks that as wages are much less in Wiltshire, Devonshire, and Cornwall, this figure is over-stated as an average.

It is not suggested that this class of evidence is at all conclusive ; but it leaves the impression that, if agricultural wages in 1760 averaged 7s., they had approximately doubled in 1812 in many parts of the country, and that the average rise cannot be put at less than two-thirds.

if wages had doubled, the price of provisions had trebled. In other words, effective earnings had diminished by a third. It is the suddenness of this advance in prices that explains, though it does not justify, the makeshift expedients for relief which were adopted by administrators of the Poor Law.

CHAPTER XV

AGRICULTURAL DEPRESSION AND THE POOR LAW
1813-37

War taxation : peace and beggary ; slow recovery of agriculture ; the harvest
of 1813 ; reality and extent of distress ; the fall of prices ; bankruptcies
of tenant-farmers ; period of acute depression, 1814-36 ; ruin of small
owners ; misery of agricultural labourers ; reduction in wages and scarcity
of employment ; allowances from the rates ; general pauperisation : the
new Poor Law, 1834, and its administration.

ENGLAND in 1815 had emerged from the Napoleonic wars victorious.
But she paid the price of victory in her huge National Debt, her
excessive taxation, her enormous Poor-Rate, her fictitious credit,
her mass of unemployed and discontented labour. Though it is
estimated that one in every six male adults was engaged in the
struggle by land or sea, the population of England and Wales had
risen from under 8¾ millions in 1792 to about 10½ millions in 1815.
Within the same period the National Debt grew from £261,735,059
to £885,186,323, and the annual expenditure, including interest
on the public debt, from under 20 millions to £106,832,260.[1] The
wealth and resources of the nation are shown by the comparative
ease with which the money was found and the increased burden met.
Yet the strain of the struggle had been intense, and on no class had
it told more severely than on agricultural labourers. If their
wages had approximately doubled, the cost of living had nearly
trebled. Of their distress the rise of the expenditure on poor-relief
affords evidence. It advanced from £1,912,241,[2] which was the

---

[1] Porter's *Progress of the Nation* (1847), p. 482.

[2] Eden's *State of the Poor*, vol. i. pp. 3½3-72. In 1776 the expenditure had
been £1,556,804. After 1785 no Returns were made till 1803, when poor-rates
stood at £4,077,891. See *Local Taxation Returns* printed by order of the House
of Commons, 1839. To the sums assessed and disbursed in relief of the poor
must be added the annual sums derived from charities appropriated to the
same object. These amounted, apart from educational charities, to £1,209,395

triennial average for 1783-4-5, to £5,418,845 in 1815. Nor did the expenditure cease to rise with the close of the war. It continued to increase, in spite of falling prices. In 1818 it had grown to £7,870,801, the highest point which it reached under the old law.[1]

For six years after the end of the war the proverbial association of " Peace and Plenty " proved a ghastly mockery to all classes of the community. To agriculturists peace brought only beggary. In the first rush of complaint, some allowance must be made for disappointment at the immediate results of the end of the war. But the evidence of commercial depression was real and widespread. The disordered state of the currency continued to injure credit, to disturb trade, to create wild speculation instead of sound business. The labour market was glutted. Discharged sailors, soldiers, and militiamen swelled the ranks of the unemployed. The store, transport, and commissariat departments were put on a peace footing. Industries to which the war had given a feverish activity languished. Thousands of spinners, combers, and hand-loom weavers were thrown out of work by the increased introduction of machinery into manufacturing processes. Continental ports were once more opened to English trade ; but money was scarce, and foreign merchandise excluded by heavy customs duties. It was soon found that home manufactures had exceeded the demand. Warehouses were overloaded, markets overstocked. Produce was unsold, or unpaid for, or bought at prices unremunerative to the producers. Only with America was increased business done. The growing imports of raw cotton were paid for by exports of British goods.

After 1821 the commercial depression began to disperse. Difficulties of the currency had been, to some extent, adjusted ; credit and confidence were reviving. Progress was for a time suspended by the financial crash of 1825. But the interruption was temporary. Trade improved, at first slowly, then rapidly. Agriculture recovered more gradually ; for a protracted period it endured an almost unexampled misery. Landlords, tenant-farmers, and labourers suffered together. It was not till 1836 that any gleam of returning

12s. 8d. a year. (See *Report of the Charity Commissioners*, 1842.) No estimate can be formed of the additional sums annually contributed by the charitable. The great increase in the Poor Rate cannot be wholly attributed to an increase in the number of paupers. It was largely due to the greater cost of provisions and to more lax administration. See Appendix II.

[1] Nicholl's *History of the Poor Law*, ed. H. G. Willink (1898), vol. ii. p. 165, and Porter's *Progress of the Nation* (ed. 1847), p. 527.

prosperity appeared. During the war, farming improvements had been stimulated by the prospect of increased profits. In peace, when once the new conditions were accepted, and some degree of confidence restored, adversity proved an efficient goad. Without improved practices there was no prospect of any profits at all. Yet down to the accession of Queen Victoria there is no substantial progress to chronicle. The characteristics of the period are a great loss of ground and a partial recovery.

The inflated prices of the war had conferred, from one point of view, a great advantage on the agriculture of the country. They brought under the plough districts which, but for their stimulus, might never have been brought into cultivation,—areas that were forced into productiveness by the sheer weight of the metal that was poured into them. Money made by farming had been eagerly reinvested in the improvement of the land. For the same purpose banks had advanced money to occupiers on the security of crops and stock which every year seemed to rise in value. Farmers had been able to meet their engagements out of loans, and wait their own time for realising their produce. Better horses were kept, better cattle and sheep bred. Land was limed, marled, or manured. Wastes were brought under cultivation ; large areas were cleared of stones in order to give an arable surface ; heaths were cleared, bogs drained, buildings erected, roads constructed. The history of Northumberland strongly illustrates these brighter aspects of a gloomy period. John Grey of Dilston,[1] " the Black Prince of the North," one of the most enterprising and skilful agriculturists of the day, played a conspicuous part in the transformation of his county. Born in 1785, and early called through the death of his father to the management of property, he lived in the midst of the agricultural revolution. When his father first settled in Glendale, the plain was a forest of wild broom. He took his axe, and, like a backwoodsman, cleared a space on which to begin his farming operations. The country was then wholly unenclosed, without roads or signposts. Cattle were lost for days in the broom forests. The inhabitants were as wild as their home,—the Cheviot herdsmen " ferocious and sullen," the rural population " uneducated, ill-clothed, and barbarous." But the character of the soil was such as to attract skill and industry. Men of the same stamp as Grey, or the Culleys, settled in the fertile vales, and by their spirited farming transformed into

---

[1] *Memoirs of John Grey of Dilston*, by Josephine E. Butler (1869).

cultivated land wide districts, like the rich valley of the Till, which before the period of war prices were wildernesses of underwood. Between 1813 and the accession of Queen Victoria falls one of the blackest periods of English farming. Prosperity no longer stimulated progress. Except in a few districts, falling prices, dwindling rents, vanishing profits did not even rouse the energy of despair. The growing demoralisation of both employers and employed, which resulted from the administration of the Poor Law, crushed the spirit of agriculturists. " Many horses die while the grass is growing." The men who survived the struggle were rarely the old owners or the old occupiers. They were rather their fortunate successors who entered on the business of land-cultivation on more favourable terms. Prices had begun to fall with the abundant harvest of 1813. The suddenness of the decline is illustrated from the contracts made on behalf of the Royal Navy. At Portsmouth in January, 1813, the price paid for wheat was 123s. 10d., in November, 67s. 10d. In February, 1813, at Deptford, flour was contracted for at 100s. 3d. per sack, in November, at 65s.[1] This rapid fall could not at that time have been due to any prospect of peace. It was rather due to over-production, which the House of Commons Committee on the Corn Trade (1814) found to have increased within the last ten years by a fourth. Besides English corn, Scottish and Irish corn were in the market. Since 1806 Irish grain had been admitted into the country free, and it poured into the western counties in considerable quantities. Deficient harvests in 1809-10-11-12 had concealed the potential yield of the increased area under corn ; its full productive power stood revealed by the favourable season of 1813. The two following harvests, 1814 and 1815, were not above the average ; but prices of wheat dropped to 74s. 4d. and 65s. 7d. per quarter respectively. As compared with 1812, the actual receipts of farmers diminished by one hundred millions, and the value of the farming stock was reduced by nearly one-half. The evidence of widespread distress is ample.[2] But it is

---

[1] *Speech of Chas. C. Western, M.P., on moving that the House should resolve itself into a Committee of the Whole House to take into Consideration the Distressed State of the Agriculture of the United Kingdom, March 7, 1816 (Pamphleteer, vol. vii. p. 508).*

[2] *E.g. 1. A Review of the present Ruined Condition of the Landed and Agricultural Interests, etc., by R. Preston, M.P. (1813).*

   2. *Letters on the Distressed State of Agriculturists, by R. Brown (1816).*

   3. *Further Observations on the State of the Nation, etc., by R. Preston, M.P (1816).*

improbable that its extent was understated. Agricultural witnesses and writers were anxious, not only to prevent any relaxation of the Corn Laws, but, if possible, to increase their stringency. The depression, therefore, " lost nothing in the telling," though its depth and reality remain unquestionable. Brougham, speaking on agricultural distress in the House of Commons, April 9, 1816, said : " There is one branch of the argument which I shall pass over altogether, I mean the *amount* of the distresses which are now universally admitted to prevail over almost every part of the Empire. Upon this topic all men are agreed ; the statements concerning it are as unquestionable as they are afflicting . . . and the petition from Cambridgeshire presented at an early part of this evening, has laid before you a fact, to which all the former expositions of distress afforded no parallel, that in one parish, every proprietor and tenant being ruined with a single exception, the whole poor-rates of the parish, thus wholly inhabited by paupers, are now paid by an individual whose fortune, once ample, is thus swept entirely away." [1]

With wheat standing at over 60s. a quarter, it is difficult to realise that the landed interests could be distressed, and it might be supposed that farmers had made enough in prosperous times to tide over a period of depression. But though the rise in prices had been enor-

4. *On the State of the Country in December* 1816, by the Rt. Hon. Sir John Sinclair (1816).
5. *Agricultural State of the Kingdom in February, March, and April,* 1816. *Being the Substance of the Replies to a Circular Letter sent by the Board of Agriculture* (1816).
6. *An Inquiry into the Causes of Agricultural Distress,* by W. Jacob, F.R.S. (1817).
7. *Observations on the Present State of Pauperism in England,* by the Rev. George Glover (1817).
8. *Speech of J. C. Curwen, M.P., in the House of Commons on May* 28th, 1816, *on a Motion for a Committee to take into Consideration the State of the Poor Laws* (1816).
9. *Two Letters on the Present Situation of the Country,* by A. H. Holdsworth, M.P. (1816).
10. *Letters on the Present State of the Agricultural Interest,* by the Rev. Dr. Crombie (1816).
11. *On Famine and the Poor Laws,* by W. Richardson, D.D. (1816).
12. *An Inquiry into Pauperism and Poor Rates,* by William Clarkson (1816).
13. *Observations . . . on the Condition of the Labouring Classes,* by John Barton (1817).
14. *Inquiry into the Causes of the Progressive Depreciation of Agricultural Labour,* by the Same (1820).

[1] " Speech on Agricultural Distress." *Speeches of Henry, Lord Brougham,* vol. i. pp. 503-4 (1838).

mous, the increase in public burdens had more than kept pace. During the ten years ending in 1792, the average price of wheat had been 47s. per quarter ; the national expenditure under twenty millions a year ; the poor-rate less than 1¾ millions ; there was also no property tax. During the ten years ending in 1812, wheat averaged 88s. a quarter. While wheat had thus not quite doubled, wages had risen by two-thirds ; the national expenditure had multiplied five-fold ; tithes had increased by more than a fourth ; a property tax had been imposed on owners and occupiers of land. The poor-rate had quadrupled ; the county-rate had risen seven-fold ; the permissive charge of 6d. in the pound for the road material of highways had been of late years habitually levied. A very large proportion of this public burden was borne by agriculturists. Upon the landed interests fell more than half the new property tax,[1] the greater part of the county-, poor-, and highway-rates, the war duties on hops and malting barley, the tax on agricultural horses, and an exceptional share of the tax on leather, which swelled the cost of every kind of harness gear. Thus the rise of the price of agricultural produce was to a great extent discounted by the growth of taxation, and it was the war, not the Corn Laws, which had given agricultural producers the monopoly of home markets.

In other respects circumstances were exceptional. During the war, the social advantages of landownership and its apparently remunerative character, as well as the large fortunes realised in recent trade, combined to give land a fancy value. New capitalists gratified both their ambitions and their speculative instincts by

[1] The Property Tax for 1814 produced Gross £15,325,720, and Net £14,545,279. It was made up thus :

Sched. A (lands, tenements and hereditaments)	£4,297,247	
Sched. B (occupiers of land) - - -	2,176,228	
Tax on houses - - - - -	1,625,939	
Total - - - - - - -		£8,099,414
Sched. C, Funded property - - - -	£3,004,861	
Sched. D, Profits on Trades and Professions -	3,021,187	
Sched. E, Naval, military, and civil lists -		
together with provincial offices - -	1,113,244	
Total - - - - - - - -		£7,139,292
Supplementary accounts, duties, penalties, etc.		87,014
Total - - - - - - - -		£15,325,720

This tax was repealed in 1816. The number of agricultural occupiers contributing to the property tax under Sched. B was 474,596, as against 152,926 assessed under Sched. D.

becoming purchasers. Their biddings forced existing owners intc ruinous competition ; they mortgaged their ancestral acres to buy up outlying properties or round off their boundaries. As much as forty-five years' purchase was given for purely agricultural land. The same spirit of competition prompted farmers to offer extravagant rents for land. Farms were put up to auction, and the tenancy fell to the highest bidder. The more prudent had left business in 1806. Many of the new men entered on their holdings with insufficient or borrowed capital. Money was still made in farming ; but, instead of being realised, it was put back into the land, where, so long as prices rose, or were even maintained, it proved a profitable investment. Among all classes, including landowners ` and farmers, a higher standard of living prevailed. Country mansions had been built, rebuilt, or enlarged, and costly improvements effected in the equipment of farms,—often by means of loans ; heavy jointures and portions had been charged on estates ; farmers and their wives had either altered their simpler habits, or brought with them into their new business more luxurious modes of life. The whole fabric rested on the continuance of the war-prices. When these began to fall, the crash came. Profits were reduced by a half ; burdens remained the same. Tenants-at-will could at least quit their holdings. But tenants occupying under long leases found themselves in a difficult position. Landlords could not meet their liabilities, unless their rents were maintained ; without reductions of rent, the bankruptcy of their tenants seemed inevitable.

In the period 1814-16 the agricultural industry passed suddenly from prosperity to extreme depression. At first farmers met their engagements out of capital. When that was exhausted, their only resource was to sell their corn as soon as it was threshed, or their stock, for what it would fetch. The great quantity of grain thus thrown on the market in a limited time lowered prices for producers, and the subsequent advance, which benefited only the dealers, suggested to landlords that no reductions of rent were necessary Farms were thrown up ; notices to quit poured in ; numbers of tenants absconded. Large tracts of land were untenanted and often uncultivated. In 1815 three thousand acres in a small district of Huntingdonshire were abandoned, and nineteen farms in the Isle of Ely were without tenants. Bankers pressed for their advances, landlords for their rents, tithe-owners for their tithe, tax-collectors for their taxes, tradesmen for their bills. Insolvencies, compositions, executions, seizures, arrests and imprisonments for debt multiplied. Farmhouses were full of sheriffs' officers. Many large farmers lost everything, and became applicants for pauper allowances. Even in Norfolk the number of writs and executions rose from 636 in 1814 to 844 in 1815 ; in Suffolk from 430 to 850 ; in Worcester from 640 to 890. In the Isle of Ely the number of arrests and executions increased from 57 in 1812-13 to 263 in 1814-15. In the same district several farmers failed for an aggregate sum of £72,500, and the creditors in hardly any instance received a dividend. Between 1815 and 1820, 52 farmers, cultivating between them 24,000 acres, failed in Dorsetshire. Agricultural improvements were at a stand-still. Live-stock was reduced to a minimum. Lime-kilns ceased to burn ; less manure was used on the land ; the least possible amount of labour was employed. The tradesmen, innkeepers, and shopkeepers of country towns suffered heavily by the loss of custom. Blacksmiths, wheelwrights, collar makers, harness makers, carpenters, found no work. At first the depression had been chiefly felt in corn-growing districts, especially on heavy land. But by 1816 it had spread to mixed and grass farms. In that year, bad seasons created a temporary scarcity ; the rise of wheat to the old prices aggravated rural distress without helping any persons except dealers, and the wealthier farmers who could afford to wait ; the potato crop, which had recently become important in England, failed ; perpetual floods in the spring and summer were succeeded by a winter of such unusual severity, that the loss of sheep in the North was enormous. Landlords, whose land was thrown upon their hands, or who had laid charges on their estates, found themselves confronted with ruin. The alternative was hard. If the mortgagee foreclosed, the estate sold for a sum which barely recouped the charges. Preston,[1] in 1816, states that " in Norfolk alone landed property to the value of one million and a half is on sale, without buyers for want of money." One property, for which " £140,000 was offered two years ago, is now on sale at £80,000." In a second pamphlet [2] he states that " some of the best estates of the kingdom are selling at a depreciation of £50 per cent. One of the finest grass farms in Somersetshire sold lately at 10 years purchase." " There

---

[1] *Review of the Present Ruined Condition of the Agricultural and Landed Interests*, by Richard Preston, M.P. (1816), (*Pamphleteer*, vol. vii. pp. 149, 167).

[2] *Further Observations on the State of the Nation* (1816), (*Pamphleteer*, vol. ix. p. 127.)

are now estates," says Glover,[1] " in the most fertile parts of England,
nay even within 50 miles of London, which are an absolute loss
to the possessor." The natural reluctance of landlords to lower
rents may have involved tenants in their fall; but by 1816
they are stated to have lost 9 millions a year by rent reductions
alone.[2]

For the next twenty years the same record of depression is continued. The attention of Parliament was continually called to
the distress of the landed interests. Petitions covered the table of
the House; innumerable pamphlets and letters demanded remunera-
ting prices for agricultural produce. Some exaggeration there
probably was, for the struggle of Free Trade against Protection had
begun. But the account which has been given of farming conditions
in the years 1814-16 was substantially confirmed by numerous
witnesses who gave evidence on the continuance of the distress
before a series of Select Committees in 1820, 1821, 1822, 1833, and
1836. Rural conditions were deplorable. Even as late as 1833, it
was stated that, in spite of rent reductions, which in Sussex amounted
to 53 per cent., there was scarcely a solvent tenant in the Wealds of
Sussex and Kent, and that many farmers, having lost all they had,
were working on the roads. Violent fluctuations in prices continued to overthrow all calculations; the wheat area alternately
expanded and contracted; the sliding scale of 1829, soon exploited
for their own profit by foreign importers, only increased the speculative character of the agricultural industry. On heavy clays less
capital and less labour were expended; wet seasons prevented

---

[1] *Observations on the Present State of Pauperism in England*, by the Rev. G.
Glover (*Pamphleteer*, vol. x. p. 384).

[2] If this statement is correct, it approximately restored the rental of *land* in
Great Britain to the figure at which it stood in 1806. In that year a subdivision for the first time was made of the classes of property the income of
which was assessed under Schedule A to the Property Tax.

Annual Income from					1806	1814
1. Property from Lands	-	-	-	-	£29,834,484	£39,405,705
2. Property from Houses	-	-	-	-	11,913,513	16,259,399
3. Amount of Tithes	-	-	-	-	2,012,064	2,732,898
4. Profits from Manors	-	-	-	-	43,521	71,672
5. Fines on Leases	-	-	-	-	72,502	216,546
6. Profits of Quarries	-	-	-	-	32,456	70,378
7. Profits of Mines	-	-	-	-	363,853	678,786
8. Profits of Iron Works	-	-	-	-	84,615	647,686
9. General Profits, etc.	-	-	-	-	477,762	65,260
					£44,834,770	£60,148,330

farmers from getting on the land, and caused the discontinuance of manure, excessive cropping, and the impoverishment, even the abandonment, of the heavier soils. To add to the difficulties of clay farmers, the rot of 1830-1, which is described as the most disastrous on record, " swept away two million sheep." Everywhere wages were lowered and men dismissed. Work became so scarce that, in spite of the fall of prices, starvation stared the agricultural labourer in the face. Distress bred discontent, and discontent disturbances, which were fostered by political agitation. While the Luddites broke up machinery, gangs of rural labourers destroyed threshing machines, or avenged the fancied conspiracy of farmers by burning farm-houses, stacks, and ricks, or wrecking the shops of butchers and bakers. In the riots of 1830-31, when " Swing " and his proselytes were at work, agrarian fires blazed from Dorsetshire to Lincolnshire.

The evidence before the Select Committee of 1836 shows that prosperity was beginning to revive. But the long period of depression left its permanent mark on the relations of landlord and tenant, as well as on the conditions of rural society. It was not merely that progress had been lost, or that much of the land was impoverished, or that farm buildings fell into ruinous condition. A great expenditure was needed to reorganise the industry, and it was the owner of the land who found the money. Necessity compelled landed proprietors to realise their position. Tenants had little capital left ; they were also more cautious of risks. Recent experience had created a profound distrust of long leases. Without security of tenure for a prolonged term of years, no man of ordinary prudence would make an outlay on the costly works which his predecessors had eagerly undertaken. It was now that the distinction becomes clearly marked between landlord's and tenant's improvements. Even in the latter class, it was already evident that, where the benefits were not exhausted at the expiration of the tenancy, compensation was payable, and that local customs afforded insufficient protection. On these new lines agriculture once more began to advance. At the accession of Queen Victoria the worst of the crisis was over. Rents had been adjusted to changed conditions. The industry had been relieved from some of the exceptional taxation. The Tithe Commutation Act of 1836 had removed a great obstacle to progress. The new Poor Law of 1834 reduced the burden of the rates, and began to re-establish the self-respect of the labourer.[1] The rapid growth of the manufacturing population not only created an increasing demand for agricultural produce, but relieved the glut of the labour-market.

To small freeholders, whether gentry, yeoman-farmers, or peasant proprietors, the Napoleonic war, with its crushing load of taxation and subsequent collapse of prices, had been fatal. The evidence before the Agricultural Committee of 1833 proves that some still held their own in every county. But it was in the first thirty years of the nineteenth century that their numbers dwindled most rapidly. Some had consulted their pecuniary interests by selling their land at fancy prices, which they took into business. Others sold and embarked their capital as tenant-farmers in hiring larger areas of land, on which they could take fuller advantage of the price of corn. Those who remained on their own estates were for the most part ruined. Many had raised mortgages to buy more land, or to improve their properties, or to put their children out in the world. Prices fell ; but the private debt, as well as the public burdens, remained. The struggle was brief ; farming deteriorated ; buildings fell out of repair ; creditors pressed ; finally the estate was sold. Even where land was free from charges, owners could not stand up against the burden of poor-rates, which was most crushing to those who employed no labour but their own. "That respectable class of English yeomanry," writes Glover[2] in 1817, " whose fathers from generation to generation have lived on the same spot and cultivated the same farms are now rapidly dwindling into poverty and decay, sinking themselves into the class of paupers." The purchasers were not men of their own class. After 1812 small capitalists no longer invested their savings in land. Their place as buyers was taken by large landowners or successful traders. In Yorkshire the number of small proprietors was dwindling ; formerly, if one freeholder went, another took his place ; but this had now ceased to be the case. The same report is made of Shropshire, Worcestershire, and Wiltshire. In Kent and Somersetshire it is stated that, though many freeholders retained their land, it was only by the practice of the most rigorous self-denial and by entirely ceasing to employ labour. Throughout

---

[1] In 1837 the expenditure dropped to the lowest point as yet recorded in the century, £4,044,741.

[2] *Observations on the Present State of Pauperism, etc.* (*Pamphleteer*, vol. x. p. 385).

the country, it is evident that most of the small landowners, who, in addition to taxes and rates, had to pay annuities or interest on mortgages, were forced to sell their properties. Everywhere large estates were built up on the ruin of small proprietors.

Morally, if not materially, no class suffered more from the prolonged period of depression than agricultural labourers. They had bitter reason to deplore the shortsighted humanity which in the last twenty years of the eighteenth century had swept away the old barriers against pauperism.[1] Where Gilbert's Act had been adopted, every man was now secure of employment from the parish or, in any case, of maintenance. In every parish, also, outdoor relief for the able-bodied poor was now compulsory on the overseers. Already in some districts men out of regular work were " on the Rounds," offering their labour from house to house, paid, if employed, partly by the householder, partly by the parish, and if unemployed, wholly by the parish. Even men in full employment were drawn within the net. When in 1795-6 the price of provisions rose to famine height, wages were supplemented by allowances from the rates. A scale of these allowances was proclaimed by the Berkshire magistrates, proportioned to the price of bread and the size of families. From the wages of the unmarried labourer, which were zero, the scale ascended, varying with fluctuations in the cost of the quartern loaf and the number of the children of the married labourer. Similar scales of allowances were adopted in many other counties. Thus able-bodied men, whether in or out of work, became dependent on the rates. That, from the first, these allowances delayed the natural rise of wages, lowered earnings by making the needs of unmarried men the most important factor, and encouraged improvident marriages, is certain. But these evils were held in check till 1813. So long as the war and the high prices continued, the demand for labour was brisk ; distress was practically confined to those who suffered from enclosures, or from the decline of local industries other than the cultivation of the land. Agricultural wages rose substantially ; employment increased owing to the extension of tillage ; even the high prices of provisions affected labourers less than might have been expected, since the provisions, in several parts of the country, were supplied to them at a lower cost than the market rates. Except for winter unemployment the allowance system was sparingly used. But during the depression

[1] See Appendix II. The Poor Law.

farmers were driven to economise in their labour bills. Wages
were greatly reduced or even ceased altogether. The replies to
the Circular Letter of the Board of Agriculture insist on the deplorable scarcity of employment. Preston [1] speaks of the daily
increase in the number of paupers, and of " a large part of the
community . . . in want of employment though willing to labour."
" At no period in the memory of man," writes Jacob,[2] " has there
been so great a portion of industrious agricultural labourers
absolutely destitute as at the present moment." It was now that
the Poor Law was most perniciously relaxed ; now also that the
demoralising system of allowances became the most conspicuous
feature in its administration.

The immediate effects of the depressed condition of agriculture
was a great reduction in the rates of wages, and in the demand for
permanent labour. Unless the farmer could lessen his costs of
production, he was rapidly sinking into bankruptcy. The Poor
Law, as it was administered in 1813-34, in two ways came to his
assistance. It enabled him to reduce wages to the lowest possible
point, because it made good the deficiency out of allowances from
the rates. Men discharged as supernumeraries were taken on again
as soon as they were on the poor-book. It also provided him with
an inexhaustible supply of cheap and temporary labour. Bound
to defray the whole cost of maintaining the able-bodied poor, the
parish gladly accepted any payment, however small, in part relief
of their liability. It became almost impossible for a farmer to
keep a man in permanent employment at reasonable wages. If
he did, he was only saving the rates for neighbours, who put their
hands into his pockets to pay their labour bills. Sometimes the
ratepayers in the parish arranged among themselves to employ
and pay a number of men proportionate to the rateable value of
their property. Sometimes the parish agreed with employers to
sell the labour of so many paupers at a given sum, and paid the
men the difference between the agreed price and the scale allowance
awarded to them according to the cost of bread and the number
of their children. Sometimes the paupers were paraded by the
overseers on a Monday morning, and the week's labour of each

---

[1] *Review of the Present Ruined Condition, etc.*, by R. Preston, M.P. (1816),
(*Pamphleteer*, vol. vii. p. 129).

[2] *Inquiry into the Causes of the Agricultural Distress*, by W Jacob, F.R.S.
(1817), (*Pamphleteer*, vol x. p. 411).

individual was offered at auction to the highest bidder. Sometimes the parish contracted for the execution of a piece of work at a given sum, and performed it by pauper labour, paying the men according to the allowance scale. If men were still unemployed, they were formed into gangs under overseers, occupied in more or less unproductive work ; it was among these men that the riots of 1830-1 are said to have originated.[1]

Against the mass of subsidised labour, free labourers could not hope to compete. It was so cheap that men who tried to retain their independence were undersold. Those who had saved money or bought a cottage, could not be placed on the poor-book ; they were obliged to strip themselves bare, and become paupers, before they could get employment. Every agency that could promote the spread of pauperism seemed brought into play. The demoralisation gradually extended from the southern counties to the North. In the most practical fashion, labourers were taught the lessons that improvidence paid better than thrift ; that their rewards did not depend on their own exertions ; that sobriety and efficiency had no special value above indolence and vice. All alike had the same right to be maintained at the ratepayers' cost. Prudence and self-restraint were penalised. The careful were unemployed, the careless supported by the parish ; the more recklessly a man married and begot children, the greater his share of the comforts of life. The effect was seen in the rapid growth of population. Among unmarried women morality was discouraged, and unchastity subsidised. The more illegitimate children, the larger the allowance from the parish ; at Swaffham a woman with five illegitimate children was in receipt of 18s. a week. The demoralisation was so complete that it threatened to overthrow the whole social fabric. Voluntary pauperism became a profession, and a paying one. Recipients considered themselves as much entitled to parish allowances as they would have been to wages that they had earned by their industry. A generation was springing up which knew no source of income but poor relief. When once the spirit of independence and self-respect was numbed, and the instincts of parental responsibility and filial obligation were weakened, a pauper's life, with its security of subsistence, its light labour, its opportunities of idleness, had attractions for the vicious and easy-

---

[1] For these varieties of the Labour Rate, the Roundsmen and Parish Employment, see the *Report on the Poor Laws* (1834), pp. 42, 31-32, 36.

going. Riots were not always protests against the existing system ; they were sometimes means of enforcing its continuance, and parochial allowances were maintained by the establishment of a reign of terror, by threats, violence, and incendiarism.[1]

Rural conditions were fast becoming intolerable. Fortunately, there still remained a leaven of agricultural labourers who resented pauper dependence as a curse and a disgrace. Fortunately, many farmers were learning by experience that cheap labour was bad labour, and that quantity was no efficient substitute for quality. Fortunately, also, there were districts which a wiser administration of the Poor Law had rescued from the general demoralisation. In four parishes, Southwell, Bingham, Uley, and Llangattock, the principle had been adopted, with marked success, of refusing relief to the able-bodied, except in well-regulated workhouses. Some sixty others had been practically depauperised (e.g. Welwyn, Leck-hampstead, and Carlisle) by stopping allowances, and exacting hard work, at low pay, under strict supervision, as a condition of parish relief. On the other hand, the parish of Cholesbury in Buckinghamshire afforded the typical illustration of the extreme consequences to which the existing system was necessarily leading. Out of 98 persons, who had a settlement in the parish, 64 were in receipt of poor relief, and the rates exceeded 24s. in the pound. Only 16 acres remained in cultivation. When able-bodied paupers were offered land, they refused it on the ground that they preferred their present position. The parish was only able to exist by means of rates-in-aid levied on other parishes in the hundred. Similar conditions prevailed elsewhere.[2] It was evident that the fund from which the rates were provided must become exhausted. Rents were already disappearing. The Poor Law had destroyed the confidence of tenants, deteriorated the moral character of the labourer, forced large areas out of cultivation, driven capital to seek investment everywhere but in land. A drastic remedy was needed. In 1832 a Commission of Inquiry was appointed to examine into existing conditions, and suggest the lines of legislative reform.

On the recommendations of this Commission was based the Act of 1834 " for the Amendment and better Administration of the Laws relative to the Poor in England and Wales." A central authority was constituted to regulate local administration. The

---

[1] *Extracts from the Information received by H.M. Commissioners as to ... the Poor Laws* (1833), p. 3.

[2] See footnote on next page.

orders issued by the new authority proceeded on the main principle of restoring the old Poor Law, without the relaxations which the legislation and practice of George III. had introduced. The workhouse test for the able-bodied was revived. If a man chose to depend for subsistence on the parish rates, instead of on his own resources, he was obliged to enter the workhouse and submit to its regulations. Out-door relief for the able-bodied was discouraged, and allowances in aid of wages were prohibited. At the same time the laws of settlement were modified, in order that labour might become more mobile and more easily transferable in obedience to the laws of demand and supply. The effect of these and other changes was soon manifest. Expenditure upon poor relief fell from £7,036,968 in 1832 to £4,044,741 in 1837. Wages rose, though for many years they remained miserably low. Landowners again poured their capital into the land ; farmers regained confidence ; agricultural progress was resumed. The evidence laid before the Select Committee of 1836 proves that signs of returning prosperity were beginning to appear, and that the distress was now practically confined to clay land.

² In the *Reminiscences of the Rev. T. Mozley* may be read an account of the state of labour in Moreton Pinckney parish, Northants, in 1832.

" There was many farmers, several yeomen, very many small freeholders, as many tradesmen as the place could find work or customers for, and not one single labourer, in the sense of an independent workman offering his labour for wages, but a multitude of paupers.

" This was the monthly apportionment of the paupers—that is, the whole labouring population, among the larger ratepayers. . . . The payments were not wages. . . . A strong man of five and twenty could not get more than a shilling a day."

CHAPTER XVI

TITHES

A SERIOUS obstacle to the progress or recovery of agriculture was presented by the incidence of tithe upon the produce of the land. Tithe-owners were sleeping partners in the cultivation of the soil. They contributed neither capital nor labour to the enterprise of the farm; they risked nothing in the venture. But they shared the profits derived from increased productiveness. While agriculture remained stationary, the burden was light. As soon as farming began to advance, and to demand a greater outlay, the grievance was acutely felt. In times of prosperity the incidence on produce discouraged improvement. In days of adversity, when every penny was important to struggling agriculturists, it retarded recovery.

Since the Tithe Commutation Act of 1836, much of the ancient law of tithes has retained only an antiquarian interest. But the long history of the payment has left an indelible mark on rural life.

Historically, tithes were a tenth part, taken yearly, of all produce of the land, of the stock nourished upon the land, and of the clear profits of the personal industry of tradesmen, artificers, millers, and fishermen. In other words, tithes were, as lawyers distinguished them by their sources, predial, mixed, or personal. Predial tithes were derived directly from the soil, such as corn, hay, beans, peas, turnips, hemp, flax, saffron, rushes, fruits, and wood of various kinds. Mixed tithes arose from the increase or produce of animals maintained by the fruits of the earth, as of cattle, sheep, pigs, poultry or their eggs, wool, milk or cheese. Personal tithes on the clear gain of the labour of man had early fallen into disuse. Manu-facturers were never liable for the payment, which only survived in such forms as a tenth of the fisherman's catch, or a tenth of the miller's clear profits on meal ground in all but ancient mills. Another classification, distinguished between great and small tithes according to the nature of the produce on which they fell. Thus great or rectorial tithes included corn, beans, peas, hay, and wood ; all the other predial tithes, together with all mixed tithes, were small or vicarial.

The legal obligation to pay tithes, as distinct from the older moral duty of giving them, dates back to a remote period of history. No real dispute arises respecting their origin, until the point is reached where the offering passed from a free-will gift into a liability enforced by legal penalties. From the fourth century onwards, throughout the Christian world, the practice of dedicating fractional parts of produce to religious objects was recognised by the faithful as a moral duty. As a matter of conscience, the gift was enjoined by Councils of the Western Church, and enforced by appeals to the rewards and punishments of religion. Thus the practice gradually acquired something of the binding force of custom. The final stage was reached when the State recognised as a civil duty the religious practice of giving tithes, and compelled payment, not by appeals to conscience, nor even by spiritual penalties, but by temporal sanctions. This last step, by which tithes passed from moral obligations into legal liabilities, was taken at different dates by the different countries of Christian Europe.

Before the landing of Augustine in England (597), and before the introduction of Christianity into this country, the moral duty of giving tithes had been enjoined on the Continent by at least one Church Council. As a matter of conscience, therefore, the first missionaries to Anglo-Saxon England preached the consecration of a tenth of produce to the service of God, and as a religious custom the practice was established by their successors among their Chris-tian followers. The appeal was the more forcible since it came from men who were believed to hold the keys of heaven and of hell. But there were as yet no divisions into parishes, no parish churches, no parochial clergy, and no parochial endowments. The cathedral, monastery, and "mother church," generally conventual, of the local-ity, were mission centres, from which radiated itinerant missioners, who preached under rude crosses the rudiments of Christianity to the inhabitants of outlying districts. Into the hands of the Bishop or monastic bodies were paid all the offerings of the faithful. The married clergy, outside the cloister, were slowly and with difficulty obtained. They were for the most part ignorant, uncouth men, recruited from the lower classes of native converts, entrusted only with the humbler offices of the ministry. They taught the Lord's Prayer and the Apostles' Creed in remote hamlets, watched by the bedsides of dying penitents, and in special cases administered the rites of baptism. It is not strange that, on earth at least, their lowly labours should have been ignored or forgotten. Very different was the fate of the monastic bodies. To them fell power, riches, credit. Kings were their nursing-fathers, queens their nursing mothers ; the wealthy nobles vied with one another in the munificence of their endowments. In comparison with existing civilisation, the monastic bodies attained a standard of wealth, refinement and culture which was at least as high as that of later times. They laid acre to acre, and field to field. For miles round their farms, barns, flocks, herds, fish-ponds, and dovecotes dotted the country. They entranced the senses by the beauty of their architecture, their music, their ritual ; they commanded respect by their learning ; they inspired awe by the austerities of their lives. They alone could offer an inviolable resting-place for the dead, since there were no parochial burial-grounds, and they practically monopolised rights of sepulture. Thus in death, as well as in life, they appealed irresistibly to the favour of the world.

Till the closing period of the Anglo-Saxon Church, there were, as has been said, no resident parochial clergy. Ecclesiastical organisation proceeded downwards, not upwards. It was provincial, diocesan, conventual, before it became local and parochial. The cathedrals of the dioceses and the conventual churches of the monasteries at first provided for the religious wants of the people. Yet the material was ready for the introduction of the parochial system. Townships suggested the necessary divisions, and village communities, on the self-sufficing system of these agrarian societies, had probably been accustomed to provide for their pagan priests. From the first the rulers of the Church felt the need of continuous local ministrations, though, probably, the earliest advances towards a parochial system were forced upon the country by external causes. From the ninth century onwards Danish invasions struck a series of staggering blows at the monastic organisation. Monasteries were the first objects of the invaders' attack ; their wealth and their defencelessness made them an easy, as well as a tempting, prey. They were sacked, pillaged, burned, and their inmates either dispersed or massacred. To save rural Christianity from extinction by a relapse into paganism it became necessary to encourage local efforts, to favour the erection of private chapels, to enlarge the powers of the rural priests or chaplains by whom they were served, even to consecrate as burial-grounds the precincts in which they stood. Thus a permanent resident clergy began to grow up on the rural estates of great nobles in connection with private chapels and oratories. With the gradual extension of this local provision for permanent religious ministrations begins the increased importance attached to the payment of tithes as parochial endowments.

Early documents confirm this explanation of the growth of parochial tithes. On the one side, the Church, backed by all her supposed power over the destinies of man, urged the consecration of tenths to the service of God. On the other hand, the earthly influence of the Crown, sometimes by royal admonitions, coupled with threats of loss of favour, sometimes by attesting and confirming the decrees of synods, sometimes even by treaties of peace with the Danes, supported the demand of the Church, and assisted in making the custom of paying tithes universal. Under this double pressure the practice grew. But it was not till 944 that King Edmund's synod at London for the first time made non-payment of tithes an ecclesiastical offence to be punished by excommunication. Henceforward the Church claimed as an ecclesiastical right what she had hitherto received, if at all, as a free-will offering. The moral duty had become a religious obligation, enforced by spiritual penalties.

The payment of tithes was not yet a legal liability, enforced by temporal sanctions. Nor were tithes, or any part of them, as yet, ecclesiastically or legally, appropriated as parochial endowments. But the times were ripening for both changes. Voluntary dedications of free-will offerings had been acted upon by the religious bodies to whom they had been made. On the faith of their continuance cathedrals had been erected and a diocesan system established ; monasteries had been founded ; manorial churches had been built and some local provision made for their service : the dim outline of a future parochial system could be discerned. By these voluntary dedications the original donors had alienated portions of their own property. If neither appeals to conscience nor threats of excommunication sufficed to obtain payment, the State might not unreasonably be asked to enforce it as a legal liability, either against the original donors or against their representatives who had inherited estates already subject to the dedication of the consecrated portion.

In the reign of Edgar the Peaceful, during the primacy of Dunstan, the payment of tithes was made a legal liability, universal in its application. At the same time a step was taken towards the appropriation of a portion to the maintenance of district churches of a particular class. At Andover, in 970, the king and his Witenagemot issued an ecclesiastical ordinance, which was to all intents and purposes an Act of Parliament. The ordinance creates no tithes. On the contrary it presupposes their existence. It regulates the times when they were to be paid, and makes their payment a legal liability, enforced by a pecuniary penalty and a power of distraint. It does not profess to give them to the clergy. The first article ran as follows :—" That God's churches be entitled to every right ; and that every tithe be rendered to the old minster, to which the district belongs ; and be then so paid, both from a thane's *inland* (*i.e.* land granted in the lord's own hands), and from *geneat-land* (*i.e.* land granted out for services), so as the plough traverses it." Undoubtedly, the law not only protects the Church in the possession of tithes already dedicated, but transforms the moral duty, religious custom, and ecclesiastical obligation into a legal liability. A reason is suggested by the passages which regulate their division. The general right of the " old minster," the mother church of the district—whether collegiate or conventual—to the local tithes was recognised. But an exception was allowed. If any landowner had built on his private estate a church with a burial-place attached, he was to assign to its support one-third of the local tithes. The remaining two-thirds were to be paid to the " mother church." If the landowner had built on his estate a church or oratory without a burial-place, the local tithes went to the " mother church," and he might provide privately for the priest of his private chapel. In other words, the old diocesan and monastic system still remained in force ; but, side by side with it, had grown up manorial churches, providing " shrift districts " with burial-grounds, and therefore claiming some more permanent support than the caprice of the builder or of his successors. They were not yet parish churches ;

but they were their original type, and in the private chapels or " field churches " of the greater landowners are seen the germs of a further extension of a parochial system.

The law of Edgar remained unaltered at the Conquest. Practically re-enacted by Canute and by Edward the Confessor, it was accepted by William the Conqueror. As years passed, district or parochial churches were multiplied by their voluntary founders in various parts of the country. Some were built by kings or great nobles as private chapels ; some by bishops, some by monastic houses, some by landowners, some by freemen on the landowners' estates. Church-building proceeded on no general system, and without any uniformity of date. There was a gradual growth under varying circumstances ; but the people, acting through the legislature in a national capacity, neither built, nor endowed, nor repaired these churches. As with the buildings, so with the endowments. They were gradually appropriated to particular churches, in different proportions, without either system or uniformity. No priest serving a district could enforce any claim to local tithes, except for the third which was appropriated only to churches with burial-grounds. Though the payment had become a legal liability, the dedication of tithes to particular parochial uses is, therefore, still unexplained. Something more remained to be done. The final steps were taken between the eleventh and thirteenth centuries.

The chief instrument by which local endowments were secured to parish churches was consecration. A founder desired to build a church on his estate, and to have it consecrated. But the bishop could refuse to consecrate, unless proper provision was made for its maintenance. Between the bishop and the founder, who in building the church was a free agent, there might be bargaining. There might also be opposition from outside. The neighbouring monastery perhaps resented the intrusion of a new church and a new priest into the field which it had regarded as its own. But at no stage, either in the bargain or in the opposition, does the national will express itself. Throughout, the founder was at first practically master of the situation. There was no compulsion on him to build a church at all. If he did, not only did he himself nominate and invest the priest, with or without the consent of the bishop, but he could delay appointing to vacancies, and thus leave the church without services. Even where local endowment had been secured to the parochial church at consecration, the system was thus incomplete. Both points were settled by ecclesiastical discipline at the close of the twelfth century, The necessity for institution by the bishop was established, and the bishop's right to appoint to vacant benefices, after a certain period of delay, was vindicated.

A further step was still required. The legal liability for the payment of tithes was satisfied if, with the exception of the third secured to the parochial churches which possessed burial-grounds, payment was made to any ecclesiastical body. A patron might increase the pittances of the poor priests at his door, or offer it to the collegiate cathedral, or heap the grain in the barn of a monastery, or sell the tithes issuing from his estate to any religious body that he chose, or even, by collusion, store the corn in the granaries of himself or his lay friends. Even after the formation of parishes had become general, and after the claim of parochial churches was commonly recognised, it was still possible, and still usual, to grant the local tithes to distant houses of religion. The same causes were at work which in Anglo-Saxon times sacrificed the secular clergy to the monasteries. Norman landlords preferred to assign their tithes to monastic bodies, with whom they were more in sympathy than with the native priests of rural districts. The increase of monasticism after the Conquest necessarily alienated a large part of the local tithes which naturally would have increased the local provision for religious services. This option on the part of land-owners is inconsistent with the theory of the endowment of parishes by an exercise of the national will, expressed in some general law. It was not till the thirteenth century, and then not by any statute or Act of Parliament, but by the growth of custom, that the land-owner's freedom of choice was limited. No doubt the growth of the custom was aided by the practice of such specific dedications of tithes to the parochial church as those of Hay and Exhall. At common law the courts presumed that the parish church was *primâ facie* entitled to the tithes which issued from the lands of the parish. By this presumption the burden of proof was thrown on tithe-payers or other claimants to show that the local tithes had been either paid to some collegiate or conventual body for so long a period as to create a prescriptive right, or had been by express grant alienated to some other religious body.

It was to custom that the parochial clergy appealed ; other claimants relied on immemorial usage or express grant. This fact is in itself of extreme importance. Had any enactment of the national assembly established the *primâ facie* right of the parochial clergy to the tithes of the parish, they would have relied, not on custom, but on statute. If, as parochial endowments, tithes were statutory in their origin, we should expect to find that they commence with the legislation by which they are alleged to be created, and that the payment was certain in practice, uniform in amount, identical in source. If, on the other hand, the endowment of parish churches with tithes originated in a series of voluntary dedications, and if the State merely protected a property which was none the less real because it began as a free-will offering, we should expect to find that customary payments preceded any recorded legislation, and were uncertain in practice, varying in amount, irregular in source. Historical facts confirm the second view. The voluntary payment of tithes in this country preceded, by upwards of three centuries, parochial organisation, as well as both ecclesiastical and secular legislation. The first secular enactment on the subject assumes the prior existence of the charge, and for more than two centuries afterwards allows tithe-payers a wide freedom in the choice of the religious body to which payment was made. When this freedom was limited, it was restricted not by legislation but by the growth of custom. Both in respect of the persons to whom tithes were due, and of the produce on which they were payable, the practice was not certain, but uncertain. The amount paid was varying, not uniform. The sources from which the payment was derived, are not identical, but irregular. If, therefore, the State endowed parochial churches with tithes, all those signs, which would naturally accompany such a national act, are conspicuously absent. On the other hand, all those signs, which naturally indicate the legislative protection of customary practices, are con-conspicuously present.

The gradual, piecemeal, and discretionary endowment of parochial churches with the tithes of the parish has left its mark on the existing organisation. It explains, for instance, as no other assumption can explain, the freedom from the payment which the " Hall," " Court," or " Manor " farm frequently enjoyed ; it lies at the root of the distinction between rectorial and vicarial tithes, and between ecclesiastical appropriators and lay impropriators ; it suggests the reason why land in one parish should be charged with tithes for the benefit of the church of another parish. Many of the old anomalies in the law of tithes have been smoothed into comparative uniformity by the Tithe Commutation Act of 1836. But the previous history of the charge renders it difficult to believe that the nation ever by a legislative action endowed parochial churches with the local tithes. The Reformation left the parochial organisation untouched. But it made an important change, which greatly embittered objections to the payment of tithes. It alienated a considerable portion of the tithes from religious uses. Rectories, together with the local tithes, might be, and often had been, " appropriated " to a monastery or other religious corporation, which appointed vicars to discharge the religious duties attached to the endowment. Originally the stipend of the vicar was arbitrary. But gradually it was recognised that the person responsible for religious ministrations in the parish ought to have some fixed determinate means of support. This was generally made by endowing the vicarage with land, or by assigning to it some portion of the great tithes, or the whole of the small tithes, or by a combination of all three methods. At the Dissolution of the Monasteries all the rectorial tithes in their possession, which had not been already allocated to the support of vicarages, passed into the hands of the Crown, and were subsequently granted out by letters patent to lay subjects. These lay grantees were called " lay rectors," or " impropriators," in order to distinguish them from the original " appropriators," who were of necessity spiritual persons or ecclesiastical corporations. When the Tithe Commutation Act was passed in 1836, and tithes of produce were commuted into rent-charges, it was found that nearly one-fourth of the annual value had thus been diverted from religious purposes into the hands of laymen.[1] There is strong evidence that the lay impropriators or their lessees, who were generally absentees, and without other interests in the parish, exacted their legal dues with a strictness which was relatively rare among clerical tithe-owners.

Tithes in themselves, and apart from their incidence, could scarcely be regarded as a legitimate grievance by either owners or occupiers of land, especially as no attack was as yet made on the religious objects to which they were devoted. No landlords could honestly believe

---

[1] The net annual value of the tithes, after a deduction of 40 per cent. from the gross value, was in 1836 estimated at £4,053,985 6s. 8½d. Of this total sum £962,289 15s. were then in the hands of lay impropriators. But this figure does not take into account the large amount of lay tithes which had been, in the course of three centuries, extinguished by purchases on the part of landowners, or bought and given back to the parochial clergy, or restored by those who, like Spelman, considered their retention by laymen a sacrilege.

that the payment robbed them of any part of the rents to which they were justly entitled. For centuries, in every transfer of land, whether by purchase or inheritance, the estimated value of tithes had been previously deducted from the value of the estate so bought or inherited. Nor could any tenant honestly complain that tithes increased the burden of his rent. Land only commands what it is worth. If 100 acres of land fetched £1 per acre, it made no pecuniary difference to the farmer whether he paid £100 to the landowner or £90 to the landowner and £10 to the tithe-owner. But the real practical grievance was the incidence of the charge upon the produce of the land. In this way tithes become a charge which was increased by good farming, or diminished by bad,—a tax on every additional outlay of money and labour,—a check upon enterprise and improvement.

Tithes in kind were admittedly out of date. Though rents and wages had long been placed on a money basis, a tithe-owner could still exact payment in the ancient fashion. As a fact, however, the Reports to the Board of Agriculture (1793-1815) prove that, at the close of the eighteenth century, comparatively little tithe was collected in kind. Especially was this the case when the tithe was in the hands of clerical owners. For this change of practice there were many reasons. Collection in kind was extremely unpopular. Where it prevailed, farmers showed their dislike to the system in various ways. Many tenants so greatly resented putting money into the pockets of tithe-owners that they preferred to lose it themselves, and refused to plough up pastures which would have been more profitable under tillage. Sometimes the tenant left his tithable land unmanured. A Hertfordshire farmer, for instance, occupied land in two parishes, in one of which a reasonable composition was paid, while, in the other, tithe was collected in kind. The result was that he farmed one part of his occupation with spirit on improved methods, and that the dung-cart never reached the other portion of his land. Sometimes the tenants made the collection as inconvenient as possible. Thus a Hampshire farmer gave notice to the tithe-owner that he was about to draw a field of turnips. When the tithe-owner's servants, horses and waggons had come on the land, the farmer drew ten turnips, gave one to the tithing-man, and said that he would let his master know when he drew any more. In a wet season the collection was often the cause of heavy loss. Notice had to be given to the tithe-owner to set out the tithe. Farmers risked a lawsuit, if they carried their crops before the

Q*—E F P P

procéss was completed. Consequently, in catchy seasons the rain often outstripped the slow progress of the tithing-man, and the crops were ruined.

The collection of tithes in kind, regulated as it was by the subtle and technical distinctions of case-made law, provoked endless bickerings, disputes and litigation. If tithe-owners were clergymen, living in their parishes, they naturally welcomed any reasonable system of payment which enabled them to live on friendly terms with their parishoners. Non-resident pluralists, or lay impropriators who let out the tithes to proctors, could better afford to defy the public opinion of the neighbourhood. But they were not always proof against business arguments. The heavy cost of collecting tithes in kind suggested the commercial prudence of adopting other arrangements. Barns must be built and repaired for the storage of produce. The weekly wages of servants must be met. Waggons and horses, with the necessary cart-sheds and stabling must be provided and maintained. The cost, not only of collecting, but of threshing, dressing and marketing corn had to be met. The net profits of a crop were thus reduced to a minimum by the duplication of expenses. Various forms of payment were therefore substituted for collection in kind. Sometimes, and especially under enclosures of open fields, tithes were extinguished by allotments of land of equivalent value. Sometimes it was considered that the increase of the area of land held in mortmain or the difficult position of clerical landowners were objections to the exchange of tithes for their equivalent in landed property, and a corn-rent was substituted. Sometimes tithes were commuted for a composition calculated on the acre or on the pound of rent paid, and either fixed for a term of years or based on an annual estimate of the value of the crops. Sometimes farmers had the option of taking the tithable portion at the surveyor's valuation or leaving it to be collected by the tithe-owner. Sometimes, in a few fortunate parishes, a *modus* had by immemorial usage taken the place of tithes. Moduses were payments of definite sums, which had been permanently fixed in amount at a time when the purchasing power of money had been far greater than it had since become. They were, therefore, advantageous to the tithe-payer. A *modus* of 1d. on every fleece shorn in the parish was no real equivalent to a tenth of the value of the wool.[1]

[1] For various methods of collecting tithes in the different counties 1793-1815, see Appendix V.

No variety in forms of payment could entirely remove the reasonable objection to a tithe of produce in kind. So long as farming remained stagnant the grievance was imperceptible. It became acute when progressive methods of agriculture were generally adopted. Here and there tithe-owners recognised the altered conditions by allowing deductions from their tithes to meet the cost of all purchased manures. But the practice was by no means general. The fair adjustment of compositions was in other ways extremely difficult. Tithable crops were of greater value to farmers, who could collect and market them at a small additional expense, than they were to tithe-owners, whose necessary outlay diminished their net profits by a half. The difference allowed a large margin for dispute. Even when compositions were reasonable, they tithed the increased produce of improved husbandry. Land, highly cultivated, might be valued at 3s. 6d. an acre; soil of the same natural quality, under slovenly management, might escape with 1s. 6d. In the case of wastes, the objection to tithes on produce was strongly felt as an obstacle to improvement. When land, which at the best had afforded only rough pasture, was reduced to cultivation, owners and occupiers risked labour and money on a venture which might succeed or fail. In either event tithe-owners were safe ; they profited by the success, and lost nothing by the failure. The legislature had endeavoured to meet the case. Under the Barren Lands Act,[1] barren heaths and waste grounds were exempted from tithes for seven years after they had been reduced for the first time to cultivation. But the decisions of the law courts deprived improvers of the benefits which they expected from the Act. Only land which was so barren that it paid no tithe *by reason of its barrenness* was held to be exempt. The initial cost of draining fen-lands, or grubbing and stubbing wood-lands, or of paring and burning moors and heaths was not to be taken into consideration. Whatever the cost at which the land had been fitted for cultivation, the only question to be asked was whether, when ploughed and sown, it was so naturally fertile as to produce a crop, or so naturally barren that it would yield nothing without an extraordinary expenditure on liming, chalking, marling, dunging, or manuring. Only in the latter case could the seven years' exemption be legally claimed.

The law of tithes needed complete revision. Its inadequacy to meet changed conditions had long been felt. The necessity for a

[1] 2 and 3 Ed. VI. c. 13.

large expenditure of capital in order to recover the ground which had been lost during a long period of disaster forced the question to the front. In 1836 the difficulty was solved. Peel in 1835 had proposed the voluntary commutation of tithe. Lord John Russell, adopting in his Bill the machinery which Peel had sketched, made commutation compulsory. When once this point was decided, party considerations were for the moment subordinated : Whig and Tory loyally co-operated to frame a workable scheme. The aim of legislators was to commute tithe of produce in kind for a variable money payment charged on the land, to make the commuted sum fluctuate with the purchasing power of money, to preserve the existing relations between the values of tithable produce and the cost of living. It never attempted to fix the payment, once and for all, at the sum which represented the value that tithe then possessed. On the contrary, it converted tithes into a corn-rent, fluctuating in value according to the septennial average of the prices of wheat, barley and oats.

The first step was to determine the value of the tithes ; the second to adjust the purchasing power of the money payment at which they were commuted.

Within a limited time tithe-owners and tithe-payers of any parish might agree upon the total sum to be paid in lieu of tithes. This agreement was first to receive the assent of the patron ; secondly, to be communicated to the bishop ; and, thirdly, to be approved and ratified by the Commissioners. If no agreement was arrived at, a local enquiry was held on the spot by the Commissioners or their assistants, who estimated the value of the tithe, taking as their basis the actual receipts of the tithe-owner during the preceding seven years ; framed their draft award ; deposited it for the inspection of interested parties ; and, finally, confirmed their award, which from that time was binding upon tithe-owners and tithe-payers.

The mode in which the purchasing power of money was intended to be preserved was as follows. The average of the gross annual value of the actual receipts of the tithe-owner was ascertained in money for the seven preceding years. The net annual value, arrived at by deducting all just expenses, was taken as the permanent commutation of the great and small tithes of the parish. This net sum was divided into three equal parts, and the average value for the seven years ending with 1835 was taken for wheat, barley and oats. It was then asked how many bushels of wheat could be bought at cost price by one of these equal portions, how many of barley by the second, how many of oats by the third. Each £100 of tithe was divided into three equal sums of £33 6s. 8d. : the septennial averages for the three grains were respectively 7s. 0½d. for a bushel of wheat ; 3s. 11½d. for a bushel of barley ; 2s. 9d. for a bushel of oats. In 1836 at those prices £33 6s. 8d. bought 94·96 bushels of wheat, or 168·42 bushels of barley, or 242·42 bushels of oats. These have been the fixed multipliers in use ever since. Each year the average prices for the last seven years are multiplied by these fixed quantities, and the result is the tithe rent charge for the coming year. It will be noticed that the charge is affected most by variations in the price of oats, and least by those of wheat.

One other point requires to be mentioned. Lord Althorp in 1833, Sir Robert Peel in 1835, Lord John Russell in 1836 were agreed that the payment should be transferred from occupiers to owners of land. Section 80 of the Act of 1836 empowered tenants to deduct the rent-charge from the rent payable to the landlord. But the section was permissive only. For mutual convenience tenants paid the rent charge direct to the tithe-owner, and their other rent to the landlord was calculated on this basis. By the Tithe Rent Charge Recovery Act of 1891 the tenant was no longer permitted to be the conduit-pipe for the payment. The liability to pay the tithe rent charge was transferred to the landowner ; the tithe-owner's remedy of distress was altered into a process through the county court ; and, instead of the corn averages absolutely determining the amount of tithe rent charges, provision was made in certain cases for a reduced payment when the charge exceeded a certain proportion of the annual value of the land.

Acting on the principles then laid down, it became the duty of the Board of Agriculture to collect the corn prices at the markets specified in the Act and at the beginning of each year to publish the average prices for the previous year and the septennial average from which the current value of £100 of tithe rent charge could be calculated. The Board itself did not publish the calculation, but tables were always produced by private enterprise. Towards the close of the century the low prices of all cereals resulted in tithe rent charge falling much below par ; for example, £100 at par which in 1875 had risen to £112 15s. 6¾d. had become as little as £66 10s. 9¼d. in 1901. Towards the end of the war, however, with the famine prices to which wheat, even though controlled, had risen, the value of the tithe rent charge began to rise steeply and would obviously become very high indeed as more of the years of scarcity became included in the septennial average. An Act was then passed in 1918 which took the current figure of £109 3s. 11d. for each £100 at par and stabilised that value until 1925, with the provision that after that year the value should be calculated on a fifteen instead of a seven years' average so as to include years of high and low prices. This was not welcome to tithe owners, who argued that for a long period they had been receiving far less than the par values, indeed, on the average from 1837 to 1918 a trifle less than £92.

However, it was avowedly temporary legislation to meet the emergency and a return to the earlier system was promised in 1925. But as that period approached it was apparent that a great increase in the charge would come into operation under the method indicated in the 1918 Act, indeed that the value of £100 at par would become over £130. An attempt was then made to put the whole vexed question of tithes upon a new basis, to commute afresh this time at a fixed, instead of a fluctuating rate, and to provide for the eventual extinguishment of a charge which had always aroused a good deal of opposition on sectarian grounds, however truly it could be argued that payment came not from the farmer but from the owner, who was cognisant of the burden when he bought the land.

By the Act of 1925 the tithe rent charge was finally stabilised at £105, to which the landowner had to add £4 10s. in order to provide a sinking fund for the eventual extinction of the charge. At the same time all ecclesiastical tithe rent charges, whether attached to benefices or owned by ecclesiastical Corporations, were vested in the Governors of Queen Anne's Bounty, which became responsible for payments to the previous beneficiaries.

Certain other legislation affecting tithe may also be recorded. It had always been a grievance among the clergy that their tithes were treated as rateable property, especially as rates even in rural parishes had been rising steeply towards the end of the nineteenth century. Relief was accorded by successive Acts of 1899, 1920, and 1922, the eventual result of which was to accord remission of rates on all tithes attached to a benefice, varying from complete remission when the income was below £300 to half remission for incomes above £500. Corporations and lay impropriators received no remissions. The 1925 Act simplified still further by removing all rates upon ecclesiastical tithes ; the rating authority received the amount due on the new tithe from the Inland Revenue Commissioners, who were in turn reimbursed from Queen Anne's Bounty.

The various Acts specified above also made a series of provisions whereby tithe rent charges could be redeemed. Under the earlier Acts redemption required agreement between tithe owner and landowner, the commutation value being fixed at twenty-four and then twenty-five years' purchase of the par value, not a very tempting proposition to the landowner when the current value had fallen below 80 as it had from 1890 onwards. The 1918 Act gave the landowner power to redeem on providing such a sum as when invested in Government securities would produce an annuity of the gross annual value of the tithe rent charge, less rates and land tax and an allowance for the cost of collection which was not to exceed $2\frac{1}{2}$ per cent. of the gross.

The Board of Agriculture had to determine the gross annual value as well as the rate of interest defined as obtainable from Government securities, and from 1918 to 1921 the formula adopted was twenty-one years' purchase of the par value less the specified deductions. Between 1921 and 1925 the following factors prevailed—118 par value $\times 17$, $108 \times 18\frac{1}{2}$, and 104×22. With the 1925 Act the gross annual value became finally defined at £105, from which a deduction of £5 was made for rates in the case of benefices (£16 for Corporations) and of £2 10s. for costs of collection. The number of years' purchase still has to vary with the current rates of interest and by 1934 had risen as high as 30. Obviously the years immediately following the war offered exceptionally favourable opportunities for redemption, for the tithe rent charge had been fixed at a relatively low figure and the rates of interest were so high that a small number of years purchase were required. Prior to the war the amount of rent charge redeemed annually lay between one and two thousand pounds, but for the years 1920–23 it averaged nearly £82,000, from which it has since fallen to something of the order of £5,000–£8,000, a figure necessarily swollen by the necessity of redemption when land has been cut up for building plots.

Though the 1925 Act had appeared to provide a not unsatisfactory settlement of a question that had long been a source of trouble, one, too, that would eventually secure the extinction of what had often been an onerous charge, like other fixed charges it was not prepared for the shock which came with the break of prices from 1929 onwards.

The larger landowner would have had no ostensible cause for complaint, tithe had become a fixed charge upon their land, they had no remedy against falling rents due to the unremunerativeness of farming. Agricultural landowners obtained little public sympathy or political consideration, in view of the accretions in land values enjoyed by the owners of the ever-increasing margin of urban land. But during the heyday of war prices no inconsiderable proportion of agricultural land had passed into the possession of the men who farmed it. They had seen its capital value fall by one-half and they had now to carry the burden of tithe when the land was earning so little. So a new agitation against the tithe sprang up from 1930 onwards, embittered in some cases by the old anti-church feeling, but obtaining its real strength from the owner occupiers. In Kent, in Essex and other of the Eastern Counties an organised opposition to the payment of tithe developed, and the distraint sales were forcibly broken up or effectively boycotted. The situation became dangerous both to public order and to the Church of England, which was placed in a most invidious position. For the moment the question still remains unsettled pending the report of a Royal Commission which was appointed in 1934, in view of which and of the large measures of relief that have been accorded to agriculture, the opposing organisations are holding their hands.

CHAPTER XVII

HIGH FARMING. 1837-1874

Condition of agriculture in 1837 ; current explanation of the distress ; preparation for a new start in farming ; legislative changes ; development of
a railway system ; live-stock in 1837 ; the general level of farming ;
foundation of the Royal Agricultural Society ; notable improvements,
1837-74 ; extension of drainage ; purchase of feeding stuffs ; discovery
of artificial fertilisers ; mechanical improvements and inventions ; Repeal
of the Corn Laws ; the golden age from 1853 to the end of 1862 ; rapid
progress in the " Fifties " ; pedigree mania in stock-breeding.

THE reign of Queen Victoria began in the midst of a transition stage
from one state of social and industrial development to another.
A complete change of agricultural front was taking place, which
necessitated some displacement of the classes that had previously
occupied or cultivated the soil. The last ten years of the nineteenth
century raised the question of whether agriculture was not passing
through another transition stage which, like its predecessor, was to
initiate another agricultural revolution and result in another dis-
ruption of rural society. This advancing change was suddenly
accelerated by the war.

Roughly speaking, the first thirty-seven years of the new reign
formed an era of advancing prosperity and progress, of rising rents
and profits, of the rapid multiplication of fertilising agencies, of
an expanding area of corn cultivation, of more numerous, better
bred, better fed, better housed stock, of varied improvements in
every kind of implement and machinery, of growing expenditure on
the making of the land by drainage, the construction of roads, the
erection of farm buildings, and the division into fields of convenient
size. So far as the standard of the highest farming is concerned
agriculture made but little advance since the " Fifties." The
last twenty-six years of the reign, on the other hand, were a period
of agricultural adversity—of falling rents, dwindling profits, con-

tracting areas of arable cultivation, diminishing stock, decreasing expenditure on land improvement.

In 1837 the farming industry had passed through a quarter of a century of misfortunes, aggravated by a disordered currency, bank failures, adverse seasons, labour difficulties, agrarian discontent. During times of adversity it has always been the practice to charge landowners, farmers, and even labourers with extravagance, to trace distress to their increased luxury, to attribute their domestic difficulties to their less simple habits. The explanation is as old as the hills. Arthur Young, writing in 1773 *On the Present State of Waste Lands*, remarks that the landed gentry were beggared by their efforts to rival their wealthier neighbours who had amassed fortunes in trade. The rural frog burst in his efforts to equal the proportions of the civic ox. " The antient prospect which afforded pleasure to twenty generations is poisoned by the pagodas and temples of some rival neighbour ; some oilman who builds on the solid foundation of pickles and herrings. At church the liveries of a tobacconist carry all the admiration of the village ; and how can the daughter of the antient but decayed gentleman stand the competition at an assembly with the point, diamonds and tissues of a haberdasher's nieces ? " Their tenants did not escape from similar charges. In 1573 Tusser had alluded to farmers with " hawk on hand " who neglected their business for sport ; in the nineteenth century it was said to be the hunting-field or the racecourse which attracted them from the farm or the market. In 1649 Walter Blith had attributed the rural depression of that day to the " high stomachs " of the farmers. So in 1816 the wiseacres of the London clubs vehemently contended that farmers had only to return from claret to beer, and their wives from the piano to the hen-house, and agricultural distress would be at an end. It was reserved for an imaginative versifier in 1801 to charge them with soaking five-pound notes instead of rusks in their port wine. Somewhat similar in tone was the outcry against labourers. " We hear," writes Borlase, the Cornish antiquary, in 1771, " every day of murmurs of the common people ; of want of employ ; of short wages ; of dear provisions. There may be some reason for this ; our taxes are heavy upon the necessaries of life ; but the chief reason is the extravagance of the vulgar in the *unnecessaries of life*." Among the tinworkers in his parish were three-score snuff-boxes at one time ; of fifty girls above fifteen years old, forty-nine had scarlet cloaks. " There is scarce a family in the parish, I mean of common labourers, but have *tea*, once if not twice a day. . . . In short, all labourers live above their conditions."

The same explanations with regard to all classes of agriculturists were repeated in 1837, and have been periodically offered ever since. The diagnosis of disease would not be so popular if it were not easy and to some extent true. It is, to say the least, inadequate. When the standard of living rises for all classes, agriculturists are not the only men who spend money more lavishly than the prudence which criticises after the event can justify. But the true explanation of the distress lay in the conditions already described. The old instrument of farming had failed ; the new had not been perfected. An agricultural revolution was in progress, which was none the less complete in its operation because it was peaceful in its processes.

In 1837 agriculture was languishing ; farming had retrograded ; heavy clay-lands were either abandoned or foul, and in a miserable state of cultivation. Indifferent pasture, when first ploughed, had produced good corn crops from the accumulated mass of elements of fertility which they had stored. But this savings bank of wealth had been soon exhausted. At peace-prices half crops ceased to be remunerative, and the newly ploughed arable area was now recovering itself from exhaustion to grass as best it could without assistance. Lighter soils had suffered comparatively little ; turnips, and the Norfolk system had helped the eastern counties to bear the stress of the storm, yet, even there, farmers had " had to put down their chaises and their nags." Much of the progress made between 1790 and 1812 had been lost. Nor was this the worst feature. The distrust which prevailed between farmers and their men had extended to tenants and their landlords. Men who had contracted to pay war rents from peace profits were shy of leases. For at least a generation confidence was shaken between landlord and tenant.

The brighter side to the picture was that, in the midst of much suffering, the ground had been prepared for new conditions. Small yeomen, openfield farmers, and commoners could never have fed a manufacturing population. They could not have initiated and would not have adopted agricultural improvements, of which some were still experimental, and of which all required an initial expenditure. It was from these classes that the most bigoted opponents of " Practice with Science " were recruited, and their contempt was heartily sincere for the innovations of the " apron-string " farmer. Socially

valuable though they were, they were becoming commercially discredited. Their disappearance was a social loss ; but it had become
an economic necessity. The land could no longer be cultivated for
the needs of a scanty, scattered population, occupied in the tillage
of the soil, or engaged in one-man handicrafts. So long as England
depended for food on her own produce,—a condition which lasted
a quarter of a century after the repeal of the Corn Laws,—it was
requisite that farming should be transformed from a self-sufficing
domestic industry into a profit-earning manufactory of bread, beef,
and mutton. Food, upon the scale that changed conditions
demanded, could only be produced upon land which had been
prepared for the purpose by the outlay of capitalist landlords and
the intelligent enterprise of large tenant-farmers.

In other respects, also, the distress of 1813-37 produced good
results. So long as war prices prevailed, prosperous years had
brought wealth to slovens, and sluggards had amassed riches in their
sleep. The collapse of prosperity spurred the energies and enterprise of both landlords and tenants, who could only hold their own
by economising the cost and increasing the amount of production.
Within certain limits, low prices and keen competition compelled
improvement. Again, though the attraction of war-prices had
driven the plough through much valuable pasture, it had also supplied
the incentive which added hundreds of thousands of acres of wastes
to the cultivated area of the country. Finally, during the era of
Protection, landlords and farmers had learned to rely too entirely
upon Parliamentary help in their difficulties. They had been prone
to expect that alterations in the protective duties would turn the
balance between the success and failure of their harvests. Now,
disappointment after disappointment had taught them the useful
lesson that they could expect no immediate assistance from legislative interference, and that, if they wanted aid, they must help
themselves.

Meanwhile legislation had been active in many useful directions.
The agricultural revolution, and the effects alike of war and peace,
had completely disorganised the labour market. Parliament co-
operated with industrial changes in redressing the balance between
demand and supply and in adapting the relations of capital and
labour to new conditions. For agricultural labourers the Poor-Law of
1834 did what the Factory legislation of 1833 had done for artisans.
The change produced immediate effect. The number of paupers

steadily diminished, and the poor-rates fell from seven millions in 1832 to four millions in 1837. New means of transport had been provided by the opening up of canals. Increased facilities of communication had been supplied by progress in the art of road-construction. Though Turnpike Trusts were proving inefficient on the great highways, the first step towards the improvement of minor roads had been taken by the Act of 1835, which substituted a rate for the old statute labour. Another legislative result of the prolonged agricultural distress had been the Tithe Commutation Act of 1836. The incidence of the charge was shifted ; it no longer operated as a check to the expenditure of capital or a discouragement to skilful and enterprising farming.

It was a period of preparation, the full significance of which was then imperfectly understood. Few persons in 1837 could have foreseen the imminence of social and industrial changes which introduced to British farming an unexampled era of prosperity, or could have foretold that new markets would not only be opened up, but brought to the doors of agriculturists. Signs of better times were indeed faintly visible. Manufacturing progress was beginning to tell upon agriculture ; steam navigation was stimulating trade ; joint-stock banks helped farmers to face their difficulties ; the new system of poor-law administration was restoring the labour market to healthier conditions ; beef, mutton, wool, barley, and oats sold briskly. Above all, the whole country was beginning to respond to the vast impulse which the introduction of railways gave to it• intelligence, its intercourse, its enterprise, its agriculture, manufacture, and commerce. Without assistance or control by the State, in the face of many difficulties and prejudices, railways were being built piecemeal by private energy and capital. They were still in their infancy. It was not till 1821 that the Act for the construction of the Stockton and Darlington Railway had been passed. The Liverpool and Manchester Railway was opened in 1830, and the line from London to Birmingham was completed in 1838. Between those two dates fifty-six Acts had been already passed for laying 1,800 miles of rails.

The era of railways had begun. The real innovations lay in the application of steam as the motive power to movable engines, the construction of new and independent lines of communication, the conveyance not only of goods but of passengers. Rail-ways to facilitate the transport of heavy weights had been in use for nearly two centuries. They seem to have been first employed in the New-castle district to convey coal from the pits to the shipping stages on the Tyne. Wooden rails, laid on continuous parallel lines, were pegged down to wooden sleepers, which were set two feet apart, the intervals being filled in with stones or ashes. On these tracks, high hopper-shaped waggons, set on solid wooden wheels, were either propelled by their own weight or drawn by horses. Log ways, thus constructed, were called in eighteenth century Acts of Parliament " dram roads." They were in fact true tram-ways,[1] though the word " tram " has been transferred from the material out of which the rails were originally constructed to the vehicle which passes over them. Successive improvements were made in their construction. Thus iron plates or iron flanges were fixed by " plate-layers " to the rails to lessen the friction at the curves or to keep the waggons on the track. About 1767 the rails began to be made entirely of iron, which were generally cast with an iron flange on the inner side. Similarly the wheels were made of cast iron, though for some years the rear wheels continued to be made of wood in order to strengthen the grip of the brake. In 1788 a still more important change was made. The projections of the flanged rail were found to be dangerous obstructions wherever lines crossed highways. To meet this difficulty, flanged wheels were introduced, and the rails were made smooth.

By the latter half of the eighteenth century, there were few collieries in the north which were not provided with their own rail-ways, often carried, in order to secure easy gradients, through hills and over valleys by means of cuttings, bridges, or embankments. They were private roads, to which the public had no access. Rail-ways laid by Canal Companies under the powers of Acts of Parliament were in a different position. Constructed by canal proprietors to feed their traffic from potteries, furnaces, collieries, and quarries,

---

[1] Whether " tram-way " is derived from the material out of which the road is contracted, or from the carriage which passes over it is doubtful. A will dated 1555 mentions the repair " of the higheway or tram " in Barnard Castle. This use of the word, like the " dram-road " of eighteenth century Acts of Parliament, suggests the log-way. On the other hand, the road may have taken its name from the application of the word " tram " in the North of England to a small carriage on four wheels, possibly gaining this meaning through the Lowland Scottish use of the word for the " shaft " of a cart. In either case, " tram " is Scandinavian in origin ; Norwegian, *tram* = a door-step of wood ; *traam* = a wooden frame, and Swedish *tromm* = a log, or a summer sledge. See Skeat's *Etymological Dictionary of the English Language*.

they were public highways, maintained, like turnpike roads, by the payments of those who used them. The Canal Companies provided no rolling stock. On payment of the stipulated tolls, any trader might transport his goods over the flanged rails in his own vehicles to the wharves. In the development of these lines, which were subsidiary to inland waterways, the lead was taken by the valley of the Severn, the Western Midlands, and South Wales. The utility of the system was at once apparent. Rail-ways multiplied rapidly, not as rivals, but as aids, to the canals which they eventually destroyed.

Numerous rail-ways, either in private hands or feeders to canals, existed at the end of the eighteenth century. The first public independent rail-way was constructed by Act of Parliamant in 1801. The Surrey Iron Rail-way connected Croydon and the mills on the Wandle with the Thames at Wandsworth. Originally intended to run to Portsmouth, it was never carried beyond Merstham. Nearly twenty years later an Act of Parliament (1821) was obtained for the construction of the Stockton and Darlington Rail-way. On this line all the stages in the transformation of the ancient rail-way into the modern type were exemplified. Hitherto speed had not been regarded as an object. Horses were generally employed, and, where steam had been introduced as the motive power, its use had been practically confined to stationary engines, placed at the top of inclines, which by means of ropes or chains drew waggons up the ascent and regulated the pace of their descent. In poetry, Erasmus Darwin [1] had anticipated the coming triumph of steam :

> " Soon shall thy arm, unconquered Steam ! afar
> Drag the slow barge, or drive the rapid car ;
> Or on wide-waving Wings expanded bear
> The flying chariot thro' the fields of air ! "

But in 1820 the vision of the " rapid car," drawn by steam, still seemed as extravagant as the dream of the " flying chariot " appeared to a later generation familiar with fast trains. The projectors of the Stockton and Darlington Railway hesitated between wooden or iron rails, between animal or steam power, between stationary or movable engines. When the line was opened in 1825, the waggons, under the advice of George Stephenson, were drawn over iron rails at an average pace of five miles an hour by steam locomotives, designed on the model of the engines which he had successfully

[1] *The Botanic Garden*, Part I. 289.

introduced at Killingworth Colliery. Goods traffic only was at first undertaken by the railway company. The conveyance of passengers was left to private enterprise ; coaches drawn by one horse ran over the rails, on payment of stipulated tolls, at intervals when the goods trains were not running. It was not till 1833 that the Company bought out the coach proprietors, and, a year later, issued notices that they proposed to provide not only carriages for goods, but " coaches " for the conveyance of passengers, drawn by steam locomotives.

Before this final stage was reached in the County of Durham, the Liverpool and Manchester Railway had been opened (1830). The project of the proposed line originated in dissatisfaction with the cost and delay of canal transport. It was directly designed not to feed but to rival the water way, and to break down a monopoly in the carriage of heavy goods. Canal companies all over the country became alive to their danger. So strong was the opposition that the first Bill was defeated. A second Bill was introduced, and passed in 1826. Like the Stockton and Darlington Company, the projectors hesitated over the choice of motive power. They were still undecided when the new line was approaching completion. To solve the problem they offered a premium of £500 for the best locomotive engine which should satisfy certain conditions. It was not to exceed £550 in price and six tons in weight ; it was also to draw three times its own weight, at a speed of ten miles an hour on level ground, The famous Rainhill trial (October 8, 1829), when Stephenson's *Rocket* won the prize, sealed the fate of canals and inaugurated the triumph of railways. Without their aid the modern organisation of industry would have been impossible. The factory, the modern farm, and the railway went hand in hand in development and were not dissimilar in their economic results.

With the ground thus prepared for a new start, but in gloom and depression, agriculturists entered on a new reign. In comparing agriculture at the end with that of the beginning of the reign of Victoria, the most striking feature was the rise in the general level of excellence. If we leave on one side the achievements of chemical science and the triumphs of mechanical invention, there were few improvements in the practices of agriculture at the end of the reign which had not been anticipated by some individuals sixty years earlier. But the knowledge which was then, at the most, confined to one or two men in the country had at the close of the century become generally diffused. The best farmers of that day could not have explained the reasons for their methods ; they farmed by experience and intuition. Judgment is still all-important ; but practice has now been reduced to principles and rules, which make the best methods more nearly common property, or at least place them within reach of all. The best arable farms in 1837 were cropped much as they are now, except that rotations were more rigid and inelastic. Pedigree barleys and pedigree wheats were already experimented upon by Patrick Shirreff, Dr. Chevallier, and Colonel Le Couteur. By the most enterprising of our predecessors all the kinds of farm produce which are raised to-day were raised seventy years ago.

Live-stock has doubtless immensely improved since the accession of Queen Victoria. Specialisation did away with " general utility " animals, and successfully developed symmetry, quality, early maturity, or yield of milk among cattle. But the value and importance of improving breeds had been thoroughly appreciated by the best farmers before 1837. Though only one herd-book—Coates's *Shorthorn Herdbook* (1822)—had begun to appear, the followers of Bakewell,—such as Charles and Robert Colling, Thomas Bates, of Kirklevington, the Booths, and Sir Charles Knightley with the Shorthorns,—Benjamin Tomkins, John Hewer, and the Prices with the Herefords,—Francis Quartly, George Turner, William Davy, and Thomas Coke of Norfolk with the North Devons,—had already brought to a high degree of perfection the breeds with which their names are respectively associated. Flockmasters, like cattle-breeders, had recognised the coming changes. Before 1837 Bakewell's methods had been extensively imitated. The Lincolns, the Border Leicesters of the Culleys, the Southdowns of Ellman of Glynde and Jonas Webb of Babraham, the Black-faced Heath breed of David Dun, the Cheviots of Robson of Belford were already firmly established ; and some of the best of the local varieties of sheep, enumerated by Sir John Sinclair in his *Address to the British Wool Society* (1791), were beginning to find their champions. Nor were pigs unappreciated. The reproach was no longer justified which, at the close of the eighteenth century, Arthur Young had directed against farmers for their neglect of this source of profit. Here, again, Bakewell had led the way. Efforts were being made to improve such native breeds as the Yorkshire Whites, the Tamworths, the reddish-brown Berkshires, or the black breeds of Essex and Suffolk.

Oxen were still extensively used for farm work. It is therefore not surprising that comparatively little attention had been paid to horses for agricultural purposes. Yet here, too, some progress was made, particularly from the point of view of specialisation. The Clydesdales were coming to the front as rivals to old English breeds. Beauty was not the strong point of the " Sorrel-coloured Suffolk Punch." Nor was he any longer suited to the pace required in the modern hunting-field or on the improved roads. But in 1837 it was recognised that his unrivalled power of throwing his whole weight into the collar fitted him pre-eminently for farm work. A similar change was passing over the Cleveland Bay. Threatened with extinction by the disappearance of coaches, he was found to be invaluable on light-soil farms. So also a definite place was assigned to another breed known to the sixteenth century. The " Large Black Old English Cart-horse," which Young calls " the produce principally of the *Shire* counties in the heart of England," was, to some extent, experimented upon by Bakewell. But the development of the breed belongs to a later date than the first half of the Victorian era, and it is as a draught-horse that the Shire has been, since 1879, patronised by Societies and enrolled in stud-books.

It has been said that while the general standard of farming was still extremely low, the best practice of individual farmers in 1837 has been little improved by the progress of seventy years. Production has been considerably increased ; but the higher averages are due to the wider diffusion of the best practices rather than to any notable novelties, and it is in live-stock that real advance is most clearly marked. If, however, we turn from the highest practice of farming to the general conditions under which it was carried on, or to the processes by which crops were cultivated, harvested, and marketed, the contrast between 1837 and 1902 is almost startling.

In 1837 the open-field system still prevailed extensively. Holdings were in general inconveniently small, though in some parts of the country farms had been consolidated. Farm-buildings, often placed at the extreme end of the holding, consisted of large barns for storing and threshing corn, a stable and yard for cart-horses, a shed for carts and waggons. But the cattle, worse housed than the waggons, were huddled into draughty, rickety sheds, erected without plan, ranged round a yard whence the liquid manure, freely diluted from the unspouted roofs, ran first into a horse-pond, and thence escaped into the nearest ditch. In these sheds the live-stock sub-

sisted during the winter months on starvation allowance. Fat
cattle, instead of being conveyed by rail quickly and cheaply, were
driven to distant markets, losing weight every yard of the way.
Long legs were still a consideration for sheep which had to plough
through miry lanes. Farm roads were few and bad. Where land
had been early enclosed, the fields were often small, fenced with
high and straggling hedges. Very little land was drained, and,
except in Suffolk and Essex, scarcely any effort had been made to
carry off the surface-water from clay soils.

Little or no machinery was employed in any operation of tillage.
In remote parts of the country, even on light soils and for summer
work, heavy wooden ploughs with wooden breasts, slowly drawn
by teams of five horses or six oxen, attended by troops of men and
boys, still lumbered on their laborious way, following the sinuous
shape of boundary fences, or throwing up ridges crooked like an
inverted S, and laid high by successive ploughing towards the crown.
In more advanced districts, less cumbrous and more effective implements of lighter draught, wheel or swing, were employed. But
not a few discoveries of real value fell into disuse, or failed to find
honour in the land of their birth, till they returned to this country
with the brand of American innovations. The mistake was too
often made of exaggerating the universal value of a new implement
in the style of modern vendors of patent-medicines. Enthusiasts
forgot that provincial customs were generally founded on common-
sense, and that farmers reasoned from actual instances which had
come within their personal experience. The boast that a two-horse
plough, with reins and one man, could, on all soils and at all seasons,
do the work of the heavy implement dear to the locality only made
the ancient heirloom more precious in the eye of its owner. It was
with antiquated implements, heavy in the draught, that most of
the soil was still cultivated. Harrows were generally primitive in
form and ineffective in operation, scarcely penetrating the ground and
powerless to stir the weeds. To keep the seed-bed firm against the
loosening effects of frost the only roller was a stone or the trunk of
a tree heavily weighted. When the bed was prepared for the crop,
the seed was still sown broadcast by hand, or, more rarely, either
dibbed or drilled. The Northumberland drill for turnips, and the
Suffolk drill for cereals, which travelled every year on hire as far as
Oxfordshire, had already attained something more than local
popularity. But corn and roots, even in 1837, were seldom either

drilled or dibbed. The advantages of both methods were still hotly denied. A man who used a drill would be asked by his neighbours when he was going to sow pepper from a pepper-caster. From the time the seed was sown and harrowed in, the infant crops waged an internecine and unaided strife with weeds. Even the hand-hoe rarely helped cereals in the struggle, for the cost was heavy, and the work, unless carefully supervised, was easily scamped.

In 1837 hand-labour alone gathered the crops. Corn was cut by scythes, fagging hooks, or sickles ; if with the first, each scytheman was followed by a gatherer and a binder ; a stooker and raker completed the party. When a good man headed the gang, with four men to each scytheman, two acres a day per scythe were easily completed. Threshed by the flail, the grain was heaped into a head on the floor of the barn. The chaff was blown away by means of the draught of wind created by a revolving wheel, with sacks nailed to its arms, which was turned by hand. Thus winnowed, the grain was shovelled, in small quantities at a time, into a hopper, whence it ran, in a thin stream, down a screen or riddle. As the stream descended, the smaller seeds were separated and removed. The wheat was then piled at one end of the barn, and " thrown " in the air with a casting shovel to the other extremity. The heavy grain went furthest ; the lighter, or " tail," dropped short. To some of the corn in both heaps the chaff still adhered. These " whiteheads " were removed by fanning in a large basket tray, pressed to the body of the fanner, who tossed the grain in the air, at the same time lowering the outer edge of the tray. By this process the whiteheads were brought to the top and extremity of the fan, whence they were swept by the hand. Lastly the corn was measured, and poured into four-bushel sacks, ready for market. The operation of dressing was slow. As the sun streamed through a crack in the barn-door, it reached the notches which were cut in the wood-work to mark the passage of time and the recurrence of the hours for lunch and dinner. The operation was expensive as well as slow, costing from six to seven shillings a quarter. Hay was similarly made in all its stages by hand, and with a care which preserved its colour and scent. The grass, mown by the scythe, fell into swathes. These were broken up by the haymakers, drawn with the hand-rake into windrows, first single, then double. The double windrows were pulled over once, put first into small cocks, then into larger which were topped up and trimmed so as to be shower proof, and finally arranged in cart-rows for pitching and loading. Women, working behind the carts, allowed scarcely a blade to escape their rakes. The farmer in 1837 had a reaper at his command, but he did not value the gift. Its sudden popularity illustrates a point which is perpetually recurring in the history of agricultural machinery. As soon as the want is created, the machine is not only discovered but appreciated. Many attempts were made to perfect a reaper. But none met with any real success till machines not only cut the corn but laid it in sheaves, till fields were enlarged, till thorough drainage was adopted, and, as a consequence, the old high-ridging system abandoned. It is a sign, and a consequence, of changes in farming that the Rev. Patrick Bell's reaper, invented in 1826, was not really appreciated till it was manufactured (1853) by Crosskill as the "Beverley Reaper." Threshing and winnowing machines were to be found on a few large farms, or travelled the country on hire, worked by horse, water, or steam power. For feeding stock, chaff-cutters and turnip-slicers were already known ; but they made their way slowly into use. Chaff was still generally cut, and turnips split, by a chopper. If cattle or sheep were unable to bite, they ran the risk of being starved.

No one who studies the agriculture of 1837 can fail to notice the perpetual contrast, often in the most glaring form, between the practices of adjoining agriculturists. A hundred farmers plodded along the Elizabethan road, while a solitary neighbour marched in the track of the twentieth century. Discoveries in scientific farming, put forward as novelties, were repeatedly found to be in practice in one district or another. The great need was the existence of some agency which would raise the general level of farming by making the best practices of the best agriculturists common knowledge. The problem was not readily solved. To diffuse scientific and practical information among agriculturists was difficult seventy years ago. Books were expensive, and those for whom they were written were often unable to read. Few of the agricultural works published before the reign of Victoria were produced by men of practical experience. Extravagant promises or incorrect science too often discounted the value of useful suggestions. What was really wanted was ocular demonstration of the superiority of new methods, or the example of men of authority who combined scientific with practical knowledge. Some of the agricultural societies already in existence were doing good work in communicating the results of experiments,

organising shows, and encouraging discoveries ; others met rather for
the consumption of meat and drink than for the discussion of their
production. The Board of Agriculture had established a strong
claim to the gratitude of farmers by providing Davy's lectures on
agricultural chemistry in 1803-13. But its dissolution in 1822 had
been one of the symptoms of distress. The foundation of the Royal
Agricultural Society of England, projected in 1837, established in
1838, and incorporated by Royal Charter in 1840, with Queen
Victoria as patron, was at once a sign of revival and a powerful
agent in restoring prosperity.

Among the founders of the Society were many of the best-known
landowners and most practical agriculturists of the day. Their
association in a common cause carried weight and authority throughout the whole country. Their recognition of their territorial duties
and enthusiasm for the general advancement of agriculture were
communicated to others, and commanded success by their sincerity.
The Society met a recognised want in the right way. It proclaimed
the alliance between practical farmers and men both of capital and
of science ; it indicated the directions in which agriculture was
destined to advance. The wise exclusion of politics, though for a
moment it threatened to endanger the existence of the new institution, eventually secured it the support of men of every shade of
political opinion. By the comprehensiveness, elasticity, and fore-
sight, with which its lines of development were traced, it has been
enabled to meet the varying needs of seventy years of change. It
has encouraged practical farming on scientific principles ; it has also
encouraged agricultural science to proceed on practical lines. It has
by premiums and pecuniary aid promoted discovery and invention ;
by its shows it has fostered competition, stimulated enterprise, and
created a standard of the best possible results, methods, processes,
and materials in British agriculture. Its *Journal* disseminated the
latest results of scientific research at home and abroad, as well as
the last lessons of practical experience. In its pages will be found
the truest picture of the history of farming in the reign of Queen
Victoria. Starting as it did under peculiarly favourable circumstances, and supported by writers like Philip Pusey and Chandos
Wren-Hoskyns, it commanded the pens of masters in the lost art of
agricultural literature—men who wrote with the knowledge of
specialists and with the forcible simplicity of practical men of the
world. Without **exaggeration** it may be said that the general

standard of excellence to which farming has attained throughout the kingdom has been to a considerable extent the work of the Royal Agricultural Society. For more than seventy years it has been the heart and brain of agriculture. The local associations which now compete with it in popularity are in great measure its own creations, and it can contemplate with pride, unmixed with envy, the sturdy growth of its own children.

From 1840 to 1901, Queen Victoria was the patron, and to Her Majesty's patronage the Society owed much of its prestige and consequent utility. It has been said that " agriculture " is " the pursuit of kings " ; yet the feeling certainly had existed that farming was beneath the dignity of gentry. Fortunately for British farming, landlords have had a truer perception of their territorial duties as well as of their pecuniary interests. In taking the lead they have made a vast outlay of private capital. Windsor, Osborne, Balmoral, Sandringham, and the home-farms of large landowners have set the fashion, and afforded the model, to hosts of agriculturists. They have helped not only to raise the standard of British farming, but also to make a costly industry a fashionable yet earnest pursuit. A detailed history, for instance, of the Windsor farms would epitomise the history of agricultural progress in the nineteenth century. Roads were laid out. Liebig's discovery that warmth is a saving of food was acted upon, and substantial buildings were erected, designed to economise the expense and labour of cattle-feeding, and at the same time to preserve manure from waste or impoverishment. Skilfully selected herds of pure-bred Shorthorns, Herefords, and Devons were formed ; quantities of food were purchased ; the soil was drained on scientific principles ; the arable land, for the most part a stiff clay, was ameliorated and enriched by high farming ; the latest inventions in implements or machinery were tested and adopted ; the grass-lands were improved by experiment and careful management ; a model dairy, designed to meet the exacting requirements of modern sanitation and convenience, was erected ; and, to supply the milk, a pure-bred herd of Jersey cattle was formed which soon became one of the most celebrated in the country.

The work which the Royal Agricultural Society was established to do was not done by it alone. Other societies, as well as associations and farmers' clubs, assisted in spreading scientific and practical knowledge of farming. Among many other useful writers on the subject the Rev. W. L. Rham, Youatt, James Johnston, Henry

Stephens, Dr. Lindley, and John Chalmers Morton, as Editor of the *Agricultural Gazette* did excellent service. The school-master was abroad, and the foundation of Cirencester Agricultural College in 1845 was a sign of the times. The need for agricultural statistics, which had long been severely felt, had been emphasised by Sir James (then Mr.) Caird in 1850-1. But it was not till 1866 that the want was supplied. Attempts had been frequently made to obtain statistical information, but without success. Fear of increased taxation closed the mouths of landowners and farmers. In 1855 a House of Lords Committee recommended the compulsory collection of statistics through the agency of the Poor-Law officials. Eleven years later (1866) the Agricultural Returns of Great Britain for the first time supplied an accurate account of the acreage, the cropping, and the live-stock of the country.

The new alliance of science with practice bore rich and immediate fruit. Science helped practical farming in ways as varied as they were innumerable. Chemists, geologists, physiologists, entomologists, botanists, zoologists, veterinaries, bacteriologists, architects, mechanics, engineers, surveyors, statisticians, lessened the risks and multiplied the resources of the farmer. Steam and machinery diminished his toil and reduced his expenses. His land was neither left idle nor its fertility exhausted. Improved implements rendered his labour cheaper, quicker, surer, and more effective. New means of transport and increased facilities of communication brought new markets to his door. Commodious and convenient buildings replaced tumble-down barns and draughty sheds. Veterinary skill saved the lives of valuable animals. The general level of agriculture rose rapidly towards that which only model farms had attained in the previous period. Sound roads, well-arranged homesteads, heavy crops, well-bred stock, skilled farmers, and high farming characterised the era which adopted the Royal Agricultural Society's rule of Practice with Science Cut off from their old resource of increasing production by adding to the cultivated area, deprived of the aid of Protection, agriculturists were compelled to adopt improved methods. The age of farming by extension of area had ended ; that of farming by intension of capital had begun.

To trace out in full detail one single point in which science has helped farmers would be the work of a separate volume. Selection and outline are all that is possible. Probably the most striking contributions which, during the period under review, were made to the progress of agriculture are the extension of drainage, the discovery of artificial manures, the increased purchase of feeding stuffs for cattle, the improvement of implements, the readier acceptance of new ideas and inventions. Such an advance was impossible in the days of pack-waggons. By the railways all that farmers had to sell or wanted to buy,—corn and cattle, coal, implements, machinery, manures, oil-cake, letters and newspapers, as well as the men themselves,—were conveyed to and fro more expeditiously and more cheaply.

Drainage was the crying need of the day both for pasture and arable land. If the land was heavy and undrained pasture, the moisture-loving plants overpowered the more nutritive herbage ; the over-wetness became in rainy seasons a danger to the stock ; the early and late growth of grass was checked ; the effect of autumn and spring frosts was more severely felt. If stiff, retentive, undrained land was under the plough, it was cultivated at greater cost, on fewer days in the year, during a season shorter at both ends, than lighter soils ; unless the seasons were favourable, it produced late and scanty harvests of corn and beans, was often unsafe for stock, could bear the introduction neither of roots nor of green-cropping, repeatedly needed bare fallows, wasted much of the benefit of manures and feeding stuff. For many years clay-farmers had been seeking for some expedient which would remedy the over-wetness of their land, and enable them to share in the profits that new resources had placed within reach of their neighbours on freer and more porous soils. It was upon them that the blow of agricultural depression from 1813 to 1836 had fallen with the greatest severity. Clay farms had fallen into inferior hands, partly because men of capital preferred mixed or grazing farms. Weaker tenants were thus driven on to the heavier land, on which they could not afford the outlay needed to make their holdings profitable. Yet their strong land, if seasons proved favourable, was still capable of yielding the heaviest crops. Some process was needed which would so change the texture of the soil as to render it more friable, easier to work, more penetrable to the rain, more accessible to air and manure, and therefore warmer and kindlier for the growth of plant life.

The usual expedient for carrying off the water from heavy soils was the open-field practice of throwing the land into high ridges, whence the rain flowed into intervening furrows, which acted as surface-drains. But this device not only stripped the land of some of its most valuable portions by washing the surface tilth into the furrows, it also robbed the soil of the fertilising agencies which rain-water holds in solution and by percolation carries downwards to the plants. For both these reasons the ancient practice of such counties as Essex and Suffolk was a great advance. In those counties trenches were cut from 2 ft. to 2½ ft. in depth at frequent intervals, filling the bottom of the cavity with boughs of thorn, heath, or alder, and the soil replaced. Sometimes, where peat or stones were easily available, they were used instead of bushes. Sometimes the filling was only intended to support the soil until a natural arch was consolidated to form a waterway. For this more temporary purpose, twisted ropes of straw or hops, or a wooden plug, which was afterwards drawn out, were generally employed. In other counties, other materials or devices were adopted. Thus in Leicestershire, a V-shaped sod was cut, the bottom end taken off, and the rest replaced. In Hertfordshire, at the lowest part of the field, a pit was sunk into a more porous stratum, filled up with stones, and covered in with earth. Many of the Suffolk and Essex drains lasted a considerable time ; but the arched waterways were apt to choke or fall in, and the depth at which they were placed was considered unsuitable for land under the plough. So little was the practice known outside these two counties that in 1841 its existence was a revelation to so enlightened an agriculturist as Philip Pusey. In tapping springs, caused by water meeting an impervious subsoil and rising to the surface, most useful work had been done by Joseph Elkington,[1] a Warwickshire farmer in the latter half of the eighteenth century. Throughout the Midland counties his services were in such request that his crow-bar was compared to the rod of Moses. In 1797 he had received £1,000 from Parliament on the recommendation of the Board of Agriculture, and an attempt was made to reduce his practice to rules. But his success was so much the result of his personal observation and experience that the attempt failed. The principles of drainage were not yet understood.

In 1823 James Smith of Deanston, then a man of 34, began to cultivate the small farm attached to the Deanston Cotton Works of which he was manager. By his system of drainage and deep

---

[1] *An Account of the most approved Mode of Draining Land according to the System practised by Mr. Joseph Elkington*, by John Johnstone, Land Surveyor, Edinburgh, 1797.

ploughing he converted a rush-grown marsh into a garden. His drains were trenches 2½ ft. deep, filled with stones and covered over, cut in parallel lines from 16 to 21 ft. apart. Agriculturists flocked to Perthshire to see with their own eyes the transformation and its causes. Smith's *Remarks on Thorough Draining and Deep Ploughing* (1831) were widely read, and in 1834 he was examined as a witness by the Committee which was then enquiring into the condition of agriculture. The value of his experience was recognised ; enquiry and discussion were excited. In 1843 Josiah Parkes, profiting by the knowledge which he had acquired in draining Chat Moss, laid down his principles of drainage. Thinking that Smith's trenches were too shallow, he advocated a depth of four feet, which would give a sufficient layer of warm mellow surface earth. On these principles millions of acres were drained, and thousands of pounds wasted where drains were laid too deep. The necessary implements were quickly perfected. But for some little time a cheap conduit remained a difficulty. Stones were not everywhere available, and, if carted and broken, their use was expensive. In 1843 John Reade,[1] a gardener by trade and a self-taught mechanic, produced a cylindrical clay-pipe. Two years later (1845) Thomas Scragg patented a pipe-making machine which enabled the kilns to work cheaply and expeditiously. The capital and the soil of the country became acquainted on an extensive scale. Within the next few years, two large public loans for drainage, repaid by annual instalments, were taken up, and treble the amount was spent by private owners or advanced by private companies. Drainage became the popular improvement by which landlords endeavoured to encourage tenants who were dismayed by the repeal of the Corn Laws. It gave clay farmers longer seasons and added to the number of the days on which they could work their land ; it increased the ease of their operations and the efficacy of their manures ; it secured an earlier seedtime and an earlier harvest, raised the average produce, and lowered the cost of working ; it enabled the occupiers of hundreds of thousands of acres to profit by past as well as future discoveries.

Drainage was a necessary preliminary to profitable manuring. On undrained land farmers could not use to full advantage the new

[1] In the *Weekly Miscellany for the Improvement of Husbandry, etc.*, for August 22, 1727, Stephen Switzer had recommended the use of pipes made of "potter's clay" for the conveyance of water, and advertised that the pipes were made by pipe-making machines which enabled the kilns to work cheaply and expeditiously.

means of wealth which agricultural chemistry was placing at their command. But while drainage, in the main, helped only one class of farmers, the benefits of manure were universal. The practice of manuring is of immemorial antiquity. But it was in the extended choice of fertilising substances, in the scientific analysis of their composition and values, in their concentration and portability, and in the greater range of time at which they could be profitably applied that a prodigious advance was made during the Victorian era.

For inland farmers in rural districts the choice of manures was practically limited to the ashes of vegetable refuse which represented the food drawn by the plant from the soil, " catch-cropping " with leguminous crops, folding sheep, and farmyard manure. " Nothing like muck " had become a proverb when there was practically " nothing but muck " to be used. On the same poverty of fertilising resources were founded the severe restrictions against selling hay, straw, and roots off farms. In another sense the proverb is true— fortunately for the fertility of the country. Rich both in organic and inorganic substances, combining both nitrogen and minerals, possessing for the loosening of clay lands a peculiar value, farmyard manure is the only substance which contains in itself all the constituent elements of fertility. Our predecessors thus commanded the most valuable of fertilising agencies, the most certain and the least capricious. But in their open unspouted, unguttered yards, in their ignorance of the importance of the liquid elements, and with their straw-fed stock, the manure was both wasted and impoverished. Nor is it only in the quantity and quality of dung, or in its collection and treatment, that farmers have the advantage to-day. Formerly distant fields suffered when no concentrated and portable fertiliser existed, and, valuable though dung is, its uses are not unlimited. In the infancy, moreover, of agricultural science, men had little knowledge of the composition of soils, the necessities of plant life, or the special demand that each crop makes on the land. It is in all these respects that modern resources are multiplied. The supply of concentrated portable manures, adapted by their varied range to all conditions of the soil, capable of restoring those elements of fertility which each particular crop exhausts, and applicable at different stages of plant life, is the greatest achievement of modern agricultural science.

It is to the great German chemist Liebig that modern agriculture owes the origin of its most striking development. In 1840 his *Chemistry in its applications to Agriculture and Physiology* [1] clearly traced the relations between the nutrition of plants and the composition of the soil. In his mineral theory he was proved to be mistaken ; but his book revolutionised the attitude which agriculturists had maintained towards chemistry. So great was the enthusiasm of country gentlemen for Liebig and his discoveries, as popularised by men like Johnston and Voelcker, that the Royal Chemical Society of 1845 was in large measure founded by their efforts. But if the new agriculture was born in the laboratory of Giessen, it grew into strength at the experimental station of Rothamsted. To Sir John Lawes and his colleague Sir Henry Gilbert (himself a pupil of Liebig) farmers of to-day owe an incalculable debt. By their experiments, continued for more than half a century, the main principles of agricultural science were established ; the objects, method, and effects of manuring were ascertained ; the scientific bases for the rotation of crops were explained ; and the results of food upon animals in producing meat, milk, or manure were tested and defined. On their work has been built the modern fabric of British agriculture.

With increased knowledge of the wants of plant or animal life came the supply of new means to meet those requirements. Artificial manure may be roughly distinguished from dung as purchased manures. Of these fertilising agencies, farmers in 1837 already knew soot, bones, salt, saltpetre, hoofs and horns, shoddy, and such substances as marl, clay, lime and chalk. But they knew little or nothing of nitrate of soda, of Peruvian guano, of superphosphates, kainit, muriate of potash, rape-dust, sulphate of ammonia, or basic slag. Though nitrate of soda was introduced in 1835, and experimentally employed in small quantities, it was in 1850 still a novelty. The first cargo of Peruvian guano was consigned to a Liverpool merchant in 1835 ; but in 1841 it was still so little known that only 1,700 tons were imported ; six years later (1847) the importation amounted to 220,000 tons. Bones were beginning to be extensively used. Their import value rose from £14,395 in 1823 to £254,600 in 1837. As originally broken in small pieces with a hammer, they were slow in producing their effect ; but the rapidity

[1] *Organic Chemistry in its applications to Agriculture and Physiology.* By Justus Liebig : edited from the manuscript of the author by Lyon Playfair, 1840.

of their action was enormously increased by grinding them to a coarse meal. Rape-dust was not known in the South of England at the beginning of the Victorian era. In 1840 Liebig suggested the treatment of bones with sulphuric acid, and in 1843 Lawes began the manufacture of superphosphate of lime, and set up his works at Bow. So far the chemists ; the next step was taken by geologists. At the suggestion of Professor Henslow, the same treatment to which bones were already subjected was applied to coprolites, and the rich deposits of Cambridgeshire and other counties, as well as kindred forms of mineral phosphates, imported from all parts of the world, were similarly " dissolved." Even Peruvian guano was subjected to the same treatment. Another important addition to the wealth of fertilising agencies was made by Odams, who about 1850 discovered the manurial value of the blood and garbage of London slaughter-houses, mixed with bones and sulphuric acid.

It is in the means of applying appropriate manures to lands which are differently composed, and to crops which vary in their special requirements, that modern farmers enjoy exceptional advantages over their predecessors. The active competition of rival manufacturers assisted the adoption of the new fertilisers. Many men, who would not listen to the lectures of professors, or read the articles of chemical experts, were worried by persistent agents for the sale of patent manures into giving them a trial. Indirectly, their use led to clean farming. A farmer who had paid £10 a ton for manure was unwilling to waste half its value on wet ill-drained land. He was less likely to allow it to fertilise weeds, and the more ready to buy a machine to distribute it carefully. Thus, as consequences of purchased fertilisers, followed the extensive use of the drain-pipe, the drill, the hand-hoe, and the horse-hoe. Yet chemical science did not at once fulfil the sanguine expectations which were formed of its capacity in the early " Fifties." The confident hope that the specific fertility extracted by a crop could be restored by a corresponding manure was scarcely confirmed by experience ; and many a farmer did himself as much harm as good by the application of fertilisers which were unsuited to his land.

Manure and drainage acted and reacted upon one another : the one encouraged the other. Previous rules of successive cropping were revolutionised ; more varied courses were gradually and universally introduced. The old exhausting system of two or three crops and a bare fallow was abandoned when land had been drained, and

fertilisers, portable, cheap and abundant, were placed at the command of the farmer. Without manure the attempt to grow roots
or clover failed ; their introduction only protracted the shift, and
aggravated the difficulty of inevitable exhaustion. Now, however,
the principle was gradually established that he who put most into
his land got most out. Farmers recognised by experience, when the
means were at their disposal, that, on the one hand, if they ruined
their land their land ruined them, and that, on the other hand,
only those who have lathered can shave. It was in readiness to
invest capital in the land that one of the chief differences between
the earlier race of agriculturists and the modern type of farmer
became most conspicuous. The main objects of the former were to
feed their families and avoid every possible outlay of cash. Hard-
living and hard-working, they rarely thought of spending sixpence
on manure, still less on cattle food to make it. They gave little to
the land and received little. The consequent loss in the national
means of subsistence can scarcely be exaggerated. Modern farmers,
on the other hand, not only purchased thousands of tons of artificial
fertilisers. They also bought for their live-stock vast quantities of
feeding-stuff, which supplemented their own produce. Roots, clover,
beans, barley-meal, hay, chaff, as well as artificial purchased food,
were supplied to the sheep and cattle, which once had only survived
the winter as bags of skin and bone. Just as guano from Peru
was turned into English corn, or bones from the Pampas into English
roots, so the Syrian locust-pod, the Egyptian bean, the Indian corn,
or the Russian linseed were converted into English meat. The
gain to the nation was immense, and to the farmer it was not small.
The return on his money was quickened. He sold his stock to the
butcher twice within the same time which was formerly needed to
prepare them once, and that less perfectly. At the same time his
command of manure was trebled in quantity and quality, and on
clay lands his long-straw muck was of special value.

The changes which have been noticed in modern farming necessitated more frequent operations of tillage, which, without mechanical
inventions would have been too costly to be possible. Here, again,
science came to the aid of the farmer, and supplied the means of
making his labour cheaper, quicker, and more certain. The Royal
Agricultural Society may legitimately pride itself on the useful part
which it has played in introducing to the notice of agriculturists
the new appliances which mechanical skill has placed at their service.

Yet, when the Society was founded, none of its promoters foresaw the importance of the mechanical department. At the Oxford show in 1839 one gold medal was awarded for a collection of implements ; three silver medals were allotted ; and a prize of five pounds was given for " a paddle plough for raising potatoes." At the show at Gloucester in 1853, 2,000 implements were exhibited. The modern system of farming had, in the interval of fourteen years, built up a huge industry employed in providing the agricultural implements that it required.

In tilling the land, sowing, harvesting, and marketing their crops, modern farmers command a choice of effective implements for which their predecessors knew no substitute. Between 1837 and 1874, ploughs in every variety, light in draught, efficient, adaptable to all sorts of soil, were introduced. Harrows suited for different operations on different kinds of land, scarifiers, grubbers, cultivators, clod-crushers, came into general use. Steam supplied its motive power to the cultivator (1851-6) and to the plough (1857). As an auxiliary in wet seasons, or in scarcity of labour, or on foul land, or to back-wardness of preparation, the aid of steam may be invaluable. But few farmers can afford to own both horse-power and steam-power, and without horses they cannot do. The time may, however, be near at hand when agriculturists may find it not only invaluable, but indispensable, to rely on an arm that never slackens, never tires, and never strikes. Corn and seed drills deposited the seed in accurate lines, and at that uniform depth which materially promotes the uniformity of sample so dear to barley growers. Rollers and land-pressers consolidated the seed-bed. Manure drills distributed fertilisers unknown to farmers in 1837. Horse-hoes gained in popularity by improved steerage gear. Crosskill's Beverley reaper was followed within the next twenty years by lighter and more convenient machines. Mowing machines, haymakers, horse-rakes, shortened the work of the hay-field. Light carts or waggons super-seded their heavy, broad-wheeled predecessors. Elevators lessened the labour of the harvesters in the yard. Threshing and winnowing machines had been invented in the eighteenth century. But in the South of England, partly perhaps from the difficulty of supplying labourers with winter work, the flail was still almost universal for threshing. From 1850 onwards, however, steam began to be applied as a motive power to machines, and within the next ten years several makers were busily competing in the manufacture of steam-driven barn-machinery, which threshed the corn, raised the straw to the loft, winnowed and dressed the grain, divided it according to quality, delivered it into sacks ready for market, and set aside the tailings for pigs and poultry. Nor did mechanical science neglect the live-stock industry, the development of which, in connection with corn-growing, was a feature of the period. Here, too, machinery economised the farmer's labour. He already knew the turnip-cutter and the chaff-cutter ; but now the same engine which superseded the flail, pumped his water, ground his corn, crushed his cake, split his beans, cut his chaff, pulped his turnips, steamed and boiled his food. Without the aid of mechanical invention farming to-day would be at an absolute standstill. No farmer could find, or if he found could pay, the staff of scarce and expensive labour without which in 1837 agricultural produce could not be raised, secured, and marketed.

The improvements which have been indicated were not the work of a day. On the contrary, during the first few years of the reign—the only period passed under Protection—progress was neither rapid nor unchecked. Farmers in general were preparing for high farming ; they had not yet adopted its practices. Whatever advance had been made between 1837 and 1846 was probably lost in the five succeeding years. Abundant materials exist for comparison. On the one side are the Reports of the Reporters to the old Board of Agriculture (1793-1815) ; and the Reports to the successive Commissions (1815-36) ; on the other, there are the Reports published in the early numbers of the *Journal of the Royal Agricultural Society*, the evidence given before the Select Committee of 1848 on tenant-right and agricultural customs, the letters of Caird to *The Times* in 1850-1, afterwards embodied in his *English Agriculture in* 1850-1, and the letters of the Commissioner to the *Morning Chronicle* during the same period. It is plain that in 1846 no universal progress had been effected ; that many landowners had made no effort to increase the productiveness of their land ; that high farming was still the exception ; that the new resources were not yet generally utilised ; and that more than half the owners and occupiers of the land had made but little advance on the ideas and practices of the eighteenth century. Another period of disaster, short but severe, forced home the necessary lessons, and ushered in the ten years, 1853-62, which were the golden age of English agriculture.

R*—E F P P

The railway manias and their collapse in 1845-7 had depressed every industry. The failure of the potato crop in 1845-6 caused appalling famine, and led to the Repeal of the Corn Laws. When in 1846 Protection was abandoned for Free Trade, an agricultural panic was the result. Caird's pamphlet on *High Farming . . . the best Substitute for Protection* (1848) pointed out the true remedy. But for the moment he preached in the wilderness. The discovery of guano and the abolition of the Brick and Timber Duties seemed no adequate set-off to the anticipated consequences of Free Trade in grain. Agriculturists predicted the ruin of their industry, and their prophecies seemed justified by falling prices in 1848-50. Many landlords and tenants had been encouraged by Protection to gamble in land. Extravagant rents had been fixed, which were not justified by increased produce. Caird calculated in 1850 that rentals had risen 100 per cent. since 1770, while the yield of wheat per acre had only risen 14 per cent.—from 23 to 26¾ bushels. In 1850 wheat stood at the same price which it had realised eighty years before (40s. 3d.). On the other hand, butter, meat, and wool had risen respectively 100 per cent., 70 per cent., and 100 per cent. The great advance which had been made was, in fact, in live-stock. Competition in farms had been reckless, and the consequences were inevitable when prices showed a downward tendency. Here and there rents were remitted, but few were reduced. Clay farmers, as before, were the worst sufferers ; dairy and stock farmers escaped comparatively lightly. But the loss was widespread. Much land was thrown on the hands of landlords, and efforts were made to convert a considerable area of arable into pasture.

From 1853 onwards, however, matters rapidly righted themselves. Gold discoveries in Australia and California raised prices ; trade and manufacture throve and expanded ; the Free Trade panic subsided ; courage was restored. The Crimean War closed the Baltic to Russian corn. During the " Sixties," while the Continent and America were at war, England enjoyed peace. The seasons were uniformly favourable ; harvests, except that of 1860, were good, fair, or abundant ; the wheat area of 1854, as estimated by Lawes, rose to a little over four million acres ; imports of corn, meat, and dairy produce supplemented, without displacing, home supplies. Even the removal of the shilling duty on corn in 1869 produced little effect. Counteracted as it was by the demand for grain from France in 1870-71, it failed to help foreign growers to force down the price of

British corn. Wool maintained an extraordinarily high price. Lincoln wool, for instance, rose from 13d. per lb. in 1851 to 27d. in 1864. Even when corn began to decline in value, meat and dairy produce maintained their price, or even advanced. Money was poured into land as the best investment for capital. Men like Mechi of Tiptree Hall, who had made fortunes in trade, competed for farms, and became enthusiastic exponents of their theories of scientific agriculture. Rentals rose rapidly ; yet still farmers made money. Holdings were enlarged and consolidated ; farmhouses became labourers' cottages ; a brisk trade was carried on in machinery. High hopes were entertained of steam. Enormous and, as has since been proved, excessive sums were spent on farm buildings. Drainage was carried out extensively, and it was now that the general level of farming rose rapidly towards the best standard of individual farmers in 1837. Crops reached limits which production has never since exceeded. and probably, so far as anything certain can be predicted of the unknown, never will exceed.

During the period from 1853 to 1874 little attention was in England paid to improvements in dairying. But in live-stock progress was great and continuous. The advance was the more remarkable as it was made in the face of outbreaks of the rinderpest, pleuro-pneumonia, and foot-and-mouth diseases. Foot-and-mouth disease had been more or less prevalent since 1839, and pleuro-pneumonia since 1840. But the scourge of rinderpest in 1865, commonly called the cattle-plague, compelled energetic action.[1] In stamping out the pest the two other diseases were nearly extinguished, so that good results flowed from a disaster which caused widespread ruin. The multiplication of shows encouraged competition ; stock-breeding became a fashion, and " pedigree " a mania among men of wealth.

It was in cattle and sheep that the improvement was most clearly marked, though neither horses nor pigs were neglected. Not only did Shorthorns, Herefords, and Devons attain the highest standard

[1] In the week ended Feb. 24, 1866, 17,875 cattle were returned as infected by the disease. The Cattle Diseases Prevention Act, dated Feb. 20, 1866, made the slaughter of diseased animals compulsory. The effect was seen at once. In the week ending March 3, 1866, 10,971 cattle were attacked, and in the week ending March 10, 10,056 were killed. At the end of April the weekly tables showed 4,442 attacked ; towards the end of May, 1,687 ; in the last week of June, 338. In the last week of the year the number had dwindled to 8. *Appendix I. to Report on Cattle Plague during the years* 1865, 1866, 1867 (Parliamentary Paper of 1868, Cd. 4060), p. 4.

of excellence in symmetry, and quality, but other breeds, now almost as well-known, were rapidly brought to perfection. Especially is this true of the Aberdeen-Angus, the Sussex, Ayrshire, and Channel Island breeds.[1] Other breeds were similarly improved by societies and the compilation of herd-books. Thus the Black cattle of South Wales and the Norfolk and Suffolk Red Polled breed have had their herd-books since 1874. In sheep the improvement was, perhaps, even more striking. The historic Leicesters, Cotswolds, and Southdowns still held their own, but other breeds made rapid strides in the popular favour. The improved Lincolns, the Oxford Downs, Hampshire Downs, and Shropshires are almost creations of the period. Between 1866 and 1874 the number of cattle in Great Britain rose from under five millions to over six millions, and sheep had increased to over thirty millions in 1874. Nor was there only an increase in numbers. The average quality was greatly improved, and good sheep and cattle were widely distributed.

[1] The Shorthorn Society was founded in 1875: the Hereford Herd-book appeared in 1846, and the Hereford Herd-book Society was incorporated in 1878: the Devon Herd-book appeared in 1851. The first volume of the Aberdeen-Angus Polled Herd-book was issued in 1862, and the second in 1872: the Sussex Herd-book Society published its first volume (1855-78) in 1879; the Ayrshire Herd-book Society was established in 1877 and published its first volume in 1878; the English Herd-book of Jersey Cattle was first issued in 1879, and the English Jersey Cattle Society was incorporated in 1883.

CHAPTER XVIII

THE GREAT DEPRESSION AND RECOVERY, 1874-1914

SINCE 1862 the tide of agricultural prosperity had ceased to flow; after 1874 it turned, and rapidly ebbed. A period of depression began which, with some fluctuations in severity, continued throughout the rest of the reign of Queen Victoria, and beyond. Depression is a word which is often loosely used. It is generally understood to mean a reduction, in some cases an absence, of profit, accompanied by a consequent diminution of employment. To some extent the condition has probably become chronic. A decline of interest on capital lent or invested, a rise in wages of labour, an increased competition for the earnings of management, caused by the spread of education and resulting in the reduction or stationary character of those earnings, are permanent not temporary tendencies of civilisation. So far as these symptoms indicate a more general distribution of wealth, they are not disquieting. But, from time to time, circumstances combine to produce acute conditions of industrial collapse which may be accurately called depression. Such a crisis occurred in agriculture from 1875 to 1884, and again from 1891 to 1899.

Industrial undertakings are so inextricably interlaced that agricultural depression cannot be entirely dissevered from commercial depression. Exceptional periods of commercial difficulty had for the last seventy years recurred with such regularity as to give support to a theory of decennial cycles.[1] In previous years, each recurring period had resulted in a genuine panic, due as much to defective information as to any real scarcity of loanable capital. The historic failure of Overend and Gurney in 1866 and the famous " Black Friday " afford the last example of this acute form of crisis. Better means of obtaining accurate intelligence, more

[1] *E.g.* 1825-6, 1836-7, 1847, 1857, 1866, 1877-8.

accessible supplies of capital, the greater stability of the Bank of England have combined with other causes to minimise the risk of financial stampedes. But, though periods of depression cease to produce the old-fashioned panic, they are not less exhausting. Their approach is more gradual ; so also is the recovery. Disaster and revival are no longer concentrated in a few months. Years pass before improvement is apparent ; the magnitude of the distress is concealed by its diffusion over a longer period. The agricultural depressions of 1875-84 and of 1891-99 had all the characteristics of the modern type of financial crisis.

In 1870 had begun an inflation of prices. The outbreak of the Franco-German War and the withdrawal of France and Germany from commercial competition enabled England to increase her exports ; the opening of the Suez Canal (1869) stimulated the shipbuilding trade ; the railway development in Germany and America created an exceptional demand for coal and iron. Expanding trade increased the consuming power of the population, and maintained the prices of agricultural produce. The wisest or wealthiest landowners refused the temptation to advance rents on sitting tenants. But in many cases rents were raised, or farms were tendered for competition. Farmers became infected with the same spirit of gambling which in trade caused the scramble for the investment of money in hazardous enterprises. In their eagerness for land they were led into reckless biddings, which raised rentals beyond reasonable limits. In 1874 the reaction began. Demand had returned to normal limits ; but the abnormal supply continued Over-production was the result. The decline of the coal and iron trade, the stoppage, partial or absolute, of cotton mills, disputes between masters and men, complications arising out of the Eastern question, the default on the Turkish debt, disturbances of prices owing to fluctuations in the purchasing power of gold and silver, combined to depress every industry. In 1878 the extent to which trade had been undermined was revealed by the failure of the Glasgow, Caledonian, and West of England Banks. One remarkable feature of the crisis was that it was not local but universal. New means of communication had so broken down the barriers of nations that the civilised world suffered together. Everywhere prices fell, trade shrank, insolvencies multiplied. In the United States the indirect consequences of the industrial collapse of 1873-4 proved to be of disastrous importance to English farming. A railway panic, a fall in the price of manufactured articles, a decline in wages drove thousands out of the towns to settle as agriculturists on the virgin soils of the West. English farming suffered from the same causes as every other home industry. In addition, it had its own special difficulties. The collapse of British trade checked the growth of the consuming power at home at the same time that a series of inclement seasons, followed by an overwhelming increase of foreign competition, paralysed the efforts of farmers. For three years in succession, bleak springs and rainy summers produced short cereal crops of inferior quality, mildew in wheat, mould in hops, blight in other crops, disease in cattle, rot in sheep, throwing heavy lands into foul condition, deteriorating the finer grasses of pastures. In 1875-6 the increasing volume of imports[1] prevented prices from rising to compensate deficiencies in the yield of corn. The telegraph, steam carriage by sea and land, and low freights, consequent on declining trade, annihilated time and distance, destroyed the natural monopoly of proximity, and enabled the world to compete with English producers in the home markets on equal, if not more favourable, terms. Instead of there being one harvest every year, there was now a harvest in every month of each year. In 1877 prices advanced, owing to the progress of the Russo-Turkish War. But the potato crops failed, and a renewed outbreak of the cattle-plague, though speedily suppressed, hit stockowners hard. The tithe rent-charge was nearly £12 above its par value. Rates were rising rapidly. Land-agents began to complain of the scarcity of eligible tenants for vacant corn-land. During the sunless ungenial summer of 1879, with its icy rains, the series of adverse seasons culminated in one of the worst harvests of the century, in an outbreak of pleuro-pneumonia and foot-and-mouth disease among cattle, and among sheep a disastrous attack of the liver rot, which inflicted an enormous loss on flockmasters. The English wheat crop scarcely averaged 15½ bushels to the acre. In similar circumstances, farmers might have been compensated for the shortness of yield by an advance in price. This was no longer the case in 1879. America, which had enjoyed abundant harvests, poured such quantities of wheat into the country as to bring down prices below the level of the favourable season of the preceding year. At the same time, American cheese so glutted the market as to create a record for cheapness. Thus,

For the growth of foreign imports of food, see Appendix VII.

at the moment when English farmers were already enfeebled by their loss of capital, they were met by a staggering blow from foreign competition. They were fighting against low prices as well as adverse seasons.

English farmers were, in fact, confronted with a new problem. How were they to hold their own in a treacherous climate on highly rented land, whose fertility required constant renewal, against produce raised under more genial skies on cheaply rented soils, whose virgin richness needed no fertilisers ? To a generation familiar with years of a prosperity which had enabled English farmers to extract more from the soil than any of their foreign rivals, the changed conditions were unintelligible. The new position was at first less readily understood, because the depression was mainly attributed to the accident of adverse seasons, and because the grazing and dairying districts had as yet escaped. Thousands of tenants on corn-growing lands were unable to pay their rents. In many instances they were kept afloat by the help of wealthy landlords. But every landowner is not a Dives ; the majority sit at the rich man's gate. In most cases there was no reduction of rents. Remissions, sometimes generous, sometimes inadequate, were made and renewed from time to time. Where the extreme urgency of the case was imperfectly realised, many old tenants were ruined. It was not till farms were relet that the necessary reductions were made, and then the men who profited were new occupiers.

If any doubt still existed as to the reality of the depression, especially in corn-growing districts, it was removed by the evidence laid before the Duke of Richmond's Commission, which sat from 1879 to 1882. The Report of the Commission established, beyond possibility of question, the existence of severe and acute distress, and attributed its prevalence, primarily to inclement seasons, secondarily to foreign competition. It was generally realised that the shrinkage in the margin of profit on the staple produce of agriculture was a more or less permanent condition, and that rents must be readjusted. Large reductions were made between 1880 and 1884, and it was calculated that in England and Wales alone the annual letting value of agricultural land was thus decreased by $5\frac{3}{4}$ millions. Yet in many cases the rent nominally remained at the old figure. Only remissions were granted, which were uncertain in amount, and therefore disheartening in effect. According to

Sir James Caird's evidence given in 1886, before the Royal Commission on Depression of Trade, the yearly income of landlords, tenants, and labourers had diminished since 1876 by £42,800,000. The worst was by no means over. On the contrary, the pressure of foreign competition gradually extended to other branches of agriculture. The momentum of a great industry in any given direction cannot be arrested in a day ; still less can it be diverted towards another goal without a considerable expenditure of time and money. Unreasonable complaints were made against the obstinate conservatism of agriculturists, because they were unable to effect a costly change of front as easily as a man turns in his bed. The aims and methods of farming were gradually adapted to meet the changed conditions. As wheat, barley, and oats declined towards the lowest prices of the century, increased attention was paid to grazing, dairying, and such minor products as vegetables, fruit, and poultry. The corn area of England and Wales shrank from 8,244,392 acres in 1871 to 5,886,052 acres in 1901.[1] Between the same years the area of permanent pasture increased from 11,367,298 acres to 15,399,025 acres. Yet before the change was complete farmers once more found themselves checkmated. The old adage " Down horn, up corn " had once held true. Now both were down together. Till 1885 the prices of fat cattle had been well maintained, and those of sheep till 1890. Both were now beginning to decline before the pressure of foreign competition. Up till 1877 both cattle and sheep had been chiefly sent in alive from European countries. Now, America and Canada joined in the trade, and the importation of dead meat rapidly increased. Consignments were no longer confined to beef and pigs' meat. New Zealand and the Republic of Argentina entered the lists. The imports of mutton, which in 1882 did not exceed 181,000 cwts., and chiefly consisted of meat boiled and tinned, rose in 1899 to 3½ million cwts. of frozen carcases. The importation of cheese rose by more than a third ; that of butter was doubled ; that of wool increased more than two-fold. Meanwhile the outgoings of the farmer were steadily mounting upwards. Machinery cost more ; labour rose in price and deteriorated in efficiency. The expenses of production rose as the profits fell.

Some attempt was made by Parliament to relieve the industry. The recommendations of the Richmond Commission were gradually

[1] For statistics of agriculture, see Appendix VIII.

carried into effect. Grants were made in aid of local taxation. Measures were adopted to stamp out disease amongst live-stock, and to protect farmers against the adulteration of feeding-stuffs, and against the sale of spurious butter and cheese. The primary liability for tithe rent-charge was transferred from occupiers to owners (1891). The law affecting limited estates in land was modified by the Settled Lands Act (1882). A Railway and Canal Traffic Act was passed, which attempted to equalise rates on the carriage of home and foreign produce. The permissive Agricultural Holdings Act of 1875, which was not incorrectly described as a " homily to landlords " on the subject of unexhausted improvements, was superseded by a more stringent measure and a modification of the law of distress (1883). A Minister of Agriculture was appointed (1889), and an Agricultural Department established.

But the legislature was powerless to provide any substantial help. Food was, so to speak, the currency in which foreign nations paid for English manufactured goods, and its cheapness was an undoubted blessing to the wage-earning community. Thrown on their own resources, agriculturists fought the unequal contest with courage and tenacity. But, as time went on, the stress told more and more heavily. Manufacturing populations seemed to seek food-markets everywhere except at home. Enterprise gradually weakened ; landlords lost their ability to help, farmers their recuperative power. Prolonged depression checked costly improvements. Drainage was practically discontinued. Both owners and occupiers were engaged in the task of making both ends meet on vanishing incomes. Land deteriorated in condition ; less labour was employed ; less stock was kept ; bills for cake and fertilisers were reduced. The counties which suffered most were the corn-growing districts, in which high farming had won its most signal triumphs. On the heavy clays of Essex, for example, thousands of acres, which had formerly yielded great crops and paid high rents, had passed out of cultivation into ranches for cattle or temporary sheep-runs. On the light soils of Norfolk, where skill and capital had wrested large profits from the reluctant hand of Nature, there were widespread ruin and bankruptcy. Throughout the Eastern, Midland, and Southern counties,— wherever the land was so heavy or so light that its cultivation was naturally unremunerative,—the same conditions prevailed. The West on the whole, suffered less severely. Though milk and butter had fallen in price, dairy-farmers were profiting by the cheapness of grain, which was ruining their corn-growing neighbours. Almost everywhere retrenchment, not development, was the enforced policy of agriculturists. The expense of laying land down to grass was shirked, and arable areas which were costly to work were allowed to tumble down to rough pasture. Economy ruled in farm management; labour bills were reduced, and the number of men employed on the land dwindled as the arable area contracted.[1]

During the years 1883-90, better seasons, remissions of rent, the fall in tithes, relief from some portion of the burden of rates, had arrested the process of impoverishment. To some extent the heavy land, whether arable or pasture, which wet seasons had deteriorated, recovered its tone and condition. But otherwise there was no recovery. Landlords and tenants still stood on the verge of ruin. Only a slight impulse was needed to thrust them over the border line. Two cold summers (1891-2), the drought in 1893, the unpropitious harvest of 1894, coupled with the great fall in prices of corn, cattle, sheep, wool, butter, and milk produced a second crisis, scarcely, if at all, less acute than that of 1879. In this later period of severe depression, unseasonable weather played a less important part than before. But in all other respects the position of agriculturists was more disadvantageous than at the earlier period. Foreign competition had relaxed none of its pressure; on the contrary, it had increased in range and in intensity. Nothing now escaped its influence. But the great difference lay in the comparative resources of agriculturists. In 1879 the high condition of the land had supplied farmers with reserves of fertility on which to draw; now, they had been drawn upon to exhaustion. In 1879, again, both landlords and tenants were still possessed of capital; now, neither had any money to spend in attempting to adapt their land to new conditions.

In September, 1893, a Royal Commission was appointed to enquire into the depression of agriculture. The evidence made a startling revelation of the extent to which owners and occupiers of land, and the land itself, had been impoverished since the Report of the Duke of Richmond's Commission. It showed that the value of produce had diminished by nearly one half, while the cost of production had rather increased than diminished; that quantities of corn-land had passed out of cultivation; that its restoration, while the present prices prevailed, was economically impossible; that its adaptation

[1] See Census Returns of Occupations, Appendix VII.

to other uses required an immediate outlay which few owners could afford to make. Scarcely one bright feature relieved the gloom of the outlook. Foreign competition had falsified all predictions. No patent was possible for the improved processes of agriculture ; they could be appropriated by all the world. The skill which British farmers had acquired by half a century of costly experiments was turned against them by foreign agriculturists working under more favourable conditions. Even distance ceased to afford its natural protection either of time or cost of conveyance, for not even the perishable products of foreign countries were excluded from English markets. Yet the evidence collected by the Commission established some important facts. It proved that many men, possessed of ample capital and energy, who occupied the best equipped farms, enjoyed the greatest liberty in cropping, kept the best stock, and were able to continue high farming, had weathered the storm even on heavy land ; that small occupiers employing no labour but their own had managed to pull through ; that, on suitable soils, market gardening and fruit-farming had proved profitable ; that, even on the derelict clays of Essex, Scottish milk-farmers had made a living. At no previous period, it may be added, in the history of farming were the advantages and disadvantages of English land-ownership more strongly illustrated. Many tenants renting land on encumbered estates were ruined, because their hard-pressed landlords were unable to give them financial help. At least as many were nursed through the bad times by the assistance of landowners whose wealth was derived from other sources than agricultural land.

When the extent of the agricultural loss and suffering is considered, the remedies adopted by the legislature seem trivial. Yet some useful changes were made. Farmers were still further protected against adulteration of cake, fertilisers, and dairy produce by the provisions of The Fertilisers and Feeding Stuffs Act (1893) and the Sale of Food and Drugs Act (1899). The Market Gardeners Compensation Act (1895) enabled a tenant, where land was specifically let for market garden purposes, to claim compensation for all improvements suitable to the business, even though they had been effected without the consent of the landlord. The Improvement of Land Act (1899) gave landowners increased facilities for carrying out improvements on borrowed money. The amendment of the Contagious Diseases of Animals Act (1896), requiring all foreign

animals to be slaughtered at the port of landing, was a valuable
step towards preventing the spread of infection. The Agricultural
Rates Act (1896) and the subsequent Continuation Acts (1901, etc.),
though they were only palliatives which did not settle the many
questions involved in the increasing burden of rates, rendered the
load of local taxation for the moment less oppressive. After all,
agriculturists received little assistance from Parliament. They had
to help themselves. Conditions slowly mended. More favourable
seasons, rigid economy in expenses, attention to neglected branches
of the industry have combined to lessen the financial strain. But the
greatest relief has been afforded by the substantial reduction in the
rents of agricultural land, which has resulted in a fairer adjustment
of the economic pressure of low prices as between owners and
occupiers.

The nadir of the great depression came in 1894–95, when the price
of wheat per imperial quarter fell to 22s. 10d. and 23s. 1d., the
lowest figures recorded for 150 years. From that time began a slow
but steady rise in prices, sufficient to counterbalance the definite
increase in wages which became manifest between 1895 and the end
of the century and indeed continued, though more gradually, until
the outbreak of war in 1914. After 1907 the price of wheat never
fell below 30s. ; but wheat had become a commodity of less importance
to farmers at large, for the acreage had declined to about one and
three-quarter millions. More than anything else milk had become
the most money-making product, for the demand was continuously
increasing with the growth of population and the industrial prosperity
of the period. Changes in the farming population were marked as
the old time arable farmers of the South and East of England, who
had persisted in their traditional but now too expensive methods of
cultivation, had to retire. About 1895, rents had really adjusted
themselves to the times, indeed on the heavy lands of Essex, where
the reductions of rent had not been rapid or large enough to save
the old tenants, farms could be had on payment of the tithe, and many
large estates took the greater part of their farms in hand rather than
let to the sort of men who offered themselves. Tempted by these
conditions, Scotchmen migrated in numbers from a country where
rents were still competitive and brought their knowledge of milk
production and their more economical methods into Essex and
Hertfordshire, and to a lesser degree into Kent and Surrey. Similarly,
the dairy farmers of the West drifted into the South and Midlands,

from Devonshire and Wales, men who had been bred to live harder and do their work more economically, if more roughly.

For this was the great lesson that was being learned, how to get the work done on the arable land with less labour. What with the turning over of arable land to grass (2½ million acres between 1872 and 1900) and economy in methods, something like a third of the labouring population left the land in the last quarter of the nineteenth century. There were other occupations to absorb the men, but none the less this forced exodus left a bitterness against farmers and landlords among the working classes that has not yet wholly disappeared.

The cheapness of land during this last decade of the nineteenth century gave to many shrewd men who had broken with tradition and learnt how to farm cheaply an opportunity of putting together exceptionally large farming businesses. S. W. Farmer of Little Bedwyn was reputed to be farming 20,000 acres at the outbreak of war, at the same time George Baylis of Wyfield Manor near Newbury was farming over 12,000 acres in Berkshire and Hampshire, growing corn and hay without any stock,[1] in Lincolnshire men like Dennis and the Worths built up great estates on potato growing out of little farms whose owners had been broken in the depression. These were examples of the success of better farming, but there were many instances where some sort of a paying return was got out of the land by turning it down to grass and reducing expenses to a minimum. Such was an estate put together near Ramsbury in Wiltshire, where about 4,000 acres of light arable land on the chalk were turned into a sheep ranch. In the early 'eighties, there was a hamlet called Snape on one part of the estate, containing a chapel, fourteen cottages, and a school attended by 44 children. In 1921 the street was grass-grown and almost obliterated, the buildings were in ruins. The working population had been reduced to a shepherd and his dog, like the owner, living elsewhere.[2]

Though the accomplishment was irregularly distributed considerable progress in the technique of farming was taking place. It was no longer possible for landowners and their agents to insist on particular methods of farming; covenants remained in the agreements, but were ignored as long as the tenant could pay his rent, so the diffusion of better methods came about by example, not by pressure from above. Indeed, in the main, landlords had accepted the

[1] See Orwin, *Progress in English Farming Systems*, No. III. 1930.
[2] See Orwin, *J. R. Ag. Soc.*, 1922, 83, p. 10.

position that there was little future for farming, that the development of their estates did not offer an outlet for their energies or capital comparable to those available elsewhere, and that their function was to be easy with their tenants in return for the sport and the social status that the ownership of land conferred. Their direct interest in agriculture was often confined to the breeding and showing of pedigree stock, the practical value of which began to be obscured almost in proportion as it became a rich man's plaything and a form of social competition. Of course, a generalisation of this kind about any of the classes engaged in agriculture, landlords, tenants, or labourers, is contradicted by a number of individuals, who worked hard at farming and managed their estates with knowledge and judgment, but none the less this period did witness the continued disappearance of the landowner as *entrepreneur*. In general the land was not sold, the possibilities of its monopoly values were too evident in a country of growing population and increasing industrial prosperity. It is true that towards the end of the first decade of the twentieth century there arose a number of land speculators who bought up embarrassed and under-rented estates and offered the farms to the selling tenants at greatly enhanced prices. The speculators had realised better than the landlords that farming was again a remunerative business and were not afraid of the odium of making the tenants pay its full value. Moreover, this was a time when considerable political attacks were being directed against the landlords, without much discrimination between the owners of agricultural land and those who were reaping "unearned increment" from ground rents in the growing urban areas. Some landowners took alarm and disembarrassed themselves of an investment which at the time was yielding an inferior rate of interest, yet carried with it heavy social obligations. Thus the Duke of Bedford sold both his Thorney and Tavistock Estates on terms favourable to the selling tenants.

It was at the very depth of the depression that a beginning was made with State-aided agricultural education, in the train of which research soon followed. But this will be dealt with elsewhere,— results had hardly begun to accrue before the new century ; the first improvements in technique came from the farmers themselves. Machinery was becoming more general upon the farm, the greatest single improvement having been the self-acting binder, the use of which began to be general about the end of the 'eighties, after the introduction of the knotting mechanism and twine. But haymaking machinery and springtined cultivators were also doing their share in reducing the costs of cultivation and the amount of labour required upon the arable land. In the more specialised industries change was at work : in hop-growing, for example, the 'eighties and early 'nineties saw the general replacement of the old poles by string and wire erections on which the bines could be trained so as to get proper exposure to sun and air, and spraying methods were evolved to deal with blight and mould. Until these improvements began, the methods of hop-growing had not altered in any substantial respect from those described by Reynolde Scot in 1574. While the use of artificial fertilisers was not growing as rapidly as in countries like Holland and Germany, that was because the acreage under arable cultivation continued to decline and the farmers who were winning through were mostly those who relied on keeping their expenses down, yet the knowledge of how to apply them appropriately was spreading. In the latter years of the nineteenth century one might still meet the landlord who forbade his tenants to put any artificials on their land or the farmer who had substituted Kainit for nitrate of soda because it was cheaper, but such instances disappeared as the new century opened out. One new fertiliser, indeed, was beginning to prove itself of immense value to the grass land which was becoming the mainstay of English farming. Basic slag was invented in 1879, but it only reached agriculture after 1885, when Wrightson and Munro demonstrated that its phosphate required no treatment with acid if only the slag was finely ground. It soon showed itself as possessing a marvellous regenerative value on old pastures, especially on the clay soils on which its application induced a speedy growth of white clover whereby not only the stock gained but the pasture continued to acquire fertility. While farmers have always been immediately appreciative of improved strains of seed, it can be said that during the years 1890–1910 their interest in the value of pure seed and good strains was being continually stimulated, though the history of actual introductions may best be dealt with under research.

In matters of live-stock the impulse towards the selection and standardisation of a pure breeding strain under the care of a Breed Society, which had been one of the chief achievements of English farming in the nineteenth century, was still active, as witness the formation of the following Societies—The Guernsey Cattle Society in 1885, the Dexter and Kerry Handbook in 1890, the Welsh Black Cattle Society in 1904, the British Holstein (now Friesian) Cattle

Society in 1909. Flock Books began for Shropshires in 1883, Oxfords in 1889, Hampshires in 1890, Lincolnshires in 1892, Romney Marsh in 1895 and many others. Though from some points of view it might be questioned whether all these new breeds were wanted, the formation of a Society did tune up the general standard in the district occupied by the breed. The chief development during the period was concerned with milk, the demand for which was continuously increasing with the growing population and industrial prosperity. The milking capacity of the various breeds received more attention; for example, during this period the Dairy Shorthorns began to be differentiated and in 1905 an Association was formed in its interests, and herds like those of Hobbs and Evens obtained a repute to rival the northern beef herds. The necessity for care and cleanliness in the preparation and despatch of milk to the public was being continually forced upon the farmers by the Health Authorities of the large towns, who had from time to time experience of milk distributed epidemics. Regulations were enforced concerning such matters as water supply and air space in cowsheds, and if at times they were uninformed and dictatorial about the unessentials, they did arouse in the dairy world the sense that success in this growing business depended upon the purity of the product. It was indeed in the 'eighties that the process of butter and cheese making, hitherto a matter of traditional and personal farm practice, were studied and standardised. At the same time the correct temperatures and acidities were determined so that the desired result could be obtained with certainty. " Creameries " and cheese factories began to be established in order to handle milk more efficiently and economically. The importations of butter from Denmark and the Baltic countries was growing rapidly and setting a standard of quality and uniformity that neither the English nor the Irish market butter could equal, however much a dairymaid here and there could turn out a " gilt-edged " product such as can never be obtained by factory methods.

However, there is little or no market in England for fine butter at an adequate price ; the English dairy farmers could get a better return by selling raw milk and abandoned the butter market to their foreign and colonial competitors. Only a few farmers in the West and South-West, Wales and its borders, continued to make butter because the rail communications precluded them from getting whole milk to market, while they could also turn the separated milk to account by calf-rearing. The Irish butter making was transformed on Danish lines, their farmers, like the Danish, being content with returns per gallon well below that expected by the English farmer. The machinery of the dairy was undergoing revolution ; barrel churns replaced the old upright churns in which a dasher worked up and down in the whole milk, the only reminder of the old shape being the metal churns in which, for a few years longer perhaps, milk will travel by rail. Rail transport again brought the necessity of milk coolers, though the customers of the small farmers, each with their own milk round, still demanded " milk warm from the cow." But the most important of these machines for the dairyman was the centrifugal separator of which really efficient types began to be available about 1890, though, as indicated above, the perfecting of this exquisite machine coincided with the decline and practical extinction of commercial butter-making in England. Among other labour saving machines that began to appear on the farms towards the close of the nineteenth century were the small oil engines to run the grinding and food preparation plant, and sheep shearing machines, the use of which grew but slowly because there were still men enough about to clip the comparatively small flocks running on the usual farm.

The commercial development of poultry rather belongs to the post-war period. Even down to the end of the century poultry-keeping still halted between the methods of the fancier and of the farmer who had a mongrel flock picking about his stack-yard. W. B. Tegetmeier, in his day an authority, was said to have an offer open of £50 for anyone who could produce an accredited balance sheet showing a year's profit on a poultry farm, excluding those dealing in stock birds or eggs. About Heathfield in Sussex there was a successful cramming industry producing birds for the table, though the crammers did no breeding, but bought young birds for fattening from as far afield as Ireland.

The period we are considering, 1890–1914, was also one of expansion and improved technique in market gardening and fruit growing, industries that were prospering in response not only to the growing population but to a change in the general dietary. Potato growing, which had proved but a treacherous foundation for the Lincolnshire Yeomen in the 'seventies and 'eighties, became one of the money-making crops for certain selected districts, like the Lothians and Ayrshire in Scotland, the silt and the warp soils of Lincoln and Yorkshire, the light soils of West Lancashire and Cheshire. Even on the gravels of Hertfordshire, where Arthur Young had found himself "living in the jaws of a wolf," potato growing brought wealth to some of the migrants from Scotland. Nothing revolution‑ ary had happened to make the industry so profitable, it was a good instance of the accumulation of a number of small improvements, each of which could be pooh‑poohed as not worth while by the old‑ time farmer. New varieties were being introduced, " Up‑to‑date " had a long run about the turn of the century ; the virtue of Scotch seed was recognised, though the reason for its success was yet undis‑ covered. Boxing the seed and planting sprouted sets became standard practice ; the fertilisers to procure large crops became understood, for the growers were substantial men willing to spend money and open to advice. Spraying with Bordeaux mixture was standardised, but did not become general in all districts because many growers preferred to gamble on intermittent appearance of blight.

Market gardening was increasing and improving its methods, though it was still dependent upon the lavish supplies of stable manure that could be obtained from the great towns. Naturally it was segregating into selected areas—the brick earths of the Thames Valley and North Kent, Bedfordshire, Huntingdon, Wisbech and the Fens, and the Vale of Evesham, where it was found that asparagus would flourish on stiffish clays, while selected areas in Cornwall could follow the earliest potato crops with a second crop of autumn and winter broccoli.

Lastly, this period saw the great development of the glass‑house cultivation in districts like Worthing, Swanley Junction and above all the Lea Valley. Fifty years ago the tomato was as great a rarity in England as an Avocado pear is to‑day ; a few were imported, a few were grown in private conservatories. In the late 'eighties market cultivation under glass began and early in the 'nineties Worthing tomatoes had established their position as superior to any importations. From that time until the outbreak of war there was no pause in the extension of the industry. Cucumbers went with the tomatoes, grapes and chrysanthemums completed the old cycle ; forced bulbs—tulips, narcissus, and iris, were later additions, as again has been the perpetual flowering carnation, for flowers have become as much a matter of general household expenditure as tomatoes or eggs.

Looking at the state of agriculture generally the early years of the century may be recalled as a time of quiet but growing prosperity for farmers. One may read in *Farmer's Glory*, that singularly faithful presentation of farming life in Wiltshire, how Mr. A. G. Street, who settled down on his father's farm in 1907, looks back to those years before the war as " the spacious days," just as the man of an older generation recalled the 'sixties and early 'seventies as the good old times. " But that large tenant farmers were doing pretty well then, there is no question. I suppose the business side of farming had its worries in those days, but it is difficult to recall any. There were good seasons and bad seasons, doubtless. I can remember wet weather in harvest time and good weather. Good luck at lambing time and bad I can also call to mind, but nothing ever seemed to make any difference in our home life. It all seemed such a settled prosperous thing."

Again I may quote my own contemporary opinion, written after a series of farming tours round the United Kingdom in 1910-12. " In the first place we must recognise that the industry is at present sound and prosperous."

CHAPTER XIX

THE WAR AND STATE CONTROL, 1914-18

Individualism in English farming; "up Horn, down Corn "; effect of the nation's commercial policy; the War 1914-18; destruction of shipping; the Milner Committee (1915); transport and food difficulties, 1917-18; necessity for increased production at home; unfavourable prospects, Dec. 1916; position compared with that of Napoleonic Wars; Government control of the industry; its aims and organisation; Corn Production Act, 1917; results of food campaign; lessons of State control.

In some foreign countries, notably in Germany and Denmark, the whole farming industry has made great and simultaneous advances within short periods of time. It has had behind it the directing force of a definite policy. During the last seventy years in England, farming has progressed on different lines. Its history is mainly that of thousands of separate enterprises, conducted as private businesses. In the slowness of its general advance, and in the variations of its standard from good to bad, it bears the stamp of individualism. Each man farms his holding according to his skill, character, tastes and capital. "The best farming is that which pays the farmer best." The idea that the nation is, or could be, vitally concerned in the use to which the land is put, or in the kind and quantity of food that it produces, seldom occurred to landowners or tenants, or, except as a political theory, to any large section of the general community. Farmers grew the produce which, from time to time, promised to yield the most certain return on their invested capital. Without State direction, or accepted leadership, the line of individual progress is the only one which they can pursue. Its main tendency has been dictated by the commercial policy of the nation. England transferred her corn-growing to distant lands, where grain can be grown more cheaply than at home. Compelled by the density of her population to buy a portion of her food abroad, she could only pay for it with the produce of her factories. The policy may have been carried too far. But its precise form was dictated by the mutual advantage of ourselves and our customers. Wheat makes a good cargo ; it is also a convenient commodity for the younger foreign nations to exchange against our manufactured merchandise. Live cattle, on the contrary, are bad travellers on ship-board, and dead meat can only be carried in chilled or frozen form. As England grew wealthy from its commerce, it became a meat-eating nation. It was ready to pay good prices for fresh beef and mutton of the best quality. Farmers as men of business followed the national lead. They obeyed the laws of demand and supply. They revised their methods. A manufacturing nation prayed " God speed the plough on every soil but our own." Farmers responded with the cry " Down Corn, up Horn," which saved them from financial ruin. Live-stock and their products became the predominant branch of the industry. The problem of feeding the nation was taken out of farmers' hands. It was solved by the commercial policy of exchanging exports of home manufactures against imports of foreign corn and meat. Farmers, therefore, chose the safest, cheapest, and least anxious method of feeding their flocks and herds. They utilised the nation's trading system to purchase foreign food-stuffs for their stock, and, secure of winter keep, dispensed as far as possible with the plough. Corn-land was laid down to grass. The tillage area in England and Wales dwindled by $3\frac{3}{4}$ million acres. In the United Kingdom (1909-13) 36 million acres were devoted to stock, and 3 million acres to such human food as wheat and potatoes Represented in cash, the net annual value to farmers of their two tillage crops was £27,000,000, and of their stock and stock products, £125,000,000. Aided by imports of concentrated feeding-stuffs, the farmers of these islands supplied our milk and three-fifths of our meat ; they also provided one-fifth of our bread. For four-fifths of its daily consumption of bread, and for two-fifths of that of meat, the United Kingdom depended on foreign producers.

The inevitable consequence of the changed system of husbandry from tillage to grass was a reduction in the number of persons maintained from the land. In 1870 a much larger population was fed from our home resources than in 1909-13. After seventy years of agricultural progress, the land filled fewer mouths, though the demand for food had nearly trebled. The reason lies in decreased tillage, not in decreased skill. In actual food value for subsistence diet, the produce of the three million acres devoted to wheat and potatoes is only fractionally inferior to that of the 36 million acres appropriated to live-stock. From 100 acres of arable, 150 persons can be maintained for a year, while from 100 acres of grass, the number is under 15. The contrast with the quantity of food per acre raised by German farmers depends on the different demands made from agriculture by Germany and by England. One country definitely aims at making itself self-supporting; the other, for commercial reasons, prefers to buy the bulk of its food abroad. English farmers were not, and to many it seemed the height of improbability that they ever would be, asked to feed the English people.

In 1914 came the War. It was thought unnecessary to change the existing system of feeding the nation. Our own, Allied, or neutral shipping brought to our ports our foreign purchases of bread, meat and feeding-stuffs. At home, farmers continued to produce our milk and their normal proportion of wheat and of fresh beef and mutton. It is true that trade was now and then disturbed by enemy cruisers, mines and submarines. In the first five months of the war, the British Empire lost 100 ships of 252,738 gross tons. Only three of these were destroyed by submarines. Even these losses were more than balanced by our gain in prizes and in new ships. The attacks were annoying ; they did not seriously aggravate difficulties. But before the war had lasted twelve months, the destruction of shipping began to arouse attention. Submarines, no longer obliged to hug their base, or to rest at frequent intervals in shallow water, were rapidly improved as weapons of attack. The figures of destruction rose from 29,376 gross tons in April, 1915, to 92,924 and 90,605 gross tons in the months of May and June respectively. For the moment the Government took alarm. A Departmental Committee was appointed (June 17) with Lord Milner as Chairman,[1] to consider and report what steps should be taken to maintain and, if possible, increase " the present production of food in England and Wales, on the assumption that the war may be prolonged beyond the harvest of 1916." Their Interim Report (July 17, 1915) is important, because upon it was based the action of the Government eighteen months later.

[1] The other members of the Committee were Lord Inchcape ; the Right Hon. F. D. Acland, M.P. ; the Hon. Edward Strutt ; Sir H. Verney, Bart., M.P. ; Mr. C. W. Fielding ; Mr. A. D. Hall ; Mr. R. E. Prothero, M.P. ; Mr. T. A. Seddon.

The Committee assumed that, in the opinion of the Government, an emergency which called for exceptional measures was likely to exist after the harvest of 1916. On that assumption they unanimously recommended (1) that, in order to encourage wheat-growing, farmers should be guaranteed for four years a minimum price per quarter of wheat, payments to be regulated by the difference between the guarantee and the *Gazette* average price for wheat for the year in which the grain was harvested ; (2) that District Committees should be set up by the County Councils, to each of which the Board of Agriculture should furnish, as a standard of endeavour, statements showing (a) the area under the plough, and (b) the acreage under wheat, oats and potatoes, in 1875 and 1914 respectively ; (3) that each District Committee should consider and report on the capacity of every farm in its district, and on the willingness of individual farmers to contribute additional food. On the receipt of these Reports, the Committee proposed to consider whether compulsion would, or would not, be necessary.

At the time, the Government decided to take no action. A portion of the Report was, however, adopted by the President of the Board of Agriculture. County Councils were asked to appoint District Committees. In many counties they were set up ; in some they were active. Meanwhile, shipping losses continued to increase. In 1915, the loss of the British Empire was 885,471 gross tons, and the total loss of the world was 1,312,216 gross tons. In 1916, the figures for the British Empire were 1,231,867 gross tons, and for the world 2,305,569 gross tons. A serious feature in the rate of loss was the rapid increase in the last three months of 1916. In October, November, December, the British Empire lost 524,574 gross tons, and the world 788,706 gross tons. The cumulative effect of these losses on the carrying capacity of shipping had become serious, especially as so large a tonnage was necessarily diverted from the transport of civilian cargoes to the conveyance of troops and war material. From the beginning of 1917 to the conclusion of hostilities, the shipping problem was of crucial importance. As soon as America joined the Allies, the cause was safe as regards money, men, munitions, and, if time allowed, ship-building. The danger-point shifted with dramatic suddenness to transport and food. On February 1, 1917, Germany had opened her intensive submarine campaign. She aimed at so large a destruction of the world's shipping as would make it impossible for the Allies to carry on the war. She met the blockade with a counter-blockade. The rate of destruction shot upwards with startling rapidity. In April alone the British Empire lost 526,447 gross tons, and the world 866,616 gross tons. It was the highest point reached in any one month. But the aggregate destruction of shipping for the year amounted to 3,660,054 gross tons and 6,078,125 gross tons for the British Empire and the world respectively. In the eleven months of 1918, owing mainly to the convoy system, the rate of destruction dropped. But it totalled for the British Empire 1,632,228 gross tons and for the world 2,528,082 gross tons. The figures tell their own tale. Against the losses must be set the gains of new ships. But it was not till July, 1918, that the corner of safety was turned, and that the monthly gain in new tonnage balanced the monthly loss in tonnage destroyed. Meanwhile, as day by day the scale of warlike operations expanded demands on transport multiplied. There was not enough tonnage left to meet essential needs. Every 5000 tons of civilian cargo carried meant the loss of 1000 American troops on the battlefield. Germany's intensive submarine campaign had seemed to some a desperate gamble. It came within measurable distance of success.

In December, 1916, a new Coalition Government came into office. It had become imperatively necessary to take careful stock of the food position of the Allies. Two and a half years of war, the drain on man-power, and the loss of territory were telling severely on the productive capacity of France and Italy. Here, as well as on the Continent, the harvest of 1916 had been bad. The area of winter-sown wheat in this country had shrunk by 60,000 acres. Exigencies of transport made America, with its short sea voyage, the main, if not the only, available source of supply. But her area of winter-sown wheat for the harvest of 1917 had declined, and it was doubtful whether she could, in the coming cereal year, meet the greatly increased demands of the Allies. We were, moreover, faced with the certainty of an acute shortage of potatoes. To these anxieties were added the difficulties of transport. As a necessity of life, bread-stuff must either be produced at home or imported from abroad. Even if sufficient wheat were on the foreign market, it could not be carried without curtailing the transport of men and munitions, essential for the successful prosecution of the war. In these circumstances, the Government determined to attempt an increase in the supply of home-grown corn and potatoes. If no addition could be made, the threatened decline might at least be arrested. Farmers were to be called on to feed the greatest possible number of the people. The decision reversed the tendencies of the last forty years. As far as was practicable, it aimed at recovering the tillage area of the 'Seventies, and reviving its system of husbandry.

The outlook in December, 1916, was not hopeful. Farmers could have doubled their output of food, if, like Munition Factories, they had commanded unlimited labour, abundant supplies of raw materials of industry, unrestricted prices for their produce, every form of priority, and protection for their men. But the nation was too much exhausted in man-power, tonnage, and finance to allow of any approach to this position. Agriculture was two years too late. Labourers, who had become efficient soldiers, could not be permanently released from the front. Even the skilled workers retained on the farms remained there subject to the paramount requirements of military service. Munitions necessarily had priority of transport over fertilisers, feeding-stuffs, or implements and agricultural machinery. The degree to which the land was depleted of labour differed from county to county and from farm to farm. But almost everywhere the shortage of skilled men was acute. Considerably more than a third had already left the land for service in the navy, army or munition factories. Ploughmen were scarce. Nearly half the steam-tackle sets were out of action, either from want of repair or from the loss of drivers. Many horses had been commandeered ; others were unshod. Harness and implements were out of order. Wide districts were denuded of such essential handicraftsmen as blacksmiths, wheelwrights, saddlers, and harness-makers. Manufacturers of agricultural machinery and implements were making munitions. Fertilisers were scarce. Potash had disappeared from the market with the declaration of war. The restricted import of phosphatic rock and iron pyrites limited the supply of super-phosphates. Exports of home-produced fertilisers were stopped; but grinding machines to grind the basic slag and men to man them were both short, and munitions made heavy demands on the supply of sulphate of ammonia. Lime kilns were closed, because the lime-burners were at the front. Discouraged by their loss of men, farmers were still further disheartened by the uncertainty of retaining those that were left. The poor corn-harvest and the failure of the potato crop were accompanied by an adverse autumn season. The winter was severe, and, as it proved, prolonged. With labour short and fertilisers scarce, with the normal channels of the supply of agricultural requisites interrupted or blocked, a decrease in food production appeared inevitable. The shrinkage in the area of winter-sown wheat seemed ominous of a general decline in arable cultivation.

The position of food supplies and transport had become so critical that it was necessary to make light of difficulties. In many respects the situation of 1801-15 was reproduced ; with contrasts as well as parallels, history repeated itself. To our ancestors, struggling in the throes of the Napoleonic War, the provision of bread became a paramount consideration. Fourteen million people in 1801, and in 1815 eighteen millions depended on the weather and the efforts of agriculturists at home. In spite of an exceptional series of bad harvests, the prodigious exertions of farmers averted actual famine, and even, in the one favourable season (1813), produced a surplus which was carried over to the two following years. To a partial extent, the position seemed likely to be repeated in 1917 and 1918. At both periods the way of safety lay in increased production on a larger arable acreage. In one material respect our ancestors were better off than ourselves. Their supply of skilled agricultural labour was never so depleted. At both periods it was imperative that we should not only grow more food, but economise in its use. Our ancestors adopted closer milling, mixed other ingredients than wheaten flour in their loaf, prohibited the sale of bread till it was twenty-four hours old. They suspended distilleries and starch manufactories. They abandoned the use of flour for hair powder. They made imitation pie-crust out of clay. Royal proclamations exhorted people to economy. By way of example Members of Parliament bound themselves to reduce their consumption by one third. Similar measures to effect economy were taken in the war with Germany. But the position had been rendered more difficult by the higher standard of living. In 1917 the country might well have revolted against hardships to which our ancestors were inured. With only a partial revelation of the facts, public opinion had to be instructed and created. In the Napoleonic War it was ready-made. Fed from home-grown food, the whole population knew from experience the need of " eating within a tether " which varied with the weather from sufficiency to scarcity. In the German War a generation had grown up which had no experience of the importance of production from their own soil or of the influence of seasons on food supplies. They could not conceive the possibility of their exclusion from foreign markets where harvests were more favourable. But the most striking difference between the two periods lies in the care which, during the recent war, was taken of the consumer. The contrast illustrates the profound change of social and political conditions and thought in the twentieth century. Our ancestors attempted no rationing or regulation of prices. They relied on the high cost of food to enforce economy and stimulate production. There was no restriction of producers' profits in the interests of consumers. Prices were allowed to find their natural level. The incentive of large gains spurred agriculturists to gigantic efforts. But, in the last two years of the German War, the appeal was as much to the farmer's patriotism as to his pocket. In the interests of consumers, flat maximum prices for agricultural produce were fixed ; the 4-lb. loaf was stabilised at 9d., partly at the expense of the farmer whose home-grown wheat was artificially cheapened ; the best and the worst qualities of home-grown beef and mutton were sold at the same price, so that the long and the short purses were in this respect placed on an equality. The general principle which guided the regulation of prices affords one of the most satis- factory contrasts in the story of agriculture during the French and the German Wars. But it did not make the problem of increasing production more easy of solution. One of the sharpest spurs to exertion was blunted.

The campaign for increased food production opened on December 20, 1916. Its course provoked acute controversies which are still discussed by agriculturists.[1] Historically, its interest lies in the experiment of State control, and in the organisation created to carry out a definite policy. But its results were so largely influenced by patriotic feeling and special circumstances that it throws little light on the question whether, in ordinary times, compulsory methods would prove successful. From the first, the Board of Agriculture recognised that the willing co-operation of farmers was essential. On the other hand, it knew that, in the urgency of the national need, it might be obliged to force its plans on a reluctant minority. Thus the three main features of the movement

---

[1] See " The Food Campaign of 1916-18," *Journal of the Royal Agricultural Society*, March 1922.

were the improvement and extension of arable cultivation, both
with the plough and the spade ; decentralisation ; and drastic
powers of compulsion, justified by the emergency of war. In its
general principles, the policy of the plough was imposed on the
agricultural industry by national necessities. Broadly speaking,
the country wanted the largest possible quantity of food in the
shortest possible time. As between grass and tillage, the only
question worth considering was, by which system of husbandry
the greatest number of people could be provided with subsistence
diet. To this there was but one answer. Arable farming feeds
at least four times as many persons as can be fed from grass of
average quality. In its main details, the policy was similarly
dictated by necessity. The interruption of sea-borne trade and
the strain on the carrying capacity of shipping threatened our food
supplies from three special directions. Our imports of corn, of concentrated feeding-stuffs, and of artificial fertilisers were imperilled.
The three danger-spots were, therefore, bread, winter meat and
winter milk, and the maintenance or restoration of exhausted
fertility. Increased output might be obtained from grading up
the cultivation of the existing arable acreage. But no adequate
results could be expected from this source only, especially as a
considerable area already needed a rest from cropping. In the
shortage of fertilisers it was necessary to release with the plough
the stored-up fertility of the grass. Our summer milk and meat
off pasture was not in danger ; but, if imported food for live-stock
was cut off, the winter supplies were in jeopardy, unless additional
fodder-stuffs could be grown on arable land at home.

The general policy could only be carried out by decentralisation.
It involved securing the adequate cultivation and cropping of
millions of acres of arable land and extending the tillage area by
ploughing up grass in sixty-one Counties and County Divisions.
Several hundreds of thousands of separate businesses could not be
treated, like factories, as controlled establishments. Even if it
had been possible, it would have been foolish. Local farmers were
the best judges, in their own districts, of insufficient cultivation
and of the most suitable land for the plough. An organisation
already existed which might be adapted for the purpose. In most
counties War Agricultural Committees had been set up on the
recommendation of the Milner Committee. They were too large
for executive action. But they were asked to appoint not more than

seven members, who, with such additions to the number as the Board might make, constituted the new County Agricultural Executive Committees. These smaller bodies, established in each County or County Division (61 in number), became the local agents of the Board of Agriculture. To them were delegated many powers which the Board exercised under the Defence of the Realm Act or retained under the Corn Production Act of 1917. They were assisted by District Committees, and, in some cases, by Parish Representatives. Each of the sixty-one Executives was provided with the necessary funds for staff and office expenses. As their work developed, they formed Sub-Committees for such branches as Survey, Cultivation, Supplies, Labour, Machinery, Horticulture, and Finance. During 1918, they were also entrusted with the responsibility of selecting recruits for the Army. The members of the Executive Committees, the majority of whom were farmers actively engaged in business, gave their time, knowledge and experience without pay or reward. Their difficult and invidious duty was performed with a discretion which reduced friction to a minimum. The Counties were further grouped in twenty-one Districts, to each of which a District Commissioner was appointed by the Board. An *ex officio* member of each of the County Executive Committees in his district, he served as a link between the Committees and headquarters. Acting under the District Commissioners were 36 Sub-Commissioners, whose special duties were to superintend the work of District Committees. By means of this network of local organisation, each parish was, as it were, connected with the central authority.

The first Cultivation of Lands Order, which vested in the County Executive Committees many of the powers of the Board, was sealed on January 19, 1917. The document was circulated to the Committees, with explanatory instructions for their guidance. It was too late in the farming year to attempt large additions to the tillage area for the harvest of 1917. The primary object set before the Committees was to assist farmers in the cultivation and cropping of their arable land, so as to secure the largest possible output of essential food. They were also asked to make a rapid survey of their districts in order to report what ploughing and sowing might be done in the coming spring. For the harvest of 1918, they were requested to make a more detailed survey, and were furnished on June 18, 1917, with *quota* of the arable acreage and crops which the Government aimed at securing in each county. The powers vested in the Committees were drastic. Where grass-land could be more profitably used in the national interest as arable, they were empowered to require it to be broken up, or to enter and plough it up themselves. Notices were to be served on occupiers specifying the grass fields to be ploughed, or the acts of cultivation to be executed, which the Committees considered necessary for the increase of food production. Failure to comply with notices rendered occupiers liable to fine or imprisonment. No appeal was allowed. Committees were further empowered to take possession of the whole or part of a badly cultivated farm, and either cultivate it themselves or arrange for its cultivation by others. The Board retained in its own hands power to determine, or to authorise the owner to determine, forthwith the tenancy of badly farmed land. By the end of January, 1917, most of the Executive Committees were established. Many had already made good progress with the preliminary survey and had begun to stimulate adequate cultivation.

The Board of Agriculture itself was not equipped to lead the movement. It had not the necessary organisation. To meet the new needs, a new branch was constituted. The Food Production Department was formed by the President, January 1, 1917, and charged with the novel functions of collecting and distributing labour, machinery, implements, fertilisers, feeding-stuffs, and other requisites of the industry and of assisting Committees to enforce their orders. It was designed to serve as a clearing-house for the requirements of individual farmers, notified through their Executive Committees. It became the pivot of the campaign. Its services were invaluable and multifarious. As the last comer in the field of national effort, agriculture found its legitimate territory already occupied by other departments, whose duties might sometimes clash with those of the farming industry. Soldier or prisoner labour was controlled by the War Office ; that of interned aliens by the Home Office ; that of civilian volunteers and public school-boys by the Ministry of National Service. Feeding-stuffs were controlled by the Minister of Food, fertilisers by the Minister of Munitions. To the latter Ministry also belonged the sanction for the manufacture or import of agricultural implements and machinery. These conditions restricted individual enterprise. But by notifying their wants through their Executive Committees to the Food Production

Department, farmers were, as far as possible, supplied with their requirements. Some idea of the scale on which the Department worked in 1918 may be gathered from the following figures of the activities of four of the principal branches. Through the Labour Division were supplied 72,247 soldiers, 30,405 prisoners of war, 3904 War Agricultural Volunteers, 15,000 public school-boys, making, with "other labour" (430), a total of 121,986 men. Through the Women's Division were provided some 300,000 part-time workers, and a Land Army, working full time, of the maximum strength of 16,000 women. The Cultivation Branch placed at the disposal of the Committees 4200 tractors ; it obtained the manufacture or import of 66 steam tackle sets, 4720 reapers and binders, 438 threshing machines ; it provided many thousands of ploughs, carts and lorries, cultivators harrows, disc harrows, land presses and rollers ; it operated 10,000 horses ; it trained ploughmen and tractor-drivers. The Supplies Branch obtained in the fertiliser year 1917-18, 232,000 tons of sulphate of ammonia, 750,000 tons of superphosphates, and 200,000 tons of basic slag ; it distributed 8700 quarters of selected seed wheat, 29,700 quarters of Irish, Scotch and Manx seed oats, and 32,800 tons of seed potatoes ; it provided 20,000 tons of binder twine. With so powerful an organisation at its back, each Executive Committee knew that it had the means in case of default by occupiers, of carrying out its orders to cultivate or plough.

Thus assisted, the farming industry was able to endure the loss of more than a third of its skilled labour, and to withstand the further shock of the calling up of many of their most experienced men—30,000 in January 1917 and upwards of 22,000 in June 1918. During the whole of 1917 and the first half of 1918, Executive Committees enforced cultivation under the extensive powers which the Board derived from two Regulations of the Defence of the Realm Act (2L and 2M). After August, 1918, the Regulations were superseded by the Corn Production Act, which had been passed in the previous year. That Act gave the Board of Agriculture powers for a period of six years to enforce the plough policy and improve the cultivation of land. In return for the acceptance of control and of the increased risks of arable farming, it guaranteed the growers of wheat and oats for the harvests of 1917-22 against substantial loss through a fall in prices. It prohibited the raising of rents during the period unless, without the operation of the guarantee, the holding could stand the rise. It provided machinery for fixing minimum rates of wages for agricultural labourers during the continuance of the Act. In the view of the Government, apart from other considerations, this latter provision was necessitated by the prosecution of the plough policy. The creation of a large subsidised force of supplementary labour,—consisting of soldiers, women, public school-boys, National Service volunteers, old-age pensioners, interned aliens, and prisoners, though farmers paid for their services the current rates of wages, flooded the labour market, upset the laws of supply and demand, and prevented agricultural workers from profiting by the scarcity of skilled labour.

For the harvest of 1917, the food campaign attempted no great increase of tillage. But approximately one million acres were added to the arable area of the United Kingdom. The production of white corn in 1917 was greater than that of 1914, 1915, and 1916 by 4,710,000 quarters, 3,837,000 quarters, and 5,827,000 quarters respectively. In weight the increase in wheat, barley and oats, and in potatoes on agricultural land, was, as compared with 1916, four million tons. At the same time, the weight of roots and hay produced in 1917 exceeded that of any of the three preceding war years. With more than a third of their skilled labour gone, the farmers of England and Wales brought under the plough an additional area of 290,000 acres (187,000 permanent grass) and sowed an increased breadth of 338,000 acres with corn or potatoes. By direction of the Board, the area under hops was reduced by half, that under mustard for seed by two-thirds, and that under bulbs or flowers considerably cut down. The land thus set free was cultivated for food crops. To the harvest of the plough must be added that of the spade. An integral part of the food production campaign was the increase in allotments. In the years 1917 and 1918 the number was raised from 530,000 to 1,400,000. As the strain on man-power and transport grew more intense, the value of the movement became more marked. The holdings were cultivated in the spare time of workers following their daily avocations, and the wide local distribution of the food produced relieved the carrying capacity of railways. Taking into account not only the increased number of allotments, but the displacement of flowers by vegetables on private and nursery gardens, the additional weight of food grown cannot be put at less than one million tons.

The increase of food was of the utmost value. Had production fallen below that of 1916—as at one time appeared probable—the consequences to the Allied cause would have been serious. Harvests in France and Italy were short ; in many districts the scarcity approached to famine. In America, if the domestic consumption had maintained the normal rate, supplies could not have met the demands of the Allies. The addition to her output made by Great Britain enabled food to be diverted from these islands to France and Italy at a critical moment. But, in the early months of 1918, wrote the Chairman of the Allied Maritime Executive, "the spectre of famine was more terrifying than at any previous period." Fresh efforts were urgently necessary. For the harvest of 1918, the Government aimed at increasing the tillage area of England and Wales cultivated for other crops than grass by 2,600,000 acres on the figures of 1916. Many delays and disappointments hindered the execution of the complete programme. Military exigencies prevented the promised supplies of labour and tractors. Owing to the protracted harvest of 1917, farm work had fallen into arrear. But the winter and spring proved more favourable. The farmers of England and Wales seized their opportunity gallantly. In some Counties the quota of tillage and cropping were obtained, if not exceeded. Everywhere, the area under cultivation for other crops than grass was so far increased that an aggregate of upwards of 1,950,000 additional acres was reached. In July, 1918, the prospects of the corn crops were magnificent. A harvest was in sight, which was compared in its promised abundance to that of 1868. In the South the crops were gathered in splendid condition ; in the North the weather broke when they were still in the fields. The incessant rain of September and October did considerable damage, and made the winning of the harvest a fine achievement on the part of farmers. Even when the loss is discounted, the addition to the food supplies of the nation was great. The following Table gives the results of the harvest of 1918 in England and Wales, as compared with 1916, and the average of the last ten years of peace.

To the corn figures must be added the grain produced on over 50,000 acres of rye, which was additional to the usual quantity cut as a green crop.

Against the increase in bread-stuffs and potatoes must be set the loss of fodder-stuffs for stock. The hay harvest was disappointing, the area under roots was diminished ; and 1,400,000 acres of pasture were ploughed. On the other hand, there was an increase of 809,000 tons of oat-straw and considerable quantities of damaged wheat and barley, not included in the following Table, were available for stock. Converting the lost fodders into their beef equivalent it is estimated [1] that from 90,000 to 110,000 tons of meat were sacrificed. Even with this deduction, the net gain in human food was large.

CROPS.	1918.	1916.	1904-13.	INCREASE.		PERCENTAGE OF INCREASE.	
				Over 1916.	Over 1904-13.	Over 1916.	Over 1904-13.
	(In Thousands of Quarters)					%	%
Wheat -	10,534	6,835	6,653	3,699	3,881	54	58
Barley -	6,085	5,181	6,212	904	-127	17	-2
Oats - -	14,336	10,411	10,572	3,925	3,764	38	36
Mixed Corn	620 [2]	— [2]	— [2]	620	620	—	—
Beans and Peas -	1,328	1,122	1,529	206	-201	18	-13
TOTAL -	32,903	23,549	24,966	9,354	7,937	49	32
	(In Thousands of Tons)						
Potatoes -	4,209	2,505	2,643	1,704	1,566	68	59

A new programme, which aimed at a further increase of one million acres in the arable area of England and Wales for the harvest of 1919, was drawn up in the early spring of 1918. It was never submitted to the Cabinet for decision. The German offensives in March and April changed the situation. A military demand was made on agriculture for thirty thousand of its most efficient workers. In July, the Board of Agriculture decided, on a careful survey of the whole position, not to enforce further additions to the arable acreage, but to rely for the maintenance or increase of production on grading-up the cultivation of existing tillages. The withdrawal of the 1919 programme terminated the policy, which as a matter of war emergency had added 2,966,000 acres to the area cultivated in the United Kingdom for other crops than grass.

So long as the campaign lasted, agriculture, under a decentralised form, was treated as a controlled factory for the production of the

---

[1] *The War Cabinet. Report for the Year* 1918, p. 235.

[2] Mixed corn is shown separately in 1918. In previous years it is shown under wheat, barley or oats.

food which the nation needed. Behind the orders of the Executive Committees stood the Government and effective means of enforcing notices to plough, cultivate or sow. But the efficacy of State control was tried under such special conditions as render the experiment comparatively valueless. In other respects the war left a more permanent mark on the industry. It added to the difficulties of landowners. Except for the favourable opportunity of sale which it afforded, they were the class which was most crippled. It largely increased the number of occupying owners. It stimulated improvement in the general standard of cultivation. It aided farmers and agricultural workers to build up strong organisations. It gave to both classes a measure of temporary prosperity. It forced the nation to realise the importance of the industry and to exercise its right to prescribe the use to which the land should be put. But it did not solve the problems of peace. On the contrary, it made them more acute.

CHAPTER XX.

AGRICULTURAL LEGISLATION SINCE THE WAR.

BEFORE reviewing the legislation dealing with agriculture since the war it is necessary to explain, however briefly, the genesis of some of the organisations which have been called upon to play a part in the current structure of agriculture. The earlier associations representative of farmers intervened but little in politics. The Royal Agricultural Society, for example, definitely excluded politics from its ambit, as did the other regional and county societies whose main activities consisted in holding annual shows of live-stock, produce, and machinery. The Farmers' Clubs, like the Farmers' Club in London, which was founded in 1842, met for purposes of discussion, and might occasionally forward resolutions to county Members of Parliament, but were not united in any common organisation. The Chambers of Agriculture had a central office and staff in London, which did focus agricultural opinion and bring it before the members of the legislature who were concerned with the land. However, the membership of both Farmers' Clubs and the Chambers of Agriculture included all classes of men connected with farming—landowners, agents, and dealers, in some cases also men who had no specific contact with the land, but who yet were active in these discussions. There developed a desire among farmers for a more definite means of stating the farmers' case, and the National Farmers' Union, which has grown out of small beginnings in Lincolnshire in 1908 into a body with branches all over England, set out to satisfy that demand. The war, which called into being so many regulations affecting the conduct of farming, brought about a rapid accession of membership, for the Union established its claim to be the one body representative of farmers with which the Government could consult. Undoubtedly it acquired this position because its membership was confined to men actually in occupation of land, as well as to the energy and capacity of its early leaders. Throughout the critical discussions of the war and almost equally disturbed post-war period the Union has retained and strengthened its position as the spokesman of the combined farmers of the country. No legislation has been passed or even proposed without previous discussion with the representatives of the Union ; its opinion has always carried great weight even when it has been cast strongly against official policies. Though the Union has been criticised for its opposition to measures which were designed for the ultimate benefit of agriculture, it must be remembered that it is essentially a " trade union," bound to put the current interests of its members in the forefront of its policy. It is concerned to protect the farmers of to-day, not to develop the agriculture of the future. At the time of writing the National Farmers' Union has 130,000 members enrolled, of whom all but a few are in occupation of land. This represents in numbers at least half of the men who can be termed farmers in England, and much more than half the land in cultivation. The central organisation in London is maintained by a *per capita* assessment on the County Branches, which in most cases call upon their members for subscriptions at a rate of 2d. per acre occupied by the member. The procedure is essentially democratic ; questions of legislation, for example, are referred to the County Branches before they are considered by the Council of the Union. While its contacts with legislation and administration alone have been touched upon, the Union is active in many other directions and has latterly been called upon to play the directive part in nominating the administrations of the various Marketing Boards.

As opposite numbers to the National Farmers' Union must be set the Trade Unions, representative of the workers in the industry. Trade Unionism in agriculture has had but a weak and chequered existence. The scattered nature of the employment, the lack of regular gatherings of the workers such as every factory affords, even the poverty of the rural worker, make combination difficult to obtain, still more difficult to maintain. The shepherds and stockmen, who are the best-paid members of the craft, will always be reluctant to join in a strike, the only ultimate weapon of unions, so great is their attachment to their animals. Again the ordinary farm in England employs but a small number of men, the average being about three workers to one employer ; in consequence no very large number of volunteers are required to enable the farms to be carried on during a strike period. This was demonstrated during the one serious strike in the Eastern Counties in 1923. From relatives, from unemployed men of farming stock, from farmers in the parts of the country unaffected, enough labour to carry on the farms was obtained. There was considerable public opinion, however, in favour of the men, who had come out against a proposed reduction of wages from the 25s a week recommended by the Conciliation Committee to the equivalent of 22s. 6d. for a week of 54 hours. Under this pressure mediators were appointed who settled on a basis of 25s. for a 50 hour week. Thus the strike effected its purpose, though it was not successful in stopping work on the farms, and indirectly it led to the re-establishment of the Wages Boards.

Of the earlier history of trade unionism Lord Ernle wrote in 1912 :

" In 1846 came the repeal of the Corn Laws. In its first effects, Free Trade was not favourable to agricultural labourers. Wages were lowered, and food prices did not fall sufficiently to counterbalance the loss. In the first impulse of despair at the anticipated effects of the loss of Protection, many tenants threw up their holdings. Much land was unlet. Employment became more and more difficult to obtain. The years 1849 to 1853, which immediately preceded the Crimean War and the era of agricultural prosperity, were a period of severe depression. Economy of production was necessarily the aim of employers. They naturally applied to their own business the Free Trade maxim, " Buy in the cheapest market ; sell in the dearest." More machinery was introduced on the land. Small farms were thrown together. There was no diminution in the number of women and children employed ; the gang system, both public and private, prevailed extensively in the Eastern Counties ; the supply of labour was still largely in excess of the demand. The competition of female and child labour continued to depress wages. It was not till some control of the gang system was established by the Gangs Act of 1867, and the employment of children regulated by the Education Acts of 1870, 1873, and 1876, that employers were deprived of the cheapest forms of labour. They were, therefore, driven to employ a larger number of adult males. But the population of rural districts still remained superabundant, in spite of a constant stream of emigration, and wages advanced little and slowly.

" Public opinion was beginning to realise the unsatisfactory position of agricultural labourers. Reports from Medical Officers of

Health on the food, housing, and sanitary conditions of rural districts aroused new feelings of sympathy. The men themselves began to entertain the idea of combining to enforce remedies. Between 1865 and 1871, in Scotland, Buckinghamshire, Hertfordshire, and Herefordshire, unions were formed. The ground was thus gradually prepared for a movement which, when once started, spread with surprising rapidity. On February 7th, 1872, a trade union of agricultural labourers was founded at Wellesbourne in Warwickshire by Joseph Arch. The tinder was ready, and the spark was struck. By the end of the following month 64 branches had been organised in Warwickshire alone. In May a Congress, attended by delegates from many parts of the country, was held at Leamington, at which the National Agricultural Labourers' Union was formed, and Arch elected to be the first president. The prime objects of the new organisation were to raise wages, shorten hours, abolish payments in kind, regulate the employment of women and children, increase the number of moderately rented allotments. In other respects it made a new departure. Up to 1878, trade unions of the ordinary type had confined themselves to the improvement of industrial conditions. From the first, the Agricultural Union included political and social reforms. It demanded not only the parliamentary franchise for agricultural labourers, but changes in the land laws, the disestablishment of the Church, enquiries into charitable endowments, the creation of peasant ownerships. It also introduced new weapons which were not employed by trade unions of the industrial type. It spent considerable sums of money in transferring labourers from congested districts to counties where there was a greater demand for labour both on land and in factories. By its aid also, in the first nine years of its existence, 700,000 persons emigrated to the British Colonies and elsewhere.[1] One feature the Labourers' Union shared in common with the trade unions, and with disastrous results. It endeavoured to combine with its other objects the work of Friendly Societies. But the attempt proved to be beyond its powers, and became one of the chief causes of its ultimate collapse.

" The immediate success of the Labourers' Union was considerable. Wages certainly rose, though no statistical evidence can be relied on to show the extent of the rise. It must be remembered that the better class of workmen had left the poorly-paid districts of the

---

[1] See the evidence of Joseph Arch before the Agricultural Commission in 1881, Qu. 58, 422. (Parliamentary Papers, 1882, vol. xiv., p. 51.)

South and East, and that employers were asked to pay more money for labour which was inferior in quality and less in result, without the advantage of the better prices for their produce which were obtainable in the industrial centres of the North and Midlands. Combination was, therefore, met by combination. Strikes proved an ineffective weapon. Employers were able to supply the places of the strikers, either from the general labourers who stood outside the National Union of Agricultural Labourers, or from the unemployed and casually employed population of the towns. In July, 1874, after a prolonged struggle of six months, the Union suffered a severe defeat. Numbers, as well as funds, began to dwindle. Inside the Union there was a split. Between it and the more recent Federal Union (1873) arose more or less open hostility. The disastrous period of 1875–1884 began to tell on the position of labourers, in regard both to amount of wages and to regularity of employment. They could no longer maintain their weekly payments, and their leaders advised against resistance to reductions. Much of the advance in wages which had been gained was lost ; but it was through the Union that the parliamentary franchise was won.

" Cynics may say that it was the parliamentary vote which gave the labourer his first real step upwards. It made him the most important of the three classes which constitute the agricultural interest, and from that moment politicians have tumbled over one another in their eagerness to secure his support. Be this as it may, there can be no doubt of his substantial progress since 1884. Most men of the class are still poorly paid ; many are precariously employed and poorly housed ; among all, poverty is chronic, and, though destitution is certainly rare, the dread of it is seldom absent. But, speaking generally, labourers in 1912 are better paid, more regularly employed, better housed, better fed, better clothed. They are better educated and more sober. Their hours of labour are shorter. They are secure of a pension for themselves and their wives in their old age. They can, if they choose, make their influence felt in the government of their parish, the administration of their county, the direction of the affairs of the nation and of the empire. Their wives and children are no longer driven by necessity to labour in the fields. What more can labourers want ? may be impatiently asked by some. Others, conscious that all is not yet well, may ask with anxiety—what more can be done ? "

As with the National Farmers' Union the war brought about a considerable strengthening of Trade Unionism in Agriculture, because the setting up of the Wages Boards provided them with a recognised function in putting forward representatives of the workers for both the Central and the County Boards. Two Unions are now in existence. The National Union of Agricultural Workers was formed in 1906 in the Eastern Counties, which have always been the home of such combination as could be formed among the agricultural workers, and at the end of that year it had but an income of £166 and a balance of £40. By the end of 1914 it had 350 branches with an income of £4500 and a reserve of £1700. It had procured the submission of a bill for the establishment of wages boards and a national minimum wage in 1913, which was, however, only to bear fruit in 1917 with the passing of the Corn Production Act. This won an immediate increase in membership, which had declined during the earlier years of the war. Numbers have fluctuated since, but have increased again since the restoration of the Wages Board, and now amount to about 40,000 enrolled in over a thousand branches, extending over almost the whole of England. One of the most important functions exercised by the Union is the nomination of members on the County Wages Committees, on the Central Wages Board, on the Council of Agriculture and on the County Agricultural Committees. But, perhaps, the chief function of the Union consists in the protection of its members in such matters as ejectment cases, the recovery of arrears of wages or underpayments, compensation under the Workmen's Compensation Act, as well as the other activities of the usual benefit societies.

The second Union catering for the agricultural workers is the Transport and General Workers' Union. In this case it is not easy to distinguish the agricultural workers among the general membership of the Union, but it includes about 380 branches composed almost exclusively of agricultural workers, and the total agricultural membership is estimated at 12,000–15,000. This Union shares with the National Union of Agricultural Workers representation on the Council of Agriculture and the nomination of members for appointment to the County Wages Committees and Agricultural Committees.

Though not called upon to exercise any statutory functions, the third of the agricultural interests, the landowners, organised an Association in 1918, confined to the owners of agricultural land in England and Wales. It now possesses nearly 12,000 members, including most of the owners of the large agricultural estates as well as a considerable number of farmer owners. The Central Landowners' Association keeps a watch over the landowners' interests in all matters of legislation affecting agriculture, and it is in a position to supply the Government Departments concerned with much information concerning the effects of the proposed measures that would be otherwise unobtainable. The Association also does a variety of advisory, legal, and educational work on behalf of its members.

The existence of these bodies representative of each of the three interests participating in agriculture—the landowners, farmers, and labourers—has ensured that in the legislation since 1918, of which a brief summary follows, the claims of all sections of the industry have received due expression.

The normal course of English agriculture had been greatly disturbed by the exigencies imposed by the war and the control of cultivation exercised by the Food Production Department. When the control was relaxed, the farming-land of the country had become, generally speaking, out of condition through the repeated growth of cereal crops and the ploughing up of grass land at a time when the supply of labour and fertilisers was inadequate. The more fortunate farmers had made considerable sums of money through the high prices that had prevailed, which, indeed, were maintained until 1921 ; on the other hand, many of them had been compelled to purchase their farms at the exaggerated values which were then ruling for land and had led to the break up of many estates. The Corn Production Act did not allow rents to be raised in response to the guaranteed prices of wheat and oats, but still more potent was the general disinclination of landowners to take any direct advantage of the national emergency. None the less the demand for farms was so great that they could be sold to produce double and treble their current rental and it is estimated that something like one-tenth of the agricultural land of the country was bought by occupiers in the years 1918–1920.

From the legislative point of view the most notable legacy of the war was the Corn Production Act, which, until 1922, guaranteed minimum prices of wheat and oats, at 60s. and 38s. 6d. per quarter respectively. The same Act initiated the Agricultural Wages Boards, which fixed minimum rates of wages, not to fall below 25s. a week. These minima, either of prices or wages rates, have never been operative, for almost as the Act was passed war conditions had forced up prices and rates far above the level specified.

The Corn Production Act was avowedly an emergency measure, and at the close of the war hopes ran high of the reconstruction of agriculture under the ægis of State, as an industry which must be developed in the interests of national safety. The beginning of a new policy was made by the Ministry of Agriculture Act of 1919, which directed the change of title from Board to Ministry and instructed the County Councils to set up Agricultural Committees, of which the majority of the members were to be nominated by the Councils though not necessarily to be members of them. The Ministry took power to nominate not more than one-third of the members, in order to secure the presence, among others, of representatives of the workers. There was also provision for the inclusion of women and representatives of science and education. There was further constituted a Council of Agriculture consisting of two representatives from each County Committee, six members of the Agricultural Wages Board and thirty-six nominated members, to include members representative of landowners, agricultural workers, women and persons connected with science and education. The Council was to meet at least twice yearly, to be addressed by the Minister and to consider such resolutions as might be submitted concerning the administration of agricultural affairs. The Council was to be a Parliament of Agriculture, though only advisory to the Minister. From the Council a small Agricultural Advisory Committee was to be selected to confer regularly with the Minister.

From the outset the National Farmers' Union took exception to the constitution of these bodies, on the ground that the nominees of the Agricultural Committees were not necessarily representative of the farmers of their area, indeed, in some cases, might be wholly representatives of the Labour Party. The Union maintained that it alone was entitled to speak for the farmers of the country, and that since the three interests concerned were the landowners, farmers, and workers, only direct representatives of these three interests should confer to advise the Minister. At a later stage the Union instructed its members to withdraw from both Council and Committee. The Council still continues to meet, but the Advisory Committee after a time fell into abeyance. This Act was followed by the Agriculture Act of 1920, which substituted for the guaranteed prices for wheat and oats of the Corn Production Act a new scale based upon the average prices of 1919, viz.: 68s. per quarter for wheat and 46s. for oats. These figures were, however, to be subject to revision from year to year by a body of Commissioners who were to adjust them in accordance with the changes in the cost of production. The Ministry was still to retain powers to enforce cultivation according to the rules of good husbandry though it could no longer call for the ploughing up of grass land. In case of default and if the necessary works of maintenance were neglected the Ministry could enter and cultivate the land under a receiver.

A very significant change was made in the law concerning landlord and tenant. A tenant became entitled to compensation for disturbance after notice to quit unless the landlord could obtain a certificate from the County Agricultural Committee that the holding was not being cultivated according to the rules of good husbandry. The compensation was to amount to one year's rent, though if the tenant could prove greater loss and expense arising from the disturbance, two years' rent could be awarded. There was also provision for arbitration in cases where either party demanded a variation of rent.

This Act had but a short lease of life, for with the break in prices of 1921 it became apparent that the Exchequer would have to face a very heavy bill. The vigour of the Scottish representatives in the discussions of 1919–20 had secured an impracticably high guarantee for oats, of which the acreage was nearly twice that of wheat. The call for economy was insistent, and though the break in the world prices was exactly the sort of emergency that the Act had been intended to counter, the Corn Production (Repeal) Act of 1921 repealed the financial provisions of both the Agriculture Act of 1920 and the Corn Production Act of 1917. Payments were made to the growers of wheat and oats in 1920–21, amounting to £15,000,000 in England and Wales, and £4,400,000 in Scotland. At the same time the Agricultural Wages Board was abolished ; instead the County Agricultural Committees were instructed to set up Conciliation Committees of employers and workmen. Naturally, this sudden denial of the promises that had been made to the agricultural community excited the utmost indignation among farmers, who, however, obtained the retention of the clauses regarding landlord and tenant and a solatium in the shape of an assignment of one million pounds to be spent within five years on education and research.

Two minor reforms may also be noted. The Seeds Act of 1920 imposed upon all vendors of seeds an obligation to declare their germinating capacity and purity, a measure which has had a valuable effect in tuning up the quality of the seed sold to farmers and market gardeners. In 1921 the Corn Sales Act made the hundredweight the standard measure by which all corn was to be sold, and so abolished the legal status of the quarter, of which two definitions had been in official use.

In 1923 the Agricultural Rates Act provided farmers with further relief of local rates, burdens which pressed unjustly upon the business of farming because farms carry a high rateable value in proportion to their turnover. The rates were to be levied on one-quarter only of the assessed value of both agricultural land and buildings. Even greater relief followed in 1928, when a general revision of the basis of rating and valuation took place. Agricultural land and buildings are now relieved of all rates, which are charged only upon the dwelling-house of the farmer.

With the advent to power of the Labour Government in 1924, the Agricultural Wages, Board was restored, though it was no longer empowered to fix a general minimum wage but, instead, to give effect to the recommendation of the County Committees. The Ministry may refer back any decision to the Committee for reconsideration. The County Committees are made up of representatives of the employers and workers, in equal numbers, together with two nominees of the Ministry and an independent Chairman. The existence of minimum wage regulations has always been a sore point with farmers, though it has not availed to check the drift of the younger men into other employment, especially as hitherto they have enjoyed no insurance against unemployment. The most cogent tie of men to the land has always been the occupation of a cottage, so general has become the shortage of rural housing.

The Land Drainage Act of 1926 gave the County Councils power to enforce the cleaning of rivers and watercourses, but this pressing question was more fully dealt with by a further Act in 1930. Thereby Catchment areas were defined and Catchment Boards were to be set up for each area, to which were transferred the powers of all internal Drainage Boards within the area. New internal Boards are to be constituted, the Catchment Board remains responsible for the main river and its outfall. The County Councils and the County Boroughs within the area become responsible for expenses of the Catchment Board not covered by the rates.

The Agricultural Credits Act of 1928 was designed to give owners and occupiers of land better access to capital. In particular, it was intended to meet the hard case of many farmers who had bought their farms at the high prices prevailing from 1918 to 1921 and who later found themselves under very onerous mortgage conditions. It recreated the old Land Improvement Company as the Agricultural Mortgage Corporation and provided it with capital furnished by the Treasury, so as to extend its power of making loans on mortgage and for improvements. In order to meet the requirements of farmers for working capital it authorised the creation of a charge on " Farming Stock " against loans made by the banks, this charge to rank before other liabilities, a privilege hitherto enjoyed by the rent. It thus created something in the nature of a " chattel mortgage " which does not involve the publication required by a Bill of Sale with its attendant danger to the credit of the borrower. As time goes on this measure may exercise a considerable influence in facilitating and safeguarding the means by which farmers obtain their short term credits, now an intricate customary affair in which merchants, auctioneers, and tradesmen, as well as the banks, bear no small part.

In the same year was also passed the Agricultural Produce (Grading and Marketing) Act, which empowered the Ministry to prescribe grades and designations for agricultural produce with specified marks. It thus provided the legal status for the National Mark scheme, which should eventually have a potent effect in standardising and raising the quality of the farmers' output.

With Dr. Addison as Minister, 1931 saw the birth of an Act of far-reaching importance, in that it aimed at the reconstruction, one might say the revolution, of the methods by which the business of agriculture is carried on.

Combination to sell the products of agriculture had never made much headway among English farmers. Individualists by temperament and tradition, the farmers in any district would be men engaged in varied sizes and types of business. Their markets were close at hand, thus removing the incentive to combination and standardisation which is supplied by working for an export market, and the dealers in these markets were in many cases financially interested in the farmers' business. Such co-operative societies as flourished were mainly concerned in the purchase of the commodities required by the farmers, and various attempts to institute collective selling had revealed the fatal sources of weakness in voluntary co-operation—the difficulty of retaining the loyalty of the partners in the enterprise. In many cases societies for the handling and sale of milk, or again for the manufacture of bacon, had started with considerable promises of support, but as soon as they attained any measure of success, had found themselves subjected to the attack of the large established commercial enterprises engaged in that business. The members of the co-operative were tempted away by the offer of prices which the society could not emulate and remain solvent. Thus rendered insecure of their supply of raw material society after society had to close down or sell to their competitors. Even more instructive had been the fate of the hop-growers, numerically a small body, not more than two thousand in all and pretty well grouped in two districts, selling, moreover, to one group of customers —the brewers, and with no alternative market. During the war and for some years after, they had been brought together under one control, which took over the whole of the hops grown and conducted the sales to the brewers. So satisfactory had the business been that when the statutory control was withdrawn it was reconstituted on a voluntary basis and obtained the adhesion of some 95 per cent. of the growers. The few who remained outside, however, enjoyed all the advantages of the control in setting a remunerative price without any of the attendant disabilities, such as delay in receiving payment, loss on the surplus grown beyond requirements, restrictions on acreage to be harvested, etc. Those outside the combine were seen to be increasing their numbers and enlarging their plantations, so that after three years the combine dissolved and its members returned to the old competitive methods of sale.

The National Farmers' Union had organised the producers of milk and entered upon a system of collective bargaining with the representatives of the wholesalers in order to fix each year a scale of prices for milk. Though this arrangement did not cover the whole of the country it did set a standard. Year by year the negotiations proved difficult and protracted, even to the point of a threatened strike of one side or other, but none the less it may be agreed that they had been effective in procuring better returns for the milk producers, even though the wholesalers appeared to hold the ruling position in the negotiations. Year by year the volume of milk was increasing without much sign of a corresponding increase of demand from the public, so that an augmented proportion of the supplies had to be directed to manufacturing ; evidently the whole structure of the milk market was becoming precarious.

It was to render collective selling practicable that the new legislation was addressed. The Agricultural Marketing Act of 1931 represented a novel conception ; if anything, English law had in the past looked askance at combinations and sanctions which might be held to be in " restraint of trade," but this was an Act to enable the majority of the producers of any agricultural commodity to compel the adhesion of all the producers. On the submission of a scheme for the more efficient production and marketing of any agricultural commodity the Minister, after enquiry and consultation with the Board of Trade, could sanction the promotion of a Marketing Board to carry out the scheme. The first step is to form a register of all producers who would proceed to the election of a Board. A poll has then to be taken to decide if the scheme was to be put in force ; the assent of not less than two-thirds of the registered producers voting is required and the supporters must also represent not less than two-thirds of the total production. Once constituted the Board becomes, in effect, the selling agency for the product. No sales can be made by others than registered producers nor on other terms than those laid down by the Board. The Board, itself, can buy and sell and conduct any manufacturing or preparation for market. In order to facilitate the preparation of schemes the Ministers can set up Reorganisation Committees to explore the situation and submit proposals. Moreover, Boards can be set up both for primary and secondary producers, e.g. a Pig Marketing Board finds its necessary complement in a Bacon Marketing Board representing the curers. For a time, the drastic nature of the policy implied in this Act roused the apprehensions of the farmers' organisations and the National Farmers' Union declared its opposition, but as some of the Reorganisation Committees got to work and as the Government made it plain that no financial assistance would be given to agriculture except through the Marketing Boards, the opposition was borne down and the National Farmers' Union threw in its lot with the new policy. A second Act in 1933 further developed and strengthened the machinery, for by that time the nation had definitely swung over to a protective policy. The Act gave power to the Board of Trade to regulate all imports of which the home production was under the control of a Marketing Board, and provided for Market Supply Committees to review the supply and demand of the commodity both from home and external sources.   At the same time the Act provided for Development Boards at the instance of any two Boards representing the primary and secondary stage of a given product, and further gave very wide powers to enforce the rationalisation of the industry. So far Marketing Boards are operating as regards Pigs and Bacon, Milk, Potatoes, and Hops ; the very intricate question of meat is under consideration, but proposals as regards fruit and vegetables have been rejected by the producers.   In so far as the Marketing Boards have been at work they have been the subject of much criticism, from the public that they have concerned themselves only with improving the returns to the farmers at the expense of the consumer, and from the farmers that the control exercised has been hampering and unbusinesslike.   Such criticisms are inevitable at the inception of new businesses on an unprecedented scale ; it is one of the fundamentals of the policy that these businesses shall be controlled by the representatives of the farmers, among whom but few can have had experience of affairs of that magnitude. Men with the capacity for management have to be found, procedure has to be learnt by trial and error.   But the Marketing Acts initiate a new era in English farming, and with this brief description of the policy they set out the account of post-war legislation may fitly close.

CHAPTER XXI

SMALL HOLDINGS

THE desire to secure access to the land for men of small means and to check the marked diminution that was taking place in agricultural employment first found expression in the Small Holdings Act of 1892.

At that time, when farmers were putting down their arable land to grass and reducing the number of workers upon their farms, there were to be seen plenty of men who, given the opportunity of obtaining a piece of land that could be worked by their own labour, were taking advantage of the smaller branches of agriculture and were making a satisfactory living. In many parts of the country where large estates had been divided and land was obtainable in small plots, even though the rental was relatively high, small holding communities had grown up, the members of which were sufficiently prosperous to create an unsatisfied demand for additional accommodation of the same type. On the face of it, land tenure of this kind was of greater value to the nation than the large farms which their occupiers could only make profitable by cheap cultivation, working for a low output at a minimum of expenditure. Both the volume of production and of labour employed was greater on the small farm. The economic gain was palpable enough when the bullock grazings of Lincolnshire were converted into 10–20 acre holdings on which a man could thrive by growing potatoes, onions, celery, and the like, or when a poor milk farm in Worcestershire, employing at the outside two men per hundred acres, would when cut up in 10–15 acre plots be eagerly sought after to produce plums and asparagus. The economic argument was reinforced by a very genuine desire to better the position of the agricultural labourers, the largest and worst paid body of working men in the country, who were being driven off the land by the changes in farming that had set in with the great depression. The idea of statutory action to secure access to the land became a part of the programme of the Liberal party of the day; Jesse Collins, one of the members for Birmingham, with his slogan of "three acres and a cow, " being the leader of the movement.

The Act of 1892 proved valueless ; on the one hand it was permissive, on the other it was designed to provide for purchase by the occupier rather than tenancy. After the Liberal government had come into power in 1904, the Small Holdings and Allotments Act was passed in 1908 under the auspices of Lord Carrington (afterwards Marquess of Lincolnshire), himself a landlord who had found satisfaction and a better rent roll by the conversion of portions of his own estates into small holdings. This Act instructed the County Councils to acquire land, voluntarily or by compulsion, to divide and adapt it by erecting buildings, drainage, etc., and then to let it to suitable occupants as small holdings, defined as holdings not exceeding 50 acres in extent or alternatively of an annual value of not more than £50.

The scheme was intended to be self-supporting ; lettings were to be at economic rents, the Government would provide loans to the County Councils from the Local Loans Fund to be amortised in eighty years, and would on approved schemes meet the legal charges of acquisition and one-half of any current losses incurred by the Council in any year. There were provisions by which the tenants could acquire the holdings by deferred payments, but these were little used ; in general, the tenants preferred an annual tenancy, though the rents were fixed so as to pay both interest and amortisation of the loan, whereby in eighty years time the Council would succeed to the full ownership of the land.

The Act was well taken up and as a rule carefully worked by the Councils, so that up to the time of the war, 12,792 statutory small holdings had been created, 70 per cent. of which were bare land holdings. The new occupiers were well chosen and as the period was one of rising prices were generally successful, the proportion of failures up to 1914 being only about 4 per cent. of the whole. None the less this scheme, designed to be self-supporting, cost the National Exchequer £394,000, chiefly for administrative expenses, the actual losses by reason of insufficient rental to cover the charges being no more than £7300.

Progress, however, under the scheme was declining when in 1914 the war put an end to such activities. Many of the earlier holdings had been small plots of land that could be acquired near a village and had incurred small costs for adaptation or for buildings. When it became necessary to buy an estate for division on which roads had to be made, drainage and water to be provided, as well as houses and farm buildings, the rents required to cover the expenditure became prohibitive or at any rate so far above current rents for holdings of similar size that they did not attract applicants. On the older holdings the buildings had generally been put up in cheaper times, indeed had often been built piecemeal by the occupier's own labour they represented a much smaller capital outlay than was required for a house and buildings which had to be designed by the County Architect, approved by the Board of Agriculture, and then erected by contractors after public tender. It was hardly possible to start with bare land and provide a small holding with house and buildings, roads, drainage, and fencing, at an economic rental of £50, even though loans could then be effected at $3\frac{1}{2}$ per cent.

As the war was drawing to its close, and while people were very conscious of the straits to which the country had been reduced because of the restricted cultivation of the land, a strong movement sprang up to provide land for ex-service men on their demobilisation. In order to implement the promises made, the Land Settlement (Facilities) Act of 1919 appropriated a fund of £20 millions for the provision of small holdings for ex-service men. The County Councils were made the authorities as before, but the Government undertook to pay the whole of the losses incurred on these new holdings up to the end of April 1926, when a final settlement would be reached.

On this basis 254,000 acres were acquired and 16,740 holdings were created, about 55 per cent. of which consisted of bare land only. By 1926 of the 24,319 men originally settled, 18,915 still remained on their holdings, and many of those who had left had moved to larger farms. The percentage of failures has been estimated as no more than 15 per cent. With the high prices of land and the cost of building and of all items of equipment, together with the high rates of interest at the time this scheme was launched, it could not be other than expensive. Roughly speaking, $15\frac{1}{4}$ millions were borrowed from the fund by the Councils for capital expenditure, and 5 millions were required to recoup them for their expenditure up to 1926. By the final settlement then reached the Ministry commuted the liabilities on the holdings thus created for an annual payment to the County Councils of about £800,000 a year for periods varying from thirty to seventy years.

Actually the rent roll of all the holdings approximates to this sum, and this is what the County Councils will ultimately inherit when all the loans are paid off. On this showing the average cost of each of these holdings to the State will be over £1000, this being the capital which has to be written off in order that the rents may constitute an economic return on the remainder of the expenditure. These figures, however, have little bearing upon the finance of small holdings— the occasion was unique and the conditions abnormal.

A further Act in 1926 removed the restriction remaining from the Act of 1908, that any scheme for the normal provision of small holdings should be self-supporting save for administrative expenses, and provided that the County Councils could embark on schemes that would involve a loss. If the scheme was approved the Ministry would repay to the Council three parts of its annual loss, leaving the Council to bear the other quarter. It was thought that as the provision of small holdings would satisfy a local demand and as the Councils would ultimately become owners of the holdings to the amortisation of which they had only contributed one-quarter of the amortisation charge, they might well be expected to adopt an active policy. This expectation was disappointed ; during the next six years, 1927–32, only 1274 holdings, amounting to 30,814 acres, were added to the list of statutory small holdings. In many of the counties there were active supporters of the movement, but they had no special representation on the Councils, where the agricultural members were generally indifferent if not hostile to small holdings, and the great mass of the urban members saw no reason to assume even one-quarter of the cost of what they considered should be a national rather than a local burden. Consideration, moreover, of such later small holdings as were provided by the machinery of the Act, under County Council management, did show that they could become very costly.

The average capital cost of these holdings was £1285 (*i.e.* total cost divided by number of holdings), but since many of them were small and bare land holdings, the expenditure on full-sized holdings that had to be equipped with house and buildings was very much greater, often over £2000. Of course, during the period building costs were high ; the house itself of the cheaper type in the North cost about £340, of the dearer type in the South £550 ; the site costs —*e.g.* drainage, water, roads and fencing, lay between £60 to £80. Money was dear during the period, the average cost of the loans was a little under 5 per cent. whereas the rents were returning only 3 per cent. on the outlay.

It may be argued that in 1935 when money could again be obtained at less than 3 per cent. the opportunity for creating small holdings on an economic basis has been revived, which may be true as regards the purchase and adaptation of little parcels of land and existing cottages, but still does not meet the case of providing a fully equipped holding *de novo* from bare land at a cost of £2000 or more. But criticism may be directed against the procedure ; a County Council whose plans have to be sanctioned by the Ministry tends to aim at a standardised holding of a substantial character that will be of value when the Council becomes the unencumbered owner in sixty-eighty years time. Thus every holding is furnished with good and true cowsteadings and pig-styes, whereas the occupier may not need them or be content to work with improvisations. Again, the accepted practice is to build a house on each holding ; if the houses were massed in a hamlet they could be built more economically in themselves and with a great saving in site costs, and again of varied sizes to meet the difference in families. Such gathering into a community would be welcome, especially to the wives. The man may want to sleep next door to his pigs, but in most cases the woman dislikes isolation and thinks of the distance her children will have to walk to school.

The case, however, for or against small holdings cannot be judged on the results of these experiments in settlement by the State ; the sample was small, the method adopted was narrow and inelastic, the course of prices during the period has been entirely abnormal. Nor does it appear that an agreed conclusion can be reached from a dispassionate consideration of such data as have been gathered on the economics of small holdings. It has been found in various districts that the gross output per acre is greater from the single man farm than from the larger holdings ; this is countered by the consideration that the small holding is usually on the better land and that the single man is forced to take up the more intensive methods of farming. It is admitted that the small farm is unsuited to corn growing or such forms of the live-stock industry as beef production or sheep farming, but it is argued that the personal attention of the small holder makes for success in the growth of fruit and vegetables, and in the keeping of cows and poultry. The answer again is that as a matter of fact the farmers who follow these special industries on a large scale are more efficient than the small holders and obtain higher returns per man and per acre, in virtue of the organisation and capital they can bring and the use of machinery which is only applicable on the larger farm. Obviously the small holder is a weak economic unit for buying and selling ; that could be remedied by co-operation did not the independence which is the characteristic virtue of the small holder render him averse to co-operation. The new Marketing Boards will, however, to a large extent, remove this difficulty. Discussion, however, tends to the conclusion that small holding is an uneconomic mode of production from the land, if judged by return from either capital or labour, even after taking into consideration the food produced for the family, but that it is kept alive by the willingness of the small holder to pay for his independence and the chance of rising by long hours of labour for himself and his family.

Discussion thus enters into the domain of the imponderables, and it becomes difficult to free it from sentimental and political prepossessions. The conception of " A bold peasantry, the country's pride " has a great appeal to townsmen, as also the life of plain honest toil it seems to imply. Then there is a current opinion among landowners and the governing classes who have contacts with the land, that in farming science and organisation are naught, that " good " farming and large scale farming cannot pay, and that only the little " dirty boot " man can muddle through.

In the face of these strong opinions, which can always be backed by specific instances, it is, perhaps, helpful to attempt a statistical examination and ascertain what in the past has been happening to small holdings in England.

We possess no numerical data which will elucidate the history of land tenure in England, but, as the earlier chapters of this book have made manifest, our earliest agriculture was carried on by a community of peasants, getting little more than a bare subsistence out of the land. Little by little their status became changed from customary tenants to that of freeholders, a process that was not effected by any one procedure or at any particular period. It was spread over the centuries, most actively from the time of the Black Death to that of the Napoleonic wars, and, as the earlier chapters set out, the changes were accompanied by a running fire of denunciation against the engrossers of land and the destroyers of the yeomen of England. It was thus a secular process of change which by the nineteenth century produced the characteristic structure of the English countryside—landlords, tenant-farmers, and labourers, instead of the peasant structure that had prevailed and still prevails in most European countries. In the main it was economic pressure that produced this revolution in our system of land tenure ; the early development of commerce and industry not only provided capital to buy up land but offered to the peasant more profitable outlets for his labour.

The countrymen who went to the wall became paid labourers in the small capitalist enterprises our tenant farms grew into. In spite of legislative efforts to keep the yeoman on the land, throughout the centuries the small farmer was being eliminated and the process was still going on during the period under consideration. So varied is the structure of farming that the process has never become complete : as late as 1930 out of 396,000 occupiers of land in England and Wales, no less than 255,000 were occupiers of less than 50 acres, but they held less than 16 per cent. of the total in cultivation. Eliminating the occupiers of less than one acre, the numbers became 182,000 small holders out of 323,000 farmers, 56 per cent. by number though only occupying 15 per cent. of the land. The larger holdings above 50 acres occupied 84 per cent. of the cultivated area and averaged in size 151 acres. The following table shows the changes in the number and acreage of small holdings in England and Wales for the last sixty years, divided into groups by size :—

NUMBER AND EXTENT OF HOLDINGS IN THOUSANDS
ENGLAND AND WALES

YEAR.	1–5 Acres.		5–20 Acres.		20–50 Acree.		TOTAL.	
	No.	Area.	No.	Area.	No.	Area.	No.	Area.
1870 – –	113	339	128	1,920	75	2,625	450	25,957
1885 – –	114	321	127	1,420	73	2,463	475	27,710
1895 – –	98	301	127	1,422	75	2,502	440	27,683
1905 – –	92	—	198	—	—	—	433	27,406
1915 – –	91	280	121	1,358	78	2,634	433	27,053
1924 – –	77	241	112	1,264	80	2,691	409	25,877
1930 – –	73	227	104	1,179	78	2,640	396	25,380
1934 – –	69	—	99	—	76	—	384	—

In the first place it will be seen that there has been a great and continuous drop in the number of small holdings from 450,000 to 384,000, about 15 per cent., i.e. the total loss has been about twice as great as the number of statutory small holdings created since 1908. Doubtless some of this diminution has been due to the growth of the towns, which would in particular absorb the occupation fields and part time holdings on their borders, but it is noticeable that until 1915 there was little or no loss of acreage by the holders of more than 5 acres.

From the trend of these figures, it can hardly be doubted that the factors leading to the secular decline in the yeomen and small farmers, which had been going on for two centuries, have been continuing in operation during the last sixty years. The losses in the 1-5 acre group may be ignored, but the reduction in the number of holdings of the 5-20 acre group are real, and the 20-50 acre group has only been kept up by the artificial creation of holdings by the County Councils. It is significant that the losses in the whole 5-50 acre group are heavy in counties like Nottingham (30 per cent.) and the East Riding of Yorkshire (20 per cent.), where ordinary arable farming predominates. But in Kent, Worcester, and Salop the number of small farms has been increased by 11, 17, and 13 per cent., these being the counties in which market gardening and fruit growing have developed. The only reasonable conclusion is that during the period under review the small mixed farm of under 20 acres had become an uneconomic proposition, only able to hold its own when the occupier was engaged in fruit growing or market gardening. The larger 20-50 acre holding was more able to support itself, but even that group had only maintained its numbers by the uneconomic provision of the statutory small holdings.

In the light of these figures it is difficult to suppose that a peasantry can be recreated in England or that English farming can be revived by a wholesale division of the large farms. Every year science, machinery, and the art of organisation are advancing and widening the gap in efficiency which separates the large from the small holding. Economic pressure will have its way in this as in preceding centuries, more rapidly, indeed, in the modern world, as science is really beginning to obtain some control over the processes of life which determine production from the land. The one chance which the recent organisation of marketing in England seems to offer to single man or family farms is through the formation of Group settlements, all the members of which are in the main devoted to a single form of output and are bound to a central processing or selling agency. Not all forms of farming can be worked on this system, but it is possible to frame schemes for pig-keeping, poultry-keeping, the growth of fruit and of vegetables for canning, possibly even milk production, which would be remunerative to the small farmer by constituting him a productive unit in the large scale business demanded by modern commerce, and yet would leave him with initiative and opportunities of increasing his business. The experiment has yet to be tried on any large scale, though one recent example, a small holding colony growing vegetables on the Ford Farm at Dagenham, is being successful in its early years. It is noteworthy that two recent enquiries, the one conducted on behalf of Lord Astor and Mr. Seebohm Rowntree (*The Agricultural Dilemma*, 1935), the other on behalf of the Carnegie Trustees (*Land Settlement*, by A. W. Menzies Kitchen, Carnegie United Kingdom Trust, 1935), both reach the conclusion that Land Settlement on a large scale cannot solve the problem of unemployment. It can, indeed, do little to overcome the drift of labourers from the land which in the last sixty years has been going on at the rate of about 10,000 a year. Comparing the 1931 census figures with those of 1871 the numbers of farmers and their dependents has changed little—326,373 to 329,028, while the labourers have dropped from 996,654 to 511,455.

The question of small holdings is now being approached from another side, as a means of aiding in the social reclamation of the inhabitants of the distressed industrial areas. A Land Settlement Association has been formed, financed initially from private sources with an equivalent grant from the Ministry of Agriculture, and to which the Carnegie Trustees have promised eventual grants amounting to £150,000. This Association is aiming at creating Group Settlements, of the 3–10 acre type, of families drawn from the depressed areas such as the Durham and the South Wales coalfields. The Settlements are to be fully co-operative in their working, but production for subsistence and the relief from a condition of neither employment nor prospect of employment, are of more importance than obtaining an economic return for the capital and labour expended. This experiment is still in its earliest stages and it is impossible to foresee what effects it may have upon English farming. It is clear that the motive is social, to provide an alternative to the purely cash relief of unemployment, and though the wild argument is sometimes advanced that since all agriculture in England has become uneconomic the subsidies now accorded (amounting to £17 millions in 1934) might better be devoted to emancipating the labourers and building up a sturdy independent population of family farmers, such a proposal ignores the high efficiency that is being attained by many English farmers and the possibilities of spreading that efficiency that lie with the schemes for orderly marketing.

The decay that has overtaken peasant farming is not confined to England or Great Britain, it is seen even more severely in the continental countries which have not advanced so far as we have in the consolidation of their peasant holdings into larger units. In those countries where the single family farm predominates and peasants may constitute from 40 to 70 per cent. of the working population, the whole economic system has been adjusted to preserve the peasant basis of the society. Protection of the most rigid character has been necessary, with the result of high food prices that have " seriously aggravated the effects of economic depression on the standards of living of the industrial population of European countries." [1]

It must not be supposed that because small holdings offer little prospect to the agricultural labourers of emerging from their dependent position and attaining a better standard of living, they will rest content with the old order of things. The census statistics reveal to what extent they have been leaving the land during the last sixty years, something like one-third of their numbers since 1871, and every one in touch with village life knows how the schoolmasters and the provident parents strive to find for their boys any other employment than in farming. What Lord Ernle wrote in 1912 still holds :—

" Labourers as a class have not formulated their general aspirations in definite form ; they are conscious of little more than discontent with a life which for them has lost its meaning and with that loss its savour. Yet perhaps a contrast between the past and the present conditions of rural villages may suggest some sort of answer. Under the old system, some of the existing evils prevailed, as well as others which are now removed. The village was not Arcadia. Its life was by no means idyllic. In one sense, at all events, it was on a lower level then than now. The peasant was too absorbed in his own surroundings to care for matters outside his own environment. He could rarely read ; he seldom thought of anything beyond his daily pursuits ; he had no ideas and few opinions except on the practical

[1] Leith Ross, League of Nations Publication, 178–97, 1935, II. xiii., p. 23.

subjects in which he was interested. For many years, under the shock of change, the mind of the agricultural labourer was even less active and even more narrow. His daily work was less varied and more monotonous ; he had lost the opportunity of practising the manifold crafts in which his grandfather had been occupied ; he toiled exclusively for a master, not for himself. He fell into a half-dazed state fatalism. Now this is changing. Labourers read, think, enquire. Their minds are awakening and curious for information. They are slowly beginning to extend their intellectual horizon beyond their own individual misfortunes or advantages, and to understand the meaning of economic laws. In this mental development, politics, honourably handled, and dealing with principles not personalities, might, and should, play an important part. A great responsibility rests on writers in the public press and on platform orators.

" The mental change in progress may account for the restlessness. It does not by itself explain the discontent. The peasant, under the old system, had a definite independent place in the community. He commanded respect for his skill, judgment, and experience in his own industries. He was not cut off by any distinctions in ideas, tastes, or habits from the classes above. On the contrary, each grade shaded almost imperceptibly into the next. To-day, the intermediate classes have disappeared. Instead of the ascending scale of peasant-labourer, the blacksmith, carpenter, wheelwright, and carrier, the small holder, the village shopkeeper, the small farmer, the larger farmer, the yeoman occupying his own land, and the squire, there are in many villages only two categories—employers and employed. The gulf is wide enough. It has been broadened by the progress of a civilisation which is more and more based on the possession of money. All the employing classes have moved on and upwards in wealth, in education, in tastes, in habits, in their standard of living. Except in education, the employed alone have stood comparatively still. The sense of social inferiority which is thus fostered has impressed the labourer with the feeling that he is not regarded as a member of the community, but only as its helot. It is from this point of view that he resents, in a half-humorous, half-sullen fashion, the kindly efforts of well-meaning patrons to do him good, the restrictions imposed on his occupation of his cottage, as well as the paraphernalia of policemen, sanitary, and medical inspectors, school-attendance officers, who dragoon and shepherd him into being sober, law-abiding, clean, healthy, and considerate of the future of his children. To his mind, it is all part of the treatment meted out to a being who is regarded as belonging to an inferior race.

" The economic side of the change further accentuates the discontent and adds a practical to the sentimental grievance. The peasant worked as hard as, or harder than, his descendant. But his industry was more interesting to him, partly because it was more manifold, partly because much of it was for himself. He had less need of money. Living more on his own produce, he could satisfy some of his wants by exchanges in kind. When he had to buy, he obtained money either by sale of his own stuff or by working for an employer. But to earn weekly wages, or to be in regular employment, was not for him an absolute necessity. He was little affected by the laws of demand and supply in relation to labour. The new commercial system, on the other hand, has made the agricultural labourer entirely dependent on employment and money wages. Instead of producing much of his own food, he has to buy it nearly all. In order to live, he must sell his labour for cash, and under the stress of new exigencies which limit his power of bargaining. Now, if the labourer loses employment, or fails to find it, he has no recourse on which to fall back. His livelihood, and, in the case of tied cottages, his home, depend on a week's notice. A change in the ownership or occupation of the land on which he works may cut him adrift. Against this uncertainty he cannot protect himself. But he may lose his wages in another way, and against this he can be partially secured out of his savings. For this reason, even his thrift is guided into new channels by the commercial spirit which necessarily controls his life. The peasant's savings went into the purchase of a pig or a calf, or into some other form of reproductive investment. They meant a step upwards in well-being. But the labourer of to-day saves for a different object. His weekly earnings, on which depend his livelihood and home, are all-important. It is against sickness, therefore, that he tries to secure himself by painful thrift. What he pays for insurance is not an investment ; the money that he puts into a Friendly Society is not capital which he can ever use. Once paid, it is gone for ever His savings do not in fact help him to put his foot on the ladder of prosperity. They only serve to protect his family and himself against a possible adversity. The great number of labourers who insure in Friendly Societies is a pathetic proof of their total dependence on weekly earnings and of their haunting dread of the loss of wages through sickness. But the popularity of Slate Clubs shows their consciousness of the disadvantages of the ordinary insurance. By dividing out at the end of each year, they are at least able to secure the use of a portion of their annual savings as capital.

" Under the older system, peasants were rarely without some real stake in the agricultural community ; they were not members of an isolated class ; they were not exclusively dependent on competitive wages for their homes and livelihood ; they were seldom without opportunities of bettering their positions ; they had not before them the unending vista of a gradual process of physical exhaustion in another's service. Under the modern commercial system, the conditions from which peasants were generally free are those under which the average agricultural labourer lives, though exceptional men may struggle out of their tyranny. They have no property but their labour. Even of that one possession, such are the exigencies of their position, they are not the masters. If they fail to sell it where they are now living, or if they lose employment by a change in the ownership or occupation of the land on which they work, they must move on. Their home is only secure to them from week to week. For all wage-earning classes, the modern conditions of industry are approximately the same. But in villages they are relieved by few of the compensations which to the eyes of country visitors appear to be offered in towns. In money wages, artisans are better paid ; they have greater chances of rising to higher rates of remuneration ; they have larger facilities for recreation and amusement ; so far as their homes are concerned, they are less directly under the thumb of their employer ; they belong to a less isolated and more numerous class ; they live in the midst of a population which is still minutely graduated in the scale of social position ; they have no excuse for imagining that laws of police, sanitation, health, and school attendance are designed and administered for the vexatious control of their social and domestic habits. Agricultural labourers believe that there is life in the towns ; they know that in the villages there is none, in which they share as a right, or which for them has any meaning. They may be indispensable, but it is only as wheels in another man's money-making machine."

No solution is yet in sight. The farmers still maintain that the imposition of minimum wage regulations has rendered farming unprofitable and that even subsidies on the present scale do not justify such increase in wage rates as will balance the rising cost of living. The reply of the workers, at any rate of the younger ones, is to leave the land at the first opportunity, and the first real revival of industry will strip the farms of their younger men, just as they get stripped locally to-day whenever some big public work gets started in a country district and calls for labourers. This is the gravest problem before the industry—to make the conditions of farming such that the workers can earn a better wage and at the same time to make the worker able to earn the better wage.

CHAPTER XXII

EDUCATION AND RESEARCH

THE development of agricultural education and research is so much a matter of the "last phase" of the policy that was adopted to pull the industry out of the great depression and of the renewal of that effort after the war, that it seems desirable to make one story of it.

Under the old dispensation such matters in England were left to private enterprise, and it was not until the setting up of the Board of Agriculture in 1889 that any State assistance was accorded to provide instruction for the farming community. Up to that time there existed one incorporated institution for education—the Royal Agricultural College at Cirencester, and one research organisation—the Rothamsted Experimental Station, which had been founded and endowed by Sir John Lawes. There were other private places of education, e.g. at Downton and Aspatria ; there was a Sibthorpian Professorship at Oxford which provided an occasional course of lectures, and the Royal Agricultural Society maintained some experimental work at Woburn correlated with the Rothamsted investigations.

The Royal Agricultural Society also conducted an annual examination in agriculture and the allied sciences, which provided a very valuable stimulus and direction to the self-education of young men taking up an agricultural career. The examination is still continued though its *raison-d'être* has largely disappeared with the widespread provision of centres of instruction leading to diploma or degree.

The earliest steps towards systematic education came with an Act of 1890 which empowered the County Councils to spend on technical instruction certain grants from the Exchequer known as the "whisky money." The more active counties initiated lecture courses and classes, and some of them combined to set up Colleges or agricultural departments within the existing Universities. To the contributions from the County Councils were now added the grants which the Board of Agriculture had been empowered to make, the original provision for which had been substantially increased. The University College of North Wales at Bangor was the first in the new field to obtain a grant from the Board of Agriculture, but during the last decade of the century other institutions for higher education were established in connection with the Universities at Cambridge, Newcastle, Leeds, Reading, and Aberystwyth, together with the South Eastern College at Wye, the Midland College and the Harper Adams College. This brought as teachers a number of young recruits from the Universities who became leaders in the scientific development of agriculture—Somerville, Middleton, and Gilchrist from Scotland, Wood and Biffen from Cambridge, Dunstan and Hall from Oxford. At first the farmers were indifferent to these new measures and new men, but the colleges soon established their position as educators of the rising generation and as centres from which lectures, classes, and experimental demonstrations were provided within their respective areas.

A great advance was, however, made possible by the Development Fund Act of 1909, a measure which agriculture owes to Mr. Lloyd George, who was Chancellor of the Exchequer at the time. By this Act the sum of two million pounds was set aside, and among the purposes to which it could be devoted were specified agricultural education and in particular research, which previously had not been recognised as a matter for State assistance. There had been some controversy as to which Department should be responsible for agricultural education—the Board of Agriculture, which had in accord with its Act administered the grants to the Colleges, or the Board of Education, but an agreement was reached in 1912 by which the Board of Agriculture became the authority for the education of persons above the age of 16 or 17, while the Board of Education continued to deal with any agricultural instruction given in schools.

The immediate result was to supplement the Colleges by the setting up of the Farm Institute scheme, under which a County Council could initiate or combine in initiating an organisation of short courses together with local lectures and classes and other methods of bringing technical instruction closer to the farmers. The scheme recognised either a Farm Institute, i.e. central buildings and a demonstration farm, or a looser organisation of instructors and demonstrators working under the direction of a County Organiser.

The Board of Agriculture made grants up to 75 per cent. of the capital expenditure in erecting and equipping the Farm Institutes, and of a varying percentage, 50 to 75, of the new current expenditure. The new scheme had barely come into operation when the war broke out, but after the Armistice the Government assigned fresh funds for the purpose ; the administration was simplified and passed wholly into the hands of the Board of Agriculture, except as regards capital grants to the Colleges. The Board repaid two-thirds of the expenditure of the Councils, and as some of the counties, often those most purely agricultural, were still unwilling to incur any expenditure on agricultural education, the grant for the salaries of agricultural organisers and horticultural superintendents was increased to 80 per cent. With minor modifications this is how the scheme for agricultural education still remains. The country is divided into provincial areas with an agricultural College or University department as its centre of higher instruction, though one or two provinces do not maintain a College. The Colleges provide the long courses of instruction over two or three years leading to a Diploma or a Degree and in some cases, e.g. Yorkshire, they are also the centres from which the extra-mural instruction, local lectures, etc., are organised. The work of the Colleges has been further strengthened by providing each of them with a group of advisory officers—chemist, entomologist, mycologist, economist, and (with some exceptions) veterinary officer, who are relieved of all but a modicum of teaching functions within the College but are available for giving advice and conducting local investigations for the benefit of farmers within the provincial area. These advisory officers also effect liaisons with the technical services of the Ministry, with the Research Institutes on the one hand, and with the agricultural organisers on the other. As regards what may be termed secondary agricultural education the County Councils still retain their autonomy, but they have established 20 Farm Institutes or Colleges giving one year courses, and with few exceptions they all possess an agricultural organiser with a team of instructors who are responsible for the direct contacts with the farming community. The scheme is elastic and the procedure varies, but the work is carried on by lecture courses, day and evening classes, discussion societies [which have proved a very stimulating method of securing the interest and co-operation of the farmers], field experiments and other organisations like the Young Farmers' Clubs. Co-operation with the colleges is ensured, as a rule, by making the organisers honorary members of the college staff and by periodical conferences to discuss the needs of the province and to arrange such matters as joint schemes for field experiments and demonstrations. Without doubt the scheme has been greatly helpful to the progress of agriculture in England, and though it may be objected that but a small proportion of even the present occupiers of land have passed through college or farm institute, some training of the kind is generally recognised as a necessary preliminary to farming. It should also be recognised that a very large number of agricultural enterprises in our Colonies are controlled from England, and that it is from our agricultural colleges they are able to obtain the managers and technical staff they require, instead of from American, Dutch, or German sources. The critics of our agricultural education deplore the relatively small numbers of ex-students of the colleges to be found farming on their own account, but they ignore the fact that farming in this country offers very few openings for young men without capital. It is the rarest thing to find a farmer, in however large a way of business, employing as assistant manager a young man trained in college or institute. Indeed the average English farm is rarely either extensive or intensive enough to provide a full time management job for even the farmer himself.

Perhaps there are too many colleges and institutes for the size of the community they are intended to serve and that better results, both educationally and economically, would be obtained with fewer schools each with larger numbers. But the farming public is very strongly localised and most of the good work that has been done has been initiated by personal contacts.

The organisation of research in agriculture followed the educational movement though with some delays owing to lack of funds. It has already been stated that England possessed the oldest agricultural research station, founded by Sir John Lawes at Rothamsted as early as 1843 and eventually provided by him with a permanent endowment. Thus at its Jubilee in 1893 Lawes and Gilbert had completed fifty years of fundamental investigations, on which the use of fertilisers all the world over had been determined. But it stood alone, and except for the work of Robert Warington, also associated with Rothamsted, of the Voelckers, father and son, as chemists to the Royal Agricultural Society, and of one or two individuals on the staff at Cirencester and Downton, and of chemists in professional practice such as Dr. Bernard Dyer, the output of research into agricultural questions was negligible. Some of the teachers at the new colleges, however, recognised that teaching must be based upon research, and investigation, however tentative, began to figure in the programmes of Cambridge, the Armstrong College at Newcastle, Reading, and Wye. Somerville started the experimental farm at Cockle Park, Wood began experiments on bullock-feeding at Cambridge where Biffen was also taking up the breeding of wheat, a question which obtained the collaboration of the Homegrown Wheat Committee of the National Association of British and Irish Millers, so providing the scientific workers with much needed information as to the qualities of wheat desired by the trade.

After the death of Gilbert in 1901, Hall, in 1902, became Director of Rothamsted, where the income from its endowment only permitted of one other trained chemist, though a donation of £10,000 from the Goldsmiths' Company and other private benefactions did within a few years enable it to recruit its present Director, Sir John Russell, and Dr. H. B. Hutchinson as bacteriologist. From one source or another the output of original work became sufficient to justify in 1904 the starting of the Journal of Agricultural Research, which has remained the main organ of publication for agricultural investigators not only in England but within the Empire.

Up to 1909 the research workers received no direct assistance, and were dependent upon such resources as their educational laboratories could supply, but with the passing of the Development Fund Act means were provided for the definite organisation of agricultural research, to which end Hall from Rothamsted became one of the original Commissioners. After some discussion as to the method to be followed, it was decided to divide the very wide field of investigation into subjects and to set up for each of the chief branches a Research Institute as the main centre for work of that description. Thus Rothamsted was recognised as the centre for the investigation of soil problems and the nutrition of the plant, at Cambridge two Institutes dealing with Animal Nutrition and Plant Breeding respectively were set up, the Dairy Research Institute was associated with the College at Reading, an Institute for investigation into Agricultural Economics was founded at Oxford, and the Fruit and Cider Institute at Long Ashton was enlarged to deal with investigations on fruit growing.

A further sum of money was allocated annually for grants in

aid of special researches which are proposed by scientific workers at such Universities or Colleges as are not in receipt of research grants.

One other part of the original research scheme was the provision of post-graduate scholarships, the object of which was to train workers for the Research Institutes by selecting each year a few graduates in the pure sciences bearing on agriculture and giving them three years' training in research in the first instance at home and then abroad. How valuable this scheme has been may be seen from the fact that among the earliest pre-war scholars were Engledow, now head of the Agricultural Department at Cambridge, Hanley, who occupies a similar position at the Armstrong College, and Neville, also head of the Department at Reading.

It should be noted that in each case the Institute is controlled by a University or other public body, the only one directly under a State Department being the Veterinary Research Station at New Haw, which was necessary to the Board of Agriculture in view of the administrative action it has to take with regard to a number of animal diseases. The need for a further independent Institute dealing with disease in animals was recognised, but a conclusion as to its location had not been reached when progress was stopped by the outbreak of war. These two principles, the division of Agricultural Research into subjects for each of which a central Institute was provided, and their independence of direct State control, has been maintained and has been found effective.

During the war most of the staffs of the Research Institutes were detached for war work of some kind or another, but at the conclusion the Board of Agriculture and the Development Commission hastened to strengthen the original scheme, for which purpose the Government of the day made a fresh allocation of two million pounds for education and research. The embargo on new expenditure consequent on the economy campaign of 1920 was relieved in the following year, when the Corn Production Acts were repealed, and as some solatium for the withdrawal of the subsidies on the growth of wheat and oats, the Government appropriated a further £850,000, to be spread over a period of five years, for education and research. This sacrifice of immediate benefits for the ultimate advancement of the industry was mainly due to Mr. R. R. Robbins, at that time Chairman of the National Farmers' Union. It may be noted, though a matter of education and not of research, that at the same time provision was made for a system of scholarships to Farm Institutes and Agricultural Colleges for the sons and daughters of agricultural workers, scholarships which cover all the costs of such students and have proved of great value in bringing forward latent ability.

The subsequent history of the research scheme is chiefly concerned with administration. In 1930, as the outcome of protracted discussions, an Agricultural Research Council was set up under the aegis of the Lord President of the Council, as are the other two bodies administering research, the Department of Scientific and Industrial Research and the Medical Research Council. The Council consists of nine members chosen for their scientific distinction with five others of more immediate contact with agriculture ; it possesses funds of its own from which to make grants and supervises the whole field of agricultural research. It would be difficult, and indeed unnecessary, to explain the complicated relationship between the Council, the two agricultural departments in England and Scotland, the Development Commission, and the Treasury and the Privy Council, but the Council now supplies for research on matters agricultural a power of initiative and enlightened criticism that before was lacking.

Rothamsted, the oldest of the research stations, remains the most extensive in the country ; in addition to its earlier reference to soil and plant nutrition problems it has become the centre for fundamental mycological and entomological work, thus dealing with the plant in disease as well as in health. It still continues to study the newer fertilisers and manurial problems, especially in relation to quality. The investigations into the bacteria of the soil have resulted in working out a process by which straw and other vegetable waste can be converted into farmyard manure without the intervention of animals, a matter of importance to market gardeners now that so little stable manure is obtainable from the cities. These investigations also explain the efficacy of the compost-making process of the Chinese cultivators, who have been able to maintain their soils indefinitely at a high pitch of fertility. Howard's scientific adaptation of these methods seems likely to play a great part in providing fertilisers for tropical and semi-tropical agriculture, and incidentally for the disposal of night soil. Another investigation with important practical results has worked out the conditions whereby lucerne becomes inoculated with its associated nitrogen fixing bacterium, and has been able to put on the market a trustworthy medium which enables proper stands of lucerne to be obtained in districts where the plant had not previously been grown.

To Rothamsted again belongs the credit of introducing the treatment of soil by heating and antiseptics, which has become standard practice among all cultivators under glass. Of less direct application, though essential to the investigator and adviser, have been the studies of the physics of the soil and again of the statistical treatment of the data obtained in agricultural and indeed in all biological experiments.

To the two Institutes at Cambridge a third has been added dealing with Animal Pathology, as well as lesser establishments for investigations on the breeding and nutrition of poultry and on virus diseases in potatoes. But great developments at Cambridge will follow from a recent benefaction from the Rockefeller Foundation which will provide for the more fundamental research in the biological sciences bearing upon agriculture. The Plant Breeding Institute under Sir Rowland Biffen has been the means of introducing to farmers two new wheats, now widely grown—Little Joss and Yeoman, the one of strength approximating to the Canadian hard wheats, the other of special usefulness to growers on the lighter soils of the Eastern Counties. Some of the later introductions both of wheat and oats have failed to obtain general recognition, indeed the long processes of trial and selection to which English cereals had been subjected in pre-scientific days had already raised their cropping powers to a point approaching the maximum.

In some respects complementary to the Plant Breeding Station and alongside it at Cambridge is the National Institute of Agricultural Botany, which serves as an intermediary between the plant breeders and the seed merchants, testing the new varieties of promise and if approved multiplying them to the point of commercial distribution. This Institute is also in charge of a scheme for accurately testing the varieties of all farm crops, to which end there are six sub-stations in other parts of the country. These trials have been of the utmost value to farmers in providing definite information on the value of new introductions as compared with the standard varieties, and in unmasking the synonyms under which old varieties are put out as specially good novelties.

Cambridge is not, however, the only station concerned with plant breeding ; at Aberystwyth under Professor Stapledon another station deals more particularly with grasses and clovers, which play an even greater part in British agriculture than all the other farm crops. The chief work of the station has been the isolation, testing, and eventual multiplication of strains of the principal grasses and clovers which possess exceptional value for grazing because of their leafy and perennial habit.

Commercial seed saving of such crops is bound to effect an unconscious selection of such types in the natural mixed population as are specifically early and prolific seed bearers, whereas the grazier does not want seed production but continued leafy growth. Some of Stapledon's selections are already being cultivated on a commercial scale so as to yield guaranteed seed, and have also been taken up in Australia and New Zealand. In a more general way Stapledon and his colleagues have done much work on the improvement of pastures, especially of hill pastures, the productive value of which could be greatly increased by a minimum of expenditure and careful management.

As regards animal nutrition T. B. Wood completed a very thorough study of the nutrition of the pig, which brought out the relationship of food to increase at the various stages in growth both as regards the composition and amount of food. These investigations indicate to the practical man the waste involved by allowing any " store " period in pig-feeding, the critical stages of the fattening process, and the progressive changes in the ration with age and the necessity of concentrated digestible foods. The other side of the nutrition question —the necessity of mineral and vitamin food accessories, is chiefly studied at the Rowett Institute in Aberdeen, between which and the Cambridge Institute there is regular collaboration. These researches have provided an exact basis for the economic feeding of pigs for either pork or bacon. The study of the fattening process in bullocks has not proceeded so far, but should none the less be the means of eliminating a great deal of waste in current practice.

The Cambridge Institute is also to be credited with some brilliant work on the physiology of reproduction in domestic animals, work, however, which does not lend itself to summary, though the importance of such investigations is apparent when the losses caused by sterility and infertility in pedigree stock are considered.

The investigations at the Dairy Research Institute at Reading have been chiefly noteworthy in supplying the scientific basis for the improvements in the methods of handling milk which have been so marked since the war. The delivery of milk in the first place to wholesalers and then to the public is now under bacteriological control, and to Reading must be attributed in a large measure the success with which the campaign for " clean milk " has met.

The investigation of animal diseases has been greatly extended since the war, and even at the time of writing further additions are being considered. Allusion has already been made to the Ministry's laboratory at Weybridge, in addition to which is the much increased staff at the Institute at Cambridge and work at the Royal Veterinary College and at the Veterinary School of Liverpool University. Such work is necessarily expensive because the experiments have to be made upon large animals, costly to buy and maintain ; it is also slow because even when the nature and sequences of the disease have been worked out there is rarely a " cure " to be suggested. Instead, further work is necessary to establish the conditions of hygiene and nutrition which will prevent or reduce the incidence of disease. Even breeding should enter into the question, though as yet little work has been attempted with the larger domestic animals.

As regards poultry, now an important item in English agriculture, the Research programme has been divided. An Institute has been set up in connection with the Harper Adams College, and while its reference is largely educational it conducts investigations into methods of management and other practical issues. Work on breeding problems is entrusted to the Professor of Genetics at Cambridge, nutrition to the Institute for Animal Nutrition there, and on disease to the Ministry's Laboratory at Weybridge.

No branch of research has achieved greater success and utilisation by the industry than that concerned with fruit growing, which is dealt with by two Institutes at Long Ashton near Bristol and East Malling, Kent. The practical results obtained will be dealt with elsewhere, but the investigations have done much to define varieties, to determine the type of growth and production to be associated with particular stocks, to put the manuring of fruit on the same experimental basis as that of wheat and other farm crops, and to bring system and economy into the spraying problem. The John Innes Horticultural Institute has contributed work on the relation-ship of varieties to ensure proper pollination, and on the laws of inheritance that govern the breeding of fruit trees. Somewhat in the same field of intensive production the Glasshouse Research Institute at Cheshunt has proved of the utmost value to the growing industry of cultivation under glass.

Investigations into agricultural machinery are conducted in an institute at Oxford, where also is placed the Institute dealing with the Economics of Agriculture. This subject, which previously had been almost entirely ignored in Great Britain, became immediately of topical interest as soon as price-fixing and control of production became necessary during the war, when the few data possessed by the Institute became the only material that had been collected without specific reference to the situation. This call for data on the economics of agriculture has prevailed down to the present day, for the agricultural situation and with it political action has been changing radically every few years. The necessity of collecting and analysing information about costs of production and the returns from different systems of farming is now seen to be essential to the direction of the industry, and that has been recognised by the extension both of the Institute and the Advisory service of the colleges, and again by the formation of a branch of economic enquiry within the Ministry itself.

From this brief outline it will be seen that England, together with certain cognate institutions in Scotland, now possesses an organisation for research into the scientific development of agriculture that in a comparatively short time has become superior to any parallel organisation in Europe. Though in magnitude it cannot compare with the service possessed by the United States, in the quality of the work done and in the reputation obtained by the workers it commands universal respect. It may even be held that the organisation is excessive considering the relatively small part that agriculture now plays in the economy of the nation, but this is far from being the case when one considers how fundamentally agricultural are the Colonies and Dominions associated with Great Britain. Already the research organisations at home are associated with the parallel services in the Dominions and Colonies by a series of Empire Bureaux which serve as clearing-houses for the exchange of information on the work accomplished at each centre and for the dissemination of results. Exchange of workers between home and overseas laboratories is arranged for, joint investigations are in progress.

As far at any rate as the Colonies are concerned Great Britain must be the training ground for the scientific men who are required not only for the Government services in agriculture, but for the great technical enterprises of a tropical and semi-tropical nature.

It is eminently the function of Great Britain to make itself the cultural centre of the Empire, as the centre where the fundamental scientific basis of all research is most intensively pursued, to which workers from all parts can resort for their final training and from time to time for the renewal of their contacts with other workers in their particular subjects.

CHAPTER XXIII

TECHNICAL PROGRESS SINCE THE WAR

BEFORE considering the changes in farming practice that have been in progress since the close of the war it is necessary to set out, however generally, the rapidly varying economic situation which has prompted farmers to successive attempts at adaptation to the altered conditions. The war period of abnormal prices came to a sudden end in 1921 with the return to the gold standard. The effect may be seen in the Price Index of Agricultural Products, which was as high as 292 in 1920, but fell to 169 in 1922. The second great break in prices came towards the end of 1929, when the price index, which had averaged 149 for the five years 1925–29, fell to 134 in 1930, 120 in 1931, and 112 in 1932, although in September 1931 the country went off the gold standard and the pound had depreciated by over 30 per cent.

The graph, page 450, and the table in Appendix show the trend from 1919 onwards of the price indices for all commodities and for agricultural produce, also for fertilisers and feeding stuffs, together with the year's average prices of wheat and fat cattle (second quality). On the same graph are also set out the weekly wage rates.

The graph shows the two great breaks in prices which began in 1921 and 1930 respectively, but also shows how closely the fluctuations in prices of agricultural produce agreed with those common to other commodities. If anything, agricultural products suffered least ; the decline in the prices of feeding stuffs and fertilisers have been significantly greater than the fall in the products into which they are converted by the farmers. Both of these requirements of the farmer were in 1935 more than 10 per cent. cheaper than in the 1911–13 period, whereas agricultural produce generally was more than 20 per cent. dearer.

The course of wage rates, however, shows a very different sequence

roughly speaking, they are now twice as high as they were in the
pre-war period. Against this may be set off the virtual extinction

GRAPH SHOWING THE COURSE OF PRICES AND WAGES, 1919-35.

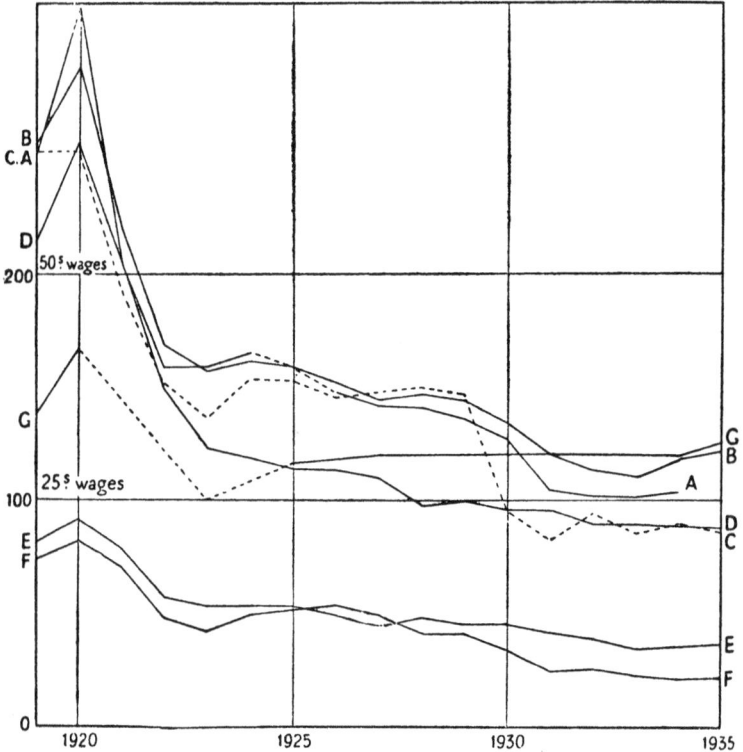

A. Index numbers of Wholesale Prices of all articles (Board of Trade).
    1913=100.
B. Index numbers of Prices of Agricultural Produce (Ministry of Agriculture).
    1911–13=100.
C. Index numbers of Prices of Feeding Stuffs (Ministry of Agriculture).
    1911–13=100.
D. Index numbers of Prices of Fertilisers (Ministry of Agriculture). 1911–13=
    100.
E. Price per live cwt. of Fat Cattle, 2nd quality (Ministry of Agriculture).
    Shillings.
F. Price per imperial quarter of English Wheat (Ministry of Agriculture).
    Shillings.
G. Minimum Weekly Wage rate for Norfolk. Shillings.

of allowances, which were considerable in some parts of the country ;
on the other hand, overtime payments have become more general.

Since labour may account for from twenty to forty per cent. of a farmer's total expenditure, the lack of parity between wage rates and the returns to be obtained for the output of the farm has been the chief source of the farmer's difficulties in the post-war period, especially as his returns are always subject to a considerable "lag" between expenditure and realisation. The farming community naturally blames the legislation, fixing minimum rates of wages, for their troubles, but without this protection the exodus of workers from the land would have been much greater. Even as it is, the younger men in the country make every effort to get employment outside farming, though the position of the agricultural workers must be regarded as considerably improved since the pre-war period, indeed as better than that of most unorganised labourers.

Seriously as farmers have been affected by wages in the more recent years, the extra cost of labour has been more than set off by the reliefs and subsidies accorded to agriculture. According to Dr. J. A. Venn, the relief of rates, protection, and direct subsidies are adding about forty millions to the income of farmers, whereas the statutory regulation of wages cannot have added more than ten millions to their expenditure. It is against the economic background thus briefly outlined that the following chapter on the technical changes in farming since the war must be read.

The course of the war had brought about a great extension of the use of tractors among farmers; the Food Production Department had imported large numbers to mitigate the shortage of men, and these were sold cheaply as soon as the Cultivation Orders ceased to operate. For a time the opinion was general that tractors were going to supply the chief motive power for the farm, but as prices fell and wages also declined, there was a sharp reaction against their use. The great value of the tractor lies in its economy of man-power: it is active for more working hours in the day, because its driver does not require the hour at each end of the day which the horsekeeper needs. A tractor, too, will run overtime as horses cannot, consequently when wages are high and oats and hay are dear the economy of the tractor is manifest. But since the war-time tractor was still an imperfect machine and its drivers comparatively unskilled, farmers found their repair bills mounting up excessively. For a time mechanical power fell into disfavour and horses again prevailed. But by degrees much better and more powerful tractors became available to farmers, who, themselves, universally had adopted motor cars for their business purposes and so had acquired a better knowledge of the care of such machines and a capacity for executing simple repairs and adjustments. The use of tractors has thus increased again, though horses have not been entirely displaced nor greatly reduced in numbers on the ordinary mixed farm. Tractors are essential because of the area they can cover at critical times, both in the preparation of the land and in harvest, but they are employed as supplements to horses.

A greater revolution in the application of power to the farm came later as the capacities of the " combine " machines used in America and Australia for harvesting cereals became known in this country. These machines not only cut but threshed and bagged the crops as they went along, the typical machine being a header which left the greater part of the straw standing. The earlier machines to be imported were not very satisfactory, because they had to deal with crops heavier and longer in straw than were general in the country for which they had been designed. The English farmer also was not prepared to leave the straw on the ground. Another objection to their use was that under English conditions the corn threshed in the field is rarely dry enough to go to the mill or into store, but has to be put through some artificial drying process. Gradually, however, machines have been constructed more suited to our conditions, in some of which the threshing and drying are dealt with as separate operations. For example, one machine cuts the crop and delivers it into a carrier. The carriers, when full, deliver to the threshing plant standing at the side of the field ; this bags the grain and passes the straw on to a baler. The grain then goes to a dryer adjoining the granary, where the grain is stored in bulk. Thus the harvesting is completed in one series of operations, there is little man-handling of the crop, no delay, and no rick building. A certain number of farms in East Anglia, Hants, and similar counties have been almost wholly mechanised for cereal growing. They have shown themselves able to produce cereals at very low costs, comparable with those of Canada and Australia (see *Studies in Power Farming*, Oxford Agricultural Economics Research Institute, 1935). Probably for really economic results 500 acres or over must be under crop, and the system has not been long enough at work to show how far the fertility and cleanliness of the land can be maintained without farmyard manure. But the experience of Mr. Baylis before the war, working with horses, demonstrated that continuous corn-growing can be maintained without visible deterioration of the land, provided that a clover crop is interpolated every five or six years. None the less, the decline in the arable acreage still continued with the high wages and low prices that prevailed. In England and Wales since the war nearly three million acres have been put down to grass or otherwise lost to arable farming, between 1919 and 1935.

The losses have been very marked with turnips and other fodder crops ; the close folding of arable land sheep which had been so characteristic a feature of farming upon the chalk and other light lands became too expensive in labour. Between 1914 and 1919 the sheep stock of Great Britain fell from 24·3 to 21·5 millions, and taking comparisons between 1912 and 1919 a group of predominantly grazing counties in England lost only 13 per cent. of their sheep, whereas a corresponding group of arable counties lost 25 per cent., and the three counties of Hants, Wilts, and Dorset, the home of the Downs folded upon green crops, lost 37 per cent. Not only are the successive cultivations for the fodder crops expensive in themselves, but the regular moving of the close fold, watering, etc., consumes much labour. Again it began to be shown that the residues from the intensive feeding of the green crops, hay, cake, etc., were by no means recovered in the succeeding crops, at least not to the extent of the values that should be charged against them. So when the sheep stock began to grow again the Downs and other arable land sheep in the southern counties of England were very generally replaced by grass sheep—by Cheviots, Exmoors, and especially by the Scottish half-breds of Cheviot—Border-Leicester parentage, which would fatten on grass alone with very little artificial feeding. The turnip crop, which had always been somewhat untrustworthy on the heavy soils and in Southern and Eastern England, because of fly in spring and mildew in the autumn, largely fell out of favour. For a time, about 1920, there was a considerable recourse to silage made either in the wooden silos imported from America or round silos of reinforced concrete ; a mixture of oats, tares, and beans being the crop most favoured for preservation as a succulent fodder for the winter feeding of milch cows. But even silage making involves a good deal of labour and to-day the silos are little used except for an excess of grass in a wet season. A considerable revolution took place in the cropping of the arable land of England ; the old rotations, whether founded on the Norfolk four-course or the Wiltshire plan of two years of green crops followed by two of corn, have given place to an alternation of a temporary ley of three or four years duration followed by a short succession of cereals. Originally this was a system designed by the Border farmers, who learned thus to utilise the capacity of wild white clover to collect nitrogen for succeeding crops, but it proved to be equally valuable to southern farmers on the second class soils where permanent pastures take years to establish and even then are rarely so productive as the temporary leys. A cheap seeds mixture is used, consisting mainly of perennial rye grass and cocksfoot, with the essential addition of wild white clover. The threatened difficulty of keeping the land clean under such an alternation of grass and cereals was found to be exaggerated if the cultivations are duly timed.

Another important development of grass farming was initiated by Warmbold of Stuttgart during the war. Having to maintain a large milk herd when artificial feeding stuffs were not available to supply the necessary protein, but synthetic nitrogen fertilisers could be obtained in quantity, he worked out a system of intensively manuring the grass-land in order to grow protein. Normally, there is a flush of rich grass in May and early June, after which the pasture loses its high-feeding value both in quantity and quality. Warmbold aimed at renewing the spring flush by repeated applications of nitrogenous fertilisers, and finally he arrived at a system of small paddocks which are successively manured and intensively eaten down in rotation in a few days, a fresh application of fertiliser being given when the stock leave each paddock. Thus the stock are always feeding upon young grass, which should never be allowed to shoot to a height of more than six inches or so. The cost of fencing and bringing water, the amount of care required to regulate the movements of the stock, the variation in production due to the weather, have prevented the widespread adoption of Warmbold's system in its full development. The two features of the system—the power of extending the period of good grazing, of obtaining an early bite in particular, and the value of the close rotational grazing of small paddocks in improving the pasture, are being realised by all thoughtful graziers and, in one form or another, are being incorporated into their practice. Another aspect of the case is receiving a good deal of attention. One of the difficulties experienced in the strict system outlined above is that in a good growing season the rate at which the grass is produced in the succession paddocks is greater than the stock can consume, so that it is desirable to cut one or more of the series before the grass gets too long, though it is as yet too short to be dealt with by the ordinary haymaking processes. Experiments at Cambridge showed that this material, if dried, could be compressed into a cake with as much as 40 per cent. of digestible protein, a concentrated feeding stuff comparable in value with cake—indeed, in some respects, superior, for it retains the vitamins and other stimulants of the young grass. This at once suggests a means of rendering a grass-land farm self-supporting throughout the year, manufacturing its own concentrates for supplementary feeding and winter production. The farmer would be able to subject the whole of his grass-land to rotational manuring and grazing without the risk of overstocking in a dry season, because there will always be a considerable margin of land that is to be cut at the right stage for drying. To realise this system, the farmer needs machinery that will catch up the short grass, and an efficient drying plant, cheap enough to be installed on a farm with two to four hundred acres of grass. One or two large scale plants, intended to produce this dried grass or lucerne fodder commercially by purchase of the raw material from farmers, have been put in operation and various patterns of smaller dryers are under trial, but the great potentialities of this method of grass-land production have not passed the experimental stage. As an alternative to drying a new process of making silage in stacks in the open field has been introduced from Sweden, the C.A.V. process in which each layer of green fodder as it is piled up in the stack is sprayed over with dilute acid. The presence of the acid changes the type of fermentation ; there is much less destruction of digestible carbohydrate, less degradation of the proteins, and the vitamins are unchanged.

Thus two types of handling grass-land have been evolved, the older one depending upon the use of basic slag and the growth of wild white clover, *i.e.* natural regeneration at small cost, and this latter intensive method in which the growth is raised to a high pitch by intensive manuring. It is, however, depressing to see how great a proportion of our most considerable agricultural asset, our grass-land, is still left uncared for, unmanured, undrained, unmanaged, and understocked.

The improvements in wheat varieties has already been dealt with under research, the parallel and indeed perhaps greater improvements in barley have almost wholly been the work of one man,

Dr. E. S. Beaven of Warminster. His work began by establishing the value of pure lines extracted from the old English barley, "Archer," but he passed on to raising a number of hybrids. At the time of writing the malting barley of the country is almost wholly either his Plumage-Archer or a similar cross—Spratt-Archer, raised under Beaven's inspiration by Hunter of the Cambridge Plant Breeding Station. Unfortunately the demand for brewing barley, of old one of the great sources of revenue on the lighter lands of the Eastern Counties, has declined with the reduction in the consumption of beer and the reluctance of brewers any longer to pay high prices for what they were accustomed to regard as "quality" in barley. Though the oat crop has become of less importance with the decline of horse traction, it is still the most extensively grown cereal. The two varieties which have most generally established themselves at the present time, Victory and Golden Rain, both originated at the Swedish Plant Breeding Station at Svälof. The Scottish varieties, Potato and Sandy, richer than any other oats in proteins and fats, have largely been displaced owing to their lower yields and to the competition of the prepared cereals with the old time Scottish oatmeal.

Among the fodder crops, marrow-stemmed kale, which had been introduced before the war, has become general in place of turnips and swedes, since it will stand a fair amount of frost. The great improvement in potato cultivation had been effected before the war and no particular change in management can be recorded, though machines for lifting the crops have come into general use. New varieties continue to be produced; Up-to-Date has almost disappeared and given place to King Edward, Majestic, Great Scot, and several varieties originated by McKelvie of Arran. It used to be supposed that potato varieties "wore out" with age, but recent research has demonstrated that the degeneration is due to the accumulation of diseases of the virus type in a plant that is propagated vegetatively. The value of Scottish and Irish grown seed is due to the fact that the aphides which transmit the virus from one plant to another are absent or rare in these cooler localities where seed potatoes are grown. In growing potatoes for seed, special care is now taken in roguing out visibly infected plants so that seed can be supplied with a guarantee of a small proportion of infection. This seed can be grown in the English potato growing areas for two or three years before the virus accumulates and the yield declines below the commercial limit. The greatest advance made in the potato industry has been the discovery towards the close of the war of varieties immune to what was then a relatively new disease of potatoes, wart or black scab (*Sinchytrium Endobioticum*). This disease had been observed many years before, but it had attracted little attention because it was only spreading slowly in allotment grounds in the industrial areas of Lancashire, the Midlands, and South Wales. But the scarcity of potatoes in 1916 induced the use of potatoes as seed from all sorts of areas, and the disease was suddenly discovered to be a menace to the maintenance of the potato crop. Its danger comes from the complete ineffectiveness of any form of treatment, but even more from the fact that the infection persists in the soil in which a diseased crop has once been grown for an indefinite period, even when the land is laid to grass. Quarantine regulations were at once applied, defining the areas in which the disease had been detected and forbidding both the sale of seed from these areas and the sale of ware except within the scheduled areas. These measures would only have delayed the spread of the disease had not the discovery of occasional healthy plants on some of the worst infected ground revealed the existence of varieties immune to the attacks of the disease. With these varieties as a basis the plant breeder got to work, immunity proved to be a heritable quality, and by degrees a series of immune varieties has been evolved which alone may be grown in the scheduled areas. There are still a few susceptible varieties possessing qualities which maintain them in cultivation ; before long only immunes will be grown, and even now this formidable disease has become of little importance.

For several years before the war, thanks mainly to the enthusiastic propaganda of the Earl of Denbigh, experiments on the growth of sugar beet were carried out in England and demonstrated that good crops could be grown here with a sugar content comparing favourably with the beet supplied to the continental factories. Capital, however, was not forthcoming to start the industry, and the first modern factory was erected at Cantley in Norfolk by a group of capitalists connected with sugar production in Holland. This company operated for four campaigns, 1912–15, but though in 1913 a grant of £11,000 from the Development Fund was spent in assisting the cultivation and transport of the crop, the losses were heavy and in 1916 the company was wound up and the factory was sold to a new company embracing both English and Dutch capital. The high price to which sugar rose during the war (90s. per cwt. as compared with 12s.–15s. in preceding years) induced the Government to purchase a large estate at Kelham, Notts, for the eventual purpose of an experiment in sugar production on a commercial scale. No attempt was made to grow sugar during the war and afterwards a portion of the estate was passed over to the Board of Agriculture for the settlement of ex-service men, but with further Government assistance a new company in 1920 took over the estate and erected a factory which began operations in November 1921. Meantime the Cantley factory resumed work a season earlier.

Almost as these factories got to work the break in prices came and that of sugar fell at the close of 1921 to about 40s. per cwt. (duty included), so that the new industry was only enabled to carry on by the remission of the heavy excise duty of 25s. 8d. per cwt. This precarious assistance was exchanged in 1924, when the Customs Duty on sugar was reduced to 11s. 8d. per cwt., for a direct subsidy for a period of ten years, for four years at the rate of 19s. 6d. per cwt. of sugar produced, 13s. for the next three years and 6s. 6d. for the remainder. At the same time the home-grown sugar was to enjoy the Empire preference of a remission of half the excise duty. With this liberal assistance progress was rapid, and by 1934 as many as 18 factories were in operation (two in Scotland) handling the beet from 403,884 acres. During the eleven years, 1924–34, the subsidies have amounted to £34,541,642 : the excise remitted is estimated at £13,350,000. As the subsidy period drew towards its end it became evident that the new industry still could not maintain itself in the face of the immense increase in sugar production throughout the world and the low price[1] to which sugar had fallen in the open market. Temporary assistance was accorded to the crops of 1934 and 1935, and at present there is a Bill before Parliament to stabilise the industry by bringing all the sugar manufacturing companies into a single corporation. It is proposed to subsidise a standard production of 560,000 tons of white sugar at the rate of 5s. 3d. per cwt., this rate being subject to variation according to variations in the cost of manufacturing and the price of sugar in the open market. The total cost to the Exchequer is not expected to exceed £2,750,000 per annum. In common with other home-produced commodities the sugar will also enjoy the remission of one-half of the excise duty.

It must be held that this great and costly experiment has so far

[1] In 1934 the open market price of raw sugar, which had been as high as 25s. 9d. in 1923, had dropped to 4s. 9d. per cwt.

failed to fulfil the expectations with which it was launched. Looking at it from the standpoint of the older agriculture, it was thought that sugar beet might find a place in English farming as a cash crop in place of the turnip break in the lighter lands of the eastern counties, on which the farmers were losing money. The turnip crop was expensive to grow and could only be realised by turning it into beef or mutton, both doubtful speculations. Even if the beet could only be sold for £10 or £12 an acre, the four and five course farmer would get a better and an immediate cash return for a crop which ensured the cleaning and manuring virtues of the turnip crop. But the English growers of sugar beet have regarded it as a crop that must produce a profit *per se*, something comparable to potatoes, which means that the beet must be sold at about 40s. per ton. What, however, the original promoters could not foresee was the drastic change in the status of the beet sugar industry. Before the war beet and cane sugar shared the world's production equally, but as wholesale destruction fell upon the chief beet-growing areas of Europe an immense stimulus was given to the production from cane. The United States developed the industry in Cuba, Hawaii, Porto Rico, and the Philippines, and put both cultivation and manufacture on a new basis of efficiency. Even more telling was the work of the Dutch sugar experts in Java, where not only was the production extended but research developed a new breed of " noble " canes with a greatly enhanced yield of sugar. Thus the cane sugar industry emerged from the war period so improved that sugar from beet could no longer hope to compete on equal terms. By excessive protection and other fiscal devices the European countries, whose agriculture was largely based upon the beet crop for which no equivalent was in sight, have restored their acreage under beet, but it has become an artificial industry maintained because upon it is based the farming system of the peasants who constitute so large a proportion of the working population. There does not appear any prospect that research will be able greatly to increase the yield of sugar from beet nor to reduce the costs of cultivation and manufacture, dependent as are both of the latter on relatively cheap labour. In England it has been disappointing to find that little improvement has been obtained in the yield per acre, which is well below that of the better continental countries ; this is probably because our farmers have never been able to give the hand labour in singling and hoeing that the European crops receive. What can indeed be said of the experiment is that it has provided a much-needed source of income for the arable land farmers during the worst of the post-war depression ; without the cash return from beet many of them must have gone to the wall. As to the future, the whole course of international trade and of world economics is in such a state of flux and the outlook so uncertain that it may be wise to preserve this form of internal production in reserve. The valid argument against it is that since we can only grow a portion of the food we consume we should not select for subsidy a form of production that can be well supplied by our Colonies, one too that considerations of national dietetics would put in a secondary position—indeed, might regard as detrimental in so far as good land which should be devoted to the more needed vegetable crops is being diverted artificially to beet.

It is not possible to point to any particular discovery or new method which characterises either the extension or the improvement of market gardening since the war. Progress has been brought about by the accumulation of a number of small modifications of technique which were already known ; in fact, by greater attention and better practice. While we possessed market gardeners whose methods were not to be excelled, much of the production of vegetables was in the hands of small men who had come into the business with little knowledge and without the guidance of a good tradition. Our private gardeners might turn out magnificent examples of vegetables, but they had been grown irrespective of cost. Their skill and practice was of little service to the commercial producers, whose methods of a generation ago were generally of a rough-and-ready description when compared with those of the French, Belgian, or Dutch market growers. The war brought many of these skilled Belgian workmen as exiles into England, and from them some of the continental technique was picked up. Two centuries ago, with the exiled Huguenots, the good tradition of French market gardening had come to the London district, where descendants of theirs are still at work ; just as a century earlier the wars in the Low Countries were the means of teaching Englishmen the new crafts on which our modern agriculture was founded.

The post-war period saw also great progress in the preparation and presentation to market of English fruit and vegetables. When plums have to be sent from California, cauliflowers from the Netherlands or Italy, apples from Tasmania, they must be carefully selected and packed ; even the style in which they are done up must be made attractive if only to compensate for the slight loss of freshness. Broccoli from Roscoff came in a simple crate, all of a size and neatly trimmed ; the Cornish broccoli with which they were in competition were thrown into a sack or barrel just as they were picked. It may be claimed that in a great measure the English growers have now learnt their lesson, the best of them, indeed, are well ahead of their foreign competitors who set the pace. The adoption of these improved methods has been largely due to the instruction and propaganda carried out by the technical staff of the Ministry. The introduction of the National Mark for certain classes of vegetables and fruit has again been effective in grading up, though it has not yet received the support it deserves either from the trade or the consumers. Latterly, the adoption of protective policies has supplied another stimulus to market gardening, especially to production of the finer kinds of forced or semi-forced vegetables, information about the growing of which has been spread about not only by the Ministry, but by experts like Mr. F. M. Secrett, who has been generous in sharing his experience. The market gardeners proper are, indeed, being forced into more skilled methods, for wherever the soil and climate are suitable farmers are treading upon their heels, putting in a field or two with the more easily grown vegetables like carrots or brussels sprouts, which are always utilisable by sheep if the market is not favourable.

The improvements in fruit growing are even more manifest, and are greatly due to the work of the two Research Stations at East Malling and Long Ashton. The newer plantations have chiefly been laid down in the eastern counties, but everywhere the better knowledge of soil and the fertiliser requirements have extended the range of soils upon which fruit can be grown commercially.

The first step has been the characterisation of varieties and stocks, and the establishment of sources from which the intending planter can obtain true material free from disease. Then has come the extension of orchards under cultivation, though the neglected grass orchards of the West Country are still too numerous. Cherry gardens must still be grassed down. The old mixed plantations of apples interplanted with plums and bush fruit at still narrower intervals are no longer in vogue, for with the mixture cultivation is impeded and a proper spraying programme cannot be carried out. Plantations must be of one kind of fruit only, including such a limited mixture of varieties as will secure pollination. Spraying methods and power-spraying plants have enormously improved. Of old, the industry was somewhat of a compromise between the easy-going methods of the farm orchard and uncommercial individualism of the private gardener ; America showed the way to organise fruit growing, but now our better fruit growers have little to learn from other countries, though there is still much that they must obtain from research.

Both fruit and vegetable growing have received a considerable stimulus through the development within the last twelve years or so of the canning industry. Twenty years ago, with the exception of a few specialised products, all canned vegetables and fruits sold in this country came from abroad, mainly from America. To-day the great bulk of the trade is English, except for such products—Citrus, Pineapple, Peaches—as cannot here be grown commercially. At the present time there are 71 canning factories at work and their output has risen to over a hundred million cans a year, representing 800,000 cwt. of fruit and vegetables. Plums, strawberries, raspberries, and loganberries constitute the bulk of the fruit, peas are the most important vegetable crop.

Space does not permit of entering into details about the development of cultivation under glass. It has been a response to the growing change in the habits of the English people, in some respects to the better wages of the working classes, but tomatoes, cucumbers, and cut flowers in their season have become almost " necessaries " in our daily life. A consequence has been the growth of businesses of which one of the most notable examples is that of Messrs. Lowe & Shawyer, who began in 1864 with two small greenhouses and in 1906 had 246,000 square feet of glass and movable lights to cover 140,000 square feet. These figures have now become 517,656 square feet and 1,140,000 square feet respectively. The establishment employs 930 permanent and 240 seasonal workers on full time, entirely in the production of cut flowers. One great item in this industry is the forcing of tulips and narcissi to provide cut flowers from Christmas onwards ; whereas the bulbs required used to be wholly imported from the Netherlands, they are now being more and more grown at home, especially on the silt lands of Lincoln. The narcissi in Cornwall and the Scilly Islands are chiefly grown for the production of cut flowers.

Turning to live stock, which must continue to be the mainstay of English farming, the devotion to pedigree, which had laid the foundation of the world-wide spread of British breeds, has led to the recognition of further local races of cattle, sheep, and pigs by the formation of further breed societies.

The older show criteria still dominate; it may be questioned whether the great Show Societies would not benefit commercial production by the adoption of some of the ideas put in practice in the shows held by good stock raising countries like the Netherlands and Sweden. One breed that has shown, perhaps, the most considerable advance has been the large black and white Dutch milk breed known in this country, first as Holstein, and now as Friesian. The foundation stock has been improved through the permitted importation on three occasions [1914, 1922, 1936] of bulls from the Netherlands and from South Africa, where the breed has been very carefully taken up. But it is less the pedigree herds of England that need grading up than the common country stock such as may be seen filling any local market. To that end the Board of Agriculture, with the assistance of the Development Commission, set on foot in 1914 a scheme for the distribution of improved sires, a method which had been widely adopted abroad and in which we had been anticipated by Ireland. A bull club has to be formed, members of which represent a certain number of cows of the breed in question, whereupon the Ministry makes a grant of one-quarter of its cost towards the purchase of a pedigree bull, for the services of which each member pays the fee necessary to cover expenses. In the purchase of the bull the Society has the guidance of the Ministry's live stock officer, who watches over the live stock generally of his province. Similar assistance is given towards the purchase of good boars and to a small extent of rams, though as a rule sheep flocks are not small enough to require the co-operative purchase of a ram.

During the post-war period there have been several agitations to continue the good work of the premium bull scheme by a system of registration of all bulls that are used for breeding and the casting of the " scrubs " who fail to reach the standard. Such proposals met with determined opposition from the dairy farmers who breed calves but do not rear them. Legislation was, however, obtained in 1930 and came into operation in 1934.

Perhaps the severest effect of the post-war depression in these latter years has been felt by the producers of beef. It is probable that public taste is changing and that beef in any case is less in demand, but the chief consumers of the better qualities of English beef had always been the working classes of the manufacturing and mining areas. London had, in the main, been content with chilled beef from the Argentine : this of late has been the prosperous part of the kingdom, while low wages and unemployment have pressed heavily upon the industrial north and midlands. The trend of beef cattle prices (p. 450) shows how indifferent the returns have been, and as regards production in Great Britain it is significant that the importation of store cattle from Ireland which averaged 404,466 for the period 1909–13 had increased in the period 1929–33 to 425,920, while the arrivals from Canada which began in 1924 have only averaged 2,289 in the latter period. It is also significant that the spread between the price of fat cattle and beef has never been greater than in these latter years.

The beef trade has become of far less importance to the English farmer than the milk trade, which has experienced a steady increase since the close of the war. The number of dairy cattle has risen from 2,340,214 in the period 1909–13 to 2,800,853 in the period 1929–33, and at the same time many milk producers in the West Country and Wales, who formerly made cheese or butter, have entered the liquid milk market. This has been made possible by the improvements in transport, particularly by the introduction of the enamel lined steel motor-driven tanks of 1500–2000 gallons capacity, which can make the journey from the collecting depot in the remote west to the depot in London within the space of a short night. Another factor increasing the volume of milk has been the improvement in the milking capacity of the cows, due on the one hand to selection based upon milk recording and on the other to more skilful rationing. Milk recording had, of course, been general among the possessors of pedigree milking breeds and had long been officially controlled in the Scandinavian countries when in 1914 a scheme was brought into force by the Ministry. A Society had to be organised embracing a certain number of cows, the owners of which had to fill in for each cow the weighings of night and morning milk. The recorder for the area visited each producer once a month and watched the two milkings through as a check of the results of the intervening days. At the end of each year the Ministry published a record of all cows reaching a certain standard and issued certificates of performance. The Ministry makes grants of £3 per herd, covering in part the salary of the Recorder and the working expenses, the rest being contributed by a levy per cow. It was soon found that the value of the scheme lay not only in better prices which recorded cows and their progeny began to command in the market, but also in the increased care it produced in the handling of the cows, for it developed a spirit of competition in their attendants. The number of cows recorded has grown from 20,000 at the close of the war to 140,000, while the average yield of the herd, per cow, has risen from 450 gallons to 540 gallons. That the scheme has not been more widely taken up, for the number of recorded cows is still but 5 per cent. of the whole, is due to its expensiveness, even though the Ministry bear so large a share of the cost. It may be possible to reduce the amount of official testing required in the light of the knowledge that has now been obtained of the regular variations in milk production that are to be expected during the period of the lactation, taking into account the age of the cow and the season of the year.

On the technical side milk production has been cheapened and improved by the widespread adoption of milking machines. They were in use before the war, but the earlier machines possessed various disadvantages ; now the current patterns under proper management are just as efficient and as unobjectionable to the cows as the best hand milking. They have this great virtue that the milk passes into a closed vessel without coming in contact with the air, a first step in maintaining a high standard of cleanliness and freedom from infection. The milking machine has rendered possible a most re-volutionary system of cow-keeping that has been worked out by Mr. A. J. Hozier, of Wexcombe, near Marlborough. The cows are kept on open pastures all the year round, and as the system was initiated on the high Wiltshire Downs at an elevation of 500–700 feet, it does not involve specially favoured climatic conditions. At milking time the cows are rounded up into a movable corral, at one side of which is the " bail," consisting of a van containing an oil engine driving the milk machine and steadings for six cows. As the cows succes_ sively take their places for milking they obtain access to their ration of concentrates ; when finished the door is lifted and they pass out of the corral into the open pasture. The bail and corral are moved across the pasture every few days. Under this system a man and a boy handle sixty cows, the cows remain remarkably free from tuberculosis and the milk is specially clean, for it never comes in contact with the air until it reaches the final receptacle. Though the bail involves some initial expense, the cost of permanent buildings is obviated. Another factor of importance is the great improvement that follows in the pasture receiving the benefit of a moving fold.

The character of the herbage on the poor thin chalk downs changes radically and after a few years of the system they will carry a succession of corn crops before being put down to pasture again. As yet it has only been employed on the drier soils which do not poach badly in wet weather, but adaptations are being made to fit it to the heavier land. The economy effected by this method can only be dsecribed as revolutionary.

Another movement of great significance to the milk industry began in 1922 with the introduction of designated grades of milk. In response to the urgent demands of the medical profession for milk that could be warranted as free from the risk of communicating bovine tuberculosis, machinery was set up by which owners could submit their herds to test so as to arrive at a segregated herd, none of the members of which reacted to the tuberculosis test. The test is renewed at regular intervals and no fresh cows or heifers can be admitted until they have passed the test. Milk from such herds has to be sold in bottles or containers : if bottled upon the farm it is designated Certified Milk ; it must notc ontain more than 30,000 bacteria per c.c., not any $B. coli$ in one-tenth of a c.c. If the milk is bulked and bottled at the depot, it is sold as Grade A (Tuberculin tested) ; the bacterial count permitted being 200,000 per c.c. Such certificated milk is inevitably more expensive, and in order to secure for general consumption milk of a better order than has usually been current, milk from herds which are submitted to a clinical examination every three months and which again contains no more than 200,000 organisms per c.c., is known as Grade A milk. " Pasteurised Milk " which has been heated to 145°–150° F. for half-an-hour is also recognised, and at present a considerable proportion of the milk delivered by the large wholesalers has been pasteurised.

The general insistence on higher standards of care in the production and handling of milk, the introduction of " clean milk " competitions, and the bacteriological examinations of the milk entering their areas carried out by the public health authorities of the large towns, has wrought a marked change for the better in the general milk supply. It is significant that whereas the night and morning milk used to be sent to market separately, the wholesalers are now content with a single daily delivery, so much better does the milk keep. With the advent of the control of the Milk Board the time is ripe for an intensive campaign for the elimination of bovine tuberculosis, which, however costly in its early stages, would eventually be profitable to both farmer and consumer. Evidence goes to show that the average working life of the milch cows in English herds is less than half of what it should be, owing to losses from tuberculosis, and as replacement is a cost in the production of milk which may rise as high as 2d.–3d. per gallon, the economic advantage to be gained by the elimination of tuberculosis will repay even a costly process of purification.

The period under review has been marked by increasing efforts to cope with disease in live stock. An attempt was made to reduce the incidents of contagious abortion, one of the major troubles of the dairy farmer, by making it an offence to expose for sale cows that had aborted, but that regulation is almost impossible of enforcement and has had little effect. Even though the agglutination test has provided a means of eliminating infected animals it has been difficult to keep valuable herds free of this disease, and the extensive research now planned will be anxiously awaited. The epidemic disease, however, which attracts the most public attention and is most noticeably costly is Foot and Mouth disease. Living as we do on the edge of the continental area in which the disease is endemic, increasing in frequency eastwards, the British Islands have always been subject to invasions of the infection. Any invasion that is detected at an early stage can be quickly cleared up by the slaughter of the affected stock and the contacts, together with the formation of a sanitary cordon, but if the affected stock gets into a market the disease may suddenly become disseminated over a wide area and contacts so multiply that months may elapse before the outbreak is under control. During this time all movements of stock in the scheduled areas except for slaughter are prohibited, hunting is stopped and even many communications may be closed. None the less the policy of slaughter has deliberately been pursued lest this debilitating disease should become endemic. Foot and Mouth belong to the class of " Virus " diseases which, whether in animals or plants, are not susceptible of cure or treatment. Immunisation affords as a rule only a temporary remedy, and these diseases are the more insidious in that they do not necessarily kill. Foot and Mouth disease, in particular, is distinguished by its singular infectivity and ease of transmission.

How the outbreaks reach Great Britain was always a matter of speculation ; the majority of outbreaks were grouped near the eastern and southern shores that are closest to the Continent ; in the west of England, Scotland, and Ireland initial outbreaks were rare though not unknown. It was thought that the wind or bird migrations might be carriers of the infection, but the almost entire freedom from outbreak during the war years indicated some articles of commerce as the most probable carriers. However, this blank period was succeeded by three of the worst outbreaks known in 1922, 1923, and 1924, which cost the country over three million pounds in compensation for slaughter. Then in 1926 infection was definitely found in some pig carcases received for curing in a bacon factory in Scotland near the source of one outbreak, and with this incrimination of fresh meat, importations from the Continent were wholly forbidden. It was impossible to prohibit the importations of Argentine chilled meat which had also become suspect, but arrangements were reached with that Government to secure the strict inspection of animals reaching the frigorificos for exportation to England, and two English inspectors spent four years in South America to assist the various Governments to reach a satisfactory control. Importations from the Continent of hay and straw even for packing have been banned, but though from time to time importations of vegetables have been suspected of initiating an outbreak the evidence has not been conclusive. Outbreaks have not ceased, but undoubtedly have been less frequent since the banning of these particular imports. In 1924 after the severe outbreak of the previous two years a research committee, in which human and animal pathologists collaborate, was set up to conduct investigations into the nature of the disease and its control, action which previously had been resisted lest the presence of the disease, even under laboratory control, should be made a reason for refusing the entry of stock from Great Britain. But though this committee has been intensively at work ever since and has arrived at a good deal of new knowledge concerning the behaviour of the viruses causing the disease (for more than one strain has been isolated) practical measures of relief are still undisclosed. Nothing has yet emerged to suggest an alternative to the policy of slaughter and isolation. It would appear that the administrative machinery for detecting and dealing with an outbreak has been so strengthened that it is rapidly stamped out without spreading. Meantime, despite all the records to show how occasional outbreaks are localised to leave the rest of the country unaffected, despite the fact that we have established an official quarantine station in which animals can be isolated under observation for a time before export, many countries and dominions still impose what amounts to an embargo on the reception of live-stock from Great Britain on the ground of Foot and Mouth disease. Regulations against disease are too often utilised by the ultra-protectionists as a most convenient means of shutting the door against imports. Yet even in 1923, during the period of the worst outbreaks, it was agreed after full enquiry to revoke a prohibition of long standing against the importation of Canadian store cattle, and that in the face of the determined opposition of all the representatives of agriculture. Neither the fears nor the expectations of the protagonists in this dispute have been realised ; no disease has been admitted and the imports of Canadian stores have been negligible.

Turning to other live-stock, the pig industry in England has undergone considerable change since the war. During the previous forty years the pig population in England had remained pretty constant, round about two millions, though during the same period it had grown from small beginnings to double that figure in Denmark, almost wholly to supply the British market. The war caused a great reduction in the stock here, but numbers were rapidly restored and with a general drive to capture a greater share of the bacon trade in 1924 they reached nearly three and a quarter millions, only to fall again to about two and a quarter millions in 1931, from which they have recovered and again passed the three million mark. Now, however, the production of bacon is under the control of Pig and Bacon Boards, and it would be premature to discuss the measures that have been taken for the improvement of this essential branch of agriculture. One effect may be noticed of the campaign of enlightenment that preceded the formation of the Boards, when the bacon manufacturers themselves were endeavouring to secure a type of pig more suited to their requirements. The bacon manufacturers insisted that many of the breeds of pigs in favour in this country, which indeed were being multiplied, specified a mistaken type, apart from such purely fanciers' points as colour, ear shape, and so forth. Their ideal bacon pig was based upon the Large White or Yorkshire breed, which, as they pointed out, was being used as the sire in the Scandinavian countries which send us the bulk of the bacon consumed here. Black pigs, again, were less desired in the bacon factory because of the tendency to discoloration in certain cuts.

This reiterated advice and the introduction of grading by the

U*—E F P P

factories receiving pigs has wrought great changes ; while some of the old breeds retain their place as foundation stock to be crossed with Large White boars, others are being discarded. That notable breed, the Berkshires, one of the most finished accomplishments of the fanciers' art, is now rarely seen except at shows. Utility is beginning to prevail over the prepossessions of the fancy and the fashion.

This is even more the case with another branch of the live-stock of the farm—poultry, an industry now counting of more value to English agriculture than wheat itself. As pointed out earlier, poultry-keeping had been the plaything of the fancier, but at the close of the war numbers of new men were attracted to a business which promised an outdoor occupation and required relatively little capital, the more so since at the time the price of eggs was very high with the depletion of stocks in England and throughout Europe. New methods, too, were under trial, intensive and semi-intensive, due to the pioneer work of men like S. G. Hanson near Basingstoke, Tom Barron in Lancashire, and J. W. Toovey in Hertfordshire, all of whom had thought out methods of reducing the labour required to handle a large stock. Egg-laying competitions began in 1920 and soon became general, some of them conducted by the County Agricultural Authorities under the auspices of the Ministry, others initiated by the popular press. These soon had the effect of sifting out the most efficient breeds for egg production and establishing foundation stocks of proved capacity. The average production of eggs per bird per annum has risen from about 120 in the earlier to 175 in the later trials. The Ministry also promoted schemes for the distribution of eggs and day-old chicks from selected flocks, which has been succeeded by the recognition of accredited poultry-breeding stations which have been inspected and found after test to be free from bacillary white diarrhœa. There are now 168 such stations distributed through 31 counties. All over England poultry began to become a common feature in the countryside either on poultry farms proper or as adjuncts on the mixed farm, and it is estimated that the production on agricultural holdings in 1935 had risen to 2956 millions of hen's eggs and 56 millions of duck's eggs from 901 and 14 millions respectively in 1913. Improved methods of grading, packing, and collective marketing also became general ; a National Mark was adopted. After a period of great development and considerable prosperity the industry began to suffer a severe setback from which it has as yet not recovered. The price of eggs fell severely, not only as part of the general slump in prices but in consequence of vastly increased imports of good quality from Australia and New Zealand. Despite the introduction of marking regulations, " Fresh Eggs, Australian produce," is a label which may be seen in any grocer's shop. At the same time disease began to assume alarming proportions in many of the English flocks. The closely confined flocks on the intensive and semi-intensive principle required a degree of cleanliness and hygienic precautions which the rank and file of poultry-keepers rarely continued to give ; the ground became infected and epidemics more severe. More recent methods incline to keeping the poultry stocks continually moving over fresh grass-land either in free range or in small pens shifted almost daily. The industry is going through a difficult period, but none the less it may be regarded to be firmly established as no inconsiderable element in English live-stock farming.

Throughout the agricultural industry a tendency towards special-isation may be noticed ; the old mixed farming is being exchanged for businesses chiefly concerned with one or two kinds of production —milk, pigs, fruit, poultry, as the case may be. Extreme cases may be noted of milk producers who do not breed, of egg producers who buy all their laying birds. It is expecting much that a man may become expert in all branches of farming, for it is no easy task to accumulate full experience of even one or two. The economic organisation of labour is facilitated when a single object is being pursued. The risk of specialisation and of having all one's ventures in a single shift is being minimised by the Marketing Boards, which have reduced the excessive fluctuations in prices.

From this brief survey it will be seen that the period under review has been one of rapid change in response to the altered conditions of prices and the high rates of wages in relation to them, and again to the new powers made available by science. In all branches of the industry surprising adaptability has been shown, success has been attained in spite of the difficult conditions. But only a minority have the capacity to meet change—farming as a whole has still to be regarded as one of the depressed industries. Since 1931, however, it has entered upon a new phase with the adoption of a protectionist policy by the nation and the grant of liberal subsidies to certain branches of agriculture. The creation of the Marketing Boards marks a new relation between the State and farmers and foreshadows an increasing measure of control of production from the land. With the advent of these novel measures the farmers' organisations have, perhaps, become more busied with political considerations than with the technical improvement of their craft ; but progressive individuals everywhere are engaged in the working out of the new methods which will be the basis of English farming in the future.

APPENDIX I

SELECT LIST OF AGRICULTURAL WRITERS DOWN TO 1700.

BEFORE THE INVENTION OF PRINTING

Latin Writers on Agriculture.—Cato, Varro, Columella, Palladius. M. Catonis Prisci *de Re Rusticâ* liber ; Marcii Terentii Varronis *Rerum Rusticarum,* libri i.-iii. ; Lutii Junii Moderati Columelloe *Rei Rusticoe* Liber Primus—Tertius—Decimus, ed. G. Merula. Palladii Rutilii Æmiliani *de Re Rusticâ* liber primus-liber xiii., ed. F. Colucia. Venice (1470 ?). 1472.

[Cato, commonly called the Censor, died B.C. 149. Varro flourished in the last century before Christ. Columella, whose agricultural writings are the most useful of the four Latin writers, flourished in the first century of the Christian era. He also wrote a treatise, *De Arboribus.* Palladius, who seems to have written in the fourth century A.D., borrows largely from Columella. The above-mentioned book is the first printed edition of any part of the Roman writers on Agriculture, though many manuscripts probably existed in the libraries of English monasteries. Frequent editions were subsequently published abroad, but none apparently in England.

Translations were also printed abroad, *e.g.* of Columella into Italian by Pietro Lauro (1544), and into French by Charles Cotereau (1555). But none seems to have been printed in England till 1725, when Richard Bradley, F.R.S., Professor of Botany in the University of Cambridge, published his *Survey of Ancient Husbandry and Gardening collected from Cato, Varro, Columella, etc.*

A manuscript translation of Palladius into English verse belonging to the fifteenth century (1420) was edited by Messrs. Lodge and Herrtage, under the title of *Palladius on Hosbondrie* for the Early English Text Society, 1873 and 1879.]

Thirteenth Century Manuscripts.—Walter of Henley's *Husbandry.* Together with an anonymous *Husbandry, Seneschaucie,* and Robert Grosseteste's *Rules.* (Ed. Elizabeth Lamond. Royal Historical Society, 1890.)

GARDINER (MAYSTER ION).

The Feate of Gardening.

[Edited from the original fifteenth century manuscript by the Hon. Alicia Amherst (Lady Rockley), *Archaeologia,* vol. liv.]

Since Lord Ernle prepared his list of early agricultural writers bibliographical research has added so many items that it has seemed desirable only to retain the more significant or more pertinent books mentioned by him. The most complete bibliography available is the Catalogue of the printed books in Agriculture between 1471 and 1840. M. S. Aslin. Rothamsted Experimental Station, 1927.

474 APPENDIX I.

PRINTED BOOKS.

CRESCENTIUS or PETRUS DE CRESCENTIIS. (Born 1233 : died 1320.)
Opus Ruralium Commodorum sive de Agriculturá. Libri xii. Augsburg,
1471.
[This work was translated into Italian, 1478 ; into French, 1486 ; into
German, 1493.]

FITZHERBERT (JOHN).
1. *The Boke of Husbondrye.* 1523.
" Here begynneth a newe tracte or treatyse mooste pftable for all
husbāde men."
[Book of Husbandry, ed. W. W. Skeat. English Dialect Society, 1882.]
2. *The Boke of Surueyeng and Improuvemēts.* 1523.
" Here begynneth a ryght frutefull mater : and hath to name the boke
of surueyeng and improuvemets."

TURNER (WILLIAM).
*Libellus de Re Herbaria Novus, in quo herbarum aliquot nomina greca,
latina, & Anglica habes, etc.* 1538.
[The first English version was entitled *The names of herbes in Greke,
Latin, Englishe, Duche and Frenche, wyth the commune names that Herbaries
and Apotecaries use.* Gathered by William Turner. 1548.
The second edition of the work appeared as *A new Herball, wherin are
conteyned the names of Herbes in Greke, Latin, etc. etc.* 1551.]

DIGGES (LEONARD).
*A Book named Tectonicon, briefly shewing the exact measuring and spedie
reckoning all maner of Land, Square Timber, etc.* 1556.
[This book continued to be republished at intervals till the close of the
seventeenth century : *e.g.* 1634, 1637, 1656, 1692.]

TUSSER (THOMAS).
A hundreth good pointes of husbandrie. 1557.
Subsequently amplified into
*Fiue hundreth good Pointes of Husbandry, united to as many of good
Huswifery.* 1578.

HERESBACH (CONRAD).
Rei Rusticae libri quatuor. Cologne, 1570.

MASCALL (LEONARD).
1. *A Book of the Art and Manner how to graff and plant all sortes of Trees,
etc.* 1572.
[The book is mainly a translation from the French, with certain Dutch
practices added, and is described as being " set forth and Englished "
by L(eonard) M(ascall). It was published in the latter half of the seventeenth century under the title of *The Country-Mans new Art of Planting
and Graffing, etc.* 1652.]
2. *The Husbandlye Ordring and Governmente of Poultrie Practised by the
Learnedste, and such as have bene knowne skilfullest in that Arte and
in our tyme.* 1581.
3. *The first Booke of Cattell : Wherein is shewed the gouernment of Oxen,
Kine, Calues, etc. : The Second booke entreating of Horses, etc. : the Third
booke entreating of the ordering of Sheepe and Goates, Hogges and Dogges,
etc.* 1591.
[The edition of 1596 seems to have been the third edition of this work.

The book was often republished under the title of *The Government of Cattell*. In the latter half of the seventeenth century it was edited and " enlarged " by Richard Ruscam, *Gent.*, and published (1680) under the title of *The Countreyman's Jewel*.]

SCOT (REYNOLDE).

A Perfite platforme of a Hoppe Garden, and necessarie Instructions for the making and mayntenaunce thereof, etc.   1574.

GOOGE (BARNABE).

Foure Bookes of Husbandrie, collected by M. Conradus Heresbachius. . . . Newely Englished and increased by B(arnabe) G(ooge), Esquire.   1577.

ESTIENNE (CHARLES) and LIEBAULT (JEAN).

L'Agriculture et Maison Rustique.   1586.
[Estienne's *Proedium Rusticum* appeared in 1554. From the French version, the book was translated by Richard Surfleet under the title of *Maison Rustique, or the Countrie Farme*, 1600. It was also published by Gervase Markham in 1616.]

MARKHAM (GERVASE).

1. *Discourse on Horsemanshippe, etc.*   1593.
2. *How to Chuse, Ride, Traine, and Diet, both Hunting Horses and running Horses, etc.*   1599.
3. *Cavalarice : or, The English Horseman, contayning all the Arte of Horsemanship, as much as is necessary for any man to understand, whether he be Horse-breeder, horse-ryder, horse-hunter, horse-runner, horse-ambler, horse-farrier, horse-Keeper, Coachman, Smith, or Sadler, etc.*   1607.
4. *The Husbandman's Faithful Orchard, etc. The Whole Art of Husbandry contained in Foure Bookes, viz. i, Of the Farme or Mansion House, etc. ii, Of Gardens, Orchards and Woods. iii, Of Breeding, Feeding, and Curing of all manner of Cattle. iv, Of Poultrie, Fowle, Fishe, and Bees, etc.*   1608.
5. *Countrey Contentments ; or, the Husbandman's Recreations, in two bookes, etc.*   1611.
[In the 1615 edition, two new treatises were added ; " the second intituled, The English Huswife " was subsequently (1638) issued as a separate work, " edited " by Countess De la Warr.]
6. *The English Husbandman, drawne into Two Bookes, and each Booke into Two Parts, etc.*   1613.
7. *Cheape and Good Husbandry, etc.*   1614.
8. *Maison Rustique ; or, the Countrey Farme, etc.*   1616.
9. *Markham's Farewell to Husbandry.*   1620.
10. *Markam's Maister-Peece.*   1623.
11. *The Inrichment of the Weald of Kent : or a Direction to the Husband-man for the true ordering, manuring, and inriching of all Grounds within the Wealds of Kent and Sussex, etc.*   1625.
12. *A Way to Get Wealth, etc.*   1628.
13. *The Art of Archerie, etc.*   1634.
14. *The Pleasures of Princes, or, Good Men's Recreations, etc.*   1635.

[The number of books, new, or revised, or republished under different titles, to which Markham put his name, makes it difficult to verify all of them at the present day. His different publishers in the seventeenth century were equally puzzled as to their properties in his works, if, as is stated, he was in 1617 compelled to sign the following memorandum : " Mem.—That I, Gervase Markham, of London, gent., do hereby promise hereafter never to write any more book or bookes to be printed of the diseases of any cattle, horse, ox, or cow, sheep, swine or goats. In witness whereof I have hereunto set my hand the 24th daie of July, 1617, Gervase Markham.'']

PLAT or PLATT (SIR HUGH).

1. *The Jewel House of Art and Nature : Conteining divers rare and profitable inventions, together with sundry new experiments in the Art of Husbandry, etc.* 1594.
2. *Sundrie New and Artificial Remedies against Famine.* . . . Written by H. P. Esq., uppon the occasion of this present Dearth. 1596.
3. *The New and admirable Arte of setting of Corne, etc.* 1600.
4. *Delightes for Ladies, to adorne their Persons, Tables, Closets and distillatories, etc.* 1602.
5. *Floraes Paradise, beautified* . . . *with sundry sorts of delicate fruites and flowers, etc.* 1608.
[Under its subsequent title of *The Garden of Eden, etc.*, this book was one of the most popular works on horticulture among seventeenth-century gardeners.]

NORDEN (JOHN).

1. *Speculum Britannioe. The first parte. An historicall* . . . *discription of Middlesex.* 1593.
2. *Speculi Britannioe Pars. Essex described* (1594). [Published by the Camden Society in 1840 from the manuscript.]
3. *Speculi Britannioe Pars. The description of Hartfordshire.* 1598.
4. *The Surveiors Dialogue, very profitable for all men to peruse, but especially for Gentlemen, Farmers, and Husbandmen, that either shall haue occasion or be willing to buy, hire, or sell Lands, etc.* Divided into five books. 1607.
[A sixth book was added, 1618, " of a familiar conference, betweene a Purchaser, and a Surveyor of Lands, etc."]
5. *Speculi Britannioe Pars Altera : or, a delineation of Northamptonshire.* 1610.

GERARDE (JOHN).

The Herball, or, Generall Historie of Plantes. 1597.

VAUGHAN (ROWLAND).

Most approved and long experienced Water Works. Containing the manner of Winter and Summer drowning of Medow & pasture. 1610.

STANDISH (ARTHUR).

1. *The Commons' Complaint. Wherein is contained two speciall Grievances. The first is the generall destruction and waste of Woods in this Kingdome.* . . . *The second Grievance is the extreame dearth of Victuals, etc.* 1611.
2. *New Directions of Experience to the Commons Complaint, etc.* 1613.

PLATTES (GABRIEL).

1. *A Discovery of infinite Treasure, hidden since the World's beginning, etc.* 1639.
[The book, with subsequent additions, was republished under the title of *Practical Husbandry Improv'd : or, A Discovery, etc.*]
2. *Recreatio Agriculturae.* 1640.
3. *The Profitable Intelligencer, etc.* 1644.
4. *Observations and Improvements in Husbandry.* 1653.
5. *The Countreyman's Recreation.* 1654.
[2, 4 and 5 are added on the authority of Donald M'Donald.]

BEST (HENRY).

Farming and Account Books.   1641.

[These records were printed by the Surtees Society (vol. xxxiii., 1857), under the title of *Rural Economy in Yorkshire in 1641, being the Farming and Account Books of Henry Best.*]

VERMUYDEN (SIR CORNELIUS).

Discourse touching the Draining of the Great Fennes, etc.   1642.

BURRELL (ANDREWES).

Briefe Relation Discovering Plainely the True Causes why the Great Levell of Fenes . . . have been drowned.   1642.

WESTON (SIR RICHARD).

A Discours of Husbandrie used in Brabant and Flanders.   1645.

[This treatise, left in manuscript by Weston to his sons as a Legacy, was piratically printed by Samuel Hartlib in 1650. The date of publication is given as 1605 ; but from Hartlib's unctuous dedication of the work to "The Right Honourable the Council of State," that date is obviously wrong. The title-page runs as follows : *A Discours of Husbandrie used in Brabant and Flanders ; shewing the wonderfull improvement of Land there ; and serving as a pattern for our practice in this COMMON-WEALTH.* London, Printed by William Du-Gard, Anno Dom. 1605. Hartlib republished the Discourse in the following year as *His Legacie.*]

HARTLIB (SAMUEL).

1. *Samuel Hartlib His Legacie ; or An Enlargement of the Discourse of Husbandry used in Brabant and Flanders ; Wherein are bequeathed to the Common-wealth of England more Outlandish and Domestick Experiments and Secrets in reference to Universall Husbandry.*   1651.

[This is a reprint and not an enlargement of Weston's work. It contains nothing by Hartlib, except some Prefaces, which are not conspicuous for honesty or sincerity. It consists of extracts from a number of letters addressed to Hartlib, and in particular of a " Large Letter " by Robert Child, who signs his full name for the first time in the edition of 1655.]

2. *The Reformed Husband-man ; or a brief Treatise of the Errors, Defects, and Inconveniences of our English Husbandry, etc.*   1651.

[This is not by Hartlib himself. It was " Imparted some years ago " to him, " And now by him re-imparted to all ingenuous English-men," etc. It is often bound up with the *Legacy.*]

3. *An Essay for Advancement of Husbandry-Learning : or Propositions for the errecting Colledge of Husbandry, etc.*   1651.

[This treatise is probably not by Hartlib, but possibly by Cressy Dymock, or by Gabriel Plattes, who in his *Discovery* had suggested a " Colledge for Inventions in Husbandry."]

4. *A Discoverie for Division or Setting out of Land, as to the best Form.* Published by Samuel Hartlib Esquire, for Direction . . . of the Adventurers and Planters in the *Fens*, etc.

BLITH (WALTER).

The English Improver ; or, a New Survey of Husbandry.   1649.

Republished in an enlarged form as

The English Improver Improved, or the Survey of Husbandry Surveyed. 1652.

G. (E.)

Waste Land Improvement ; or, Certain Proposals, etc.   1653.

[" Robberies, thefts, burglaries, rapes, and murders receive their nourishment and encouragement " from the wastes lying unenclosed.]

MOORE (ADAM).
Bread for the Poor. . . . Promised by Enclosure of the Wastes and Common Grounds of England. 1653.

MOORE (JOHN).
The Crying Sin of England of not caring for the Poor, wherein Inclosure is . . . arraigned, convicted, and condemned by the Word of God, etc. 1653.
[This tract is answered by " Pseudomisius " in *Considerations concerning commonfields and inclosures* . . . partly to answer some passages in another discourse . . . by Mr. J. M., under this title *The Crying Sinne of England*, etc. 1654.]

LEE (JOSEPH).
Εὐταξία τοῦ Ἀγροῦ *or a Vindication of a Regulated Enclosure.* 1656.
[To this work and to " Pseudomisius," John Moore replied in *A Scripture Word against Inclosure, viz. such as doe un-people townes and un-corne fields, etc.* (1656). To this, " Pseudomisius " replies in *A Vindication of the Considerations concerning Commonfields and Inclosures ; or, a Rejoynder unto that Reply which Mr. Moore hath pretended to make unto those Considerations* (1656).]

DUGDALE (SIR WILLIAM).
The History of Imbanking and Draining. 1662.

FORTREY (SAMUEL).
England's Interest considered in the Increase of Trade. 1663.
[The author argues in favour of enclosures, mainly on the ground that they enable occupiers to apply the land to the use for which it is best suited.]

YARRANTON (ANDREW).
1. *The Great Improvement of Lands by Clover, or the Wonderful Advantage by right management of Clover.* 1663.
2. *England's Improvement by Sea and Land, etc.* 1677.

EVELYN (JOHN).
1. *Sylva ; or, a Discourse of Forest Trees.* 1664.
2. *Terra ; a Philosophical Discourse of Earth, relating to the Culture and Improvement of it for Vegetation, etc.* 1675.
3. *Pomona : a Discourse concerning Cyder.* 1679.

WORLIDGE (JOHN).
1. *Systema Horticulturoe, etc.* 1667.
2. *Systema Agriculturoe: The Mystery of Husbandry discovered.* 1669.
3. *Treatise of Husbandry, etc.* 1675.
4. *Vinetum Britannicum ; or the Treatise on Cyder.* 1676.
5. *Apiarium ; or a Discourse on Bees.* 1676.
6. *The Most Easie Method for Making Cyder.* 1687.

B[ALGRAVE] (J[OSEPH]).
The Epitome of Husbandry. Comprising all necessary directions for the improvement of it. 1669.
[Other editions were published in 1670, 1675, 1685. The book is based on Fitzherbert's *Book of Husbandry*.]

REEVE (GABRIEL).

*Directions left by a Gentleman to his sonns ; for the Improvement of Barren
and Healthy Land, etc.*   1670.
[This is to a great extent a repetition of Sir Richard Weston's *Discours.*]

MEAGER (LEONARD).

1. *The English Gardener ; or, a sure guide to young planters, and gardeners,
in three parts, etc.*   1670·
[The tenth edition of this book was published in 1704.]
2. *The Mystery of Husbandry : or, arable, pasture and woodland improved,
etc.*   1697.

SMITH (JOHN).

1. *England's Improvement Revived ; in a Treatise of all manner of Husbandry & Trade by Land and Sea, etc.*   Experienced in thirty years
Practice, and digested into six Books.   1670.
2. *Profit and Pleasure United ; or, the Husbandman's Magazine, etc.*   1704.
[It is uncertain whether the John Smith who wrote *England's Improvement Revived* was also the J. Smith, *Gent.*, who wrote *Profit and Pleasure
United.*   The two works have been attributed to the same author ; but
probably they were by different writers.]

PLOT (ROBERT).

A Natural History of Oxfordshire.   1677.

HOUGHTON (JOHN), F.R.S.

A Collection of Letters for the Improvement of Husbandry and Trade.
                                                           1681-83 and 1691-1703.
[The materials collected by Houghton were rearranged and republished
in 4 vols. in 1727 by Richard Bradley, F.R.S.]

MOORE (SIR JONAS), F.R.S.

1. *History or Narrative of the Great Level of the Fens, etc.*   1685.
2. *England's Interest : or the Gentleman and Farmer's Friend.*   1703.

APPENDIX II

THE POOR LAW FROM 1601 TO 1834

THE Poor Law suggests two distinct lines of enquiry—(1) the collection of the funds for Poor Relief ; (2) the expenditure of the money when collected. In other words, there are two questions—how was the money raised ? and how was it spent ?   Only a slight outline of the complicated subject can be here attempted.

1. THE COLLECTION OF THE FUNDS FOR POOR RELIEF.

A variety of rates, statutory or otherwise, were collected in mediaeval times.   Fixed sums were required for various purposes, and their payment was locally apportioned to individuals in each district or parish according to their ability to pay.   Some of the rates were assessed on particular persons in proportion to the benefits they individually received : some were raised for purposes of more general utility on all the inhabitants of the wider areas which were benefited by the expenditure.   Some originated in feudal tenures. A part of the national taxes even was raised as a local rate.   Thus the subsidies granted to the Crown under the name of " tenths and fifteenths " were apportioned in fixed sums to the inhabitants of each district.   A tenth of the capital value of the movables in cities, boroughs, and ancient demesnes, and a fifteenth of movables in the rest of the country, were thus raised.   In 1334, a searching inquisition was made, in order to levy the tax with the utmost accuracy and precision.   The assessment of this year remained for nearly two centuries the basis of future demands.   Till the reign of Henry VII., when another elaborate assessment was made, the grant of tenths and fifteenths meant a grant of the sums produced and apportioned in 1334.   On this basis was also levied money needed for many purposes of local government, not covered by other rates.   The required sums, represented by some fraction of the valuation of 1334, were apportioned direct upon the contributors, according to their estimated ability to pay.   As guides to relative means, records were kept in which were entered the size of the houses in which contributors lived, or the acreage of the land that they farmed.   In effect these records resembled valuation lists.   Their existence possibly facilitated the eventual transfer of liability from inhabitants in respect of the income which they enjoyed from all sources to persons in respect of the annual value of the immovable property that they occupied.

The history of the origin of Poor Rates is, however, entirely different from that of other rates, although, when once the contribution to the funds for poor relief became a legal liability, they naturally were influenced by the characteristics of the rates already in existence.

The indigent, in early times, were relieved by personal charity, which religion enforced as a Christian duty.   The contribution of money for the poor was an exercise of the will, measured in amount by the means of the giver. It was a moral income-tax from which no person escaped, according to his ability and substance. Whether the payment was urged as a Christian obligation, or enforced as a duty which no one had the right to refuse, or demanded by the civil power as a legal liability, it preserved this universal character, universal both as to persons and as to sources of income. Gradually the optional charity passed into compulsory alms-giving, and finally into local taxation. Because this last stage was reached in the latter years of the reign of Elizabeth, it is usual to say that the Poor Law dates from 1601. In principle there was undoubtedly an important change. But in practice it had come to matter little whether the universal obligation to contribute, which attached to all persons according to their ability, was enforced by religious, or by legal, penalties. For agriculturists, the really serious change came later, when the universal income-tax raised from all inhabitants became a rate levied upon occupiers in proportion to the annual value of the immovable property which they occupy. It grew still more important, when practically all other rates were levied according to the annual value of property rateable to the Poor Rate.

The charitable relief of the poor was never a national, but always a local burden. Hence some law of settlement was necessary. When in 1388 (12 Ric. II. c. 7) " beggars impotent to serve " were confined to the places where they happened to be at the passing of the Act, local provision was obviously needed for their support. For a time, the alms of the faithful provided adequate funds. When, however, not only " beggars impotent to serve," but " valiant vagabonds," and " able-bodied vagrants " were saddled on their birthplace or last permanent abode,[1] their maintenance imposed a heavy burden on parochial charity. Voluntary effort begins to flag and compulsion to be applied. The machinery is still the alms-box ; the donors are still all inhabitants according to their general ability. But if any parish fails to provide sufficient funds to succour the impotent poor, or to keep " sturdy vagabonds and valiant beggars " to continual labour, the defaulting parish was to forfeit 20s. a month.[2]

Charity did not, however, provide the whole funds. In the legislation of Edward VI. a distinction begins to be drawn between the actual support of the poor and the expenses incidental to their maintenance. Thus the money administered in direct relief was still to be raised by charitable donations. But the costs of removing the poor to their birthplace or last permanent abode, or the initial sums expended in providing " convenient houses " for the impotent,[3] or the expenses of building Houses of Correction,[4] are to be compulsorily raised by means of existing methods of taxation.

The growing burden of poor relief, the inadequacy of the voluntary principle, perhaps, also, the relaxation of any sense of the moral obligation of charity, are shown in Elizabethan legislation. Alms-giving is abandoned for compulsory provision. In the Act of 1572 (14 Eliz. c. 5) the voluntary principle of charity survives only in name. The justices are to number the poor, calculate the weekly sums required for their support " by their good discretions," tax and assess " all and every inhabitant," register the names of the taxpayers, and the amount of their taxation, appoint collectors, as well as overseers of the poor. If any person, thus taxed and assessed, " obstinately refuses to give towards the help and relief of the poor," or " wilfully discourages others from so charitable a deed," he is to be brought before the justices, and committed to gaol until he is " contented " with their order. Finally, in 1597 (39 Eliz. c. 3) the appeal to charity is practically thrown

---

[1] E.g. 19 Hen. VII. c. 12 (1503-4) ; 22 Hen. VIII. c. 12 (1530-1).
[2] 27 Hen. VIII. c. 25 (1535-6).
[3] 1 Ed. VI. c. 3 (1547), and 3 and 4 Ed. VI. c. 16 (1549).
[4] 18 Eliz. c. 3 (1575-6).

aside altogether. The churchwardens, and four "substantial householders, to be called overseers of the poor," are empowered to raise the necessary funds by taxation of "every inhabitant and every occupier of lands " in the parish, "according to the ability of the same parish." If anyone refuses to contribute as he is assessed, distress is to be levied, and in default of distress imprisonment is to be inflicted. The better known Act of 1601 (43 Eliz. c. 2) is a repetition of its predecessor, except that it gives a more precise definition of the sources from which the money is to be raised. The overseers are to raise the funds " by taxation of every inhabitant, parson, vicar, and other, and of every occupier of lands, houses, tithes impropriate, or propriations of tithes, coal mines, and saleable underwood."

Under the Act of Elizabeth, not only every occupier, but every inhabitant was required to contribute, and the measure of his liability was, not the annual rental value of the immovable property which he occupied, but his ability to pay. In modern practice, inhabitants, as such, except parsons and vicars, escape payment, and the accepted criterion of ability to pay is rental value. It was nearly two centuries and a half before these two changes were completed.

In the seventeenth century, the annual rent, which traders, manufacturers, or farmers paid for their shops, factories, or farms, was probably the best guide to an estimate of their business profits. It was at least more satisfactory than the estimate which their neighbours might form of those profits. Very few years had, therefore, passed before rental value was generally accepted as practical evidence of an occupier's ability to pay. The other change was slower and more gradual. Inhabitants continued liable till a much later date. The earnings of labour, whether fees, salaries, or wages, went first. The rent of landlords escaped next.[1] Personal property and the profits of stock-in-trade were not fully relieved till 1840 (3 and 4 Victoria, c. 89). That Act provides that it shall not be lawful for the overseers of any parish, " township, or village to tax any inhabitant thereof, as such inhabitant, in respect of his ability derived from the profits of stock-in-trade or any other property, for or towards the relief of the poor." Since 1840, this measure has been renewed from year to year.

2. The Expenditure of the Funds raised for Poor Relief.

From mediaeval times two distinct classes of the poor had to be dealt with —the impotent, and the able-bodied. Each class fell into two subdivisions ; the impotent—into children, or the aged, sick, and infirm ; the able-bodied— into sturdy rogues and vagabonds, or the honest poor, " willing to worcke " but unable to find employment. Early legislators seem to have recognised that poverty is a relative condition implying the want of comforts habitual to any particular class, and that destitution is the absolute state of wanting the necessaries of life. They accepted as a moral duty the direct relief of the destitute ; they did not encourage public aid to the unconditional relief of poverty. They also discriminated between voluntary and involuntary indigence. They distinguished those who are indigent owing to their own conduct from those who are reduced to want by causes for which they themselves are not responsible. Thus the impotent poor were to be succoured and relieved. The able-bodied were treated on different lines, and each of the two divisions into which the class falls was differently handled. Untold misery might have been saved, if the original principles of poor relief had been more strictly maintained ; their gradual relaxation, culminating in the years 1795-1834, sums up the history of the Poor Law from Elizabeth to William IV.

The relief of the impotent poor affects agricultural labour only indirectly. For this reason it will not be specially discussed here. It will be enough to

[1] Sir Anthony Earby's case, 1633.

say that the Act of Elizabeth in 1601 (43 Eliz. c. 2) followed previous legislation by directing money to be raised for the relief of the impotent poor—" of lame, impotent, blind, and such other among them being poor and not able to work," for the provision for them of "convenient houses of dwelling," and for "the putting out of . . . children to be apprentices."

From the first, the problem of dealing with the able-bodied presented the greatest difficulty to legislators. The necessity of discriminating between the two classes of the able-bodied was early felt. It was forced into prominence, partly by the increase of mendicancy fostered by indiscriminate almsgiving, partly by the industrial changes, which, during the Tudor period, affected both agriculture and manufacture. Sturdy rogues and vagabonds were punished, under penal laws of such ferocious severity that they defeated themselves—with whipping, branding, imprisonment, and transportation. For less hardened offenders labour was used both as a test and a penalty; for them Houses of Correction were created; rewards were offered for their apprehension; they were harried by laws of settlement from parish to parish until they reached their place of birth. Towards the able-bodied poor, who were willing to work, a different policy was adopted. It was at first scarcely more lenient. For them a living at least, but not wages, was to be provided, on condition that they earned their food as slaves. Labour was the test of their necessities. In 1547 (1 Ed. VI. c. 3) every "city, town, parish, or village " was required to provide work for its able-bodied poor, or " to appoint them to such as will find them work " for meat and drink. Elizabethan legislation proceeded on more humane lines. In 1575, and again in 1597 and 1601,[1] "a convenient stock of flax, hemp, wool, thread, iron," and other stuff was to be provided in every parish "to set the poor on work." The material was to be wrought up at the home of the needy able-bodied person, finished at a given time, and paid for according to skill. Those who either refused to work, or spoilt or embezzled the material, fell into the class of vagabonds, and were consigned to a House of Correction. The stock was to be replenished by the sale of the manufactured goods, so that the system might become self-supporting.

The Elizabethan Poor Law was imperfectly administered. Political disorders increased the disorganisation of the system. Overseers failed to collect the rates; the stock was not uniformly provided; the vagrant population had greatly increased. Stanley,[2] the ex-highwayman, writing probably in 1605, imagined that there were then "not so few as 80,000 idle vagrants that prey upon the common-wealth." It is improbable that their number had decreased during the Civil Wars or under the Commonwealth. It was against this class, and especially against squatters, that the Act of 1662 "for the better Relief of the Poor " (14 Car. II. c. 12), commonly known as the "Act of Settlement " was directed.[3] The principle on which it proceeded was as old as the Anglo-Saxons. As the first step towards progress and order, every man was required within 40 days to have a settled domicile, and to be enrolled in some fixed community. Stranger and outlaw were synonymous terms. Throughout the Poor Law legislation from Richard II. to Elizabeth, the same principle had been enforced for the removal of the poor to their birthplace or last permanent abode. If relief was to be treated as a parochial, and not as a national, burden, a settlement was necessary. But the Act of Charles II. indisputably made the law more rigid, and imposed new fetters on the mobility of labour. It recites that people wandered from parish to parish, endeavouring " to settle themselves where there is the

---

[1] 18 Eliz. c. 3; 39 Eliz. c. 3; 43 Eliz. c. 2.

[2] *Stanleye's Remedy: or the Way how to Reform Wandring Beggars, Theeves, Highway Robbers, and Pickpockets.* London, 1646.

[3] *Report to the Poor Law Board,* by G. Coode, 1851.

best stock, the largest commons or wastes to build cottages, and the most woods for them to burn and destroy, . . . to the great discouragement of parishes to provide stocks, where it is liable to be devoured by strangers." To these wanderers were applied the regulations for the removal of vagabonds. Any person "*likely to be chargeable to the parish,*" unless he had acquired a settlement by a residence of 40 days, might be removed to his birthplace. On this Act was built up a mass of settlement law, which occupied the time of Sessions and Assizes, wasted the money of ratepayers, and, worst of all, fettered labour to one spot. But the policy of the legislature seems to have been to encourage and facilitate the relief of those who satisfied the settlement test. The cost of the Poor Law increased so rapidly that, in order to check what the Act of 1691 (3 and 4 Wm. and Mary, c. 11) styles the frivolous pretences of the overseers, controlling powers were given to the vestry and the justices. The control of the magistrates only led to larger expenditure. To reduce the growing cost a test of destitution was authorised. By the Act of 1722 (9 Geo. I. c. 7) churchwardens and overseers were empowered to provide houses for the maintenance of the poor, to contract for the employment of the inmates, and to apply the surplus of their earnings to the reduction of the rates. If any applicant refused to enter the house, he was not entitled to relief of any kind from the rates. The offer of a living was made entirely conditional on his entering the workhouse.

The Act of 1722 effectively checked the spread of pauper relief wherever it was adopted. But the legislation of George III. moved in the opposite direction of increased laxity. No doubt the great rise in the price of provisions from 1765 to 1774,[1] the disturbance of trade by the wars of 1756-63 and 1774-83, the invention of the steam-engine and the spinning-jenny, the rapid growth of population, presented the old problems of unemployment and poverty in an acuter form. As civilisation advanced, humanitarian sentiment asserted new claims, and social legislation occupied a larger share of the attention of Parliament. The need for detailed information respecting the cost of Poor Relief led to the Act of 1776, which required overseers to furnish returns of their assessments and expenditure. From these returns it appeared that the actual outlay in that year was £1,530,800. During the early years of the reign, numerous amendments were passed in the administration of the Poor Law. But the most important changes were made by the Act of 1782 " for the Better Relief and Employment of the Poor," usually known as " Gilbert's Act " (22 George III. c. 83). In any parish, or union of parishes, which adopted the provisions of the Act, the management of the poor was vested in a visitor and guardians. Poor relief was thus taken out of the hands of the overseers, whose duties were restricted to collecting and accounting for the rates. For the impotent poor, houses were to be provided which were in effect almshouses, though they unfortunately inherited the traditions of the old Houses of Correction and workhouses. The most serious alterations affected the able-bodied. The workhouse test of destitution, or of voluntary pauperism, was partially discontinued. In any parish where poor persons, able and willing to work but unable to find employment, applied to the guardian, he was obliged, under a penalty, to find them work conveniently near the residence of the applicants, to receive their earnings, to apply the money to their maintenance, to make up any deficiency out of the rates, to hand over any surplus to the earners. Alone among wage-earners, the able-bodied poor had not to exert themselves to find work or conduct their own bargains for wages. They were secure of a living, not if they worked their best, but if they worked hard enough and well enough to escape punishment. Nothing depended on their own

[1] For 1755-64, the average price of wheat was 37s. 6d. a quarter ; for 1765-74, 51s. a quarter. The price rose 35 per cent.

characters, skill, or industry, provided only that they kept out of the class of sturdy rogues and vagabonds, or, as they were now called, "the idle and disorderly."

Such measures as the encouragement of Friendly Societies, the foundation of Savings Banks, the establishment of industrial schools were designed by the Legislature to improve the condition of the labouring classes and to provide means of escape from poverty. Their remedial effects were necessarily slow : they could afford no relief to the hardships and privations into which wage-earners were suddenly plunged by the exceptional rise in the price of provisions in 1795-6. The labouring classes must have been brought to the verge of famine, unless the advance in the cost of necessaries was met by a corresponding rise in wages, or unless wages were supplemented by some form of charitable allowances. In these circumstances, legislators and county magistrates unfortunately turned for their immediate remedies to permanent alterations in the Poor Law. In 1795 a tardy attempt was made to remedy the worst abuse of the laws of settlement. The removal of any persons was prohibited until they had actually become chargeable to the parish (35 Geo. III. c. 101). In the same year the Berkshire magistrates, by what was from their place of meeting known as the Speenhamland Act, endeavoured to fix a " fair wage " by using the rates to supplement earnings in proportion to the price of bread and the size of families (see Chapter XIV.). Other counties adopted similar scales of supplementary allowances out of the rates. At the time the expedient was of doubtful legality ; but in the following year (1796) Parliament confirmed its principle. It sanctioned indiscriminate outdoor relief (36 Geo. III. c. 23) by completely abandoning the workhouse test of destitution. That part of the Act of 1722 (9 Geo. I. c. 7) which had permitted relief to be made conditional on entry into the workhouse was repealed, on the ground that it prevented " an industrious poor person from receiving such occasional relief as is best suited to his peculiar case," and held out " conditions of relief injurious to the comfort and domestic situation and happiness of such poor persons." Overseers were authorised to give occasional relief in cases of temporary illness or distress at the houses of the industrious poor unconditionally, although the recipients refused to enter any house provided for their maintenance.

Justices were also empowered, at their discretion, to order money grants to be given to the industrious poor in their own homes. The consequences of these successive relaxations of the Poor Law were not at the time visible. During the greater part of the war they were mainly used for the relief of winter unemployment. Substantial advances in wages, the progress of manufactures, the increased demand for labour created by the larger area under tillage combined to relieve distress. It was during the period of depression 1813-36 that the full effects were revealed. Both in manufacturing towns and agricultural districts employment had become scarce. Employers, hard-pressed by falling prices, took advantage of the relaxed Poor Law to reduce their expenses, by throwing on the ratepayers the greater part of their labour bills. A single justice was further empowered, at his discretion, to order relief to be given to poor persons in their own homes for one month ; two justices might extend the order for two months, " and so on from time to time, as the occasion shall require." [1]

Within the next forty years the consequences of these relaxations of the Poor Law were fully developed. They are summarised in the Report of the Parliamentary Committee of 1817, and with greater detail by the Poor Law Inquiry Commissioners, appointed in 1832, whose Report was published two

---

[1] These powers were still further enlarged in 1815 (55 Geo. III. c. 137). A single justice might make an order for three months, and two justices for six months, " and so on from time to time as the occasion shall require."

years later.[1] The general principle of this latter Report, on which was based the Poor Law Amendment Act of 1834 (4 and 5 Wm. IV. c. 76), was that outdoor relief of able-bodied paupers was the root-evil of the existing system, and that their position must be made less eligible than that of independent labourers. Relief must, therefore, not be given in the home of the recipient. Destitution, or, as the Report expresses it, indigence, not poverty, is to be regarded as the proper object of a Poor Law, and the workhouse affords the only reliable test by which the two conditions can be discriminated. A man is only destitute when he is ready to accept the restraint of a workhouse rather than rely on his own resources. Parliament refused to adopt the extreme course of absolutely and universally prohibiting outdoor relief to able-bodied persons or to their families, as the Commissioners recommended. It is significant that no power was given to dissolve the Gilbert Unions, in which the overseers were compelled by statute to employ able-bodied applicants for relief in work conveniently near their own homes. Nor was Parliament altogether convinced that workhouses were essential to properly administered Poor Relief, and it showed its hesitation by not arming the Commissioners with powers to compel their provision. In other respects the main principles of the Report were adopted.

The Act of 1834 did not itself attempt to frame a new system of Poor Relief. It rather aimed at correcting abuses by substituting for every variety of practice the uniform adoption of improvements which experience had proved to be salutary. Instead of elaborating its own code of rules, it conferred on a newly constituted public body ample discretionary powers. It altered the existing law of settlement and of bastardy. But the really important change which it effected was the creation of a Central Authority, consisting of three Poor Law Commissioners, empowered to issue orders regulating every detail of the local administration of Poor Relief. To this Board were transferred the powers of all the unskilled and irresponsible authorities of 15,000 parishes. Existing incorporations under Gilbert's Act or local Acts were not to be dissolved. But with these exceptions, the Commissioners were specially empowered to group parishes into unions for the management of workhouses common to the united parishes, thus spreading the cost over a larger area, minimising the influence of local interests, enabling each union to employ competent paid officials, and facilitating the classification of the inmates. Power was reserved to two justices to order outdoor relief in cases where one of the justices making the order certified, of his own knowledge, that the recipient was, from old age or bodily infirmity, unable to work. In all other cases, relief to the able-bodied was to be regulated by the Commissioners' orders, and any relief administered contrary to their regulations was declared unlawful, and was to be disallowed. In pursuance of these wide powers, the Poor Law Commissioners, appointed under the Act, entrusted the ordinary administration of relief within each union to the "relieving officers," under the direction of the Board of Guardians, subject to the following, among other, regulations : (1) Except in cases of sickness or accident, no relief is to be given in money to any able-bodied pauper, who is in employment, nor to any part of his family. (2) If any able-bodied male pauper applies to be set to work by the parish, one-half at least of the relief is to be in kind. (3) No relief is to be given by payment of house-rent, or by allowance towards the same. The first regulations were temporary, designed as the first stage in a stricter administration of the law. But the attitude of Parliament towards the complete prohibition of outdoor relief and the universal establishment of workhouses faithfully reflected the feeling of the country. Round these two points raged a prolonged contest.

---

[1] Feb. 24, 1834. It had been preceded by the publication of *Extracts from Information received by His Majesty's Commissioners*, etc. (March 19, 1833).

The workhouse test was denounced as inhuman ; the houses were condemned as Bastilles of the poor ; the classification of the inmates was resisted as contrary to nature. Though the Commissioners for some time steadily persevered in their policy, it cannot be said that a complete victory rested with them or with their successors (1847), the Poor Law Board. The points which Parliament left open in 1834 are still those on which administrators of the Poor Law are divided.

APPENDIX III.

THE CORN LAWS.

A. Prices of Wheat, 1646-1926 (p. 488).

B. The Principal Acts relating to the Corn Trade (p. 490).

C. The Assize of Bread (p. 496).

D. Exports and Imports of Corn, 1697-1801 (p. 500).

E. Bounties paid on Exports of Corn, 1697-1765 (p. 500).

APPENDIX III.—CORN LAWS

A. PRICES OF WHEAT, 1646-1926.[1]

(i) STATEMENT OF THE ANNUAL AVERAGE PRICE OF WHEAT PER IMPERIAL QUARTER AT ETON, FROM THE YEAR 1646 TO 1770.

Years.	Average Prices.		Years.	Average Prices.		Years.	Average Prices.	
	s.	*d.*		*s.*	*d.*		*s.*	*d.*
1646 -	44	0	1688 -	42	1	1730 -	33	5
1647 -	67	5	1689 -	27	6	1731 -	30	0
1648 -	77	10	1690 -	31	8	1732 -	24	4
1649 -	73	3	1691 -	31	1	1733 -	25	11
1650 -	70	2	1692 -	42	8	1734 -	35	6
1651 -	67	2	1693 -	61	11	1735 -	39	4
1652 -	45	4	1694 -	58	7	1736 -	36	11
1653 -	32	5	1695 -	48	6	1737 -	34	9
1654 -	23	9	1696 -	65	0	1738 -	32	5
1655 -	30	6	1697 -	55	0	1739 -	35	2
1656 -	39	4	1698 -	62	7	1740 -	46	5
1657 -	42	8	1699 -	58	7	1741 -	42	8
1658 -	59	6	1700 -	36	7	1742 -	31	1
1659 -	60	6	1701 -	34	5	1743 -	22	9
1660 -	51	8	1702 -	26	11	1744 -	22	9
1661 -	64	1	1703 -	33	0	1745 -	25	2
1662 -	67	9	1704 -	42	7	1746 -	35	9
1663 -	52	3	1705 -	27	6	1747 -	31	10
1664 -	37	1	1706 -	23	9	1748 -	33	10
1665 -	45	2	1707 -	26	1	1749 -	33	10
1666 -	33	0	1708 -	37	11	1750 -	29	8
1667 -	33	0	1709 -	71	11	1751 -	35	2
1668 -	36	7	1710 -	71	6	1752 -	38	3
1669 -	40	7	1711 -	49	6	1753 -	40	10
1670 -	38	1	1712 -	42	5	1754 -	31	8
1671 -	38	6	1713 -	46	9	1755 -	31	0
1672 -	37	6	1714 -	46	1	1756 -	41	4
1673 -	42	8	1715 -	39	4	1757 -	55	0
1674 -	62	10	1716 -	44	0	1758 -	45	9
1675 -	59	2	1717 -	41	10	1759 -	36	4
1676 -	34	9	1718 -	35	6	1760 -	33	5
1677 -	38	6	1719 -	32	0	1761 -	27	7
1678 -	54	0	1720 -	33	10	1762 -	35	9
1679 -	55	0	1721 -	34	4	1763 -	37	2
1680 -	41	3	1722 -	33	0	1764 -	42	8
1681 -	42	8	1723 -	31	9	1765 -	49	6
1682 -	40	3	1724 -	33	10	1766 -	44	5
1683 -	36	7	1725 -	44	5	1767 -	59	1
1684 -	40	3	1726 -	42	1	1768 -	55	5
1685 -	42	8	1727 -	38	6	1769 -	41	10
1686 -	31	1	1728 -	49	11	1770 -	44	10
1687 -	23	0	1729 -	42	10			

[1] The figures for 1646-1770 are taken from *Returns Relating to the Importation and Exportation of Corn*, etc. ; Parliamentary Paper 177 (1843), p. 17. Those for 1771-1910 are taken from *Agricultural Statistics*, 1910, vol. xlv. pt. 3 [Cd. 5786, pp. 232-35]. Those for 1911-26 are taken from *Agricultural Statistics* for the years in question.

(ii) STATEMENT OF THE ANNUAL AVERAGE PRICE OF BRITISH WHEAT PER IMPERIAL QUARTER IN ENGLAND AND WALES, FROM 1771 TO 1935.

Years.	Average Prices. s. d.	Years.	Average Prices. s. d.	Years.	Average Prices. s. d.
1771 -	48 7	1826 -	58 8	1881 -	45 4
1772 -	52 3	1827 -	58 6	1882 -	45 1
1773 -	52 7	1828 -	60 5	1883 -	41 7
1774 -	54 3	1829 -	66 3	1884 -	35 8
1775 -	49 10	1830 -	64 3	1885 -	32 10
1776 -	39 4	1831 -	66 4	1886 -	31 0
1777 -	46 11	1832 -	58 8	1887 -	32 6
1778 -	43 3	1833 -	52 11	1888 -	31 10
1779 -	34 8	1834 -	46 2	1889 -	29 9
1780 -	36 9	1835 -	39 4	1890 -	31 11
1781 -	46 0	1836 -	48 6	1891 -	37 0
1782 -	49 3	1837 -	55 10	1892 -	30 3
1783 -	54 3	1838 -	64 7	1893 -	26 4
1784 -	50 4	1839 -	70 8	1894 -	22 10
1785 -	43 1	1840 -	66 4	1895 -	23 1
1786 -	40 0	1841 -	64 4	1896 -	26 2
1787 -	42 5	1842 -	57 3	1897 -	30 2
1788 -	46 4	1843 -	50 1	1898 -	34 0
1789 -	52 9	1844 -	51 3	1899 -	25 8
1790 -	54 9	1845 -	50 10	1900 -	26 11
1791 -	48 7	1846 -	54 8	1901 -	26 9
1792 -	43 0	1847 -	69 9	1902 -	28 1
1793 -	49 3	1848 -	50 6	1903 -	26 9
1794 -	52 3	1849 -	44 3	1904 -	28 4
1795 -	75 2	1850 -	40 3	1905 -	29 8
1796 -	78 7	1851 -	38 6	1906 -	28 3
1797 -	53 9	1852 -	40 9	1907 -	30 7
1798 -	51 10	1853 -	58 3	1908 -	32 0
1799 -	69 0	1854 -	72 5	1909 -	36 11
1800 -	113 10	1855 -	74 8	1910 -	31 8
1801 -	119 6	1856 -	69 2	1911 -	31 8
1802 -	69 10	1857 -	56 4	1912 -	34 9
1803 -	58 10	1858 -	44 2	1913 -	31 8
1804 -	62 3	1859 -	43 9	1914 -	34 11
1805 -	89 9	1860 -	53 3	1915 -	52 10
1806 -	79 1	1861 -	55 4	1916 -	58 5
1807 -	75 4	1862 -	55 5	1917 -	75 9
1808 -	81 4	1863 -	44 9	1918 -	72 10
1809 -	97 4	1864 -	40 2	1919 -	72 11
1810 -	106 5	1865 -	41 10	1920 -	80 10
1811 -	95 3	1866 -	49 11	1921 -	71 6
1812 -	126 6	1867 -	64 5	1922 -	47 10
1813 -	109 9	1868 -	63 9	¹ 1923 -	42 2
1814 -	74 4	1869 -	48 2	¹ 1924 -	49 3
1815 -	65 7	1870 -	46 11	¹ 1925 -	52 2
1816 -	78 6	1871 -	56 8	¹ 1926 -	53 3
1817 -	96 11	1872 -	57 0	¹ 1927 -	49 3
1818 -	86 3	1873 -	58 8	¹ 1928 -	42 10
1819 -	74 6	1874 -	55 9	¹ 1929 -	42 2
1820 -	67 10	1875 -	45 2	¹ 1930 -	34 3
1821 -	56 1	1876 -	46 2	¹ 1931 -	24 8
1822 -	44 7	1877 -	56 9	¹ 1932 -	25 4
1823 -	53 4	1878 -	46 5	¹ 1933 -	22 10
1824 -	63 11	1879 -	43 10	¹ 1934 -	20 9
1825 -	68 6	1880 -	44 4	¹ 1935 -	22 2

¹ Converted from the ascertained price per cwt.

B. PRINCIPAL ACTS OF PARLIAMENT AFFECTING THE CORN TRADE.

1360.   34 Edw. III. c. 20.—Prohibiting exportation of corn, except by the King's license for the supply of Calais and Gascony.

1393.   17 Ric. II. c. 7.—Permitting the export of corn, except to the King's enemies, subject to the power of the King's Council to restrain exportation in the interests of the nation.

1436.   15 Hen. VI. c. 2.—Permitting the exportation of corn, without the royal license, when the price of wheat at the place of shipment did not exceed 6s. 8d. per quarter, and at proportionate prices for other grains.

1463.   3 Edw. IV. c. 2—Prohibiting the importation of foreign corn, when the price of wheat at the place of import did not exceed 6s. 8d. per quarter, and at proportionate prices for other grains.

1533.   25 Hen. VIII. c. 2.—Prohibiting the export of corn without the royal license.

1551-2.   5 and 6 Edw. VI. c. 14.—Persons engrossing corn (*i.e.* buying corn to sell again) were subjected to heavy penalties, and, on a third offence, to the pillory, forfeiture of goods, and imprisonment. Persons were, however, permitted to engross corn, provided that they did not forestall it, or regrate it (*i.e.* hold it for a rise), when the price of corn did not exceed 6s. 8d. per quarter. Farmers buying corn for seed were compelled to sell an equivalent amount, or forfeit double what they had bought.

1554.   1 P. and M. c. 5.—Restoring freedom of exportation when the price of wheat did not exceed 6s. 8d. per quarter, and of other grains in proportion.

1562-3.   5 Eliz. c. 12.—Corn-badgers, *i.e.* persons buying corn at open fairs and markets, were required to take out licenses and to give security not to engross, forestall, or buy otherwise than at open fairs and markets.

1562-3.   5 Eliz. c. 5.—Freedom of exportation from ports specially licensed by the Crown extended when the price of wheat did not exceed 10s. per quarter and of other grains in proportion.

1570.   13 Eliz. c. 13.—Providing for the annual settlement of the average prices by which exportation was regulated. The Lord President and the Council in the North, the Lord President and the Council in Wales, and the Justices of Assize, within their respective jurisdictions, were yearly, upon conference, had with the inhabitants of the country, on the cheapness and dearth of all kinds of grains, to determine the averages for the year, and permit or prohibit the exportation of grain.

Corn could be exported freely to any foreign country subject to a customs duty of 1s. per quarter of wheat and other grains in proportion, when no proclamation was issued to the contrary.

1593.   35 Eliz. c. 7.—Permitting exportation of grain subject to a customs duty of 2s. per quarter of wheat, and other grains in proportion, when the prices of wheat did not exceed 20s. per quarter and other grains in proportion.

1604.   1 Jac. I. c. 25.—Raising the limit of price for the export of wheat to 26s. 8d. per quarter and other grains proportionately.

1623.   21 Jac. I. c. 28.—Raising the export limit for wheat to 32s. per quarter and for other grains in proportion.

1660.   12 Car. II. c. 4.—The export of corn was permitted whenever the prices at the port of exportation did not exceed, for wheat, 40s. per quarter; rye, pease, and beans, 24s.; barley and malt, 20s.; oats, 16s. The poundage on exportation amounted to 1s. per quarter for wheat, 4d. for oats, and 6d.

for other grain. The import rates, from poundage, were as follows, subject
to an allowance of 5 per cent. for discount :

						s.	d.
Wheat and buck-wheat, when not exceeding 44s.							
at the port of importation,	-	-	-	per quarter	2	0
Ditto, exceeding 44s.,	-	-	-	-	„	0	4
Rye, beans, barley, and malt, when not exceed-							
ing 36s.,	-	-	-	-	-	„	1	4
Ditto, exceeding 36s.,	-	-	-	-	„	0	3
Oats and pease,	-	-	-	-	-	„	0	2½
					per last of		
Meal of wheat or rye,	-	-	-	-	12 barrels	3	0

1663. 15 Car. II. c. 7.—Corn was allowed to be exported when the prices
did not exceed the following sums per quarter at the places of shipment :
Wheat, 48s. ; rye, pease, and beans, 32s. ; barley, buck-wheat, and malt,
28s. ; oats, 13s. 4d. The import duties were increased to the sums undermentioned, on corn, when not exceeding the same prices at the port of
importation :

	s.	d.
Wheat, when not exceeding 48s. per quarter, - - -	5	4
Rye, pease, and beans, when not exceeding 32s. per quarter,	4	0
Barley and malt, when not exceeding 28s. per quarter, -	2	8
Oats, when not exceeding 13s. 4d. per quarter, - - -	1	4
Buck-wheat, when not exceeding 28s. per quarter, - -	2	0

1670. 22 Car. II. c. 13.—Intituled " An Act for the Improvement of
Tillage," corn might be exported, although the prices exceeded the sums
fixed by 15 Car. II. c. 7, paying poundage as before, viz. 1s. per quarter on
wheat, 4d. on oats, and 6d. on other grain. Rates of duty on importation
were imposed, in lieu of poundage, on corn when under certain prices at the
port of importation, as follows :

	s.	d.
On wheat, when not exceeding 53s. 4d. per quarter, - -	16	0
Ditto, exceeding 53s. 4d., and not above 80s. per quarter, -	8	0
Rye, pease, and beans, when not exceeding 40s. per quarter,	16	0
Barley, malt, and buck-wheat, when not exceeding 32s. per		
quarter, - - - - - - - - -	16	0
Oats, when not exceeding 16s. per quarter, - - -	5	4
Wheat-flour—as Wheat.		

When exceeding those prices, poundage as before was chargeable, viz. 4d.
on wheat, 3d. on rye, beans, barley, and malt, 2½d. on oats and pease.

1689. 1 William and Mary, c. 12.—Bounties were granted on the exportation of the following articles, ground or unground :

	s.	d.
Wheat, 5s. per quarter, when the price was at or under -	48	0
Rye, 3s. 6d. per quarter, when the price was at or under -	32	0
Barley and malt, 2s. 6d. per quarter, when barley was at or		
under - - - - - - - - -	24	0

1707. 5 Anne, c. 29.—Bounties were allowed on the export of the
following :

	s.	d.
Beer or bigg, 2s. 6d. per quarter, when the price was at or		
under - - - - - - - - -	24	0
Malt made of wheat, 5s. per quarter, - - - -	48	0
Oatmeal, 2s. 6d. per quarter, when oats were - - -	15	0

1774. 13 Geo. III. c. 43, from 1st January.

DUTIES ON IMPORTATION.

Whenever the prices of middling British corn at the port of importation were at or above the following sums, the undermentioned rates were to be paid :

		Prices.		Duty.
Wheat (and buck-wheat),	-	48s.	-	6d. per qr. on wheat. 2d. per cwt. on flour.
Rye, pease, and beans,	- -	32s.	-	3d. per qr.
Barley, beer or bigg,	- -	24s.	-	2d. per qr.
Oats, - - - -	-	16s.	-	2d. per qr.

When the prices were under those sums, the former scale of duties became chargeable ; in which case, corn and flour were allowed to be warehoused at the principal ports, paying, on delivery for home consumption, such duty as might be due at that time.

EXPORT.

The export to foreign parts of corn or meal, flour, malt, and bread and biscuit made therefrom, was prohibited when British corn was at or above the following sums, per quarter, at the port of exportation : wheat, 44s. ; rye, pease, and beans, 28s. ; barley, beer or bigg, 22s. ; oats, 14s.

BOUNTIES ON BRITISH CORN, GROUND OR UNGROUND, EXPORTED IN BRITISH SHIPS.

When the prices of middling corn were under the following sums at the port from which exported, viz. :

Wheat, - - -	44s.	A bounty allowed of 5s. per quarter on wheat, and malt made therefrom.
Rye, - - -	28s.	A bounty allowed of 3s. per quarter on rye.
Barley, beer or bigg,	22s.	A bounty allowed of 2s. 6d. per quarter on barley, beer or bigg, and on malt made therefrom.
Oats, - - -	14s.	A bounty allowed of 2s. per quarter on oats, and 2s. 6d. per quarter on oatmeal, at the rate 276 lbs. to the quarter.

1791. 31 Geo. III. c. 30, from 15th November.—In order to ascertain the average home price of corn, the Act directed that the maritime counties of England should be divided into twelve districts, and that the prices, ascertained separately for these districts, should regulate the duties and bounties in each. The export of wheat was prohibited, when wheat was at or above 46s. per quarter. The export of other corn or meal was prohibited at proportionate prices. The duties on imported wheat were as follows :

From Ireland, or British Colonies or Plantations in North America.	From other parts.	s.	d.
Under 48s. per qr., - -	Under 50s., - - - -	24	3
At or above 48s. but under 52s.,	At or above 50s. but under 54s.,	2	6
At or above 52s., - - -	At or above 54s., - - -	0	6

The importation of other species of grain was regulated by proportionate duties at proportionate prices.

1796. 36 Geo. III. c. 21.—Bounties granted (for the first time) on corn imported at certain ports in British or neutral vessels, from 24th September, 1795, to 30th September, 1796, viz. :

Wheat and flour from any port of Europe, south of Cape Finisterre, from the Mediterranean or Africa :

20s. on every quarter of wheat weighing not less than 440 lbs.,   -   -   -   -
16s. on every quarter of wheat weighing not less than 424 lbs.,   -   -   -   -
6s. on every cwt. of wheat-flour,   -   -

Until the quantity of such wheat and flour together amounts to 400,000 quarters.

Wheat and flour from other parts of Europe (not British Dominions) :

15s. on every quarter of wheat weighing not less than 440 lbs.,   -   -   -   -
12s. on every quarter of wheat weighing not less than 424 lbs.,   -   -   -   -
4s. 6d. on every cwt. of wheat-flour, -   -

Until the quantity amounts to 500,000 quarters.

Wheat and flour from the British Colonies or the United States of America :

20s. on every quarter of wheat weighing not less than 440 lbs.,   -   -   -   -
16s. on every quarter of wheat weighing not less than 424 lbs.,   -   -   -   -
6s. on every cwt. of wheat-flour,   -   -

Until the quantity amounts to 500,000 quarters.

On wheat and flour exceeding the quantities to which the beforementioned bounties are limited :

10s. on every quarter of wheat weighing not less than 440 lbs.
8s. on every quarter of wheat weighing not less than 424 lbs.
3s. on every cwt. of wheat-flour.

INDIAN CORN AND MEAL.

5s. for every quarter of corn,   -
1s. 6d. for every cwt. of meal,   -

Until the quantity together amounts to 500,000 quarters.

3s. for every quarter of corn,   -
1s. for every cwt. of meal,   -

Exceeding the limited quantity of 500,000 quarters.

RYE.

10s. for every quarter weighing not less than 400 lbs. until the quantity amounts to 100,000 quarters.

6s. for every quarter exceeding the limited quantity of 100,000 quarters.

N.B.—On warehoused corn delivered out within three months, three-fifths of the bounty granted on the importations from the Mediterranean, allowed. (2½ cwt. of wheat-flour deemed equal to one quarter of wheat, and 3½ cwt. of Indian meal to one quarter of Indian corn.)

1804. 44 Geo. III. c. 109, from 15th November.—The plan for ascertaining the average prices, as laid down by the Act of 1791, was altered, and it was now directed that importation and exportation should be regulated in England by the aggregate average of the 12 maritime districts into which it was divided, and in Scotland by the aggregate average of the four Scotch districts. The averages were to be taken four times a year. In 1805, by 45 Geo. III. c. 86, the plan was again altered, and it was arranged that the prices both in England and Scotland should be regulated by the average prices of the 12 English maritime districts.

1814. 54 Geo. III. c. 69, from 17th June.—Corn, grain, meal, and flour (bread and biscuit added by Treasury Order), to be exported at all times without payment of duty, and without receiving any bounty.

1815. 55 Geo. III. c. 26, from March 23.—Foreign and colonial corn, meal, or flour might be at all times imported, and warehoused, without payment of duties ; but it could only be taken out of warehouse for home consumption, or entered for the like purpose on importation, whenever the prices of British corn should be at or above the following sums, and then duty free :

	For Corn of the British Colonies in North America.	For Corn not of the British Colonies in North America.
Wheat, - - - -	67s. per quarter.	80s. per quarter.
Rye, pease, and beans, - -	44s. ,,	53s. ,,
Barley, beer or bigg, - -	33s. ,,	40s. ,,
Oats, - - - - -	22s. ,,	27s. ,,

1822. Table of duties payable by Act 3 George IV. c. 60, on wheat when admitted for home consumption :

From the British Colonies or Plantations of North America.	From other parts.	Duty.	Additional for the first three months.
When the average prices of British Wheat should be	When the average prices of British Wheat should be		
Under 67s. per quarter, -	Under 80s. per quarter, -	12s.	5s.
At or above 67s., but under 71s., - - - -	At or above 80s., but under 85s., - - - -	5s.	5s.
At or above 71s., - -	At or above 85s., - -	1s.	—

The importation of other grain was regulated by proportionate prices and duties.

1825. 6 Geo. IV. c. 64 (temporary).—The prohibitions and restrictions upon the importation of wheat, the produce of the British possessions in North America, and the duties payable thereon, under the Acts of 1815 and 1822, were suspended from 22nd June, 1825, until the end of the then next session of Parliament, during which period such wheat might be entered for home consumption, whatever might be the average price of British corn, on payment of a duty of 5s. per quarter. (The provisions of this Act expired in July, 1827.)

1828. 9 Geo. IV. c. 60.—The duties on wheat imported from any foreign country were as follows : When the average price of wheat is

66s. and under 67s.,	the duty to be 20s. 8d. per quarter.
67s. ,, 68s.	,, 18s. 8d. ,,
68s. ,, 69s.	,, 16s. 8d. ,,
69s. ,, 70s.	,, 13s. 8d. ,,
70s. ,, 71s.	,, 10s. 8d. ,,
71s. ,, 72s.	,, 6s. 8d. ,,
72s. ,, 73s.	,, 2s. 8d. ,,
at or above 73s.	,, 1s. 0d. ,,

The duties on wheat imported from any British Possession out of Europe were as follows : When the average price of wheat is

Under 67s. per quarter, the duty to be 5s. 0d.
At or above 67s. „   „   0s. 6d.

The importation of other grain, foreign or colonial, was regulated by proportionate prices and duties.

1842. 5 and 6 Vict. c. 14, from 29th April.—The Act of 1828 repealed and another scale of duties imposed. The prices for the regulation of the duty to be made up and computed on Thursday in each week from the returns received of the sales of corn the produce of the United Kingdom during the preceding week ending Saturday, and a certificate thereof to be transmitted to the officers of the Customs at the respective ports on the same day.

Rates of duty on wheat " Imported from any foreign country " (i.e. not being the produce of and imported from any British Possession out of Europe).

Prices			Duty per qr.		Prices.			Duty per qr.	
			s.	d.				s.	d.
When under 51s.,	-		- 20	0	When 62s. and under 63s.,			- 10	0
51s. and under 52s.,			- 19	0	63s.	„	64s.,	- 9	0
52s.	„	55s.,	- 18	0	64s.	„	65s.,	- 8	0
55s.	„	56s.,	- 17	0	65s.	„	66s.,	- 7	0
56s.	„	57s.,	- 16	0	66s.	„	69s.,	- 6	0
57s.	„	58s.,	- 15	0	69s.	„	70s.,	- 5	0
58s.	„	59s.,	- 14	0	70s.	„	71s.,	- 4	0
59s.	„	60s.,	- 13	0	71s.	„	72s.,	- 3	0
60s.	„	61s.,	- 12	0	72s.	„	73s.,	- 2	0
61s.	„	62s.,	- 11	0	73s. and upwards,			- 1	0

Rates of duty on wheat, the produce of and imported from any British Possession out of Europe.

When under 55s.,	-		- 5	0	When 57s. and under 58s.,		- 2	0
55s. and under 56s.,			- 4	0	58s. and upwards,		- 1	0
56s.	„	57s.,	- 3	0				

These duties ceased to apply to wheat and flour, *the produce of Canada,* imported after 10th October, 1843. See 1843.

The importation of other grain, foreign or colonial, was regulated by proportionate prices and duties.

1843. 6 and 7 Vic. c. 29.—In consideration of a duty of 3s. per quarter having been imposed by the Legislature of Canada on wheat imported into that province from other places than the United Kingdom or British Possessions, the duty on wheat and wheat flour, the produce of Canada, imported into the United Kingdom after 10th October, 1843, and during the continuance of the duty of 3s. in Canada, is to be at all times 1s. per quarter on wheat, and on flour, for every 196 lbs. a duty equal to that payable on 38½ gallons of wheat (or 4¼d. per cwt.).

1846. 9 and 10 Vict. c. 22.—From Feb. 1, 1849, the following fixed rates of duty on corn, meal and flour came into operation in lieu of those previously levied under the same Act :

Wheat, barley, beer or bigg, oats, rye, pease, and beans,   1s. per qr.
Wheatmeal and flour, barley meal, oatmeal, rye, meal
and flour, pea meal and bean meal,   -   -   -   4¼d. per cwt.

1869. The duties on corn were entirely repealed from the 1st June by 32 and 33 Vic. c. 14.

C. THE ASSIZE OF BREAD.[1]

The Assize of Bread, like the Assize of Ale, formed part of the system by which, in the interests of consumers, prices of food and drink were regulated. The Assize is defined in one of the most famous of mediaeval statutes—the *Assisa Panis et Cervisiæ*. But though the name is familiar, the date of the statute is uncertain. Probably passed towards the end of the reign of Henry III. (1266 ?), that part of it which referred to bread, though revised and altered by subsequent legislation, remained in force for London till 1815, and for the rest of the country till 1836.

It was the duty of Justices of the Peace to " set the Assize," in other words, to adjust the weight, quality, and price of bread to the current prices of wheat, with the addition of an allowance for the labour and skill of the baker. The method, by which this adjustment was effected, was extremely complicated. It started with the legal liability of the baker to make 418 lbs. of bread out of every quarter of wheat. The first step was to ascertain the average price fetched by a quarter of wheat in the public markets of the neighbourhood. The next step was to add to this price the discretionary allowances for the expenses and skill of the baker. These two sums, added together, represented the total sum for which the 418 lbs. were to be sold. The last step was to calculate the exact weight of bread which each penny would buy, in order that the whole 418 lbs. might realise the ascertained sum. The table, thus calculated, was called the Assize of Bread : in it were given the weights of the loaves which were to be sold at the customary prices. The Assize was periodically proclaimed, and to sell bread above the price or below the weight set out in the current table was a penal offence.

The amount of bread to be made from each quarter of wheat remained unaltered down to 1710 (8 Anne, c. 11), when it was reduced, it is said accidentally, from 418 lbs. to 417 lbs. The allowance of the baker on each quarter of wheat varied more widely. In 1497 it was 2s. But the Church Rate, Education, Sanitary, Police, and Poor Rates, had not then to be taken into consideration. In 1620 these payments entered into the calculation. In that year the white bakers of London petitioned that the allowance should be raised from 6s. to 8s., owing to their necessary expenses for food and clothing, and " the teaching at school " of their children, their " duties to the parson, the scavengers, for the poor, for watching and warding," etc. The items of the allowances at the earlier period are sometimes quaint, e.g. : [2]

" Furnace and wood, - - - - - - - -	6d.
The Miller, - - - - - - -	4d.
Two journeymen and two apprentices, - - - -	5d.
Salt, yeast, candle, and sackbands, - - - - -	2d.
Himself, his house, his wife, his dog and his cat, - - -	7d.

In all   - 2s. 0d."

The average price of a quarter of wheat from 1453 to 1497 is said to have been, in modern money, 14s. 1d.[3] Taking this figure as an illustration, the method of " setting the Assize " may be thus exemplified. The addition of the discretionary allowance of 2s. to the price (14s. 1d.) of the quarter of wheat gives as the total 16s. 1d. The justices had to calculate the weight of each penny loaf, so that the whole 418 lbs. of bread might be sold to realise

---

[1] The Assize of Bread is fully treated in G. Atwood's *Review of the Statutes and Ordinances of Assize*, 1202-1797 (1801), and in an article by Sidney and Beatrice Webb in *The Economic Journal* for June, 1904, pp. 196-218.

[2] Quoted by Sidney and Beatrice Webb in "The Assize of Bread" (*Economic Journal*, June, 1904, p. 197).

[3] Adam Smith, *Wealth of Nations*, M'Culloch's edition (1850), p. 117.

16s. 1d. The table of weights and prices, so computed and proclaimed, was the Assize of Bread. Besides his allowance, the baker had for his profits the offals, and his " advantage bread," consisting of the additional amount which he could make out of the quarter of wheat over and above the fixed 418 lbs. of bread. The flour which the baker obtained from the miller was to be worked up into three different·qualities of bread. Their proportionate values were fixed. Thus, when one penny would buy 1 lb. of " bread treet," or household bread, it would approximately buy ¼ lb. of " bread of the whole wneat " or wheaten bread, or ½ lb. of " wastell," or white bread. Put in another way, when the finest white bread cost one penny a pound, the pound of wheaten and of household bread could be bought at the approximate prices of three-farthings and of one halfpenny respectively.

It is difficult to say whether Assizes of Bread were in the Stewart period very generally set. They were certainly much more frequently proclaimed in towns than in country districts where bread was usually baked at home. By the end of the seventeenth century, the practice seems to have fallen into disuse, even in corporate towns. An attempt to revive it was made by the Government in 1710 in the interests of consumers. The statutory allowance of bakers was raised to twelve shillings per quarter, and, as has been already stated, the amount of bread to be made from the quarter of wheat was reduced from 418 lbs. to 417 lbs. A still more important change was necessitated by the intervention of a new class of trader between the baker and the wheat-grower. The industrial organisation had become more complicated than it was in the Middle Ages. Bakers no longer bought their quarters of wheat direct from the farmer, carried them to the mill, paid the miller for grinding, and carried away the product in the form of flour and offals. Now millers themselves bought the wheat from the growers, ground it into flour, separated it into different qualities, and sold them at different prices to the bakers. These changed conditions were most ineffectively met by empowering the justices, at their option, to calculate their tables on the prices either of wheat or of flour. The local prices, thus settled by the justices and returned to the Custom House officers, supplied the statistics by which, under the Corn Laws, the bounties, the prohibitions of exports, and the duties on imports were to a great extent regulated.

The Act of 1710 remained unaltered till 1758. If no other proof existed, it might be concluded from this fact, that the prices of wheat remained low. When the prices of food rose, public discontent was generally expressed in a demand for some change in the laws by which they were regulated. Practices which, though irritating, were tolerated in days of cheapness, became in times of scarcity burdensome beyond endurance. Complaints were always numerous, but mostly from the trade. It was, for instance, alleged that country districts were unprotected against frauds by neglect of the practice of setting Assizes which had proved beneficial in towns ; that the informers who profited by the penalties under the Act were mischievously active ; that bakers could not, owing to their dependence on the millers, comply with the regulations ; that the wheat prices were improperly taken ; that the best white bread could not be produced at the prices fixed in the tables ; that the lower quality of bread was largely adulterated by the use not only of alum but of burnt bones, chalk, lime, and whiting ; that the poorest classes refused to eat any bread except that made from the whitest flour, and sacrificed the nourishment of wheat to an absurd fashion. The last complaint illustrates the tendency of history to repeat itself. The pamphlets of the day exhaust the subject of the " whole meal " agitation of 1911 ; but, in the early part of the eighteenth century, the demand for white bread was a sign of an improved standard of living.

The Act of 1710 remained in force till 1758. Among the bakers it was unpopular; there is less evidence that it dissatisfied consumers. When prices rose, it was human, since natural causes were beyond control, to blame the method of setting the Assize. The year 1757 was one of scarcity; by the Bristol tables the penny loaf weighed only 4 oz. 14 dwt. There was an immediate demand for a change in the law. In 1758 a new Act was passed (31 Geo. II. c. 29), codifying the existing law, and introducing alterations which seem to have been unduly favourable to bakers. Bread was to be made of two qualities only, wheaten and household; the number of pounds to be sold per quarter was reduced from 417 to 365; bakers were allowed to choose whether they would sell in their shops "assized" loaves, which varied in weight but were fixed in price, or "prized" loaves, varying in price but fixed in weight. The Statute fixes by a table of proportion the weight of the 1d. loaf of wheaten and of household bread, regulated by the price of wheat and the baker's allowance. It also settles in the same way the price of the "prized" loaf. By a later Act the wheaten bread was to be stamped with W and the household with H. The "prized" loaf was the "peck loaf," made from 2 gallons of wheat or 14 lbs. of flour, and weighing 17 lbs. 6 oz., and its subdivisions—the half peck loaf (8 lbs. 11 oz.); the quarter peck or quartern (4 lbs. 5½ oz.); and the half quartern (2 lbs. 2¾ oz.). To prevent fraud, no baker was allowed to sell in the same shop "assized" as well as "prized" loaves. The Assizes were to be periodically proclaimed according to the fluctuations in prices, and were not to remain in force more than 14 days.

The Act of 1758 fell on troublous times. The period 1765-74 was one of scarcity. The high prices of bread were attributed to the malpractices of bakers and millers. It was alleged that the changes in the law had allowed them unusual opportunities for making excessive profits. It is possible that this was the case. No assay of flour was attempted. Consequently, millers were able to return their product as being of superior quality, though they continued to supply the inferior grades, and bakers conformed to the Assize by adulterating the standard of both the legalised classes of bread. An agitation was begun, which resulted in the Act of 1773 (13 Geo. III. c. 62). The number of pounds of bread to be sold out of each quarter of wheat was restored to its former basis; regulations were made prescribing the method in which flour was to be dressed; the old standard wheaten bread, which, "according to the antient order and custom of the realm," had existed "from time immemorial," was again legalised, and was to be stamped S.W. There were, therefore, once more three qualities of bread—wheaten, standard wheaten, and household. The proportionate cost of the three kinds was also regulated. The same weights of wheaten, standard wheaten, and household were to be sold respectively at 8d., 7d., and 6d. No attempt was made, except in the case of standard wheaten, to define the quality of the different breads. Standard wheaten was to consist of the whole produce of the grain, the bran or hull only excepted, and the flour of which it was made was to weigh three-fourths of the wheat from which it was ground.

The new Act may have been as easily evaded as the old. But the fall in prices after 1774 cheapened bread, and the contented consumer probably attributed his relief to the success of the new law. The extraordinary rise in prices which took place from 1794 to 1812 revived the whole question of the efficacy of Assizes of Bread in aggravated form. In 1812 at Grantham the weight of assized bread to be bought for one penny was only 4 oz. 6 dwt., bringing the cost of the quartern loaf to 1s. 4d. The rise was so unprecedented, that millers and bakers were suspected of every variety of misdemeanour. Bakers especially were charged with reducing the weight, and raising the price of bread beyond the limits justified by the advance in the price of wheat. There was a general demand for the enforcement of the

Assize of Bread. In some districts, justices prohibited the sale of any kind of bread except the standard wheaten. In others, they set the table of prices so low that bakers refused to bake. The Assize was either disregarded, or public subscriptions were raised to induce bakers to continue their trade. The regulations naturally failed to lower prices, though some effect may have been produced in compelling bakers to follow their variations. Among other devices to reduce the price of wheat was the authority given to bakers in 1795 (36 Geo. III. c. 22) to make and sell bread, stamped with M, which was mixed with other ingredients than corn. But the people, for whose relief the mixed bread was designed, resolutely refused to touch it. They rejected even standard wheaten and household bread, and demanded the finest and whitest bread. It was in vain that members of Parliament, Privy Councillors, magistrates, aldermen, and vestrymen endeavoured to set the fashion by eating the coarser qualities themselves. The people clung to their improved standard of living, and bakers could only satisfy the tastes and pockets of their customers by the production of white bread which was artificially whitened by wholesale adulteration.

Efforts were made to improve the system of setting the Assize. In consequence of the Report of a Select Parliamentary Committee, an amending Act was passed in 1813 (53 Geo. III. c. 116). But the feeling was becoming more and more general that regulations affecting prices of food were mischievous, that legislation was powerless, and that, where laws failed, free competition might succeed. Another Select Committee was appointed to consider the " Laws relating to the Manufacture Sale and Assize of Bread." On their Report in 1815, an Act was passed (55 Geo. III. c. 49), which applied only to London and a metropolitan area of ten miles round. Bakers were permitted to sell loaves of specified weight at any price they chose. In provincial towns the Assize still lingered. In 1821 a Report in favour of complete freedom of trade was presented to the House of Commons by a Committee appointed to consider the " Regulations relative to the Making and the Sale of Bread." The immediate result of their Report was the Bread Act of 1822 (3 Geo. IV. c. 106), which finally abolished all regulations of weight or price in London. Its remoter effect was the application of a similar Act to the provinces. In 1836 (6 and 7 Wm. IV. c. 37) the Assize of Bread was at last, after an existence of nearly six centuries, finally abolished.

D EXPORTS AND IMPORTS OF CORN (1697-1801).

KIND OF GRAIN.	NUMBER OF QUARTERS EXPORTED.			Total Exports 1697-1801.	NUMBER OF QUARTERS IMPORTED.			Total Imports 1697-1801
	1697-1731.	1732-1766.	1767-1801.		1697-1731.	1732-1766.	1767-1801.	
Wheat, Flour, etc.,	3,592,163	11,540,216	3,063,649	18,196,028	124,417	291,773	10,541,300	10,957,490
Barley, Malt, etc., -	7,467,129	10,259,675	2,614,385	20,341,189	35,340	49,270	1,957,838	2,042,448
Oats and Oatmeal,	217,490	364,779	736,777	1,319,046	432,514	1,050,320	14,918,447	16,401,281
Rye and Ryemeal, -	1,098,885	1,437,727	173,179	2,709,791	178,224	22,205	1,122,641	1,323,070
Peas and Beans, -	690	25,274	666,096	692,060	185	9,569	1,184,304	1,194,058
TOTALS, -	12,367,367	23,627,671	7,254,086	43,258,114	770,680	1,423,137	29,724,530	31,918,347

E. BOUNTIES PAID UNDER 1 WM. AND MARY, C. 12, AND SUBSEQUENT ACTS OF PARLIAMENT ON EXPORTS OF CORN BETWEEN THE YEARS 1697 AND 1765.

	£ s. d.
1697-1705,	£289,670 14 0
1706-25,	1,371,032 4 0
1726-45,	1,769,756 4 2
1746-65,	2,628,503 4 7
1697-1765	£6,058,962 6 9

APPENDIX IV.

TABLES OF ESTIMATES BY GREGORY KING, CHARLES DAVENANT AND W. COULING.

TABLE 1.—*A Scheme of the Income and Expence of the several Families of England, calculated for the year 1688.*[1]

Number of Families.	Ranks, Degrees, Titles, and Qualifications.	Heads per Family.	Number of Persons.	Yearly Income per Family.		Yearly Income in General.	Yearly Income per Head.		Yearly Expense per Head.			Yearly Increase per Head.			Yearly Increase in General.
				£	s.	£	£	s.	£	s.	d.	£	s.	d.	£
160	Temporal lords,	40	6,400	3,200	0	512,000	80	0	70	0	0	10	0	0	64,000
26	Spiritual lords,	20	520	1,300	0	33,800	65	0	45	0	0	20	0	0	10,400
800	Baronets,	16	12,800	880	0	704,000	55	0	49	0	0	6	0	0	76,800
600	Knights,	13	7,800	660	0	390,000	50	0	45	0	0	5	0	0	39,000
3,000	Esquires,	10	30,000	450	0	1,200,000	45	0	41	0	0	4	0	0	120,000
12,000	Gentlemen,	8	96,000	280	0	2,880,000	35	0	32	0	0	3	0	0	288,000
5,000	Persons in greater offices and places,	8	40,000	240	0	1,200,000	30	0	26	0	0	4	0	0	160,000
5,000	Persons in lesser offices and places,	6	30,000	120	0	600,000	20	0	17	0	0	3	0	0	90,000
2,000	Eminent merchants and traders by sea,	8	16,000	400	0	800,000	50	0	37	0	0	13	0	0	208,000
8,000	Lesser merchants and traders by sea,	6	48,000	198	0	1,600,000	53	0	27	0	0	6	0	0	288,000
10,000	Persons in the law,	7	70,000	154	0	1,540,000	22	0	18	0	0	4	0	0	280,000
2,000	Eminent clergymen,	6	12,000	72	0	144,000	12	0	10	0	0	2	0	0	24,000
8,000	Lesser clergymen,	5	40,000	50	0	400,000	10	0	9	4	0	0	16	0	32,000
40,000	Freeholders of the better sort,[2]	7	280,000	91	0	3,640,000	13	0	11	15	0	1	5	0	350,000
120,000	Freeholders of the lesser sort,[2]	5½	660,000	55	0	6,600,000	10	0	9	10	0	0	10	0	330,000
150,000	Farmers,	5	750,000	42	10	6,375,000	8	10	8	5	0	0	5	0	187,500
15,000	Persons in liberal arts and sciences,	5	75,000	60	0	900,000	12	0	11	0	0	1	0	0	75,000

[1] *Political and Commercial Works of Charles Davenant*, collected and revised by Sir C. Whitworth, vol. ii. p. 184 (London, 1761).

[2] Gregory King gives 140,000 as the number of "lesser" freeholders.

TABLE 1.—Continued.

Number of Families.	Ranks, Degrees, Titles, and Qualifications.	Heads per Family.	Number of Persons.	Yearly Income per Family. (£ s.)	Yearly Income in General. (£)	Yearly Income per Head (£ s.)	Yearly Expense per Head. (£ s. d.)	Yearly Increase per Head. (£ s. d.)	Yearly Increase in General. (£)
50,000	Shopkeepers and tradesmen, -	4½	225,000	45 0	2,250,000	10 0	9 0 0	1 0 0	225,000
60,000	Artisans and handicrafts, -	4	240,000	38 0	2,280,000	9 10	9 0 0	0 10 0	120,000
5,000	Naval officers, - - -	4	20,000	80 0	400,000	20 0	18 0 0	2 0 0	40,000
4,000	Military officers, - -	4	16,000	60 0	240,000	15 0	14 0 0	1 0 0	16,000
500,586		5¼	2,075,520	68 18	34,488,800	12 18	11 15 4	1 2 8	3,023,700
50,000	Common seamen, - -	3	150,000	20 0	1,000,000	7 0	7 10 0	Decrease 0 10 0	Decrease 75,000
364,000	Labouring people and out-servants, -	3½	1,275,000	15 0	5,460,000	4 10	4 12 0	0 2 0	127,500
400,000	Cottagers and paupers, -	3¼	1,300,000	6 10	2,000,000	2 0	2 5 0	0 5 0	325,000
35,000	Common soldiers, -	2	70,000	14 0	490,000	7 0	7 10 0	0 10 0	35,000
849,000	Vagrants ; as gipsies, thieves, beggars, etc., -	3¼	2,795,000	10 10	8,950,000	3 5	3 9 0	0 4 0	562,500
			30,000		60,000	2 0	4 0 0	2 0 0	60,000
	So the general account is								
500,586	Increasing the wealth of the kingdom, - - -	5¼	2,675,520	68 18	34,488,800	12 18	11 15 4	1 2 8	3,023,700
849,000	Decreasing the wealth of the kingdom, -	3¼	2,825,000	10 10	9,010,000	3 3	3 7 6	0 4 6	622,500
1,349,586	Neat totals, - -	4 3/13	5,500,520	32 5	43,491,800	7 18	7 9 3	0 8 9	2,401,200

TABLE 2.—The Land of England and Wales and its Products in 1688. (*Natural and Political Observations and Conclusions upon the State and Condition of England*, 1696, by Gregory King ; ed. Chalmers, 1804, p. 52.)

	Acres.	Value per Acre.	Rent.
		£ s. d.	£
Arable land,[1] · · · ·	9,000,000	0 5 6	2,480,000
Pasture and meadow, · ·	12,000,000	0 8 8	5,200,000
Woods and coppices, · ·	3,000,000	0 5 0	750,000
Forests, parks, and commons, ·	3,000,000	0 3 8	570,000
Heaths, moors, mountains, and barren land, · · ·	10,000,000	0 1 0	500,000
Houses and homesteads, gardens and orchards, churches and churchyards, · ·	1,000,000 {	The land The buildings	450,000 2,000,000
Rivers, lakes, meres, and ponds,	500,000	0 2 0	50,000
Roads, ways, and waste land, ·	500,000		
In all, ·	39,000,000	about 6 2	12,000,000

TABLE 3.—*An Estimate of the Live Stock of England and Wales in* 1688. (*Davenant on Trade*, ed. Whitworth, vol. ii. p. 219.)

	Yearly breed or increase.	The whole Stock.	Value of each besides the skin.	Value of the Stock.
			£ s. d.	£
Beeves, sterks, and calves, · ·	800,000	4,500,000	2 0 0	9,000,000
Sheep and lambs, ·	3,600,000	12,000,000	0 7 4	4,440,000
Swine and pigs, ·	1,300,000	2,000,000	0 16 0	1,600,000
Deer and fawns, ·	20,000	100,000	2 0 0	200,000
Goats and kids, ·	10,000	50,000	0 10 0	25,000
Hares and leverets,-	12,000	24,000	0 1 6	1,800
Rabbits and conies,	2,000,000	1,000,000	0 0 5	20,833
	7,742,000	19,674,000	0 0 0	15,287,633

TABLE 4.—1827.

(Select Committee on Emigration, 1827. Evidence of Mr. W. Couling. *Sessional Papers*, 1827, vol. v., p. 361.)

Territorial divisions.	Arable land and gardens.	Meadows, pastures, and marshes.	Uncultivated improveable wastes.	Unimproveable wastes.	Total acreage.
	Acres	Acres	Acres	Acres	Acres
England, -	10,252,800	15,379,200	3,454,000	3,256,400	32,342,400
Wales, -	890,570	2,226,430	530,000	1,105,000	4,752,000
Total, -	11,143,370	17,605,630	3,984,000	4,361,400	37,094,400

APPENDIX V.

COLLECTION OF TITHES (1793-1815.)

The following variations in the methods of collecting tithes are mentioned in the Reports to the Board of Agriculture on the respective counties.

1. Eastern and North-Eastern Counties.

Bedford. A corn-rent generally adopted instead of an allotment of land when enclosures are made. Few enclosed parishes continue tithable. Half the parishes are Vicarages, and the great tithes are in the hands of lay impropriators. In about 10 parishes tithes are collected in kind.

Cambs. Much tithe collected in kind, the hirer paying from 3s. to 5s. 4d. per acre. Compositions average the same rate or higher. One-tenth of the prime cost of purchased manures allowed by several titheowners. On enclosures $\frac{1}{5}$th of arable, $\frac{1}{8}$th of pasture, and $\frac{1}{9}$th of fen, all fenced at the cost of proprietors, allotted in lieu of tithe.

Essex. Average for great and small tithes is a composition of 3s. 9d. per pound rent.

Herts. Titheowners generally moderate because the light soils could only produce heavy crops under spring-dressings brought from a distance at great cost. In 1813 there is no instance of tithe being taken in kind.

Hunts. Instead of land being allotted in lieu of tithes on enclosures, a corn-rent varying with the price of corn is generally arranged.

Lincoln. Compositions average from $\frac{1}{5}$th to $\frac{1}{7}$th of the rent of arable land, $\frac{1}{7}$th to $\frac{1}{8}$th of meadow, $\frac{1}{9}$th of rich pasture. Tithes exchanged for land on enclosure. Much land unploughed in order to escape tithes.

Norfolk. Very little tithe collected in kind. Compositions average 3s. 6d. an acre (1793) or 4s. 9d. (1803) on arable land and 1s. 6d. for grass.

Suffolk. Some tithe taken in kind. Compositions, sometimes by the acre, sometimes by the pound rent, vary in amount, but are generally much under real value.

2. South-Eastern and East Midland.

Berks. Great tithes compounded at 5s. in the pound rent; small at 1s. 3d.

Bucks. Lay impropriators less careful of interests of parish than clerical titheowners. Average of composition 4s. 6d. an acre. Out of 204 parishes 82 are tithe free, *i.e.* extinguished by allotments of land in lieu; 30 partly and chiefly tithe free; three pay a corn-rent, one a *modus*. The remaining 114 are tithable. Only one tithe collected in kind. Compositions moderate.

Hants. Much tithe taken in kind; composition from 4s. 6d. to 7s. on full improved rent.

Kent. Much tithe let out to proctors who collect in kind.

Middlesex. Tithe taken in kind, or annually compounded for; some parishes pay a *modus*.

Northants. Enclosed land generally tithe free. Average of tithes reckoned at from 3s. to 3s. 6d. per acre over the whole open-field farm, including the portion which is annually fallowed. The loss to the occupier, if tithe taken in kind, estimated at 5s. to 6s. per acre.

Notts. Tithes taken in kind, or compounded for: in new enclosures, land generally allotted in lieu of tithe: *moduses* charged on particular products; lands originally belonging to religious houses generally tithe-free.

Oxford. Tithe averages $\frac{1}{4}$th of rent: on enclosures $\frac{1}{5}$th of arable and $\frac{1}{9}$th of pasture allotted in lieu.

Rutland. One-third of the land subject to tithes: on enclosures $\frac{1}{5}$th of arable and $\frac{1}{9}$th of greensward allotted in lieu. Clergy more reasonable in collection than lay impropriators.

Surrey. Common opinion is that land tithe-free and worth 20s. an acre is worth 13s. if subject to tithe. Tithes mean more than loss to a farmer; they cramp his energies.

Sussex. Compositions average 4s. 6d. an acre for wheat; 2s. 6d. for barley, oats, beans, or pease; pasture and meadow 2s. These compositions are "generally allowed to be moderate and very fair."

Warwick. Lay impropriators more rigorous in exactions than clergy. Tithes generally compounded for at the rate of 6s. to 12s. per acre of tillage, and from 1s. 6d. to 5s. 6d. for meadow and pasture.

3. West Midland and South-Western Counties.

Cornwall. Great or sheaf tithes generally in the hands of laymen who farmed them out to proctors. In the hands of the clergy, small tithes were generally compounded at 1s. to 1s. 3d. in the pound rent, and the great tithes at from 2s. 6d. to 3s. 6d. in the pound rent. Lay tithes were either taken in kind, or valued and agreed in the field at harvest.

Dorset. Tithes averaged on pasture and arable land 3s. 6d. an acre; on commons 8d., on downs 4d. A very low *modus* was common. Great tithes were often in lay hands, and compositions were frequently 5s. or 6s. in the pound rent.

Gloucester. Tithes were generally compounded, but on yearly valuations. The average on arable land was 6s. an acre, and on grass 2s. 6d. to 3s. When land was allotted in lieu of tithes, commissioners generally allotted $\frac{1}{5}$th of the arable and $\frac{1}{9}$th of the pasture, the land, so allotted to the titheowner, to be fenced, and the fences repaired for 7 years, at the cost of the proprietors. Very little tithe was taken in kind "at least among the clergy."

Hereford. Very little tithe taken in kind. Compositions averaged from 3s. 6d. to 4s. in the pound rent.

Shropshire. Very little tithe taken in kind. Compositions generally fair and equitable, either by annual valuations, or an agreed sum for a term of years: averaged not more than 2s. in the pound rent.

4. North and North-Western Counties.

Cheshire. Lay tithes generally let on lease for 21 years. Some lay owners collect a portion in kind and let the remainder, or have the tithes valued in the field, and give farmers the option of taking or leaving the tithable produce at the valuation. Hay was either tithed in kind or valued as above. Sometimes a trifling *modus* was paid.

Cumberland. Tithes generally taken in kind. A few parishes pay a *modus* ; others are tithe-free in consideration of a portion of the common being allotted to the impropriator.

Derby. Survey and valuation made annually before harvest. The usual rates charged are : hay, 2s. 6d. to 4s. 6d. an acre ; wheat, 12s. to 14s. ; oats, 7s. to 10s. 6d. ; barley, 10s. Often surveyors fix a gross sum for the parish, which is paid by the principal occupier, and the proportions are adjusted among the farmers themselves.

Lancashire. Tithes in many places collected in kind ; corn, $\frac{1}{11}$th of the crop ; hay, compounded for at 5s. per acre, and at 6s. for clover in the first year.

Northumberland. Some tithes collected moderately, others rigorously ; some let the tithes at a fair rent for a term of years ; others value and let every year.

Staffordshire. Tithes diminish the wheat-growing area : farmers decline to sow corn, and prefer to graze their land, because of the large sums demanded by the tithingmen.

Westmoreland. Some tithes collected in kind, or farmers have the option of taking at valuation.

Yorkshire, West. Great tithes often taken in kind ; but much reduced by grazing.

Yorkshire, North. Tithes in lay hands being extinguished by purchases made by landowners ; in clerical hands, tithes being diminished by enclosures and allotments of land in lieu.

APPENDIX VI.

THE AGRICULTURAL POPULATION.

TABLE I.

Census Returns of those who were engaged on Farms in England and Wales in 1851, 1861, 1871, 1881, 1891, 1901, 1911.

Occupations.	1851.	1861.	1871.	1881.	1891.	1901.	1911.
Farmers and Graziers, -	249,431	249,735	249,735	223,943	223,610	224,299	208,761
Relatives of Farmers and Graziers assisting on Farms, - - -	111,604	92,321	76,466	75,197	67,287	107,783	97,689
Farm Bailiffs or Foremen, -	10,561	15,698	16,476	19,377	18,205	22,662	22,141
Agricultural Workers o. all Classes on Farms, -	965,514	983,824	962,348	870,798	780,707	620,986	643,117
Totals, - -	1,337,110	1,341,578	1,305,025	1,189,315	1,089,809	975,730	971,708

TABLE II.

NUMBER OF PERSONS AGED 12 YEARS AND OVER, IN ENGLAND AND WALES, RETURNED AT THE OCCUPATIONAL CENSUS AS BEING IN AGRICULTURAL OCCUPATIONS IN 1921 AND 1931.

	1921.		1931.	
	Males.	Females.	Males.	Females.
Land and Estate Agents and Managers (not Auctioneers and Estate Agents)	1,831	35	2,301	51
Farmers	244,653	19,440	230,879	17,367
Farmers' Sons, Daughters and other Relatives assisting in the work of the Farm	80,257	15,384	72,593	8,189
Gardeners, Nurserymen, Seedsmen, Florists	192,560	6,156	216,569	4,402
Agricultural and Forestry Pupils (not at Colleges)	6,382	341	1,697	141
Farm Bailiffs and Foremen	22,462	217	16,588	114
Foresters and Woodmen	10,525	14	12,401 (a)	53 (a)
Agricultural Machine Tractor Proprietors, Managers, Foremen	1,553	39	1,393	28
Drainage Superintendents, Foremen, etc.	121	3	—	—
Shepherds	11,240	42	10,298	25
Agricultural Machine, Tractor Drivers, Attendants	9,525	61	6,103	25
Agricultural Labourers, Farm Servants :				
Distinguished as in charge of Cattle	59,382	10,603	62,342	6,461
Distinguished as in charge of Horses	113,616	313	69,754	119
Not otherwise distinguished	376,331	21,349	334,590	11,164
	549,329	32,265	466,686	17,744
Gardeners' Labourers	28,156	4,344	50,090	3,521
Land Drainers, Drainage Labourers	834	2	—	—
Labourers in Woods and Forests	918	15	(b)	(b)
Estate Labourers	3,550	1	5,081	9
Pea and Fruit Pickers	1,327	3,370	40	171
Other Agricultural Occupations	6,075	1,323	23,178	3,838
Total engaged in Agricultural Occupations	1,171,298	83,052	1,115,897	55,678

(a) Includes labourers in woods and forests.   (b) Included in foresters and woodmen.

TABLE III.

CENSUS OF POPULATION, 1931.

Agricultural Industry showing Males and Females, aged 14 years and over, classified by Industrial Status.

ENGLAND AND WALES.

INDUSTRIES.	Employers, Directors, Managers.		Operative Employees.		Working on Own Account.		Out of Work (All Classes)		Total.	
	Male.	Female.	Male.	Female.	Male.	Female.	Male.	Female.	Male.	Female.
Farming (not Fruit or Poultry) and Stock-rearing -	138,074	10,996	529,957	25,527	82,516	5,114	42,210	1,439	792,757	43,076
Poultry Farming - -	2,814	490	8,771	1,461	11,186	2,311	624	46	23,395	4,308
Market Gardening and Fruit Farming - -	10,139	494	23,076	2,919	17,050	377	5,004	206	55,269	3,996
Flower and Seed Growing and Nursery Gardening	4,316	195	34,902	5,522	2,755	136	3,086	285	45,059	6,138
Other or undefined Gardening	757	14	5,567	101	19,794	178	2,312	16	28,430	309
Forestry :										
Government - - -	29	—	2,133	68	—	—	252	9	2,414	77
Other - - -	114	—	6,134	18	623	—	488	1	7,359	19
Other Agricultural Industries	326	13	1,654	95	2,494	80	386	5	4,860	193
Total Agricultural Industries	156,569	12,202	612,194	35,711	136,418	8,196	54,362	2,007	959,543	58,116

APPENDIX VII.

Annual Values of Imports of the Undermentioned Articles, 1866-1935.

Years.	Group 1. Beef, Mutton, Pork, Bacon, Hams, Eggs, Butter, Cheese, etc.	Group 2. Live Cattle, Sheep, and Pigs.	Group 3. Fruit (Raw), Nuts (Edible), and Vegetables (Raw).	Group 4. Wheat and Wheat Flour.	Group 5. Oats, Indian Corn, Barley, Meal, Hops, Rice, Sugar, etc.	Total of the Groups.
	£	£	£	£	£	£
1866-70 (averages),	15,044,181	4,528,203	2,469,991	22,628,516	32,398,540	77,069,431
1871-75 (averages),	23,332,813	5,613,583	4,352,181	30,953,009	44,321,935	108,573,521
1876,	28,039,220	7,326,288	4,619,238	32,380,726	47,663,103	120,018,575
1877,	31,447,611	7,260,119	5,526,991	27,919,526	48,984,551	121,138,798
1878,	31,227,346	6,012,564	6,776,858	40,694,419	55,587,495	140,298,682
1879,	33,983,462	7,453,309	6,885,756	34,217,641	50,596,046	133,136,214
1880,	33,645,153	7,075,386	7,182,739	39,970,120	49,130,312	137,003,710
1881,	39,838,081	10,239,295	8,298,583	39,327,820	52,133,506	149,837,285
1882,	39,650,098	8,525,256	5,976,106	40,736,754	49,950,198	144,838,412
1883,	35,442,109	9,271,956	6,502,354	44,921,565	50,845,848	146,983,832
1884,	40,792,229	11,983,754	6,753,148	43,799,259	54,192,407	157,520,797
1885,	39,736,081	10,504,877	6,519,290	30,065,577	43,090,450	129,916,275
1886,	38,110,303	8,734,754	6,009,119	33,736,358	42,162,070	128,752,604
1887,	36,101,454	7,142,397	6,178,176	26,137,681	37,359,579	112,919,287
1888,	38,193,960	6,149,048	6,327,895	31,365,802	36,814,160	118,850,865
1889,	40,339,136 [1]	7,727,694	6,703,486	31,526,720	42,290,236	128,587,272 [1]
1890.	46,060,430	10,359,832	6,884,787	31,054,410	47,365,165	141,724,614
	48,957,436	11,216,311	7,514,692	32,668,132	43,943,599	144,290,170

Year					Total
1891,	49,503,337	9,246,398	8,311,209	39,633,091	47,693,979	154,388,014
1892,	53,745,375	9,362,135	8,473,694	37,125,355	46,790,767	155,497,326
1893,	54,857,020	6,351,704	8,076,013	30,831,538	47,444,542	147,560,817
1894,	55,474,902²	9,089,883	9,311,281	26,755,178	45,091,495²	145,722,739
1895,	56,872,963	8,966,252	8,897,720	30,210,189	41,409,406	146,356,530
1896,	59,043,670	10,438,699	9,426,560	30,906,862	44,169,176	153,984,967
1897,	63,586,303	11,380,092	10,463,736	32,963,159	40,854,545	159,247,835
1898,	66,304,331	10,385,676	11,358,409	37,692,699	47,429,261	173,170,376
1899,	71,645,720	9,515,005	11,329,661	32,982,199	48,754,483	174,227,068
1900,	78,001,601	9,622,319	12,299,415	33,448,477	49,943,653	183,315,465
1901,	83,921,119	9,426,803	11,468,502	33,422,891	51,940,142	190,179,457
1902,	84,768,553	8,269,175	13,070,870	36,005,440	46,229,883	188,343,921
1903,	86,292,555	9,755,185	15,319,994	39,663,843	46,992,151	198,023,728
1904,	82,839,570	10,324,420	14,984,553	41,525,016	48,367,906	198,045,465
1905,	85,486,700	9,944,859	13,872,842	41,324,776	48,980,558	199,609,735
1906,	92,738,461	9,889,127	14,016,565	39,493,398	47,752,293	203,889,844
1907,	91,273,987	8,273,640	15,516,140	44,040,630	52,500,197	211,604,594
1908,	92,060,674	6,671,810	14,657,722	45,370,558	49,265,676	208,026,440
1909,	90,775,615	5,579,028	14,536,386	51,642,611	54,947,389	217,481,029
1910,	97,050,856	4,028,672	14,039,110	49,671,789	54,476,585	219,267,012
1935,	145,960,607	5,567,905	39,682,310	33,683,926	37,223,176	262,119,540

¹ The Imports of Condensed or Preserved Milk, not having been separately distinguished until 1888, are not included in the above figures for the preceding years.

² Groups 1 and 5 are not made up after 1893. The totals are obtained by adding together the figures, taken from *Agricultural Statistics*, 1910, vol. 45, pl. 3. [Cd. 5786, 1911], page 307 *et seqq.*, which constitute the various headings, viz. :

Group 1. Beef, Mutton, Pork, Bacon, Hams, Meat (not enumerated), Rabbits, Poultry, Game, Eggs, Fish, Lard (not including Imitation Lard), Butter, Margarine, Cheese, Preserved or Condensed Milk.

Group 5. Barley, Oats, Maize, Rye, Buckwheat, Beans, Peas, Oatmeal, Maizemeal, other Meal, Rice, Sago, Sago Flour, other farinaceous substances, Sugar, Molasses, Malt, and Hops.

APPENDIX

AGRICULTURAL

TABLE I.—*Total Acreage in England and Wales under each kind*

	1866.		1871.	
	England.	Wales.	England.	Wales.
	Acres.	Acres.	Acres.	Acres.
Total Area (in Statute Acres), - - -	32,590,397	4,734,486	32,590,397	4,734,486
Abstract of Acreage under all kinds of Crops, Bare Fallow, and Grass, - -	22,236,737	2,284,674	23,717,660	2,604,817
Corn Crops—				
Wheat, - - - - - -	3,126,431	113,862	3,312,550	126,334
Barley or Bere, - - - - -	1,877,387	146,323	1,964,210	169,751
Oats, - - - - - - -	1,503,990	251,893	1,454,144	253,672
Rye, - - - - - - -	50,570	2,452	57,775	2,338
Beans, - - - - - -	492,586	3,534	512,929	4,071
Peas, - - - - - -	314,206	3,010	382,104	4,534
Total of Corn Crops, - - -	7,365,170	521,074	7,683,692	560,700
Green Crops—				
Potatoes, - - - - - -	311,151	44,266	391,531	51,853
Turnips and Swedes, - - - -	1,610,610	62,442	1,592,933	69,833
Mangold, - - - - - -	254,081	3,864	351,523	7,380
Cabbage, K. Rabi, and Rape, - -	159,539	1,329	174,643	1,062
Carrots, - - - - - -	15,598	295	18,634	341
Vetches, Lucerne, and any other Crop (except Clover or Grass), - -	408,933	27,069	368,281	6,072
Vetches or Tares, - - - -				
Lucerne, - - - - - -				
Beet Root (except Sugar Beet), - -				
	After 1881 these items appear separately.			
Sugar Beet, - - - - -				
Miscellaneous Green Crops, - -				
Total of Green Crops, - - -	2,759,912	139,265	2,897,545	136,541
Clover, Sainfoin, and Artificial and other Grasses under Rotation, - - -	2,296,087	256,722	2,694,370	375,086
Permanent Pasture, Meadow or Grass, not broken up in Rotation (exclusive of Heath or Mountain Land), - - -	8,998,027	1,257,721	9,881,833	1,494,465
Bare Fallow or Uncropped Arable Land, -	760,979	109,878	484,249	37,843
Hops, - - - - - - -	56,562	14	60,022	7
Flax, - - - - - - -	—	—	15,949	175
Small Fruit,* - - - - -	* Not separately distinguished before 1888.			

VIII.

STATISTICS, 1866-1935.

of Crop, Bare Fallows, Grass, Hops, Flax, and Small Fruit.

	1881.		1891.		1901.		1911.	
	England.	Wales.	England.	Wales.	England.	Wales.	England.	Wales.
	Acres.	Acres.	Acres.	Acres.	Acres.	Acres.	Acres.	Acres.
	32,597,398	4;721,823	32,509,322	4,779,343	32,550,698	4,776,779	32,394,302	4,749,651
	24,663,937	2,784,963	25,113,343	2,887,791	24,694.252	2,823,062	24,478,426	2,770,397
	2,641,045	90,026	2,192,393	61,590	1,617,721	47,019	1,804,045	38,487
	2,029.499	142,318	1,772,432	117,101	1,635,426	101,907	1,337,513	86,800
	1,627,004	243,544	1,672,835	234,055	1,831,740	208;773	1,841,136	206,037
	32,372	1,905	38,081	1,086	49,649	1,460	39,962	366
	417,789	2,538	337,493	1,884	237,361	1,215	299,846	1,608
	213,249	1,984	201,648	1,533	152,185	1,596	166,182	712
	6,960,958	482,315	6,214,882	417,249	5,524,082	361,970	5,488,684	334,010
	347,733	42,440	354,606	38,238	415,105	31,979	402,505	26,667
	1,478,682	66,356	1,367,960	70,607	1,144,035	61,934	1,066,625	57,947
	330,385	7,386	345,405	7,883	386,044	9,811	438,916	11,154
	138,341	1,096	145,764	1,902	161,727	4,194	139,513	5,386
	13,855	454	14,115	485	† probably	among	‡ 10,441	273
	863,957	6,818			"Misc. Gre	en Crops "	below.	
			214,225	2,145	147,186	1,344	102,204	532
			17,559	222	43,165	370	52,757	366
			2,875	41 ⎞	† This total probably	‡ "Carrots	" deducted	
				⎟	includes	" Carrots."	from this total—see	
				3 ⎬			above.	
			169	⎟	104,003	847	117,680	775
			67,712	659 ⎠				
	2,681,953	124,550	2,530,450	122,185	2,401,265	110,479	2,330,641	103,100
	2,548,952	331,401	2,762,021	324,744	2,862,658	400,268	2,327,265	281,512
	11,655,825	1,815,413	13,085,117	2,012,432	13,457,660	1,941,365	13,903,494	2,046,109
	744,896	31,271	409,972	10.069	329,002	7,882	318,909	4,764
	64,943	—	56,145	—	51,127	—	33,056	—
	6,410	13	1,787	4	630	6	416	30
			52,969	1,108	67,828	1,092	76,287	902

TABLE L.

	1914.		1915.	
	England.	Wales.	England.	Wales.
	Acres.	Acres.	Acres.	Acres.
Total Area (in Statute Acres), • • •	32,388,998	4,750,155	32,387,409	4,750,155
Abstract of Acreage under all kinds of Crops, Bare Fallow, and Grass, • •	24,367,509	2,746,495	24,310,744	2,742,356
Corn Crops—				
Wheat, • • • • • • •	1,770,470	37,028	2,121,519	48,651
Barley or Bere, • • • • • •	1,420,346	84.425	1,151,544	80,178
Oats, • • • • • • •	1,730,091	199,535	1,888,568	199,479
Rye, • • • • • • •	52,348	1,551	46,907	774
*Mixed Corn, • • • • •	—	—	—	—
Beans, • • • • • • •	292,612	1,404	265,286	1,229
Peas, • • • • • • •	168,233	608	128,886	495
Total of Corn Crops, • • •				
Green Crops—				
Potatoes, • • • • • •	436,172	25,449	436,940	26,459
Turnips and Swedes, • • • •	989,523	55,571	881,103	50,753
Mangold, • • • • • •	421,336	11,031	402,262	11,461
Cabbage, K. Rabi, and Rape, • •	131,860	6,545	128,286	5,450
Carrots, • • • • • •	10,534	234	9,016	229
Vetches or Tares, • • • • •	123,185	545	109,243	387
Lucerne, • • • • • •	53,343	311	52,705	297
Sugar Beet, • • • • • •	2,318	16	2,346	3
Miscellaneous Green Crops, • •	133,774	1,126	132,458	1,194
Total of Green Crops, • • •	2,302 045	100,828	2,154,359	96,233
Clover, Sainfoin, and Artificial and other Grasses under Rotation, • • • •	2,121,541	259,810	2,101,761	260,604
Permanent Pasture, Meadow or Grass, not broken up in Rotation (exclusive of Heath or Mountain Land), • •	14,061,042	2,054,708	14,038,071	2,049,322
Bare Fallow or Uncropped Arable Land, •	335,208	5,529	305,200	4,439
Hops, • • • • • • • •	36,661	—	34,744	—
Flax, • • • • • • • •	581	42	625	39
Small Fruit, • • • • • •	76,331	1,027	73,274	913

* Mixed Corn. The areas were in years previous to 1918 apportioned among Wheat, Barley and Oats.

Continued.

1916.		1917.		1918.	
England.	Wales.	England.	Wales.	England.	Wales.
Acres.	Acres.	Acres.	Acres.	Acres.	Acres.
32,387,409	4,750,155	32,387,409	4,750,155	32,387,409	4,750,153
24,317,993	2,756,091	24,322,870	2,758,611	24,262,040	2,725,47:
1,862,211	49,997	1,854,870	63,615	2,460,695	95,96●
1,244,639	87,437	1,364,630	95,166	1,394,861	105,948
1,862,502	222,172	2,012,719	246,190	2,414,561	365,502
52,840	636	55,138	875	101,199	233
—	—	—	—	113,799	27,720
235,080	1,177	209,311	1,281	247,787	2,894
112,068	615	130,193	806	149,230	874
5,369,340	362,034	5,626,861	407,933	6,882,132	599,137
399,586	28,392	473,342	34,645	596,607	37,225
885,477	52,682	921,553	50,821	858,516	52,302
366,818	11,319	376,912	11,930	388,077	13,215
125,571	7,070	112,141	5,650	105,453	6,524
9,999	226	15,220	298	11,656	196
88,484	630	78,207	557	61,449	550
53,895	272	49,907	306	39,885	184
143	8	269	6	658	18
169,997	1,500	65,849	526	66,479	474
2,099,970	102,069	2,093,400	104,739	2,128,780	110,683
2,311,267	279,043	2,226,061	273,484	1,875,363	219,864
14,015,840	2,007,143	13,868,721	1,966,654	12,798,361	1,790,511
416,953	4,933	351,526	3,772	405,176	3,542
31,352	—	16,946	—	15,666	—
853	58	2,427	80	18,307	97
72,418	811	71,211	731	65,053	626

TABLE II.—*Agricultural*

	1866.		1871.	
	England.	Wales.	England.	**Wales.**
Horses—Used solely for Agriculture,	No. No Return before 1869	No. No Return before 1869	No. 733,257 [1]	No 68,708 [1]
Unbroken horses, - - - ⎫ Mares kept solely for breeding, - ⎬ Unbroken horses 1 year and above, - ⎭ „ „ under 1 year, -			229,583 [1]	48,468 [1]
Total, - - -			962,840	117,176
Cattle—Cows and heifers in milk or in calf,	1,290,529	222,546	1,460,693	250,551
Other cattle : 2 years old and above, Ditto, 1 year and under 2, - - Ditto, under 2 years old, - - Ditto, under 1 year, - - -	946,975 1,069,530	133,560 185,295	981,295 1,229,076	118,005 228,082
Total, - - -	3,307,034	541,401	3,671,064	506,588
Sheep—1 year old and above, - - - Under 1 year old, - - - Ewes kept for breeding, - -	10,620,190 4,504,351	1,287,809 380,854	11,144,527 6,385,880	1,885,839 820,576
Total, - - -	15,124,541	1,668,663	17,530,407	2,706,415
Pigs, - - - - - - - Sows kept for breeding, - - - - Other Pigs, - - - - - -	2,066,299	191,604	2,073,504	225,456

[1] Including Ponies. [2] Including Mares kept for breeding.

Live Stock.

	1881.		1891.		1901.		1911.	
	England.	Wales.	England.	Wales.	England.	Wales.	England.	Wales.
	No. 772,087[1]	No. 72,208[1]	No. 796,869	No. 80,645	No. 843,624[2]	No. 91,876[2]	No. 843,632[2] Stallions[3] 6,323	No. 93,117[2] Stallions[3] 1;495
	322,016[1]	65,559[1]	300,674 45,507	57,015 12,526	223,430 94,860	41,504 21,244	195,573 87,053	36,443 20,367
	1,094,103	137,767	1,143,050	150,186	1,161,914 Other horses,	154,624 -	1,132,581 125,741	151,442 11,175
					Total, -		1,258,322	162,597
	1,621,249	260,480	1,917,078	294,599	1,887,414	In Milk, 280,899 In Calf but not in Milk,	1,630,089 478,461	237,355 47,025
	1,103,389	132,753	1,114,301	126,832	1,059,495 944,205	94,472 179,914	1,038,708 1,037,107	89,607 174,470
	1,435,447	262 112	1,838,836	337,878	900,421	187,793	989,661	191.814
	4,160,085	655.345	4,870,215	759,309	4,791,535	743,078	5,173,976	740,271
	9,818,852 5,564,004	1,771,213 695.732	10,783,208 7,091,514	2,139,367 1,094,569	3,420,482 6,282,168 5,845,407	845,693 1,217,963 1,364,078	3,359,815 6,419,615 5,960,099	795,578 1,285,567 1,509,976
	15,382,856	2,466,945	17,874,722	3,233,936	15,548,057	3,427,734	15,739,529	3,591,121
	1,733,280	191,792	2,461,185	270,082	268,909 1,573,224	35,446 177,525	333,786 2,080,942	41,797 194,514
				Total, -	1,842,133	212,971	2,414,728	236,311

[3] Above 2 years old used, or intended to be used, for service.

TABLE II.

	1914.		1915.	
	England.	Wales.	England.	Wales.
	No.	No.	No.	No.
Horses—Used solely for Agriculture, • •	712,743	78,554	656,166	72,913
Stallions, - - - - - •	6,165	1,335	6,370	1,319
Unbroken horses, 1 year and above, -	178,315	34,753	168,888	33,190
„ „ under 1 year, - -	82,289	19,817	80,522	18,720
Total, • • -	979,512	134,459	911,946	126,142
Other Horses, - - - • -	260,425	25,151	226,444	22,650
Total of Horses, -	1,239,937	159,612	1,138,390	148,792
Cattle—				
Cows and Heifers in milk • • -	1,658,996	248,620	1,633,272	248,402
Cows in calf but not in milk, - • -	240,269	24,414	231,176	25,997
Heifers in calf, - - - • -	285,350	26,571	269,259	26,397
Other cattle—				
2 years old and above, - • • -	878,733	73,598	922,227	72,101
Under 2 years old, • • • -	995,401	179,549	1,101,556	195,667
Under 1 year old, • • • -	1,060,696	205,747	1,123,457	214,643
Total of Cattle, - -	5,119,445	758,499	5,280,947	783,207
Sheep—				
One year and above, • • • -	2,415,106	736,445	2,758,152	723,495
Under 1 year, - • • • -	5,916,938	1,352,871	5,763,344	1,405,848
Ewes kept for breeding, • •	5,319,921	1,518,413	5,303,000	1,568,738
Total of Sheep, - -	13,651,965	3,607,729	13,824,496	3,698,081
Pigs—				
Sows kept for breeding, • • -	306,736	33,648	267,497	30,666
Other Pigs, - - - - • -	1,953,215	187,882	1,939,088	182,776
Total of Pigs, - -	2,259,951	221,530	2,206,585	213,442

Continued.

1916.		1917.		1918.	
England.	Wales.	England.	Wales.	England.	Wales.
No.	No.	No.	No.	No.	No.
691,737	81,033	712,956	83,082	734,381	88,047
6,805	1,452	6,963	1,446	6,519	1,188
184,081	34,833	193,900	35,092	187,582	32,268
88,504	21,303	84,063	20,299	80,508	19,562
971,127	138,621	997,882	139,919	1,008,990	141,065
226,633	23,185	212,663	22,358	203,078	22,699
1,197,760	161,806	1,210,545	162,277	1,212,068	163,764
1,606,975	248,479	1,591,308	240,135	1,623,350	234,845
225,373	26,677	242,338	29,199	299,426	35,664
291,200	30,978	328,260	33,554	355,981	28,700
982,513	85,252	1,010,682	83,088	929,435	71,334
1,171,714	202,375	1,148,103	205,219	1,138,675	199,836
1,125,890	218,355	1,103,790	211,472	1,077,336	205,903
5,403,665	812,116	5,424,481	802,667	5,424,203	776,282
2,818,250	778,711	5,247,246	1,624,783	4,963,696	1,523,079
5,870,783	1,436,272	2,832,990	730,530	2,529,520	631,197
5,382,921	1,664,186	5,422,108	1,312,200	5,490,320	1,337,363
14,071,954	3,879,169	13,502,344	3,667,513	12,983,536	3,491,639
253,754	29,243	226,693	27,598	262,371	27,169
1,723,712	161,232	1,509,829	154,421	1,279,182	128,344
1,977,466	190,475	1,736,522	182,019	1,541,553	155,513

TABLE III.—*Average under Crops and Grass ; and number of Live-*

	1919.	1920.	1921.
	Acres.	Acres.	Acres.
Total Area (excluding Water), -	37,136,005	37,136,005	37,136,005
Total Acreage under Crops and Grass [1]	26,747,953	26,507,011	26,144,071
Arable Land - - - - -	12,308,876	12,019,745	11,618,236
Permanent Grass, - - - -	14,439,077	14,487,266	14,525,835
Wheat, - - - - - -	2,221,105	1,874,652	1,976,004
Barley or Bere, - - - -	1,509,716	1,637,222	1,435,620
Oats, - - - - - -	2,564,326	2,271,703	2,148,943
Mixed Corn, - - - - -	142,661	147,477	135,484
Rye, - - - - - -	106,518	95,594	78,831
Beans, - - - - - -	284,626	257,142	246,819
Peas, - - - - - -	163,479	165,677	142,598
Potatoes, - - - - -	475,376	544,615	577,800
Turnips and Swedes, - - -	983,398	991,408	895,010
Mangolds, - - - - -	396,051	385,866	374,771
Cabbage,[2] - - - - -	51,582	62,101	57,955
Kohl-Rabi,[2] - - - - -	9,629	11,020	9,882
Rape,[2] - - - - -	93,233	100,277	82,026
Vetches or Tares, - - - -	76,962	121,732	103,686
Lucerne, - - - - -	38,761	44,501	47,174
Hops, - - - - - -	16,745	21,002	25,133
Small Fruit, - - - - -	58,699	58,814	72,587
Clover, Sainfoin, and Grasses under rotation, - - - - -	2,258,435	2,448,363	2,549,034
Other Crops - - - - -	207,045	213,983	172,197
Bare Fallow, - - - - -	650,439	566,596	506,682
	No.	No.	No.
Horses used for Agricultural Purposes, - - - - - -	814,198	788,939	822,739
Stallions being used for Service,[2] -	7,204	6,892	7,169
Unbroken Horses :			
1 Year and above - - -	223,623	228,593	225,508
Under 1 Year, - - - -	104,005	97,298	92,269
Other Horses, - - - - -	237,794	244,006	236,902
Total of Horses, - -	1,386,824	1,365,728	1,384,587
Cows and Heifers in milk, - -	1,943,666	1,827,733	1,876,114
Cows in calf, but not in milk, - -	292,201	243,009	251,818
Heifers in calf, - - - -	317,526	282,086	373,465
Bulls being used for Service,[2] -	88,978	82,020	78,742
Other Cattle :			
2 Years and above, - - -	1,078,103	1,095,830	922,774
1 Year and under 2, - - -	1,271,392	1,109,002	893,517
Under 1 Year, - - - -	1,202,583	907,125	1,120,294
Total of Cattle, - - -	6,194,539	5,546,805	5,516,724
Ewes kept for Breeding, - - -	5,764,300	5,108,452	5,336,536
Rams and Ram Lambs to be used for Service,[3] - - - -	156,747	156,106	165,829
Other Sheep :			
1 Year and above [3] - - -	3,411,297	2,848,743	2,685,058
Under 1 Year, - - -	5,791,969	5,269,372	5,644,090
Total of Sheep, - - -	15,124,313	13,382,673	13,831,513
Sows kept for Breeding, - - -	250,752	289,546	335,893
Boars being used for Service,[2] -	18,582	20,933	24,134
Other Pigs. - - - - -	1,529,134	1,683,445	2,145,439
Total of Pigs, - - -	1,798,468	1,993,924	2,505,466

[1] Not including rough grazings.   [2] After 1921 fodder only, cabbage for human consumption not included.   [3] In 1926 numbered together.

Stock in England and Wales for each of the Years from 1919 to 1926.

1922.	1923.	1924.	1925.	1926.
Acres.	Acres.	Acres.	Acres.	Acres.
37,136,005	37,136,005	37,136,005	37,136,005	37,136,005
26,025,793	25,943,261	25,876,797	25,755,486	25,675,000
11,310,515	11,181,137	10,928,673	10,682,053	10,548,000
14,715,278	14,762,124	14,948,124	15,073,433	15,127,000
1,966,917	1,740,257	1,544,804	1,499,628	1,592,000
1,364,048	1,326,947	1,314,072	1,317,810	1,150,000
2,163,905	1,977,633	2,037,948	1,868,176	1,861,000
125,224	116,850	134,530	124,424	114,800
84,532	73,217	58,642	50,062	48,000
284,928	234,993	241,694	190,813	214,200
173,611	141,415	171,162	131,162	119,200
561,177	466,653	452,242	403,241	499,500
821,128	862,015	832,464	806,461	766,400
422,641	402,897	389,660	359,073	338,500
136.776	139,972	137,171	127,639	{ 84,200 / 10,200 / 66,800
136,179	86,943	112,300	88,498	97,100
50,630	57,880	64,615	54,020	47,000
26,452	24,893	25,897	26,256	25,000
74,746	63,698	73,515	68,352	69,600
2,302,645	2,600,263	2,547,687	2,573,724	} 2,502,000
209,898	227,041	229,570	232,873	
404,068	435,592	355,599	403,204	417,900
No.	No.	No.	No.	No.
805,094	798,122	782,494	773,200	760,500
6,074	5,459	4,707	4,800	} 125,500
224,487	201,380	176,715	143,736	
83,890	66,323	54,801	44,875	41,000
220,950	209,995	213,481	197,629	201,700
1,340,495	1,281,279	1,232,198	1,164,240	1,128,700
1,933,986	1,974,546	2,014,241	2,035,061	2,064,900
288,634	269,021	281,556	299,657	294,500
299,321	371,230	367,405	378,454	389,500
82,547	80,949	81,942	86,042	} 1,053,000
840,366	937,525	904,817	974,844	
1,167,060	1,108,190	1,084,386	1,177,687	1,221,100
1,110,747	1,081,531	1,159,982	1,211,585	1,229,400
5,722,661	5,822,992	5,894,329	6,163,330	6,252,400
5,428,146	5,505,245	5,993,581	6,397,091	6,752,300
152,189	155,456	173,577	180,624	} 2,904,500
2,137,705	2,369,228	2,384,203	2,691,691	
5,719,980	5,805,604	6,291,834	6,705,388	7,202,100
13,438,020	13,835,533	14,843,195	15,974,794	16,858,900
302,046	388,545	449,022	316,454	301,100
22,821	26,224	32,266	24,478	} 1,899,100
1,974,069	2,196,837	2,747,042	2,303,424	
2,298,936	2,611,606	3,228,330	2,644,356	2,200,200

TABLE III.—

Acreage under Crops and Grass ; and Number of Live Stock in

	1927.	1928.	1929.	1930.
	Acres.	Acres.	Acres.	Acre .
Total Area (excluding Water), -	37,134,967	37,134,967	37,132,571	37,132,571
Total Acreage under Crops and Grass [1]	25,590,330	25,505,252	25,437,679	25,380,447
Arable Land - - - - -	10,310,087	10,108,745	9,947,758	9,832,949
Permanent Grass for Hay, - -	4,317,347	4,500,163	4,695,863	5,050,936
„ „ not for Hay, -	10,962,896	10,896,344	10,794,058	10,496,562
Rough Grazings - - - -	5,125,951	5,177,958	5,282,884	5,293,644
Wheat, - - - - - -	1,635,980	1,395,543	1,330,209	1,346,150
Barley, - - - - - -	1,048,926	1,185,003	1,120,282	1,020,225
Oats, - - - - - -	1,751,146	1,762,716	1,854,408	1,778,597
Mixed Corn, - - - - -	109,236	119,303	140,445	130,695
Rye, - - - - - -	35,952	30,506	34,370	43,996
Beans, - - - - - -	201,663	169,971	157,050	175,869
Peas, - - - - - -	118,752	114,008	132,502	134,321
Potatoes, - - - - -	513,947	489,019	518,808	424,660
Turnips and Swedes, - - -	716,314	722,263	699,376	671,436
Mangold, - - - - -	305,584	298,395	299,174	288,286
Sugar Beet, - - - - -	222,566	175,734	229,918	347,257
Cabbage for fodder, Kohl-rabi, and Rape,[2] - - - - - -	128,964	124,942	124,717	134,414
Vetches or Tares, - - - -	80,725	67,665	67,879	75,076
Lucerne, - - - - - -	43,563	37,104	35,783	39,781
Hops, - - - - - -	23,004	23,805	23,986	19,997
Small Fruit, - - - - -	69,154	64,721	64,942	66,209
Orchards, - - - - -	248,705	248,379	248,353	246,979
Clover, Sainfoin, and Grasses under rotation, for Hay, - - -	1,586,173	1,567,943	1,523,753	1,595,045
„ „ not for Hay, -	874,719	868,957	845,725	828,421
Other Crops, - - - - -	204,718	205,099	201,709	203,544
Bare Fallow, - - - -	423,443	468,135	325,389	294,048
	No.	No.	No.	No.
Horses used for Agriculture, - -	746,208	732,510	706,765	682,828
Stallions being used for service, -	3,174	2,995	2,845	3,323
Unbroken Horses :				
1 Year and above, - - -	103,711	93,291	89,274	84,784
Under 1 Year, - - - -	40,214	38,161	37,583	37,907
Other Horses on Agricultural Holdings,	183,914	171,503	162,806	152,511
Total of Horses, - -	1,077,221	1,038,460	999,273	961,353
Cows and Heifers in Milk, - -	2,096,387	2,066,483	2,054,073	2,033,381
Cows in Calf but not in Milk, - -	307,059	301,774	293,715	288,768
Heifers in Calf, - - - -	387,257	355,231	364,763	352,790
Bulls being used for service, -	88,405	82,112	80,271	82,816
Other Cattle :				
2 Years and above, - - -	971,470	925,980	918,878	888,403
1 Year and under 2, - - -	1,226,498	1,174,956	1,143,640	1,117,075
Under 1 Year, - - -	1,198,164	1,119,897	1,102,254	1,086,543
Total of Cattle, - -	6,275,240	6,026,433	5,957,594	5,849,776
Ewes kept for Breeding, - - -	6,962,142	6,836,912	6,717,258	6,810,727
Rams and Ram Lambs to be used for service, - - - - -	196,351	191,491	187,209	192,163
Other Sheep :				
1 Year and above, - - -	2,625,282	2,379,277	2,266,762	2,213,668
Under 1 Year, - - -	7,288,500	6,981,930	6,934,224	7,099,285
Total of Sheep, - -	17,072,275	16,389,610	16,105,453	16,315,843
Sows kept for Breeding, - - -	392,934	380,063	307,144	315,655
Boars being used for service, - -	25,524	26,222	21,990	20,787
Other Pigs, - - - - -	2,273,056	2,564,758	2,037,409	1,973,799
Total of Pigs, - -	2,691,514	2,971,043	2,366,543	2,310,241

[1] Not including rough grazings.
[2] After 1921 fodder only, cabbage for human consumption not included.

Continued.

England and Wales for each of the Years from 1927 to 1935.

1931.	1932.	1933.	1934.	1935.
Acres.	Acres.	Acres.	Acres.	Acres.
37,132,688	37,132,688	37,132,688	37,132,688	—
25,283,320	25,206,082	25,119,648	25,030,494	24,956,724
9,582,428	9,366,506	9,249,886	9,249,627	9,397,864
4,777,843	4,542,497	4,603,764	4,822,651	4,642,704
10,923,049	11,297,079	11,265,998	10,958,216	10,916,156
5,316,031	5,357,297	5,397,776	5,424,004	5,419,922
1,196,701	1,287,944	1,660,389	1,759,448	1,772,332
1,029,157	960,542	751,349	860,598	791,653
1,652,005	1,580,377	1,494,797	1,402,017	1,418,625
121,927	114,373	104,548	96,399	93,899
32,513	24,642	19,777	17,513	20,459
158,171	153,614	153,416	151,865	154,014
132,358	126,318	136,336	142,981	132,972
446,772	504,275	518,934	487,558	462,796
620,955	580,517	555,194	520,391	497,638
270,664	229,712	238,079	246,405	251,020
233,219	254,983	364,068	396,348	367,304
125,714	109,399	117,571	136,938	169,618
64,164	48,609	53,274	52,630	53,803
46,193	39,407	34,844	34,212	
19,528	16,531	16,895	18,037	18,251
62,023	59,535	59,979	61,033	59,485
244,778	247,304	249,574	254,857	262,389
1,726,444	1,538,265	1,261,559	1,289,606	1,388,628
854,822	872,243	812,750	783,198	945,428
215,341	210,685	214,357	222,537	272,300
356,985	433,888	457,903	339,933	287,661
No.	No.	No.	No.	No.
666,538	655,104	645,611	596,326	586,394
2,829	2,804	3,819	5,254	7,205
82,328	80,735	80,793	82,475	86,522
37,793	39,104	41,228	43,780	47,035
149,006	139,326	131,137	157,729	146,696
938,494	917,073	902,588	885,564	873,852
2,043,077	2,116,616	2,179,007	2,213,918	2,232,113
321,720	352,039	358,177	363,933	382,229
425,336	402,813	417,991	417,310	436,816
90,837	91,315	95,506	96,787	85,850
845,332	847,593	900,948	944,575	923,484
1,130,396	1,241,753	1,356,867	1,369,075	1,313,733
1,208,345	1,305,886	1,311,692	1,254,557	1,166,919
6,065,043	6,358,015	6,620,188	6,660,155	6,541,144
7,262,519	7,634,366	7,767,937	7,308,355	7,124,852
216,057	224,988	226,769	218,206	225,491
2,567,761	2,728,393	2,197,230	1,484,412	1,531,219
7,702,898	7,907,671	7,897,942	7,516,069	7,595,438
17,749,235	18,495,418	18,089,878	16,527,042	16,477,000
402,424	425,204	405,643	450,354	494,132
25,720	29,265	28,671	30,849	39,779
2,354,864	2,730,089	2,634,834	2,838,955	3,279,462
2,783,008	3,184,558	3,069,148	3,320,158	3,813,373

APPENDIX IX.

Table of Agricultural Wages in England, 1768-70, 1824, 1837, 1850-1, 1860, 1869-70, 1872, 1882, 1892, 1898, 1910. It must, however, be noticed that this Table does not include Payments for Piece or Task Work; the occupation of Cottages, with or without gardens, free or at rents below the letting value; harvest earnings, overtime money, or any extra allowances in Kind or Cash. Wages of those having the care of animals are not included.

Wages for 1920 and 1926 were fixed by tribunals with compulsory powers. They also apply to ordinary Labourers only, but of not less than 21 years of age. They include allowances in Kind (milk, potatoes, cottage, board and lodging), where these are provided by the employer. They are minimum, not necessarily actual, rates, and they are paid on the shortened hours, generally of 48 hours all the year round, or of 50 hours per week (summer) and 48 hours (winter). In Cheshire, Leicester, Rutland, Shropshire, Staffordshire, Derby, the wages are for a 54 hours week all the year round. In no case are harvest or overtime wages included. The average minimum rate in operation in England and Wales during October, 1926, was 31s. 8d.

Divisions and Counties.	Average Weekly Wages of ordinary Agricultural Labourers in England.												
	1768-70.	1824.	1837.	1850-1.	1860.	1869-70.	1872.	1882.	1892.	1898.	1910.	1920.	1926.
	s. d.	s. d.	s. d.	s. d.	s. d.	s. d.	s. d.	s. d.	s. d.	s. d.	s. d.	s. d.	s. d.
(i) Eastern and North-Eastern													
Bedford	7 3	8 7	9 6	9 0	10 3	11 1	12 0	12 6	12 0	12 6	13 6	46 0	30 6
Cambridge	—	—	9 6	—	10 0	11 0	—	12 6	12 0	12 0	12 8	46 0	30 0
Essex	7 6	9 4	10 4	8 0	11 3	11 4	15 9	12 6	11 6	12 0	13 9	46 0	30 0
Hertford	7 5	9 0	9 6	9 0	10 0	11 3	12 3	13 6	11 6	12 6	14 8	46 6	31 6
Huntingdon	7 0	—	9 6	8 6	10 9	10 6	12 6	12 0	12 6	12 6	13 8	46 0	30 6
Lincoln	8 6	10 2	12 4	10 0	13 0	13 11	—	14 3	14 6	14 3	15 6	48 6	33 6
Norfolk	7 0	9 1	10 4	8 0	10 7	10 5	13 4	12 6	12 0	11 6	12 4	46 0	30 0
Suffolk	7 6	8 3	10 4	7 11	10 7	10 10	13 0	12 6	12 0	10 6	12 9	46 0	30 0
East Riding	—	11 8	12 0	12 0	13 6	—	—	15 0	15 6	16 0	16 11	49 0	35 0
Average	7 6	9 5	10 4	9 1	11 1	11 3	13 2	13 0	12 7	12 8	14 0	46 7	31 2

Average Weekly Wages of ordinary Agricultural Labourers in England.

Divisions and Counties.	1768-70	1824	1837	1850-1	1860	1869-70	1872	1882	1892	1898	1910	1920	1926
(ii) South-Eastern and East Midland.													
Berks,	7 6	—	—	7 6	10 0	10 7	14 0	12 0	10 6	11 6	13 2	46 0	30 0
Bucks,	8 0	—	8 0	8 6	—	—	—	14 0	—	13 6	14 8	46 0	31 0
Hants,	8 0	8 6	9 6	9 0	12 0	10 10	13 8	12 6	11 6	12 0	13 9	46 0	30 0
Kent,	—	11 9	12 0	12 0	12 0	14 3	15 2	16 0	14 6	14 6	16 4	47 0	32 6
Leicester,	—	—	10 0	—	13 0	13 2	16 6	13 6	14 3	15 0	15 9	46 0	34 0
Middlesex,	6 6	—	10 6	—	—	—	—	15 6	—	16 0	17 10	48 6	33 0
Northants,	9 0	8 1	9 0	9 0	11 0	11 10	16 0	—	14 0	13 6	14 1	46 0	30 0
Notts,	7 0	10 3	12 0	10 0	12 9	13 4	—	16 6	15 0	15 0	17 3	46 0	32 0
Oxford,	—	—	9 0	9 0	—	—	16 0	13 3	12 0	11 6	12 0	46 0	30 0
Rutland,	9 0	10 8	9 0	—	12 9	12 6	14 4	—	—	14 6	15 9	46 0	32 6
Surrey,	8 6	9 7	10 6	9 6	12 9	13 9	13 0	15 0	15 0	15 0	16 4	47 6	32 3
Sussex,	8 0	8 10	10 7	10 6	11 8	12 2	15 0	13 6	12 0	14 0	14 10	46 6	31 0
Warwick,	—	—	10 0	8 6	10 9	12 0	—	14 0	11 6	14 0	14 4	46 0	30 0
Average,	7 11	9 8	10 0	9 5	11 9	12 5	14 10	14 1	13 0	13 10	15 1	46 5	31 5
(iii) West Midland and South-Western.													
Cornwall,	—	7 6	10 0	—	11 0	11 0	13 0	14 6	14 0	14 0	14 6	46 0	31 0
Devon,	6 9	6 11	8 0	7 6	9 2	10 4	11 4	13 0	13 6	13 0	13 9	46 0	32 6
Dorset,	6 9	9 3	7 6	7 0	9 4	9 6	10 0	11 0	10 0	11 0	11 11	46 0	30 0
Gloucester,	—	7 1	9 0	—	9 5	10 0	14 4	13 0	10 6	12 6	11 11	46 0	30 0
Hereford,	—	7 1	8 0	—	9 0	10 3	12 0	13 0	—	12 0	13 3	46 0	31 0
Monmouth,	—	10 10	10 6	—	11 8	13 1	17 5	12 0	—	15 0	16 6	50 0	32 0
Shropshire,	—	8 2	9 0	—	10 0	11 6	14 3	13 0	14 0	14 0	14 8	46 0	32 0
Somerset,	7 0	8 7	8 8	7 0	10 0	10 0	14 4	13 0	11 0	12 6	13 6	46 0	32 0
Wilts,	—	7 2	8 8	—	9 6	10 7	11 5	12 0	10 0	11 0	12 9	46 0	30 0
Worcester,	—	8 2	9 6	—	10 0	11 11	13 0	13 6	12 6	14 0	14 2	46 0	30 0
Average,	6 10	8 2	8 10	7 2	9 11	10 10	13 1	12 10	11 10	12 11	13 9	46 5	31 2

Average Weekly Wages of ordinary Agricultural Labourers in England.

Divisions and Counties.	1768-70. s. d.	1824. s. d.	1837. s. d.	1850-1. s. d.	1860. s. d.	1869-70. s. d.	1872. s. d.	1882. s. d.	1892. s. d.	1898. s. d.	1910. s. d.	1920. s. d.	1926. s. d.
(iv) North and North-Western.													
Cheshire, - -	—	10 8	13 0	—	11 8	14 0	15 9	18 0	15 0	—	17 0	52 0	35 0
Cumberland, - -	6 6	12 3	12 0	13 0	15 0	15 3	18 7	18 6	18 0	17 0	18 4	48 0	31 0
Derby, - -	—	10 10	12 0	—	12 0	13 9	—	16 9	16 0	17 6	18 8	46 0	36 0
Durham, - -	6 6	11 6	12 0	11 0	14 3	16 2	20 6	17 9	—	18 0	20 0	50 6	32 0
Lancashire, -	6 6	—	—	13 6	—	14 8	15 7	17 6	18 0	19 0	18 10	47 6	37 4
Northumberland, -	6 0	11 5	12 0	11 0	14 0	15 8	19 2	16 6	17 0	17 0	19 4	50 6	33 0
Stafford, - -	6 4	10 8	12 0	9 6	12 6	13 0	14 0	13 6	16 0	15 0	15 11	46 0	31 6
Westmoreland, -	—	12 0	12 0	—	14 3	16 4	19 6	18 0	—	17 0	18 4	48 0	31 0
Yorkshire, - North Riding,	6 6	10 3	12 0	11 0	13 6	15 10	16 9	16 6	15 6	16 0	16 11	49 0	33 0
Yorkshire, - West Riding,	6 0	12 6	12 0	14 0	13 6	15 4	15 6	16 6	16 0	16 0	16 11	49 0	36 0
Average, - -	6 4	11 4	12 1	11 10	13 5	15 0	17 4	16 9	16 5	16 11	18 0	48 8	33 7

[1] The materials from which these Tables are compiled are, for 1768-70, Arthur Young's *Tours*; for 1824, the very defective and unreliable *Abstract of Returns prepared by order of the Select Committee on Labourers' Wages* (1825); for 1837, Purdy's paper in the *Journal of the Statistical Society* for 1860 (vol. xxiv.), apparently based on Reports from the Assistant Commissioners to the Poor Law Commission; for 1850-1, from Sir James Caird's *English Agriculture in 1850-1*; for 1860, from the *Returns of the average rate of Weekly Earnings of Agricultural Labourers* laid before Parliament in 1861, and also contained in Purdy's above-mentioned paper; for 1869-70, and for 1872, from the *Returns of the average rate of Weekly Earnings of Agricultural Labourers in the Unions of England and Wales*, presented to Parliament, and published in 1869, 1870, 1871, 1873; for 1882, on the *Reports from Assistant Commissioners to the Royal Commission on Agriculture* (1882); for 1892, from the *Report of the Royal Commission on Labour* (1891-94); for 1898 and 1910, from the *Reports on the Wages and Earnings of Agricultural Labourers* to the Board of Trade (Labour Department), and especially the important Report of Mr. Wilson Fox; for 1920, from *Standard Time Rates, etc.* (Ministry of Labour), 1921 (Cmd. 1253), pp. 196-7; for 1926, from Report of Wages Board (Ministry of Agriculture).

For convenience of reference the material has been arranged under the headings adopted for agricultural statistics.

The actual value, in individual cases, of the Returns of Wages, before the establishment of Wage Boards, is doubtful; but they probably represent some general approximation to the truth.

Other materials will be found in Sir R. Giffen's Presidential Address on the Progress of the Working Classes in the last half-century, *Journal of the Statistical Society* (vol. xlvi.); *The Agricultural Labourer*, by T. E. Kebbel (eds. 1870 and 1887); *The Romance of Peasant Life*, etc., by F. G. Heath (1872); *The English Peasantry*, by the same (1874); *British Rural Life and Labour*, by the same (1911); *Rural England*, by H. Rider Haggard (1906); *A History of the English Agricultural Labourer*, by W. Hasbach, tr. Ruth Kenyon (1908), etc., etc.

Average minimum Weekly Rates of Wages for ordinary Male Agricultural Workers from information supplied by the Ministry of Agriculture.

Divisions and Counties.	1919.		1920.		1925.		1926.		1927.		1928.		1929.		1930.		1931.		1932.		1933.		1934.		1935.	
	s.	d.	s.	d.	s.	d.	s.	d.	s.	d.	s.	d.	s.	d.	s.	d.	s.	d.	s.	d.	s.	d.	s.	d.	s.	d.
East and North East Group	35	8	42	9	30	10	31	4	31	6	31	6	31	6	31	7	31	2	30	8	30	5	30	8	31	8
South East and East Midland	35	6	42	6	31	3	31	6	31	6	31	6	31	6	31	6	31	6	31	3	30	9	31	1	31	9
West Midland and South West	35	0	42	1	31	0	31	1	31	1	31	1	31	1	31	2	31	2	30	10	30	2	30	5	31	6
North and North West	37	10	44	6	34	8	34	9	34	7	34	6	34	6	34	6	34	4	33	9	32	1	31	1	32	8

From October 1921 to the beginning of 1925 no statutory regulation of wage rates was in force and the agreements reached by the Conciliation Committees were in most cases voluntary and without legal effect. In the agricultural counties the wage rates in 1923 and 1924 had sunk to 25s. The averages given above are not weighted for the number of workers in each county.

INDEX NUMBERS OF PRICES SINCE 1918

APPENDIX X.

TABLE OF INDEX NUMBERS OF PRICES SINCE 1918.

Years.	A	B	C	D
1919 - - - - - -	255	258	268	215
1920 - - - - - -	313·9	292	273	259
1921 - - - - - -	202·2	219	151	220
1922 - - - - - -	158·8	169	146	147
1923 - - - - - -	159·1	157	136	123
1924 - - - - - -	166·2	161	154	119
1925 - - - - - -	159·1	159	152	114
1926 - - - - - -	148·1	151	125	113
1927 - - - - - -	141·4	144	139	110
1928 - - - - - -	140·3	147	154	98
1929 - - - - - -	136·5	144	139	100
1930 - - - - - -	119·5	134	96	101
1931 - - - - - -	104·1	120	83	96
1932 - - - - - -	101·6	[1] 114 112	95	90
1933 - - - - - -	100·9	[1] 111 107	85	90
1934 - - - - - -	104·1	[1] 119 114	91	90
1935 - - - - - -	106·1	[1] 123 117	87	88

[1] Includes allowances for payment under the Wheat Act 1932 and the Cattle Industry (E. P.) Act, 1934.

A. Board of Trade Index Numbers of wholesale prices of all articles, 1913 = 100.

B. Ministry of Agriculture Index Numbers of Prices of Agricultural Produce, 1911–13 = 100.

C. Ministry of Agriculture Index Numbers of Prices of Feeding Stuffs, 1911–13 = 100.

D. Ministry of Agriculture Index Numbers of Prices of Fertilisers, 1911–13 = 100.

INDEX

Abb's Court, Lord Halifax at, 172.
Aberdeen-Angus Polled Herd-Book, The, 376.
Abortion, contagious, 467.
Abraham man, the, 76.
Acorns, bread made from, 84.
Acreage of England and Wales, 145; in 1688, 503; in 1866-1911, 463, 512; unenclosed, in 1696, 152; in eighteenth century, 154.
Acts of Parliament affecting the Corn Trade, 1360-1869, 490 *seq.*
Actus, the Roman, 279.
Addington, Stephen, his *Inquiry into the Reasons for and against inclosing Open Fields*, 303.
Adulteration of food-stuffs, 382, 384.
" Advantage-bread," 497.
Affers, 12.
Agistment, on commons, 27, 158.
Agriculture, under Elizabeth, 91 *seq.*; under James I., 112; advance of, in eighteenth and nineteenth centuries, 148; fashionable under George II., 173; George III., 207; in 1837, 349; (1837-1846), 373; (1853-1862), 373; (1863-1874), 375; (1874-1885), 377 *seq.*; (1883-1890), 383; (1891-1899), 377, 383; in 1837 and 1912, 349; advance since the "fifties," 349; "apron-string" farmers, 351; commercial basis, organisation of, on a, 48, 205, 290; Department of Agriculture, 382; changes in modern times, 365-368; extravagance blamed for distress, 350; depression from 1814, 322 *seq.*; reorganisation of, 325.
Agriculture Act, of 1920, 416, 417.
Agriculture, The Annals of, and Robert Bakewell, 185; started by Arthur Young, 196; George III. a contributor to, 196, 207; on enclosure, 307.
Agriculture, Board of, established by Pitt in 1793, 196, 209; Reports

to, 225 *seq.* (and see Reports); Report of 1795, on uncultivated lands, 152; Sir Humphry Davy's lectures in 1803, 216, 362; and the Enclosure Bill of 1800, 251; on open-fields, 157; on voluntary enclosure, 162; *Communications to the Board of Agriculture*, 314; dissolved in 1822, 362; recommended potatoes, 135.
Agricultural Committee, of 1833, 326; of 1836, 325, 326.
Agricultural Commission of 1879 (Duke of Richmond's), 380.
Agricultural Customs, Select Committee on, of 1848, 373.
Agricultural Gazette, The, 364.
Agricultural Holdings Act, of 1875, 382.
Agricultural Population, Census Returns, 507.
Agricultural Returns, the (1866), 364.
The Agricultural State of the Kingdom in February, March, and April, 1816 . . . being the Substance of the Replies to a Circular Letter sent by the Board of Agriculture, etc., 320.
Agricultural Statistics, 512.
Agricultural Workers, National Union of, 414.
Aire, the, 117, 277.
Alcester, the needle industry at, 310.
Aldsworth, 232.
Ale, Assize of, 496.
Allen, William, his *Landlord's Companion*, 262.
Allowances, Parish, 327, 485.
Althorp, Lord, 208; on tithes, 345.
America, corn from, after 1815, 271, 273; the Civil War in, 374; increase of trade with, after 1815, 317; industrial collapse in 1873-4, 378; influx of wheat from, in 1879, 379; innovations from, 359; our war with, 268.
Amherst, Hon. Alicia. See **Mrs. Evelyn Cecil.**

For Product Safety Concerns and Information please contact our EU
representative GPSR@taylorandfrancis.com
Taylor & Francis Verlag GmbH, Kaufingerstraße 24, 80331 München, Germany